Convergence in Ergodic Theory and Probability

Editors

V. Bergelson
P. March
J. Rosenblatt

Walter de Gruyter · Berlin · New York 1996

Editors

VITALY BERGELSON
Mathematics Department
The Ohio State University
Columbus, OH 43210, USA

PETER MARCH
Mathematics Department
The Ohio State University
Columbus, OH 43210, USA

JOSEPH ROSENBLATT
Mathematics Department
University of Illinois
Urbana-Champaign
Urbana, IL 61801, USA

Series Editors:

Gregory R. Baker, Karl Rubin
Department of Mathematics, The Ohio State University, Columbus, Ohio 43210-1174, USA

Walter D. Neumann
Department of Mathematics, The University of Melbourne, Parkville, VIC 3052, Australia

1991 Mathematics Subject Classification: Primary: 28-06, 60-06
Secondary: 28D05, 28D10, 28D20, 60B15, 60F05, 60F10, 60G10, 60J10, 60J15
Keywords: Ergodic theory, probability theory, measure-preserving transformations, Fourier transforms, random walks, Markov chains, large deviations, martingales, and distributions

♾ Printed on acid-free paper which falls within the guidelines of the ANSI to ensure permanence and durability.

Library of Congress Cataloging-in-Publication Data

Convergence in ergodic theory and probability / editors, V. Bergelson, P. March, J. Rosenblatt
p. cm. – (Ohio State University Mathematical Research Institute publications, ISSN 0942-0363 ; 5)
Papers from the Conference on Convergence in Ergodic Theory and Probability, held at Ohio State University, in June 1993. ISBN 3-11-014219-8 (acid-free paper)
1. Convergence – Congresses. 2. Inequalities (Mathematics) – Congresses. 3. Probabilities – Congresses. 4. Ergodic theory – Congresses. I. Bergelson, V. (Vitaly), 1950– . II. March, P. (Peter), 1955– . III. Rosenblatt, J. (Joseph), 1946– . IV. Conference on Convergence in Ergodic Theory and Probability (1993 : Ohio State University) V. Series.
QA295.C67 1996
515'.42–dc20 96-21518
CIP

Die Deutsche Bibliothek – Cataloging-in-Publication Data

Convergence in ergodic theory and probability / ed. V. Bergelson ... – Berlin ; New York : de Gruyter, 1996
(Ohio State University, Mathematical Research Institute publications ; 5)
ISBN 3-11-014219-8 100550177
NE: Bergelson, Vitaly [Hrsg.]; International Mathematical Research Institute <Columbus, Ohio>: Ohio State University ...

Printed in Germany. Printing: Arthur Collignon GmbH, Berlin.
Binding: Lüderitz & Bauer GmbH, Berlin. Cover design: Thomas Bonnie, Hamburg.

Dedication

This volume of articles is dedicated to Professor Louis Sucheston. His commitment to research and teaching in mathematics and his persistent efforts to develop the ergodic theory and probability groups at The Ohio State University have had a great impact on the mathematical community as a whole and the personal lives of the editors of this volume in particular. We are all indebted to him and grateful for his support of our work.

Preface

This volume is a sample of the mathematics that was presented and discussed at the Conference on Convergence in Ergodic Theory and Probability, held at The Ohio State University, Fawcett Center for Tomorrow, in June, 1993.

The conference was the third in a recent series of such meetings, the first being held at The Ohio State University in the Spring of 1988 (organized by Louis Sucheston and Gerald Edgar) and the second being held at Northwestern University in the Fall of 1989 (organized by Alexandra Bellow and Roger Jones).

The organizers of the meeting which gave rise to this volume would like to express their appreciation for the financial support given this conference by The Ohio State University Mathematical Research Institute and the Mathematics Department at The Ohio State University, the College of Mathematical and Physical Sciences at The Ohio State University, the National Science Foundation, and the Department of Defense (Army). Without this generous help, this meeting would not have been possible.

The organizers would also sincerely like to thank the de Gruyter Publishing Company and their editor, Dr. Manfred Karbe, who have made the publication of this volume of articles a reality. Can there ever be creation without communication?

Vitaly Bergelson
Peter March
Joseph Rosenblatt

Contents

J. Aaronson and B. Weiss
A $\mathbb{Z}^d$ Ergodic Theorem with Large Normalising Constants 1

M. A. Akcoglu, J. R. Baxter, and S. E. Ferrando
A Repeated Filling Scheme for Upcrossings 15

E. Akin, J. Auslander, and K. Berg
When is a Transitive Map Chaotic? 25

M. Atlagh et M. Weber
Une Nouvelle Loi Forte des Grands Nombres 41

J. R. Baxter and N. C. Jain
An Approximation Condition for Large Deviations and Some Applications 63

V. Bergelson and I. J. Håland
Sets of Recurrence and Generalized Polynomials 91

N. H. Bingham
The Strong Arc-Sine Law in Higher Dimensions 111

M. Boshernitzan, R. L. Jones, and M. Wierdl
Integer and Fractional Parts of Good Averaging Sequences in Ergodic Theory 117

R. E. Bradley
The Uniqueness of Induced Operators 133

N. Etemadi
On Convergence of Partial Sums of Independent Random Variables 137

R. A. Fefferman
Harmonic Analysis of Operators Associated with a Multiparameter Group of Dilations 145

S. R. Foguel
Markov Matrices 157

N. E. Frangos and P. Imkeller

Existence and Continuity of the Quadratic Variation of Strong Martingales 179

C. D. Fuh and T. L. Lai

Convergence Rate in the Strong Law of Large Numbers for Markov Chains 185

H. Furstenberg and B. Weiss

A Mean Ergodic Theorem for $\frac{1}{N}\sum_{n=1}^{N} f(T^n x)g(T^{n^2} x)$ 193

R. L. Jones and M. Wierdl

Convergence of Ergodic Averages 229

M. Lacey

Bourgain's Entropy Criteria 249

S. P. Lalley

Self-Intersections of Closed Geodesics on a Negatively Curved Surface: Statistical Regularities 263

D. Li and M. B. Rao

A Note on the Law of the Iterated Logarithm for Weighted Sums of Independent Identically Distributed Random Variables 273

M. Lin and J. Olsen

Besicovitch Functions and Weighted Ergodic Theorems for LCA Group Actions 277

V. Losert

A Remark on the Strong Sweeping Out Property 291

D. Maharam

Ergodic Invertible Liftings 295

G. Peškir and M. Weber

The Uniform Ergodic Theorem for Dynamical Systems 305

M. A. Pinsky

Pointwise Fourier Inversion and Related Jacobi Polynomial Expansions 333

P. Révész

The Distribution of the Particles of a Branching Random Walk 345

B.-Z. Rubshtein

On a Random Ergodic Theorem for Bistochastic Operators 365

B.-Z. Rubshtein
A Central Limit Theorem for Conditional Distributions 373

Y. Sagher and N. Xiang
Complex Methods in the Calculation of Some Distribution Functions 381

Y. Sagher and K. Zhou
Exponential Integrability of Rademacher Series 389

D. Schneider and M. Weber
Weighted Averages of Contractions Along Subsequences 397

Y. Sun
On Uniform Distribution of Sequences 405

R. Wittmann
Ergodic Theorems for Exit Laws 421

Q. Zhang
On Skew Products of Irrational Rotations with Tori 435

A $\mathbb{Z}^d$ Ergodic Theorem with Large Normalising Constants

Jonathan Aaronson and Benjamin Weiss

Abstract. We show, *inter alia* that for a non-integrable, non-negative stationary, ergodic $\mathbb{Z}^d$ random field, there are no normalising constants for the pointwise ergodic theorem.

0. Introduction

Let $(X, \mathcal{B}(X), m)$ be a Lebesgue probability space, let $\{T_g : g \in \mathbb{Z}^d\}$ be a free, probability preserving action of $\mathbb{Z}^d$ on X, and suppose that $f: X \to \mathbb{R}$ is integrable. The pointwise ergodic theorem (**[Kr]**) states that

$$\frac{1}{n^d} S_n f \longrightarrow E(f|\mathcal{I}(T)) \text{ a.e.}$$

where

$$S_n f := \sum_{g \in \mathbb{Z}^d,\ \|g\| \leq n} f \circ T_g \quad (\|(n_1, \ldots, n_d)\| := \max_{1 \leq j \leq d} |n_j|)$$

and $E(f|\mathcal{I}(T))$ denotes the conditional expectation of f with respect to the σ-field $\mathcal{I}(T)$ of T-invariant sets.

It will follow from our main theorem that if T is ergodic, and $f \geq 0$ is nonintegrable, then there is no sequence of constants $b_n > 0$ such that

$$\frac{1}{b_n} S_n f \longrightarrow 1 \text{ a.e.} \tag{1}$$

This is the statement in the abstract, as a *stationary* $\mathbb{Z}^d$ *random field* is a stochastic process of $\{Y_g : g \in \mathbb{Z}^d\}$ of form $Y_g = f \circ T_g$ where f and T are as above.

Our main theorem is

Theorem A **($\mathbb{Z}^d$-actions alternative).** *Let $(X, \mathcal{B}, m, T)$ be a free, ergodic, probability preserving $\mathbb{Z}^d$-action on a Lebesgue probability space, and suppose that $f: X \to \mathbb{R}$ is measurable. Let $a: [0, \infty) \to [0, \infty)$ be continuous, strictly increasing, and satisfy $a(x + y) \leq a(x) + a(y)$, and $a(x) = o(x)$ as $x \to \infty$.*

Then either, $\limsup_{n \to \infty} \frac{a(|S_n f|)}{n^d} = \infty$ *a.e., or* $\frac{a(|S_n f|)}{n^d} \to 0$ *a.e. as $n \to \infty$.*

The authors would like to thank IMPA, Rio de Janeiro for inspiration and hospitality while this work was conceived.

As a corollary, we obtain a version of the ergodic theorem with large normalising constants (Feller's theorem):

Theorem F. *Let* $(X, \mathcal{B}, m, T)$ *be a free, ergodic, probability preserving* $\mathbb{Z}^d$*-action on a Lebesgue probability space, and suppose that* $f: X \to \mathbb{R}$ *is measurable. Let* $b: \mathbb{N} \to [0, \infty)$ *satisfy* $\frac{b(n)}{n} \uparrow \infty$ *as* $n \uparrow \infty$.

Then either, $\limsup_{n\to\infty} \frac{|S_n f|}{b(n^d)} = \infty$ *a.e., or* $\frac{|S_n f|}{b(n^d)} \to 0$ *a.e. as* $n \to \infty$.

As far as we know, these are the first multi-dimensional versions of these results. There are previous results for $d = 1$.

Theorem F, in case $\{f \circ T_g : g \in \mathbb{Z}\}$ are independent, identically distributed random variables, is due to W. Feller **[Fe]**. The general stationary one dimensional results were obtained in **[Aa2]**. The nonexistence of a sequence of constants satisfying (1) was obtained by Chow and Robbins **[C-R]** in the independent case, and in general in **[Aa1]**.

In §1, we consider the multi-dimensional independent case, also proving a lemma which is needed for the proof of Theorem A.

This proof is carried out in §2, the main tool being a "random Kac inequality" (Theorem 2.1). The Kac formula (obtained by M. Kac in **[Kac]**) says that if T is an ergodic measure preserving transformation of $(X, \mathcal{B}, m)$, $A \in \mathcal{B}_+$ (here, and throughout, $\mathcal{B}_+$ denotes the collection of sets in $\mathcal{B}$ with positive measure: $\mathcal{B}_+ = \{A \in \mathcal{B} : m(A) > 0\}$), and $\varphi = \varphi_A : A \to \mathbb{N}$ is the *first return time* function: $\varphi(x) = \min\{n \geq 1 : T^n x \in A\}$, then

$$\int_A \varphi dm = m(X).$$

No such formula is available for $\mathbb{Z}^d$ actions. We attempt to provide a substitute with Theorem 2.1.

To complete this introduction, we deduce Theorem F from Theorem A, and, afterwards in Proposition 0, we show a little more than the nonexistence of a sequence of constants satisfying (1) from Theorem F (as in **[C-R]**).

Proof of Theorem F. Suppose that $\limsup_{n\to\infty} \frac{|S_n f|}{b(n^d)} < \infty$ on some set of positive measure, and in particular suppose that $A \in \mathcal{B}_+$, $M > 1$ are such that

$$S_n f \leq M b(n^d) \text{ on } A \;\; \forall\, n \geq 1.$$

Setting $a = b^{-1}$ we have $\frac{a(x)}{x} \downarrow$ as $x \uparrow$, and therefore $a(x + y) \leq a(x) + a(y)$. Also, on A, for $n \geq 1$, we have

$$a(S_n f) \leq a(M b(n^d)) \leq M n^d,$$

whence, by Theorem A

$$\frac{a(S_n f)}{n^d} \to 0 \quad \text{a.e.}$$

and hence

$$\frac{S_n f}{b(n^d)} \to 0 \quad \text{a.e.}$$

(since $a(S_n f) \leq \epsilon n^d$ implies that $S_n f \leq b(\epsilon n^d) \leq \epsilon b(n^d)$). □

Proposition 0. *Suppose that $\{T_g : g \in \mathbb{Z}^d\}$ is a free, ergodic probability preserving action of $\mathbb{Z}^d$ on X, and suppose that $f: X \to \mathbb{R}_+$ is measurable and $E(f) = \infty$. Let $b_n > 0$ be a sequence of constants, and $F \in \mathcal{B}_+$ satisfy*

$$\limsup_{n\to\infty} \frac{S_n f}{b_n} < \infty \text{ on } F.$$

Then $\exists\, n_k \to \infty$ such that

$$\frac{S_{n_k} f}{b_{n_k}} \longrightarrow 0 \text{ a.e.}$$

Proof. By the pointwise ergodic theorem we have that

$$\frac{b_n}{n^d} \longrightarrow \infty.$$

Now define

$$b(n^d) = n^d \max_{1\leq k\leq n} \frac{b_k}{k^d},$$

and

$$b(k) = \frac{kb(n^d)}{n^d} \text{ for } (n-1)^d < k < n^d.$$

It follows that $b(n^d) \geq b_n$, and $\frac{b(n)}{n} \uparrow$ as $n \uparrow$. Also, $\exists\, n_k \to \infty$ such that $b(n_k^d) = b_{n_k}$.

Suppose that $A \in \mathcal{B}_+$, $M > 1$ and

$$\limsup_{n\to\infty} \frac{S_n f}{b_n} < M \quad \text{a.e. on } A.$$

Since $b_n \leq b(n^d)$, also

$$\limsup_{n\to\infty} \frac{S_n f}{b(n^d)} < M \quad \text{a.e. on } A.$$

By Theorem F,

$$\frac{S_n f}{b(n^d)} \to 0 \quad \text{a.e. , and } \frac{S_{n_k} f}{b_{n_k}} \to 0 \quad \text{a.e.}$$ □

1. The independent case

In this section we prove

Theorem 1.1. *Suppose that* $\{Y_g : g \in \mathbb{Z}^d\}$ *are independent, non-negative, identically distributed random variables, and that* $a: [0, \infty) \to [0, \infty)$ *is continuous, strictly increasing, and satisfies* $a(x + y) \leq a(x) + a(y)$, *and* $a(x) = o(x)$ *as* $x \to \infty$.

Then $\limsup_{n\to\infty} \frac{a(S_n)}{n^d} < \infty$ *a.e. iff* $E(a(X)) < \infty$. *Here* $S_n := \sum_{g\in\mathbb{Z}^d,\ \|g\|\leq n} Y_g$.

The proof of Theorem 1.1 will use

Lemma 1.2. *Suppose that* $(X, \mathcal{B}, m, T)$ *is a free, ergodic, probability preserving* $\mathbb{Z}^d$*-action on a Lebesgue probability space, and suppose that* $f: X \to \mathbb{R}$ *is measurable. Let* $a: [0, \infty) \to [0, \infty)$ *be continuous, strictly increasing, and satisfy*

$$a(x + y) \leq a(x) + a(y), \quad \textit{and} \quad a(x) = o(x)$$

as $x \to \infty$.

If $\int_X a(|f|)dm < \infty$, *then*

$$\frac{a(|S_n f|)}{n^d} \to 0 \textit{ a.e. as } n \to \infty.$$

Lemma 1.2 will also be used elsewhere in the sequel. For a one dimensional version of Lemma 1.2, see (**) in **[Aa2]**.

Proof of Lemma 1.2. We first show that we can assume without loss of generality that $a(0) = 0$. To do this, we exhibit $\tilde{a}: [0, \infty) \to [0, \infty)$ continuous, increasing, and such that $\tilde{a}(x + y) \leq \tilde{a}(x) + \tilde{a}(y)$, $\tilde{a}(0) = 0$, and, $a \equiv \tilde{a}$ on $[m, \infty)$ for some $m > 0$.

To do this, choose $m > 0$ such that $a(m) = \alpha m$ where $0 < \alpha < 1$, and define

$$\tilde{a}(x) = \begin{cases} \alpha x & 0 \leq x \leq m\,, \\ a(x) & x \geq m\,. \end{cases}$$

It is straightforward to verify the required properties.

Fix $\epsilon > 0$ with the intention of proving that

$$\limsup_{n\to\infty} \frac{a(|S_n f|)}{n^d} \leq \epsilon \quad \text{a.e.}$$

Now, we claim that there are measurable functions $g, h: X \to \mathbb{R}_+$ such that

$$|f| = g + h,\ \sup g < \infty, \text{ and } \int_X a(h)dm < \epsilon.$$

This is because

$$a\Big(|f| 1_{[|f|\geq M]}\Big) \to 0 \quad \text{a.e. as } M \to \infty,$$

whence by the dominated convergence theorem (as $a(|f|)$ is integrable)

$$\int_X a\left(|f|1_{[|f|\geq M]}\right)dm \to 0 \text{ as } M \to \infty,$$

and for suitably large M we can set

$$g = |f|1_{[|f|<M]}, \; h = |f|1_{[|f|\geq M]}.$$

Now, we have that

$$\begin{aligned}\frac{a(|S_n f|)}{n^d} &\leq \frac{a(S_n g)}{n^d} + \frac{a(S_n h)}{n^d} \\ &\leq \frac{a(Mn^d)}{n^d} + \frac{S_n a(h)}{n^d} \\ &\longrightarrow \int_X a(h)dm < \epsilon\,. \end{aligned}$$

□

Proof of Theorem 1.1. In order to complete the proof of this theorem, we must show that $\limsup_{n\to\infty} \frac{a(S_n)}{n^d} < \infty$ a.e. implies $E(a(X)) < \infty$. To do this, note that, by the Hewitt–Savage 0-1 law, $\limsup_{n\to\infty} \frac{a(S_n)}{n^d}$ is a.e. constant. Thus, $\limsup_{n\to\infty} \frac{a(S_n)}{n^d} < \infty$ on some set of positive measure implies that for some constant M,

$$\limsup_{n\to\infty} \frac{a(S_n)}{n^d} < M \quad \text{a.e.}$$

whence,

$$\sum_{g\in\mathbb{Z}^d} 1_{[a(Y_g)>M\|g\|^d]} < \infty \quad \text{a.e.}$$

and by the Borel–Cantelli lemma

$$\begin{aligned}\infty &> \sum_{g\in\mathbb{Z}^d} P([a(Y_g) > M\|g\|^d]) \\ &\asymp \sum_{n\geq 0} n^{d-1} P([a(X)^{\frac{1}{d}} > Mn]) \\ &\asymp E(a(X))\,. \end{aligned}$$

Here (and throughout) if $a_n, b_n > 0$ $(n \geq 1)$ then $a_n \asymp b_n$ means that $\exists\, M > 1$ such that $M^{-1} \leq \frac{a_n}{b_n} \leq M \; \forall\, n \geq 1$.

□

2. Random Kac inequality and proof of Theorem A

In this section, we prove our main theorem, using

Theorem 2.1 (Random Kac Inequality). *Let T be a free, ergodic, probability preserving $\mathbb{Z}^d$-action on the Lebesgue probability space $(X, \mathcal{B}(X), m)$, and let $(\Omega, \mathcal{B}(\Omega), P, \sigma)$ be the $\mathbb{Z}^d$-dyadic odometer.*

Given $A \in \mathcal{B}(X)$, $\exists\, \varphi\colon A \times \Omega \to \mathbb{N}$ such that

$$\text{for a.e. } (y, s) \in X \times \Omega,\ \exists\, (x, r) \in A \times \Omega, \qquad g \in \mathbb{Z}^d \ni T_g x = y, \sigma_g r = s,\ \|g\| \le \varphi(x, r) \tag{1}$$

and

$$\int_{A \times \Omega} \varphi^d dm < \infty.$$

First, we prove the theorem assuming the random Kac inequality. Suppose

$$\limsup_{n \to \infty} \frac{a(S_n f)}{n^d} < \infty \quad \text{a.e.}$$

and M is such that

$$A := \{x \in X : \frac{a(S_n f(x))}{n^d} \le M \ \forall\, n \ge 1\}$$

has positive measure. Let $\varphi\colon A \times \Omega \to \mathbb{N}$ be as in the random Kac inequality, and set

$$g(x, r) = \begin{cases} S^T_{\varphi(x,r)} f(x) & x \in A, \\ 0 & \text{else.} \end{cases}$$

Then

$$\int_{X \times \Omega} a(g) dm < \infty \ \because\ a(g) \le M \varphi^d 1_{A \times \Omega},$$

and by Lemma 1.2,

$$a(S_n^{T \times \sigma} g) = o(n^d) \quad m \times P\text{-a.e.}$$

The theorem is now established by the following

Claim. $S_{2n}^{T \times \sigma} g(x, r) \ge S_n^T f(x)$ $\forall\, n$ large.

Proof.

$$S_{2n}^{T \times \sigma} g(x, r) = \sum_{\|g\| < 2n} g(T_g x, \sigma_g r) = \sum_{\|g\| < 2n} \sum_{\|h\| < \varphi(T_g x, \sigma_g r)} f(T_{g+h} x)$$

It is therefore sufficient to prove that

$$\{T_{g+h} x : \|g\| < 2n,\ \|h\| < \varphi(T_g x, \sigma_g r)\} \supset \{T_c x : \|c\| < n\} \ \forall\, n \text{ large.}$$

We deduce this from the random Kac inequality as follows:

Since $\int_X \varphi^d dm \times p < \infty$ we have that

$$\sum_{g\in\mathbb{N}^d} m \times p([\varphi \geq \epsilon\|g\|]) < \infty \ \forall \epsilon > 0$$

whence

$$\limsup_{g\to\infty} \frac{\varphi \circ (T \times \sigma)_g}{\|g\|} = 0 \quad \text{a.e.}$$

$\therefore \exists N\colon A \times \Omega \to \mathbb{N}$ such that

$$\varphi(T_g x, \sigma_g r) \leq \frac{\|g\|}{2} \quad \forall \ \|g\| \geq N(x, r).$$

Suppose $n > N(x, r)$, $\|c\| < n$. By the random Kac inequality, $\exists\, d \in \mathbb{Z}^d$ such that $T_{c+d}x \in A$, and $\|d\| < \varphi(T_{d+c}x, \sigma_{c+d}r)$. Write $g = d + c$, and $h = -d$. We show that $\|g\| < 2n$ and $\|h\| < \varphi(T_g x, \sigma_g r)$.

Clearly $\|h\| = \|d\| \leq \varphi(T_{c+d}x, \sigma_{c+d}r) = \varphi(T_g x, \sigma_g r)$. To see that $\|g\| \leq 2n$, note that $\|g\| = \|c + d\| \leq n + \varphi(T_g x, \sigma_g r)$, so that if $\|g\| > N(x, r)$ (e.g. if $\|g\| > 2n$), then $\|g\| \leq n + \frac{\|g\|}{2}$, whence $\|g\| \leq 2n$. □

Proof of the random Kac inequality. For simplicity of notation, we restrict the proof to the case $d = 2$. The interested reader is invited to reconstruct the general proof from this.

We begin by recalling some facts about the $\mathbb{Z}^2$ dyadic odometer $(\Omega, \mathcal{B}, p, \sigma)$. Here

$$\Omega = \{0, 1\}^{\mathbb{N}} \times \{0, 1\}^{\mathbb{N}}, \quad \sigma_{1,0}r = (\tau u(r), v(r)), \quad \sigma_{0,1}r = (u(r), \tau v(r))$$

where τ is the one dimensional dyadic adding machine.

For $n \geq 1$, let

$$B_n(r) = \{r' \in \Omega : u_k(r') = u_k(r),\ v_k(r') = v_k(r),\ \forall\, k \geq n + 1\}.$$

Note that the $B_n(r)$ are equivalence classes of an equivalence relation which we denote by $\mathcal{R}_n$. Explicitly

$$B_n(r) = \{\sigma_b r \ :\ b \in -a(n, r) + Q_n\}$$

where

$$a(n, r) = \left(\sum_{k=0}^{n} 2^{k-1} u_k(r), \sum_{k=0}^{n} 2^{k-1} v_k(r)\right), \text{ and } Q_n = [0, 2^n - 1]^2 \cap \mathbb{Z}^2.$$

Write $D_n(r) = -a(n, r) + Q_n$. For $(\alpha, \beta) \in \{0, 1\}^2$ set

$$\begin{aligned} a_n^{(\alpha,\beta)}(r) &= a(n + 1, r) + 2^n(\alpha, \beta),\ D_n^{(\alpha,\beta)}(r) \\ &= -a_n^{(\alpha,\beta)}(r) + Q_n,\ B_n^{(\alpha,\beta)}(r) \\ &= \sigma_{D_n^{(\alpha,\beta)}(r)}\{r\}. \end{aligned}$$

It follows that

$$a(n,r) \in \{a_n^{(\alpha,\beta)}(r) : (\alpha,\beta) \in \{0,1\}^2\},\ D_{n+1}(r) = \bigcup_{(\alpha,\beta)\in\{0,1\}^2} D_n^{(\alpha,\beta)}(r),$$

and

$$B_{n+1}(r) = \bigcup_{(\alpha,\beta)\in\{0,1\}^2} B_n^{(\alpha,\beta)}(r).$$

Now define

$$\overline{B}_n(x,r) = T_{D_n(r)}\{x\},\ \overline{B}_n^{(\alpha,\beta)}(x,r) = T_{D_n^{(\alpha,\beta)}(r)}\{x\}.$$

It follows that

$$\overline{B}_{n+1}(x,r) = \bigcup_{(\alpha,\beta)\in\{0,1\}^2} \overline{B}_n^{(\alpha,\beta)}(x,r).$$

Next, let

$$\ell(x,r) = \min\{\ell : A \cap \overline{B}_n^{(\alpha,\beta)}(x,r) \neq \emptyset\ \forall\,(\alpha,\beta) \in \{0,1\}^2,\ n \geq \ell\}.$$

The claim is that $\ell(x,r) < \infty$ a.e. because, by the ergodic theorem, for any $A \in \mathcal{B}(X)_+$,

$$\forall\, M > 1, \text{ for a.e. } x,\ \exists\, N = N(M,x) \ni \sum_{\|g\|\leq n} 1_A(T_{g+h}x) \geq 1 \qquad (2)$$

$$\forall\, n \geq N,\ \|h\| \leq Mn,$$

whence, for $2^n \geq N(36,x)$,

$$\sum_{g\in D_n^{(\alpha,\beta)}(r)} 1_A(T_g x) \geq 1\ \forall\ \alpha,\beta = 0,1$$

since

$$D_n^{(\alpha,\beta)}(r) = a_n^{(\alpha,\beta)}(r) + Q_n,\ \|a_n^{(\alpha,\beta)}(r)\| \leq 2^{n+2} = 8 \cdot 2^{n-1};$$

and $\ell(x,r) \leq N+1$.

Set $D(x,r) := D_{\ell(x,r)}(r)$, and $\overline{B}(x,r) := \overline{B}_{\ell(x,r)}(x,r)$.

Lemma 2.2. $\ell(T_g x, \sigma_g r) = \ell(x,r)\ \forall\ g \in D(x,r)$.

Proof. Since $g \in D(x,r)$, we have that $(r, \sigma_g r) \in \mathcal{R}_n\ \forall\, n \geq \ell(x,r)$. It follows that

$$\overline{B}_n(T_g x, \sigma_g r) = B_n(x,r)\ \forall\, n \geq \ell(x,r),$$

whence $\ell(T_g x) \leq \ell(x)$. On the other hand, if $\ell(T_g x, \sigma_g r) < \ell(x,r)$, then

$$A \cap \overline{B}_{\ell(x,r)-1}^{(\alpha,\beta)}(T_g x, \sigma_g r) \neq \emptyset\ \forall\ (\alpha,\beta) \in \{0,1\}^2,$$

whence since

$$\{\overline{B}_{\ell(x,r)-1}^{(\alpha,\beta)}(T_g x, \sigma_g r) : (\alpha,\beta) \in \{0,1\}^2\} = \{\overline{B}_{\ell(x,r)-1}^{(\alpha,\beta)}(x,r) : (\alpha,\beta) \in \{0,1\}^2\}$$

we have $A \cap B^{(\alpha,\beta)}_{\ell(x,r)-1}(x,r) \neq \emptyset\ \forall\ (\alpha,\beta) \in \{0,1\}^2$, whence $\ell(x,r) \leq \ell(x,r) - 1$, a contradiction. □

Lemma 2.3. $\forall\ g \in \mathbb{Z}^2$, $\begin{cases} \text{either} & \overline{B}(T_g x, \sigma_g r) = \overline{B}(x,r) \\ \text{or} & \overline{B}(T_g x, \sigma_g r) \cap \overline{B}(x,r) = \emptyset \end{cases}$.

Proof. It follows from Lemma 2.2 that if $g \in D(x,r)$, then $\overline{B}(T_g x, \sigma_g r) = \overline{B}(x,r)$ or, in other words, writing, for $y = T_g x \in \overline{B}(x,r)$, $s = s(y) = \sigma_g r$ (the $g \in \mathbb{Z}^2$ involved is uniquely determined by the freeness of the action), $\overline{B}(y,s) = \overline{B}(x,r)\ \forall y \in \overline{B}(x,r)$. Now suppose that $g \in \mathbb{Z}^2$, and $y \in B(x,r) \cap B(T_g x \sigma_g r)$, then, evidently, $B(T_g x, \sigma_g r) = B(y,s) = B(x,r)$. □

Now, given $(x,r) \in X \times \Omega$, let $b_{(x,r)} \in Q_{\ell(x,r)}$ be lexicographically smallest such that $y(x,r) = T_{b_{(x,r)} - a(\ell(x,r),r)} x \in A$. It is clear from the definition of ℓ that $y(x,r) \in B(x,r)$. Now define

$$\varphi(T_b x, \sigma_b r) = \begin{cases} 2^{\ell(x,r)} & b = b_{(x,r)} - a(\ell(x,r),r)\,, \\ 0 & \text{else.} \end{cases}$$

By Lemma 2.3, $\forall\ (x,r) \in X \times \Omega$, $\varphi : \{(T \times \sigma)_g(x,r) :\ g \in \mathbb{Z}^2\} \to \mathbb{N}$ is well defined, and for $k \in \mathbb{N}$,

$$\{(x,r) \in X \times \Omega :\ \varphi(x,r) = 2^k\} = \{(x,r) \in X \times \Omega : \ell(x,r) = k, \\ \text{and } x = y(x,r)\} \in \mathcal{B}(A \times \Omega).$$

To show $\int_X \varphi^2 dm dp < 36$, we show that

$$\limsup_{N\to\infty} \frac{1}{N^2} \sum_{\|g\|\leq N} \varphi(T_g x)^2 \leq 36 \quad \text{a.e.}$$

Claim. For a.e $(x,r) \in X \times \Omega\ \exists\ n_{(x,r)}\ \ni\ \forall\ g \in \mathbb{Z}^2,\ \|g\| \geq n_{(x,r)} :\ \varphi(T_g x, \sigma_g r) \leq 2\|g\|$.

Proof. By (2), for a.e. $x \in X$, $\exists\ N$ such that

$$\sum_{g \in b + Q_n} 1_A(T_g x) \geq 1\ \ \forall\ 2^n \geq N,\ \|b\| \leq 2^{n+2}.$$

For $g \in \mathbb{Z}^2$, we have that

$$\overline{B}_n(T_g x, \sigma_g r) = T_{a'_n} \overline{B}_n(x,r) \text{ where } \|a'_n\| \leq \|g\|,$$

whence

$$\overline{B}_n^{(\alpha,\beta)}(T_g x, \sigma_g r) = T_{a'_{n+1}} \overline{B}_n^{(\alpha,\beta)}(x,r).$$

Suppose that $2^n \geq \|g\| \geq N$, then

$$\overline{B}_n^{(\alpha,\beta)}(T_g x, \sigma_g r) = T_{a'_{n+1} + D_n^{(\alpha,\beta)}(r)}\{x\} = T_{a'_{n+1} + a_n^{(\alpha,\beta)}(r) + Q_n}\{x\}$$

where $\|a'_{n+1} + a_n^{(\alpha,\beta)}(r)\| \le 2^{n+2}$ and so

$$\#\overline{B}_n^{(\alpha,\beta)}(T_g x, \sigma_g r) \cap A \ge 1 \ \forall\ \alpha, \beta = 0, 1$$

whence $\ell(T_g x, \sigma_g r) \le n$. The conclusion is that $\varphi(T_g x, \sigma_g r) \le 2^n$ whenever $N \le \|g\| \le 2^n$, in particular $\varphi(T_g x, \sigma_g r) \le 2\|g\|$ whenever $N \le \|g\|$. □

Now let $(x, r) \in X \times \Omega$, and $n_{(x,r)}$ be as in the claim, then $\forall\ N \ge n_{(x,r)}$,

$$\begin{aligned}\sum_{\|g\| \le N} \varphi(T_g x, \sigma_g r)^2 &= 4\#\Bigg(\bigcup_{g \in \mathbb{Z}^2,\ \overline{B}(T_g x, \sigma_g r) \cap T_{[-N,N]^2}\{x\} \ne \emptyset} \overline{B}(T_g x, \sigma_g r)\Bigg) \\ &\le 4\#T_{[-3N,3N]^2}\{x\} \\ &= 36N^2.\end{aligned}$$

□

3. Counterexamples

In this section, we show that generalization of Theorem 1.1 to an arbitrary non-negative, stationary, ergodic $\mathbb{Z}^d$ random field is impossible.

Theorem 3.1. *Let T be a free, ergodic, probability preserving $\mathbb{Z}^d$ action on a standard probability space X. Suppose $a(x + y) \le a(x) + a(y)$ and $a(x)/x \to 0$ as $x \to \infty$. There is a measurable function $f: X \to [0, \infty)$ such that*

$$E(a(f)) = \infty, \text{ but } a(S_n f) = o(n^d) \text{ a.e. as } n \to \infty.$$

Proof. Start with $\epsilon, \delta > 0$, $K > 0$, and choose $M \in \mathbb{N}$ such that $a(M)\epsilon > K$. Next, choose $N \in \mathbb{N}$ such that $\frac{a(M(2n)^d)}{n^d} < \delta$ for every $n \ge N$.

By **[K-W]**, $\exists\ A \in \mathcal{B}(X)$ such that $\{T_a A : \|a\| \le 2N\}$ are disjoint, and

$$m\Bigg(\bigcup_{\|a\| \le 2N} T_a A\Bigg) = 4\epsilon,$$

and set

$$f = M \sum_{\|a\| \le N} 1_A \circ T_a.$$

This function $f: X \to \mathbb{N}$ satisfies

$$\int_X a(f) dm = a(M)\epsilon > K,$$

and

$$\frac{a(S_n f(x))}{n^d} < \delta\ \forall\, n \ge 1,\ \ x \notin B := \bigcup_{\|a\| \le 2N} T_a A, \text{ and } \forall\, n \ge N,\ x \in X.$$

To see this, note that for $x \notin \bigcup_{\|a\| \le 2N} T_a A$, $S_n f(x) = 0\ \forall n \le N$, and for $n \ge N$, $S_n f(x) \le (2n)^d M$, whence

$$\frac{a(S_n f(x))}{n^d} \le \frac{a(2^d n^d M)}{n^d} < \delta, \quad \forall n \ge N.$$

To prove the theorem, fix $\epsilon_k = \delta_k = \frac{1}{2^k}$, $K_k = k$, construct

$$f_k \colon X \to [0, \infty), \quad A_k, B_k \in \mathcal{B}, \quad M_k, N_k \in \mathbb{N}$$

accordingly, and set

$$F = \sum_{k=1}^{\infty} f_k.$$

Clearly

$$\begin{aligned} \int_X a(F)dm &\ge \int_X a(f_k)dm \quad \forall\, k \ge 1 \\ &\ge k \to \infty \text{ as } k \to \infty. \end{aligned}$$

Let

$$\kappa(x) = \min\{k \ge 1 : x \notin B_r\ \forall\, r \ge k\}$$

which is finite a.e. because $\sum_{k=1}^{\infty} m(B_k) < \infty$. It follows that for $\ell \ge \kappa(x)$, $n \ge N_\ell$,

$$\begin{aligned} a(S_n F) &\le \sum_{k=1}^{\infty} a(S_n f_k) \\ &= \sum_{k=1}^{\ell} a(S_n f_k) + \sum_{k=\ell+1}^{\infty} a(S_n f_k) \\ &\le \ell a(S_n f_\ell) + \sum_{k=\ell+1}^{\infty} \frac{n^d}{2^k} \\ &\le \frac{\ell+1}{2^\ell} n^d. \end{aligned}$$

□

4. Concluding remarks

Suppose that T is an ergodic probability preserving $\mathbb{Z}^d$ action on $(X, \mathcal{B}, m)$, and suppose that $f \colon X \to \mathbb{R}$ is measurable.

Let $a \colon [0, \infty) \to [0, \infty)$ be continuous, strictly increasing, and satisfy $a(x + y) \le a(x) + a(y)$, and $a(x) = o(x)$ as $x \to \infty$.

In this section, searching for a suitable generalization of Theorem 1.1, we'll try to characterise the situation

$$\limsup_{n\to\infty} \frac{1}{n^d} a\Big(|S_n f|\Big) = \infty \quad \text{a.e.}$$

Let us define a *Kac function* for a set $A \in \mathcal{B}_+$ to be a measurable function $\varphi: A \to \mathbb{N}$ satisfying

$$\bigcup_{n=1}^{\infty} \bigcup_{\|g\| \leq n} T_g[\varphi = n] = X \mod m,$$

and

$$\int_A \varphi^d dm < \infty.$$

When $d = 1$, the first return time function to a set $A \in \mathcal{B}_+$ is a Kac function for that set by Kac's formula (see §1 and **[Kac]**).

The proof of the random Kac inequality can be adapted to show that if T has the $\mathbb{Z}^d$ dyadic odometer as factor, then every set of positive measure has a Kac function. Indeed, this last is true in general and will be proved in a future article.

This means that there is no loss of generality if we restrict ourselves to those $\mathbb{Z}^d$ actions T with the property that every set of positive measure has a Kac function. The methods of §2 now establish

Theorem 4.1. *If $\exists A \in \mathcal{B}_+$, and a Kac function $\varphi: A \to \mathbb{N}$ such that*

$$\int_A a\left(\max_{1 \leq n \leq \varphi} |S_n f|\right) dm < \infty,$$

then

$$\frac{1}{n^d} a\left(|S_n f|\right) \to 0 \text{ a.e. as } n \to \infty.$$

Otherwise,

$$\limsup_{n \to \infty} \frac{1}{n^d} a\left(|S_n f|\right) = \infty \text{ a.e.}$$

References

[Aa1] J. Aaronson, On the ergodic theory of non-integrable functions and infinite measure spaces, Israel J. Math. 27 (1977), 163–173.

[Aa2] J. Aaronson, An ergodic theorem with large normalising constants, Israel J. Math. 38 (1981), 182–188.

[C-R] Y. S. Chow, H. Robbins, On sums of independent random variables with ∞ moments, Proc. Nat. Acad. Sci. U.S.A. 47 (1961), 330–335.

[Fe] W. Feller, A limit theorem for random variables with infinite moments, Amer. J. Math. 68 (1946), 257–262.

[Kac] M. Kac, On the notion of recurrence in discrete stochastic processes, Bull. Amer. Math. Soc. 53 (1947), 1002–1010.

[K-W] Y. Katznelson, B. Weiss, Commuting measure preserving transformations, Israel J. Math. 12 (1972), 161–172.

[Kr] U. Krengel, Ergodic theorems, de Gruyter Stud. Math. 6, Walter de Gruyter, Berlin-New York 1985.

School of Mathematical Sciences
Tel Aviv University
69978 Tel Aviv, Israel
aaro@math.tau.ac.il

Institute of Mathematics
Hebrew University of Jerusalem
Jerusalem, Israel
weiss@math.huji.ac.il

A Repeated Filling Scheme for Upcrossings

M. A. Akcoglu, J. R. Baxter, and S. E. Ferrando

1. Introduction

A general upcrossing inequality, that holds in particular in the setting of the Ratio Ergodic Theorem, was proved by Bishop **[1]**, **[2]**, **[3]**.This result implies the Ratio Ergodic Theorem, and also gives some information about the rate of convergence. Since the Maximal Ergodic Lemma is essentially a statement about the first upcrossing of a particular interval, the upcrossing inequality also implies the Maximal Ergodic Lemma. In the present paper we will give a simple proof of Bishop's theorem in the basic case, that is, in the case of the Ratio Ergodic Theorem setting. Our proof will be based on a variant of the filling scheme construction used originally by Chacon and Ornstein to prove the Maximal Ergodic Lemma **[4]**. We will also strengthen Bishop's result in one direction, since we allow a more general form of upcrossing, as explained below.

The existence of a filling scheme construction in this setting is not surprising, since the work of Rost (cf. **[9]**, **[7]**, **[8]**) has given a general criterion for the existence of a filling scheme connecting two distributions. Bishop's proof of the upcrossing inequality uses a combinatorial argument together with a technique similar to Garsia's proof (cf. **[5]**) of the Maximal Ergodic Lemma. An examination of Bishop's proof shows that (just as in the case of Garsia's proof) the result of Rost is applicable, in the sense that by combining Rost's theorem with the facts proved by Bishop one can deduce the existence of an appropriate filling scheme. However, to prove the existence of the filling scheme in this way would not be shorter than the simple direct argument that we give here.

We now introduce some notation. Let $(X, \mathfrak{F}, \mu)$ be a complete σ-finite measure space. We will study $L^1(X, \mathfrak{F}, \mu)$, and will denote this space more briefly by L^1. Let T be a positive L^1-contraction, that is, let T be a linear map from L^1 to itself such that $||T\varphi||_1 \leq ||\varphi||_1$ for every $\varphi \in L^1$, and such that $T\varphi \geq 0$ for every $\varphi \in L^1$ with $\varphi \geq 0$. The transformation T will be considered to be fixed throughout this paper. The Ratio Ergodic Theorem states that for any nonnegative φ and θ in L^1 the functions ξ_n defined by

$$\xi_n = \frac{\varphi + T\varphi + \cdots + T^{n-1}\varphi}{\theta + T\theta + \cdots + T^{n-1}\theta} \tag{1}$$

converge almost everywhere on $\{\theta > 0\}$. We wish to consider a bound for the upcrossings of the sequence of functions ξ_n defined by (1). Before stating this bound, we will give a somewhat abstract definition of the notion of a sequence of upcrossings.

Definition 1. Let f, g, p be nonnegative functions in L^1. Let versions be chosen for the L^1-functions $T^j(f-g)$, $T^j(f-g-p)$, $j=0,1,\ldots$. Let u_r, v_r, $r=1,2,\ldots$ be elements in $\{-1,0,1,\ldots,\infty\}$, such that

$$u_r < v_r \text{ if } u_r < \infty, \tag{2}$$

$$v_r < u_{r+1} \text{ if } v_r < \infty. \tag{3}$$

We denote the sequence $(u_r, v_r)_{r=1,2,\ldots}$ by ζ. For a given point x, the sequence ζ will be called a sequence of upcrossing times at x for f, g, p if

$$\sum_{j=0}^{u_r} T^j(f-g)(x) \le \sum_{j=0}^{v_r} T^j(f-g-p)(x) \tag{4}$$

for all r with $v_r < \infty$, and

$$\sum_{j=0}^{u_{r+1}} T^j(f-g)(x) \le \sum_{j=0}^{v_r} T^j(f-g-p)(x) \tag{5}$$

for all r with $u_{r+1} < \infty$.

To connect this definition with the usual definition of an upcrossing, let φ and θ be nonnegative functions in L^1, and let α, β be nonnegative numbers with $\alpha < \beta$. Let $f=\varphi$, $g=\alpha\theta$, $p=(\beta-\alpha)\theta$. Let versions be chosen for the L^1-functions $T^j\varphi$, $T^j\theta$, and define versions

$$T^j(f-g) = T^j\varphi - \alpha T^j\theta, \quad T^j(f-g-p) = T^j\varphi - \beta T^j\theta.$$

Let x be a point. Let u_r, v_r, $r=1,2,\ldots$, be a sequence of elements in $\{-1,0,1,\ldots\}$ such that (2) and (3) hold, such that if ξ_n is defined as in (1) then $\xi_{u_r}(x) \le \alpha$ for all r with $0 \le u_r < \infty$ and $\xi_{v_r}(x) \ge \beta$ for all r with $v_r < \infty$. It is easy to see that in this case u_r, v_r will also be a sequence of upcrossing times at x for the functions f, g, p, in the sense of Definition 1. Accordingly, a bound for the number of upcrossings according to Definition 1 will certainly give a bound for the number of upcrossings of an interval in the context of the Ratio Ergodic Theorem.

Definition 1 is a special case of the definition of upcrossings in the more general setting used by Bishop. For simplicity we will not deal with Bishop's most general case. However, we will widen the definition of an upcrossing sequence, as follows.

Definition 2. Let f, g, p be nonnegative functions in L^1. Let versions be chosen for the L^1-functions $T^j(f-g)$, $T^j(f-g-p)$, $T^j(f-p)$, $j=0,1,\ldots$. Let u_r, v_r, $r=1,2,\ldots$ be elements in $\{-1,0,1,\ldots,\infty\}$, such that

$$u_r < v_r \text{ if } u_r < \infty, \tag{6}$$

$$v_r \le u_{r+1} \text{ for all } r. \tag{7}$$

We denote the sequence $(u_r, v_r)_{r=1,2,\ldots}$ by ζ. For a given point x, the sequence ζ will be called a sequence of generalized upcrossing times at x for f, g, p if

$$\sum_{j=0}^{u_r} T^j(f-g)(x) \leq \sum_{j=0}^{v_r-1} T^j(f-g-p)(x) + T^{v_r}(f-p)(x) \tag{8}$$

for all r with $v_r < \infty$, and

$$\sum_{j=0}^{u_{r+1}} T^j(f-g)(x) \leq \sum_{j=0}^{v_r-1} T^j(f-g-p)(x) + T^{v_r}(f-p)(x) \tag{9}$$

for all r with $u_{r+1} < \infty$.

It is easy to verify that an upcrossing sequence at a point x in the sense of Definition 1 is also an upcrossing sequence at a point x in the sense of Definition 2. From now on we will concentrate on finding a bound for the upcrossings of Definition 2.

We will prove the following theorem (cf. Theorem 6 in Chapter 8 of **[2]**).

Theorem 1. *Let f, g, p be nonnegative functions in L^1. Let $\mathfrak{U}$ be the set of all sequences $\zeta = (u_r, v_r)_{r=1,2,\ldots}$ satisfying* (6) *and* (7). *For each $\zeta \in \mathfrak{U}$, let*

$$\text{length}(\zeta) = \sup\{\mathrm{r} : \mathrm{v_r} < \infty\},$$

with $\text{length}(\zeta) = 0$ *if $v_r = \infty$ for all r. Let*

$$\omega(x) = \sup \text{length}(\zeta),$$

where the supremum is taken over all $\zeta \in \mathfrak{U}$ such that ζ is a sequence of generalized upcrossing times at x for the functions f, g, p. This defines ω up to equivalence. The function ω is measurable, and satisfies the inequality

$$\int \omega p \, d\mu \leq \int_{\{\omega \geq 1\}} f \, d\mu. \tag{10}$$

The measurability of ω follows easily from the fact that we could have defined the same function ω using the set $\mathfrak{U}_0$ in place of $\mathfrak{U}$, where $\mathfrak{U}_0$ is the set of elements in $\mathfrak{U}$ with finite length, so that $\mathfrak{U}_0$ is countable.

We will prove Theorem 1 in the Section 3. Our proof has the same structure as the proof of the Maximal Ergodic Lemma in **[4]**, in the sense that we first define the filling scheme and collect some intuitively obvious facts about it, in Section 2, and then complete the proof of Theorem 1 by simply writing down the definition of an upcrossing and expressing it in terms of quantities associated with the filling scheme.

The following example illustrates the more general definition of upcrossing given in Definition 2.

Example. Let φ and θ be nonnegative functions in L^1. Let versions be chosen for the L^1-functions $T^j\varphi$, $T^j\theta$, $j = 0, 1, \ldots$. Let $c > 0$ be a real number. Let $f = \varphi$, $g = \varphi$, $p = c\theta$. For any x, let $(w_r)_{r=1,2,\ldots}$ be a sequence taking values in

$\{0, 1, 2, \ldots, \infty\}$, such that $w_r < w_{r+1}$ if $w_r < \infty$, and such that the sequence (w_r) consists of exactly those integers $n \geq 0$ with

$$T^n\varphi(x) \geq c \sum_{j=0}^{n} T^j\theta(x). \tag{11}$$

Let $u_r = w_r - 1$, $v_r = w_r$ for all $r = 1, 2, \ldots$. Let $\zeta = (u_r, v_r)_{r=1,2,\ldots}$. Then ζ is a sequence of generalized upcrossing times at x for f, g, p so the function ω is an upper bound for the number of times that (11) holds.

Theorem 1 implies in particular that the integral of ωp is finite, so that ωp is finite almost everywhere for every choice of $c > 0$. Thus

$$\lim_{n\to\infty} \frac{T^n\varphi}{\sum_{j=0}^{n} T^j\theta} = 0 \text{ a.e. on } \{\theta > 0\}.$$

This is a well-known fact which is usually established in the course of proving the Ratio Ergodic Theorem (cf. Lemma 2.3 in Section 3.2 of **[6]**).

Remark. As another illustration of the use of generalized upcrossing times, we note that Theorem 1 implies the main combinatorial result (Lemma 1) in **[1]**.

Consequences of filling

Although the upcrossing inequality for the Ratio Ergodic Theorem appears similar to that for submartingales, it might be noted that there is extra information in the filling scheme. Let's consider again the upcrossings of an interval (α, β) as described following Definition 1. Here the filling scheme statement of the upcrossing inequality is that the function $(\beta - \alpha)\omega\theta$ can be completely filled by the mass of the function φ (see (31) below). A standard argument then shows that the ratio limit using φ and θ is greater than or equal to the ratio limit using $(\beta - \alpha)\omega\theta$ and θ. In particular if there are at least k upcrossings everywhere on the space then the ratio limit using φ and θ is at least $k(\beta - \alpha)$. Similarly if there are at least k upcrossings on most of the space then the ratio limit using φ and θ is nearly $k(\beta - \alpha)$ on most of the space. Statements of this sort are of course not generally true for submartingales.

2. A filling scheme

We continue to use the same space L^1 and operator T as in Section 1. Let f, g, p be given nonnegative functions in L^1. We will define an alternating filling scheme. Intuitively, the function p will represent a hole, which we alternately fill, at each point, with mass from f and from g. As usual, the operator T is the shovel that we use to move the mass.

We will define our alternating filling scheme soon, by inductively defining sequences of functions φ_n, θ_n, π_n, κ_n, ρ_n, σ_n, $n = 0, 1, \ldots$. The functions φ_n, θ_n, π_n, κ_n

are nonnegative L^1-functions. The functions ρ_n, σ_n are bounded measurable functions taking nonnegative integer values. Our construction of these functions takes place in steps, labelled by the time n. Each step n has three parts, which we refer to as turns. During the first turn of step n, we attempt to fill a copy of the hole using the mass of f, if this is appropriate. During the second turn of step n, we attempt to fill a copy of the hole using the mass of g, if this is appropriate. During the third turn of step n, we move the unused mass of both f and g, using the operator T. The function φ_n represents the unused mass of f just before step n. The function π_n represents the unfilled portion, just before step n, of the copy of the hole that we will try to fill with f-mass during the first turn of step n. The value $\rho_n(x)$ is the number of times that a copy of the hole has been completely filled at x with f-mass prior to step n. The functions θ_n, κ_n, and σ_n have similar interpretations using g instead of f.

For convenience we will also define some other functions which are expressible in terms of the functions already mentioned.

We now begin the inductive definition of all these functions. It should be noted that we construct φ_{n+1}, θ_{n+1}, π_{n+1}, κ_{n+1}, ρ_{n+1}, σ_{n+1} during step n, since these functions refer to conditions just before step $n+1$. We also construct functions e_n, h_n, f_n, g_n during step n. These functions describe results of operations during step n.

Initial definitions

Let

$$\varphi_0 = f,\ \theta_0 = g,\ \pi_0 = \kappa_0 = p,\ \rho_0 = \sigma_0 = 0.$$

Step n. Suppose that we have already defined φ_n, θ_n, π_n, κ_n, ρ_n, σ_n for some $n = 0, 1, \ldots$.

First turn. On $\{\rho_n = \sigma_n\}$ let $e_n = \varphi_n \wedge \pi_n$. On $\{\rho_n = \sigma_n\}^c$ let $e_n = 0$. On $\{\rho_n = \sigma_n\} \cap \{e_n = \pi_n\}$ let $\rho_{n+1} = \rho_n + 1$, $\pi_{n+1} = p$. On $(\{\rho_n = \sigma_n\} \cap \{e_n = \pi_n\})^c$ let $\rho_{n+1} = \rho_n$, $\pi_{n+1} = \pi_n - e_n$.

Second turn. On $\{\rho_{n+1} = \sigma_n + 1\}$ let $h_n = \theta_n \wedge \kappa_n$. On $\{\rho_{n+1} = \sigma_n + 1\}^c$ let $h_n = 0$. On $\{\rho_{n+1} = \sigma_n + 1\} \cap \{h_n = \kappa_n\}$ let $\sigma_{n+1} = \sigma_n + 1$, $\kappa_{n+1} = p$. On $(\{\rho_{n+1} = \sigma_n + 1\} \cap \{h_n = \kappa_n\})^c$ let $\sigma_{n+1} = \sigma_n$, $\kappa_{n+1} = \kappa_n - h_n$.

Third turn. Let $f_n = \varphi_n - e_n$, $g_n = \theta_n - h_n$, and let $\varphi_{n+1} = Tf_n$, $\theta_{n+1} = Tg_n$.

It is easy to see by induction that the sequences (ρ_n) and (σ_n) are nondecreasing, and that

$$\sigma_n \le \rho_n \le \rho_{n+1} \le \sigma_n + 1 \tag{12}$$

for all n. In particular $\rho_n \le n$ for all n.

It is also obvious that for all n we have

$$0 \le e_n \le \pi_n \le p, \quad 0 \le h_n \le \kappa_n \le p. \tag{13}$$

Again by induction we have at once that

$$\{\rho_n = \sigma_n\} \subset \{\kappa_n = p\}, \ \{\rho_{n+1} = \sigma_n + 1\} \subset \{\pi_{n+1} = p\}. \tag{14}$$

For each $n = 0, 1, \dots,$ the usual easy induction shows the following standard filling scheme fact:

$$T^n f = f_n + \sum_{k=0}^{n} T^{n-k} e_k, \tag{15}$$

$$T^n g = g_n + \sum_{k=0}^{n} T^{n-k} h_k. \tag{16}$$

In the same way, we also have

Lemma 1. *For every n,*

$$\int f \geq \sum_{j=0}^{n} \int e_j + \int f_n.$$

We next must justify another intuitively obvious fact, namely that the functions ρ_n and σ_n measure the number of times that the holes have been filled. This follows at once from the next lemma, which is also easily proved by induction.

Lemma 2. *For all $n = -1, 0, 1, \dots,$*

$$\sum_{k=0}^{n} e_k = \rho_{n+1} p + (p - \pi_{n+1}), \tag{17}$$

and

$$\sum_{k=0}^{n} h_k = \sigma_{n+1} p + (p - \kappa_{n+1}). \tag{18}$$

Another property of the filling scheme

The properties given so far are essentially the same as those of the standard filling scheme. We will need another fact, expressed in (21) and (22) below, in our proof of the upcrossing inequality. Equations (21) and (22) involve the relation between the filling using f and the filling using g, and so do not correspond to properties of the ordinary filling scheme. However, they are still very easy to prove and are intuitively obvious when looked at the right way.

We first define two new quantities. For each $n = 0, 1, \dots,$ let

$$\alpha_n = p + \sum_{k=0}^{n-1} (h_k - e_k), \quad \beta_n = \sum_{k=0}^{n} e_k - \sum_{k=0}^{n-1} h_k. \tag{19}$$

We also define $\alpha_n = p$ and $\beta_n = 0$ for all negative integers n.

As a consequence of the definition we have

$$\alpha_n - e_n + \beta_n = p, \quad \alpha_{n+1} + \beta_n - h_n = p \tag{20}$$

for all $n \geq 0$.

From Lemma 2, (14) and (20) we see that whenever $n \geq 0$ and $\rho_n = \sigma_n$ we have

$$\alpha_n = \pi_n, \quad \beta_n = p - (\pi_n - e_n), \tag{21}$$

and whenever $n \geq 0$ and $\rho_{n+1} = \sigma_n + 1$ we have

$$\alpha_{n+1} = p - (\kappa_n - h_n), \quad \beta_n = \kappa_n. \tag{22}$$

In particular, we have established that

$$0 \leq \alpha_n \leq p, \quad 0 \leq \beta_n \leq p \tag{23}$$

for all n.

These equations have a simple intuitive meaning, since we could have constructed our filling scheme in a slightly different way, which we now describe. In our previous description, we had essentially one hole at each point. When this hole was filled with mass from either f or g, it was immediately emptied, the mass being discarded, and filling began again using the mass of the other function. Now we will suppose that we have two holes at each point, instead of one. We use one hole for f-mass and one for g-mass. We begin with the f-hole empty and the g-hole full. As we fill the f-hole, whenever mass is put into the f-hole we remove and discard an equal amount from the g-hole. Clearly, when the f-hole is filled the g-hole is empty. We continue in this way, next filling the g-hole, and removing an equal mass from the f-hole whenever mass is put into the g-hole. At any moment during the construction one hole is being filled and the other is being emptied, at the same rate. The quantities connected with the hole being filled correspond to the quantities in the earlier construction. The size, just before step n, of the empty part of the hole for f-mass is α_n, and the size, just before the second turn of step n, of the empty part of the hole for g-mass is β_n.

3. Linking filling and upcrossings

At this point we have finished proving the facts about the filling scheme that we need. Now we give the crucial argument for Theorem 1.

Lemma 3. *As in Theorem* 1, *let* $\mathfrak{U}$ *be the set of all sequences* $\zeta = (u_r, v_r)_{r=1,2,\ldots}$ *satisfying* (6) *and* (7). *For any* $\zeta \in \mathfrak{U}$, *let* $H(\zeta)$ *be the set of* x *such that* ζ *is a sequence of generalized upcrossing times at* x *for* f, g, p.

Let $\zeta \in \mathfrak{U}$, $\zeta = (u_r, v_r)_{r=1,2,\ldots}$. *For* $r = 1, 2, \ldots$,

(i) *if* $v_r < \infty$ *then*

$$\rho_{v_r+1} \geq r \text{ almost everywhere on } \mathrm{H}(\zeta); \tag{24}$$

and

(ii) *if* $u_{r+1} < \infty$,

$$\sigma_{u_{r+1}+1} \geq r \text{ almost everywhere on } \mathrm{H}(\zeta). \tag{25}$$

Proof. We shall prove by induction on $n = -1, 0, 1, \ldots$ that

(a) inequality (24) holds for all r such that $v_r \leq n$;
(b) inequality (25) holds for all r such that $u_{r+1} \leq n$.

Let $n = -1$. We see by (6) that $v_r \leq n$ never holds, and we see by (7) that $u_{r+1} \leq n$ never holds. Thus (a) and (b) are true for $n = -1$.

Assume that (a) and (b) hold for $n = -1, 0, \ldots, m-1$, for some $m \geq 0$. We will prove that they also hold for $n = m$.

Suppose that r is such that $v_r = m$. Let $u_r = \ell$. Then $\ell < m$. If $r > 1$, then by our inductive assumption $\sigma_{\ell+1} \geq r-1$ almost everywhere on $H(\zeta)$. The same inequality is clearly true for $r = 1$. Thus for all r, $\rho_{\ell+1} \geq r-1$ almost everywhere on $H(\zeta)$. We wish to show that $\rho_{m+1} \geq r$ almost everywhere on $H(\zeta)$. On $\{\rho_{\ell+1} \geq r\}$ this is certainly true. On $\{\rho_{\ell+1} = r-1\} \cap \{f_j > 0\}$, for any $j = \ell+1, \ldots, m$, we have from the definition of ρ_{j+1} that if $\rho_j = r-1$ then $\rho_{j+1} = r$, and we are done. Thus we need only consider ρ_{m+1} on the set

$$A \equiv \{\rho_m = r-1\} \cap \{f_{\ell+1} = 0\} \cap \cdots \cap \{f_m = 0\} \cap H(\zeta).$$

On A, combining (8) with (15) and (16), we have

$$\sum_{j=0}^{m} T^j p \leq \sum_{j=\ell+1}^{m} \sum_{k=0}^{j} T^{j-k} e_k - \sum_{j=\ell+1}^{m-1} \sum_{k=0}^{j} T^{j-k} h_k. \tag{26}$$

Exchanging the order of the sums, using the definition of α_n and β_n, and integrating both sides of the inequality over A, we have

$$\sum_{i=0}^{m} \int (T^{*i} 1_A) p \leq \sum_{i=0}^{m} \int (T^{*i} 1_A)(\beta_{m-i} + \alpha_{\ell+1-i} - p). \tag{27}$$

Since $0 \leq \alpha_n, \beta_n \leq p$ for all n, we see that

$$(T^{*i} 1_A)\beta_{m-i} = (T^{*i} 1_A) p$$

almost everywhere for each $i = 0, 1, \ldots, m$. In particular, $\beta_m = p$ almost everywhere on A. But since $\rho_m = r-1$ we then have $e_m = \pi_m$ by (21) and hence $\rho_{m+1} = \rho_m + 1 = r$ by the definition of the filling scheme. Thus $\rho_{m+1} \geq r$ in all cases and (a) holds for $n = m$, as claimed.

Let us now consider (b). Suppose that r is such that $u_{r+1} = m$. Let $v_r = \ell$. Then $\ell \leq m$. By (a) for $n = m$, we have $\rho_{\ell+1} \geq r$ almost everywhere on $H(\zeta)$. Thus using (12) we know that $\sigma_\ell \geq r-1$ almost everywhere on $H(\zeta)$. We wish to show that $\sigma_{m+1} \geq r$ almost everywhere on $H(\zeta)$. On $\{\sigma_\ell \geq r\}$ this is certainly true. On $\{\sigma_\ell = r-1\} \cap \{g_j > 0\}$, for any $j = \ell, \ldots, m$, we have from the definition of σ_{j+1} that if $\sigma_j = r - 1$ then $\sigma_{j+1} = r$, and we are done. Thus we need only consider values on the set

$$B \equiv \{\sigma_m = r - 1\} \cap \{g_\ell = 0\} \cap \cdots \cap \{g_m = 0\} \cap H(\zeta).$$

On B, combining (9) with (15) and (16), we have

$$\sum_{j=0}^{\ell} T^j p \leq - \sum_{j=\ell+1}^{m} \sum_{k=0}^{j} T^{j-k} e_k + \sum_{j=\ell}^{m} \sum_{k=0}^{j} T^{j-k} h_k. \tag{28}$$

Exchanging the order of the sums, using the definition of α_n and β_n, and integrating both sides of the inequality over B, we have

$$\sum_{i=0}^{\ell} \int (T^{*i} 1_B) p \leq \sum_{i=0}^{m} \int (T^{*i} 1_B)(\alpha_{m+1-i} - p + \beta_{\ell-i}). \tag{29}$$

If $i > \ell$ then $\beta_{\ell-i} = 0$, so $\alpha_{m+1-i} - p + \beta_{\ell-i} \leq 0$. Thus we have

$$\sum_{i=0}^{\ell} \int (T^{*i} 1_B) p \leq \sum_{i=0}^{\ell} \int (T^{*i} 1_B)(\alpha_{m+1-i} - p + \beta_{\ell-i}). \tag{30}$$

It follows that

$$(T^{*i} 1_B)\alpha_{m+1-i} = (T^{*i} 1_B) p$$

holds almost everywhere for each $i = 0, 1, \ldots, \ell$. In particular, $\alpha_{m+1} = p$ almost everywhere on B. We then have $h_m = \kappa_m$ by (22) and since $\sigma_m = r - 1$ we then have $\sigma_{m+1} = \sigma_m + 1 = r$ by the definition of the filling scheme. Thus $\sigma_{m+1} \geq r$ in all cases, proving (b) for $n = m$ and proving the lemma. □

As immediate corollaries, we see first that for any $\zeta \in \mathfrak{U}$,

$$\lim_{n\to\infty} \rho_n(x) \geq \text{length}(\zeta),$$

almost everywhere on $H(\zeta)$, and hence that

$$\lim_{n\to\infty} \rho_n(x) \geq \omega \tag{31}$$

almost everywhere. Combining this fact with Lemma 2 gives

$$\int \omega p \, d\mu \leq \int_{\{\omega \geq 1\}} (\sum_{n=0}^{\infty} e_n) \, d\mu.$$

Let $\tilde{f} = 1_{\{\omega \geq 1\}} f$. Since $f \leq p$ on $\{\omega < 1\}$, it is easy to see from the definitions that if we perform our filling scheme with f replaced by $\tilde{f}$ and p replaced by $\tilde{p} \equiv p - 1_{\{\omega<1\}} f$ then on $\{\omega \geq 1\}$ the functions e_n are unchanged. Hence by Lemma 1 we have

$$\int \omega p \, d\mu \leq \int \tilde{f} \, d\mu,$$

proving Theorem 1.

References

[1] Bishop, E., An upcrossing inequality with applications, Michigan Math. J. 13 (1966), 1–13.

[2] Bishop, E., Foundations of Constructive Analysis, McGraw-Hill, New York 1967.

[3] Bishop, E., A constructive ergodic theorem, J. Math. Mech. 17 (1968), 631–639.

[4] Chacon, R. V., Ornstein, D. S., A general ergodic theorem, Illinois J. Math. 4 (1960), 153–160.

[5] Garsia, A. M., Topics in almost everywhere convergence, Markham, Chicago 1970.

[6] Krengel, U., Ergodic Theorems, de Gruyter Stud. Math. 6, Walter de Gruyter, Berlin 1985.

[7] Meyer, P. A., Travaux de H. Rost en théorie du balayage, in: Séminaire de Probabilités V, Université de Strasbourg 1970, pp. 237–250, Lecture Notes in Math. 191, Springer-Verlag, Berlin-Heidelberg-New York 1971.

[8] Meyer, P. A., Le schéma de remplissage en temps continu, in: Séminaire de Probabilités VI, Université de Strasbourg 1971, pp. 130–150, Lecture Notes in Math. 258, Springer-Verlag, Berlin-Heidelberg-New York 1972.

[9] Rost, H., On the stopping distributions of a Markov process, Invent. Math. 14 (1971), 1–16.

Department of Mathematics
University of Toronto
Toronto, Ontario M5S 1A1, Canada
akcoglu@math.toronto.edu
ferrando@math.toronto.edu

Department of Mathematics
University of Minnesota
Minneapolis, MN 55455, U.S.A.
baxter@math.umn.edu

When is a Transitive Map Chaotic?

Ethan Akin, Joseph Auslander, and Kenneth Berg

In their Monthly paper of April 1992, Banks, Brooks, Cairns, Davis, and Stacey **[3]** showed that, for a transitive map, sensitivity to initial conditions follows from the assumption that the periodic points are dense. To understand their lovely proof, we pushed and prodded it, as one does, in an attempt to generalize the result. While we succeeded in a modest way (Theorem 2.5 below), we were struck by the paucity of examples of transitive systems that are not sensitive. Transitive equicontinuous systems are in fact minimal systems, and it is well known that they occur only as translations on compact groups, preserving an invariant metric. For example, the unit circle K is a compact multiplicative subgroup of the complex numbers, multiplication by a fixed $\zeta \in K$ is an isometry on K and, if ζ is not a root of unity, every orbit is dense. Auslander and Yorke showed some time ago **[2]** that all nonequicontinuous minimal maps are sensitive and they also raised the question which inspired the title of this paper: Among transitive maps which are not minimal, do any exhibit the sort of orbital stability which we call almost equicontinuity?

The answer turns out to hinge upon a seemingly unrelated question. For which systems is every point recurrent (every orbit returns arbitrarily close to its initial state)? Since for minimal maps every point is transitive, and hence recurrent, the issue is to find transitive nonminimal maps for which every point is recurrent. Weiss and Katznelson **[6]** constructed examples which satisfy an even stronger condition, later designated uniform rigidity by Glasner and Maon **[4]** in their investigation of several related conditions. Recently Glasner and Weiss **[5]** answered our question in the affirmative, revealing a close connection between uniform rigidity and what we call almost equicontinuity. We further explore this connection and show, for example, that among the uniformly rigid examples of Weiss and Katznelson some are, and some are not, almost equicontinuous.

I. Recurrence and transitivity

We begin by reviewing (and occasionally modifying) some of the standard concepts and definitions of topological dynamics. A topological dynamical system (X, f) consists of a topological space X and a continuous map f of X into itself. We will assume that X is a compact, metrizable space although many of the results hold in a more general setting.

The dynamics is introduced by iterating the map. As usual, $f^0 = 1_X$, the identity map of X, and f^n for $n > 0$ is the n-fold composition $f \circ f \circ \cdots \circ f$. For $x \in X$ we define the orbit

$$\mathcal{O}f(x) = \{f^n(x) \mid n \in \mathbb{N}\} \subseteq X \tag{1}$$

where $\mathbb{N} = \{0, 1, 2, \ldots\}$.

In topological dynamics we are interested in asymptotic, or long term, properties of the system, so we consider certain "extensions" of an orbit. The first of these is the set of limit points of the orbit sequence:

$$\omega f(x) = \bigcap_{N \geq 0} \overline{\bigcup_{n \geq N} \{f^n(x)\}} \subseteq X. \tag{2}$$

Equivalently $y \in \omega f(x)$ iff for every neighborhood U of y and every $N \in \mathbb{N}$ there is an $n > N$ such that $f^n(x) \in U$. Since X is metrizable, $y \in \omega f(x)$ iff there is a sequence $\{n_k\}$ in $\mathbb{N}$ such that $n_k \to \infty$ and $f^{n_k}(x) \to y$. Since X is compact, $\omega f(x) \neq \emptyset$. We observe that the orbit closure $\overline{\mathcal{O}f(x)} = \mathcal{O}f(x) \cup \omega f(x)$.

We identify a function g with its graph $\{(x, g(x)) \mid x \in X\}$. For example, the identity function 1_X is the diagonal subset of $X \times X$. In general, a relation on X is a subset $R \subseteq X \times X$ with $R(x) = \{y : (x, y) \in R\}$. We call $\mathcal{O}f := \cup_{n \in \mathbb{N}} f^n$ the orbit relation. We observe that the relation notation $\mathcal{O}f(x)$ is consistent with our previous definition. The limit point relation ωf is defined by $\omega f = \{(x, y) \mid y \in \omega f(x)\}$.

Although each $\omega f(x)$ is closed in X, the relation ωf is not generally closed in $X \times X$. We define:

$$\Omega f = \bigcap_{N \geq 0} \overline{\bigcup_{n \geq N} f^n} \subseteq X \times X. \tag{3}$$

Thus, $y \in \Omega f(x)$ iff for every neighborhood U of y, every neighborhood V of x and every $N \in \mathbb{N}$ there is an $n > N$ and $\bar{x} \in V$ such that $f^n(\bar{x}) \in U$. Again, for metrizable X, $y \in \Omega f(x)$ iff there is a sequence $\{n_k\}$ in $\mathbb{N}$ and a sequence $\{\bar{x}_k\}$ in X such that $n_k \to \infty$, $\bar{x}_k \to x$ and $f^{n_k}(\bar{x}_k) \to y$. Clearly Ωf is a closed relation containing ωf and thus containing $\overline{\omega f}$. The containment may be proper.

A subset A of X is called *positively invariant* if $f(A) \subseteq A$, *negatively invariant* if $f^{-1}(A) \subseteq A$, and *invariant* if $f(A) = A$. A set is positively invariant iff its complement is negatively invariant. An invariant set is positively invariant but it need not be negatively invariant unless f is injective. A set that is both positively and negatively invariant, such as X, need not be invariant unless f is surjective. In the important case when $f: X \to X$ is a homeomorphism, a set is invariant iff it is both positively and negatively invariant. If $A \subseteq X$ is a closed positively invariant set then $(A, f|_A)$ is called a *subsystem* of (X, f). It is clear that each $\omega f(x)$ and each $\Omega f(x)$ is positively invariant and, since X is compact, they are in fact invariant.

In this paper we will be concerned with the related notions of recurrence and transitivity. We say that x is a *recurrent point* for f if $x \in \omega f(x)$. If x satisfies $x \in \Omega f(x)$ then x is said to be *non-wandering*. The system (X, f), or the map f, is called *topologically transitive* if $\omega f(x) = X$ for some $x \in X$, and such a point x is called a *transitive point*. Clearly a topologically transitive map on a compact space is surjective. (Note that topological transitivity requires more than the existence of a dense orbit, which is sometimes taken as the definition; in particular a transitive point is recurrent.) In case (X, f) is topologically transitive, we write Trans_f for the set of transitive points of (X, f).

The system (X, f) is said to be *minimal* if every point is a transitive point ($\mathrm{Trans}_f = X$). A closed positively invariant subset $Y \subset X$ is called minimal if the subsystem $(Y, f|_Y)$ is minimal.

The following result collects several elementary facts about compact (metrizable) systems. We will outline some key points of the proof. See the book of Akin **[1]** (Theorem 4.12) for further details.

Theorem 1.1. *Let (X, f) be a compact system* $\mathbb{P}$.

(1) *The following conditions are equivalent:*

a. *(X, f) is topologically transitive.*

b. *$\Omega f(x) = X$ for all $x \in X$.*

c. Trans_f *is a countable intersection of open dense sets.*

d. *Every open, non-empty negatively invariant subset of X is dense.*

e. *Every closed, proper positively invariant subset of X has empty interior.*

(2) *The following conditions are equivalent:*

a. *(X, f) is minimal.*

b. *X has no proper nonempty open negatively invariant subsets.*

c. *X has no proper nonempty closed positively invariant subsets.*

d. *X has no proper closed nonempty invariant subsets.*

e. *$\overline{\mathcal{O}f(x)} = X$ for every $x \in X$.*

Most of these results are known and not difficult to prove. See the book of Akin **[1]** (Theorem 4.12) for details. We sketch the implication b $\to$ c in (1), which is not quite standard. If $y \in X$, $N \in \mathbb{N}$, and $\varepsilon > 0$, let $V = V(y, N, \varepsilon)$ consist of all $x \in X$ such that $d(f^k(x), y) < \varepsilon$ for some $k > N$. Clearly V is open and it follows easily from b that it is dense. Moreover Trans_f is the intersection of the sets $V(y, N, \varepsilon)$, for positive integers N, positive rationals ε and y in a countable dense subset of X. The implication c $\to$ b follows from the Baire Category Theorem. The rest of the proof of (1) (a $\to$ b, d $\iff$ e, a $\to$ e, and d $\to$ b) is straightforward, as are the proofs of the equivalences in (2).

The usual definition of a minimal system is the condition c of (2) in the above theorem. A straightforward Zorn's Lemma argument shows that any nonempty closed positively invariant set contains a minimal set. A point x is called a *minimal point* if it is contained in a minimal set Y or, equivalently, if $(\overline{\mathcal{O}f(x)}, f)$ is a minimal system. Obviously a minimal point is a recurrent point. It's well known that the converse is false, and we'll see some examples illustrating this later.

It is useful to define the following sets of integers. If $x \in X$ and $A \subseteq X$ let $R(x, A) = \{n \in \mathbb{N} \mid f^n(x) \in A\}$. Clearly $\mathbb{N} \setminus R(x, A) = R(x, X \setminus A)$. If U is a neighborhood of x then $R(x, U)$ is called the set of *return times* of the point x to the neighborhood U. Clearly a point x is recurrent iff $R(x, U)$ is infinite for every neighborhood U of x.

Definition. A subset $T \subseteq \mathbb{N}$ is called *relatively dense* if there is a positive number L such that $[n, n+L] \cap T \neq \emptyset$ for all $n \in \mathbb{N}$.

Lemma 1.2. *Suppose that U is an open subset of X such that $R(x, U)$ is infinite but not relatively dense. Then there is a point $y \in \overline{U} \cap \omega f(x)$ such that $U \cap \omega f(y) = \emptyset$.*

Proof. For $n = 1, 2, \ldots$ there is an increasing sequence $\{k_n\}$ of positive integers such that $f^{k_n}(x) \in U$ and $f^{k_n+j}(x) \notin U$ for $j = 1, 2, \ldots, n$. By compactness we may assume that $f^{k_n}(x) \to y \in \overline{U}$. Clearly $y \in \omega f(x)$. Now let $j > 0$. Then for all $n > j$, $f^{k_n+j}(x) \notin U$ so $f^j(y) \notin U$. Since this holds for all $j > 0$ it follows that $\omega f(y) \cap U = \emptyset$. □

Lemma 1.3. (1) *Let $K \neq \emptyset$ be a closed positively invariant subset of $\omega f(x)$. Let E be a closed set disjoint from K. Then $R(x, E)$ is not relatively dense.*

(2) *Suppose that Y is a minimal subset of $\omega f(x)$ such that $R(x, U)$ is relatively dense for every open set U containing Y. Then Y is the unique minimal set contained in $\omega f(x)$.*

Proof. Let $U = X \setminus E$. For any positive integer ν there is, since K is positively invariant, an open set V containing K such that $f^j(V) \subseteq U$ for $0 \leq j \leq \nu$. Since $f^n(x) \in V$ for some n, the result follows.

Let K be a minimal subset of $\omega f(x)$. If $K \neq Y$ then $K \cap Y = \emptyset$. There is thus an open set U containing Y such that $\overline{U} \cap K = \emptyset$. By (1), $R(x, \overline{U})$, and thus $R(x, U)$, is not relatively dense contrary to our assumptions. □

The above lemma may be used to establish the following classical theorem, whose proof we omit. It will also be used later, in the proof of Theorem 4.1.

Theorem 1.4. *Let (X, f) be a compact dynamical system and let $x \in X$. The point x is minimal iff $R(x, U)$ is relatively dense for every neighborhood U of x.*

For a transitive system (X, f) we have already seen that the transitive points are residual (Theorem 1.1). This next theorem shows that non-minimal systems have a dense set of non-transitive points.

Theorem 1.5. *If the compact dynamical system (X, f) is transitive but not minimal then the set of non-transitive points is dense.*

Proof. Let V be a non-void open set. Since the set of transitive points is dense, we can choose $x \in V \cap \mathrm{Trans}_f$. The point x is not minimal so there is a neighborhood U of x such that $R(x, U)$ is not relatively dense. Clearly we may choose U so that $\overline{U} \subseteq V$. By Lemma 1.2 there is a point $y \in \overline{U}$ such that $\omega f(y) \cap U = \emptyset$. In particular, $y \notin \mathrm{Trans}_f$. □

II. Equicontinuity and almost equicontinuity

We call a dynamical system (X, f) *equicontinuous* if the family of mappings $\{f^n \mid n \in \mathbb{N}\}$ is uniformly equicontinuous. In terms of a metric d for X, this means that for every $\varepsilon > 0$ there is a $\delta > 0$ such that whenever $x, y \in X$ with $d(x, y) < \delta$, $d(f^n(x), f^n(y)) < \varepsilon$ for $n = 0, 1, 2, \ldots$. The definition of equicontinuity is independent of the choice of metric since X is compact.

Equicontinuity is a strong form of stability and indeed is one way in which the overworked expression "stability" can be made precise. It takes the form of continuity of a certain map in function space. First a few definitions. Let $X^{\mathbb{N}}$ denote the set of all sequences $\boldsymbol{x} = (x_0, x_1, \ldots)$ in X with the product topology (the topology of pointwise convergence). This same set of points, with the topology of uniform convergence, will be denoted by $C(\mathbb{N}, X)$. A metric D for $C(\mathbb{N}, X)$ can be defined in terms of a metric d on X by $D(\boldsymbol{x}, \bar{\boldsymbol{x}}) = \sup d(x_i, \bar{x}_i)$. Later we will let D denote the uniform metric on other function spaces. For example, $C(X, X)$ will denote the space of continuous mappings from X into itself and $D(g, h) := \sup\limits_{x \in X} d(g(x), h(x))$.

Define $\mathcal{O}_f(x) = (x, f(x), f^2(x), \ldots)$. We can regard $\mathcal{O}_f$ as a map from X to $X^{\mathbb{N}}$ or to $C(\mathbb{N}, X)$. With the product topology on the range, the *orbit map* $\mathcal{O}_f$ is clearly continuous; this is not in general the case for the uniform topology.

Given a metric d it is useful to introduce the metric d_f on X:

$$d_f(x, y) = \sup_{n \in \mathbb{N}} d(f^n(x), f^n(y)).$$

Clearly convergence with respect to d_f implies convergence with respect to d but the converse is not generally true.

Theorem 2.1. *The following are equivalent:*

a. *The system (X, f) is equicontinuous.*
b. *The map $\mathcal{O}_f : X \to C(\mathbb{N}, X)$ is continuous.*
c. *The metrics d and d_f are equivalent (and therefore uniformly equivalent).*
d. *The topology on X defined by the metric d_f is compact.*
e. *There is a metric d^* equivalent to d such that $d^*(f(x), f(y)) \leq d^*(x, y)$ for all $x, y \in X$.*

Proof. Conditions b and c are clearly reformulations of a. That c $\to$ e (choose $d^* = d_f$), and e $\to$ a (choose $\delta = \varepsilon$ in the definition of equicontinuity) are obvious. Clearly, c $\to$ d. Finally, d $\to$ e since the identity map $(X, d_f) \to (X, d)$ is continuous and thus, if (X, d_f) is compact, a homeomorphism. □

The notion of equicontinuity can be localized in the obvious way. That is, $x \in X$ is called an *equicontinuity point* (or (X, f) is *equicontinuous at* x) if for every $\varepsilon > 0$ there is a $\delta > 0$ such that whenever $y \in X$ satisfies $d(x, y) < \delta$ then $d(f^n(x), f^n(y)) < \varepsilon$ for all $n \in \mathbb{N}$. If every $x \in X$ is an equicontinuity point then the compactness of X implies that the system (X, f) is equicontinuous.

The proof of the following proposition is immediate.

Proposition 2.2. *The following are equivalent:*
a. x *is an equicontinuity point.*
b. *The orbit map* $\mathcal{O}_f : X \to C(\mathbb{N}, X)$ *is continuous at* x.
c. *For every* $\varepsilon > 0$ *there is a* $\delta > 0$ *such that* $d(x, y) < \delta$ *implies* $d_f(x, y) \leq \varepsilon$.

The next result is less obvious.

Theorem 2.3.
(1) *Suppose* x *is an equicontinuity point for* f. *Then*
a. $\omega f(x) = \Omega f(x)$.
b. *If* x *is non-wandering, then* x *is recurrent.*
c. *If* x *is a limit of a sequence of minimal points, then* x *is a minimal point.*

(2) *Let* $f \times f$ *denote the product map on* $X \times X$. *The following are equivalent:*
a. x *is an equicontinuity point for* f.
b. (x, x) *is an equicontinuity point for* $f \times f$.
c. $\omega(f \times f)(x, x) = \Omega(f \times f)(x, x)$.
d. $\Omega(f \times f)(x, x) \subseteq 1_X$ *(the diagonal of* $X \times X$*).*

Proof. (1) a. Suppose $y \in \Omega f(x)$. Then there are sequences $\{x_k\}$ and $\{n_k\}$ with $n_k \to \infty$ and $x_k \to x$ such that $f^{n_k}(x_k) \to y$. Let $\varepsilon > 0$ and let $\delta > 0$ correspond to $\frac{\varepsilon}{2}$ in the definition of equicontinuity at x. Choose k so large that $d(x, x_k) < \delta$ and $d(f^{n_k}(x_k), y) < \delta$. By the choice of δ we have $d(f^{n_k}(x), f^{n_k}(x_k)) \leq \frac{\varepsilon}{2}$. Then $d(f^{n_k}(x), y) < \varepsilon$ and it follows that $y \in \omega f(x)$.

b. This is an immediate consequence of a.

c. Let U be a neighborhood of x. We show that the return time set $R(x, U)$ is relatively dense. Let $\varepsilon > 0$ be such that U contains $B_{3\varepsilon}(x)$, the open 3ε ball around x, and choose $\delta > 0$ corresponding to ε in the definition of equicontinuity at x. (In particular $\delta \leq \varepsilon$.) Let z be a minimal point with $d(x, z) < \delta$. By Theorem 1.4 $R(z, B_\varepsilon(z))$ is relatively dense. If $k \in R(z, B_\varepsilon(z))$, then $d(x, f^k(x)) \leq d(x, z) + d(z, f^k(z)) + d(f^k(z), f^k(x)) < 3\varepsilon$ (since x is an equicontinuity point). Therefore $R(z, B_\varepsilon(z)) \subset R(x, U)$ and so the latter set is relatively dense. Since U is arbitrary, x is a minimal point, again by Theorem 1.4.

(2) a $\to$ b and c $\to$ d are immediate, b $\to$ c follows from (1)a, and d $\to$ a is an easy exercise. □

Recall that for any map between metric spaces, the set of continuity points of the map is a G_δ (a countable intersection of open sets). If we apply this result to the map $\mathcal{O}_f$ we obtain from Theorem 2.3 that the set of equicontinuity points of (X, f) is a G_δ. We get sharper results by explicitly describing this G_δ. To this end we introduce for $k = 1, 2, \ldots$ the sets $G_k = \{x \in X \mid$ there exists a neighborhood U of x such that $x_1, x_2 \in U \to d(f^n(x_1), f^n(x_2)) \leq \frac{1}{k}$ for all $n \geq 0\}$. Note that $x \in G_k$ iff x has a neighborhood whose d_f diameter is $\leq \frac{1}{k}$. Also, it is easy to see that each G_k is an open negatively invariant set and that $\cap G_k$ is the set of equicontinuity points of f.

As an application of this representation of the set of equicontinuity points, we prove (in a somewhat different way) a result by Auslander and Yorke [**2**, Corollary 2] which establishes, for minimal systems, a dichotomy between sensitivity and equicontinuity. We first define two terms that are central to this paper.

Definition. A topologically transitive system is called *almost equicontinuous* if there is at least one equicontinuity point.

A topologically transitive system is called *sensitive* if there is an $\varepsilon > 0$ such that whenever U is a non-void open set there exist $x, y \in U$ such that $d(f^n(x), f^n(y)) > \varepsilon$ for some $n \in \mathbb{N}$. Equivalently, every nonempty open set has d_f diameter greater than ε

The term *chaotic*, used in the literature with various meanings, was used in [**2**] to denote what we have here called *sensitive*.

Theorem 2.4. *Let (X, f) be topologically transitive.*

If (X, f) is almost equicontinuous then the set of equicontinuity points coincides with the set of transitive points (and so the set of equicontinuity points is a dense G_δ). In particular, a minimal almost equicontinuous dynamical system is equicontinuous.

If (X, f) has no equicontinuity points then it is sensitive. In particular a minimal system is either equicontinuous or sensitive.

Proof. Let $x \in \mathrm{Trans}_f$. Suppose that every G_k is nonempty. Then (since G_k is open) there is an $n_k > 0$ such that $f^{n_k}(x) \in G_k$. Since G_k is negatively invariant, $x \in G_k$. Thus $x \in \cap G_k$, which is the set of equicontinuity points of f.

Furthermore, since f is transitive $\Omega f(x) = X$ for all $x \in X$. If x is an equicontinuity point $\omega f(x) = \Omega f(x)$ by the previous theorem so $\omega f(x) = X$.

Each G_k is open and negatively invariant and their intersection is the set of equiconit-inuity points. If there are no equicontinuity points then some G_k is empty by the argument of the first paragraph. It then follows that every nonempty open set has d_f diameter greater than $\frac{1}{k}$ and so (X, f) is sensitive. □

The construction and analysis of (necessarily non-minimal) almost equicontinuous topologically transitive systems that are not equicontinuous will be carried out in the next two sections.

The next result is the generalization of the theorem of Banks, et al, which we mentioned in the introduction. It appears as Theorem 1.3 in [**5**].

Theorem 2.5. *Let (X, f) be a topologically transitive system which is not minimal. Suppose that the set $\mathcal{M}$ of minimal points is dense. Then (X, f) is sensitive.*

Proof. If (X, f) is not sensitive then there is an equicontinuity point x, and x is the limit of a sequence of minimal points. By Theorem 2.3 (1c), x is a minimal point. But the equicontinuity point x is a transitive point, so (X, f) is minimal. □

Our next theorem relates the metric d_f to the set of transitive points. It will be used in the next section. There is no assumption of almost equicontinuity.

Theorem 2.6. *The metric* d_f *is complete, and* Trans_f *is closed with respect to* d_f.

Proof. Let $\{x_k\}$ be a d_f Cauchy sequence in X. Since $d \le d_f$, $\{x_k\}$ is a d Cauchy sequence so $\{x_k\}$ converges in the metric d to some $x \in X$. We will show that, in fact, $\{x_k\}$ converges to x with respect to d_f. Let $\varepsilon > 0$. Since $\{x_k\}$ is d_f Cauchy, there is an $m \in \mathbb{N}$ such that if $k, k' \ge m$, then $d_f(x_k, x_{k'}) \le \varepsilon$. Thus for each fixed $n \in \mathbb{N}$, $d(f^n(x_k), f^n(x_{k'})) \le \varepsilon$ for $k, k' \ge m$. Now let $k' \to \infty$. Then $d(f^n(x_k), f^n(x)) \le \varepsilon$ for all $k \ge m$. Since this holds for all $n \in \mathbb{N}$ we have $d_f(x_k, x) \le \varepsilon$ for $k \ge m$. Therefore $\{x_k\}$ d_f converges to x. Thus d_f is complete.

Now let $\{x_k\}$ be a sequence in Trans_f which d_f converges to $x \in X$. We will show that $x \in \mathrm{Trans}_f$. Let $y \in X$, let $M \in \mathbb{N}$ and let $\varepsilon > 0$. Choose $k \in \mathbb{N}$ such that $d_f(x, x_k) < \frac{\varepsilon}{2}$, and, using transitivity, choose $m > M$ such that $d(f^m(x_k), y) < \frac{\varepsilon}{2}$. Then $d(f^m(x), y) \le d(f^m(x), f^m(x_k)) + d(f^m(x_k), y) \le d_f(x, x_k) + d(f^m(x_k), y) < \varepsilon$. □

III. Uniform rigidity

Now for something completely different. In this section we will focus on a strong form of recurrence. We say that the compact system (X, f) is *uniformly rigid* if there is a sequence $\{n_k\}$ such that $n_k \to \infty$ and f^{n_k} converges uniformly to 1_X, the identity map of X.

The notion of uniform rigidity was introduced by Glasner and Maon **[4]** who constructed minimal nonequicontinuous examples and also investigated some related notions, which are mentioned below. Somewhat earlier, Weiss and Katznelson **[6]** (without explicitly referring to the property) constructed a nonminimal example. We will analyze this example in Section 4.

In this section, we develop some consequences of uniform rigidity (Theorem 3.4). This is obviously an extremely strong property, and there is room for a number of weaker properties. Recall that the point x is recurrent if $x \in \omega f(x)$. If X is topologically transitive, clearly any $x \in \mathrm{Trans}_f$ is recurrent. The system (X, f) is called *pointwise recurrent* if every point is recurrent. Obviously a minimal system is pointwise recurrent; the construction of transitive nonminimal examples takes some work. Other conditions which have been studied **[4]** include *weak rigidity* – every finite product $f \times \cdots \times f$ is pointwise recurrent, and *rigidity* – the identity map 1_X is the pointwise limit of some sequence f^{n_i}.

Lemma 3.1. *Suppose* (X, f) *is pointwise recurrent. Then the map* f *is surjective and every positively invariant set is both invariant and negatively invariant.*

Proof. Let A be any closed positively invariant set. For each $a \in A$, $a \in \omega f(a) \subseteq A$, so A is the union of omega limit sets. As we pointed out in Section 1, every omega limit set is invariant and so A is invariant. Applying this result to $A = X$ we see that f is surjective. To show that A is negatively invariant, suppose that $f(x) \in A$. Then $x \in \omega f(x) = \omega f(f(x)) \subseteq A$. □

Let $C = C(X, X)$ denote the set of all continuous maps of X to itself, with the topology of uniform convergence. A metric D for $C(X, X)$ is given by: $D(f, g) = \sup_{x \in X} d(f(x), g(x))$.

Lemma 3.2.
(1) *(X, f) is uniformly rigid iff for every $\varepsilon > 0$ there is an $n \geq 1$ such that $D(f^n, 1_X) < \varepsilon$.*
(2) *If f is a homeomorphism and $n \in \mathbb{N}$ then $D(f^n, 1_X) = D(f^{-n}, 1_X)$.*

Proof. (1) is an easy consequence of the definition. As for (2), note that if $f_1, f_2, g: X \to X$ with g surjective then $D(f_1 \circ g, f_2 \circ g) = D(f_1, f_2)$. If we put $f_1 = f^n$, $f_2 = 1_X$ and $g = f^{-n}$ we obtain $D(1_X, f^{-n}) = D(f^n, 1_X)$. □

Lemma 3.3. *Let (X, f) be a compact dynamical system such that $f \times f$ is pointwise recurrent. Then*
(1) *f is a homeomorphism.*
(2) *If $x \in X$, $\omega f^{-1}(x) \subseteq \omega f(x)$.*
(3) *f is an isometry for the metric d_f*

Proof. By Lemma 3.1, f is surjective and the assumption that $f \times f$ is pointwise recurrent clearly implies that f is injective.

If $x \in X$, $\omega f(x)$ is positively invariant and therefore, by Lemma 3.1, negatively invariant. Since $x \in \omega f(x)$ it follows that $\omega f^{-1}(x) \subseteq \omega f(x)$.

If $x, y \in X$, $(f^{n_i}(x), f^{n_i}(y)) \to (x, y)$ for some sequence $\{n_i\}$ with $n_i \to \infty$. It follows that $d_f(f(x), f(y)) \geq d(x, y)$. But clearly

$$d_f(x, y) = \sup\{d(x, y), d_f(f(x), f(y))\}$$

so $d_f(x, y) = d_f(f(x), f(y))$. □

Theorem 3.4. *Suppose (X, f) is uniformly rigid. Then:*
(1) *The restriction of f to any subsystem is uniformly rigid. Any finite Cartesian product $f \times \cdots \times f$ on X^k is uniformly rigid. Any power f^n on X is uniformly rigid.*
(2) *f is a homeomorphism and (X, f^{-1}) is uniformly rigid. In fact $\{f^{-n_k}\}$ converges uniformly to 1_X whenever $\{f^{n_k}\}$ does.*
(3) *If $x \in X$ then $\omega f(x) = \omega f^{-1}(x)$, so* $\mathrm{Trans}_f = \mathrm{Trans}_{f^{-1}}$.
(4) *The metrics d_f and $d_{f^{-1}}$ coincide on X, and f is an isometry for this metric. If $\{f^{n_k}\}$ converges uniformly to 1_X with respect to the given metric d then it also does so with respect to the metric d_f.*

Proof. The first two statements are immediate. To prove the third, note that $f^{k_i} \to 1_X$ uniformly implies that $f^{nk_i} \to 1_X^n = 1_X$ uniformly $\mathbb{P}$. (2) follows from (1) of Lemma 3.3 and (1) of Lemma 3.2.

By (2) of Lemma 3.3, $\omega f^{-1}(x) \subseteq \omega f(x)$. But f^{-1} is also uniformly rigid, so $\omega f(x) \subseteq \omega f^{-1}(x)$.

By (3) of Lemma 3.3, f is a d_f isometry. It follows that f^{-1} is a d_f isometry. Then for all $k \in \mathbb{N}$, $d_f(x, y) = d_f(f^{-k}(x), f^{-k}(y))$ and therefore $d_f(x, y) = \sup_{k \in \mathbb{N}} d_f(f^{-k}(x), f^{-k}(y)) \geq \sup_{k \in \mathbb{N}} d(f^{-k}(x), f^{-k}(y)) = d_{f^{-1}}(x, y)$. Therefore $d_f \geq d_{f^{-1}}$. Since f^{-1} is also uniformly rigid we have $d_{f^{-1}} \geq d_f$ as well. The proof of the final statement in (4) follows easily from the definitions. □

We now bring together the two parts of our story. The next lemma will be crucial in accomplishing this.

Lemma 3.5 (Hinge Lemma). *Let (X, f) be a topologically transitive system and let $x \in \mathrm{Trans}_f$. Let $\varepsilon, \delta > 0$. Then the following are equivalent:*
(H1) *If $y \in X$ and $d(x, y) < \delta$ then $d(f^k(x), f^k(y)) \leq \varepsilon$ for all $k \in \mathbb{N}$.*
(H2) *If $n \in \mathbb{N}$ and $d(x, f^n(x)) < \delta$ then $d(f^k(x), f^{n+k}(x)) \leq \varepsilon$ for all $k \in \mathbb{N}$.*
(H3) *If $n \in \mathbb{N}$ and $d(x, f^n(x)) < \delta$ then $d(y, f^n(y)) \leq \varepsilon$ for all $y \in X$.*

Proof. We prove H1 → H2 → H3 → H2 → H1.
H1 → H2: Apply H1 to $x = x$ and $y = f^n(x)$.
H2 → H3: We see from H2 that $d(y, f^n(y)) \leq \varepsilon$ for all $y = f^k(x)$. By continuity the inequality holds for all $y \in \omega f(x) = X$.
H3 → H2: Apply H3 to $x = x$ and $y = f^k(x)$.
H2 → H1: Fix $k \in \mathbb{N}$. We see from H2 that $d(f^k(x), f^k(y)) \leq \varepsilon$ for each $y = f^n(x)$ that satisfies $d(x, y) < \delta$. Any y satisfying $d(x, y) < \delta$ is the limit of some sequence $f^{n_i}(x)$ with $d(x, f^{n_i}(x)) < \delta$, and continuity implies $d(f^k(x), f^k(y)) \leq \varepsilon$. □

Note that in this lemma ε and δ are fixed positive numbers, although in the applications they will be quantified away. The proof was carried out by "swinging" via (H2) between (H1), which is an equicontinuity type condition and (H3) which is related to uniform closeness of f^n to 1_X; this is the reason for calling it the "Hinge Lemma".

Recall that in Section 2 we defined a system to be almost equicontinuous if it is topologically transitive and if there are points of equicontinuity (which necessarily coincide with the set of transitive points). Our next theorem gives several conditions which are equivalent to almost equicontinuity. The resulting observation that almost equicontinuity implies uniform rigidity was earlier proved in [**5**, Lemma 1.2].

Theorem 3.6. *Let (X, f) be a topologically transitive system. Then the following are equivalent:*
(1) *(X, f) is almost equicontinuous.*
(2) *For every transitive point x, the convergence $f^{n_i}(x) \to x$ implies the uniform convergence $f^{n_i} \to 1_X$.*
(3) *For some transitive point x, the convergence $f^{n_i}(x) \to x$ implies the uniform convergence $f^{n_i} \to 1_X$.*

(4) *There is a transitive point* x *such that for every* $\varepsilon > 0$ *there exists* $\delta > 0$ *such that* $d(f^n(x), x) < \delta$ *implies* $d(f^{n+k}(x), f^k(x)) \leq \varepsilon$ *for all* $k \in \mathbb{N}$.
(5) *The topologies induced by the metrics* d *and* d_f *on* Trans_f *agree.*
(6) *The restriction of the orbit map* $\mathcal{O}_f$ *to* Trans_f *defines a continuous map from* Trans_f *to* $C(\mathbb{N}, X)$.

Moreover, these conditions imply that (X,f) is uniformly rigid.

Proof. (1) → (2) follows from the implication (H1) → (H3) in the Hinge Lemma. (If $\varepsilon > 0$ choose $\delta > 0$ to correspond to ε in the definition of equicontinuity at the transitive point x.)

(2) → (3) since (X, f) is topologically transitive.

(3) → (4) follows from (H3) → (H2) in the Hinge Lemma.

(4) → (1): We use the Hinge Lemma again, this time the implication (H2) → (H1). Let x be the transitive point whose existence is assumed in (4). It is sufficient to show that x is an equicontinuity point. Let $\varepsilon > 0$, and let $\delta > 0$ correspond to ε as in (4). Then (H2) of the Hinge Lemma is satisfied for this ε and δ. So for every $\varepsilon > 0$ there is a $\delta > 0$ so that (H1) holds, which is exactly equicontinuity at x.

We have shown that the first four statements are equivalent. We will now show that (1) → (6) → (5) → (4).

(1) → (6): Recall that almost equicontinuity is equivalent to the property that the transitive points are the points of continuity of the orbit map $\mathcal{O}_f$. Thus $\mathcal{O}_f$ is certainly continuous when restricted to Trans_f.

(6) → (5): To prove (5) it is sufficient to show: if $\{x_j\}$ is a sequence in Trans_f, $x \in \mathrm{Trans}_f$, and $d(x, x_j) \to 0$ then $d_f(x, x_j) \to 0$. But this follows immediately from the assumption of (6), namely the continuity of the orbit map $\mathcal{O}_f$ on Trans_f.

(5) → (4): First recall that if $x \in \mathrm{Trans}_f$ then so is $f^n(x)$ for all $n \in \mathbb{N}$. Now fix $x \in \mathrm{Trans}_f$, and let $\varepsilon > 0$. Since the metrics d and d_f are equivalent on Trans_f there is a $\delta > 0$ such that whenever $y \in \mathrm{Trans}_f$ with $d(x, y) < \delta$ then $d_f(x, y) < \varepsilon$, so $d(f^k(x), f^k(y)) < \varepsilon$ for all $k \in \mathbb{N}$. In particular, if $y = f^n(x)$ where $d(f^n(x), x) < \delta$, then $d(f^k(x), f^{n+k}(x)) < \varepsilon$ for all $k \in \mathbb{N}$ which proves (4).

Finally, since a transitive point is recurrent, uniform rigidity follows from (2). □

The following corollary summarizes properties of almost equicontinuous systems. It is a direct consequence of the theorems in this section and Theorem 2.6.

Corollary 3.7. *If the system* (X, f) *is almost equicontinuous, then* f *is a homeomorphism and* (X, f) *is uniformly rigid. The system* (X, f^{-1}) *is also almost equicontinuous,* $\mathrm{Trans}_f = \mathrm{Trans}_{f^{-1}}$, *and a point is a transitive point iff it is an equicontinuity point (for either* f *or* f^{-1} *). The set of transitive points is residual, and on it the metric* $d_f = d_{f^{-1}}$ *is complete and induces the same topology as* d. *The restriction of* f *to* Trans_f *is an isometry with respect to* d_f.

After this extensive and delicate analysis of almost equicontinuous nonminimal systems, it is time to show that this class is nonempty, and this is what we do in the final section. For an alternative construction see **[5**, Prop 1.5**]**.

IV. The examples of Weiss and Katznelson

We will now construct a class of examples of uniformly rigid systems, some of which are almost equicontinuous (and not equicontinuous). The construction we present below was outlined by Weiss and Katznelson **[6]**, who were, in turn, inspired by a much earlier construction of Nemitsky **[7]**.

The examples are all subsystems of the so called Bebutov system, which is defined by the shift map S on $I^{\mathbb{N}}$. ($I = [0, 1]$, the closed unit interval, $I^{\mathbb{N}}$ is provided with the product topology – the topology of pointwise convergence – and the shift map is defined by $S(\alpha)(n) = \alpha(n+1)$ for $\alpha \in I^{\mathbb{N}}$ and $n \in \mathbb{N}$.) A metric which generates the product topology is given by

$$d(\alpha, \beta) = \sup_{n=0}^{\infty} \frac{|\alpha(n) - \beta(n)|}{2^n}. \tag{1}$$

It will also be convenient to consider the uniform topology on $I^{\mathbb{N}}$, with the metric D given by

$$D(\alpha, \beta) = \sup_{n \in \mathbb{N}} |\alpha(n) - \beta(n)|. \tag{2}$$

It follows that

$$D(\alpha, \beta) = \max\{d(\alpha, \beta), D(S(\alpha), S(\beta))\}. \tag{3}$$

Now for the construction. Let $L \geq 2$ and let a_0 be any function from $[-1, 1]$ to I satisfying

$$a_0(-1) = a_0(1) = 1 \text{ and } |a_0(s_1) - a_0(s_2)| \leq L|s_1 - s_2| \text{ for } s_1, s_2 \in [-1, 1]. \tag{4}$$

Extend a_0 periodically to $a_1 \colon \mathbb{R} \to I$, so $a_1(s) = a_0(s)$ for $|s| \leq 1$ and $a_1(s+2) = a_1(s)$ for $s \in \mathbb{R}$. Note that a_1 satisfies the same Lipschitz condition as a_0.

For any $p > 0$ define a_p by $a_p(s) = a_1(\frac{s}{p})$. Then (for $s, s_1, s_2 \in \mathbb{R}$):

$$0 \leq a_p(s) \leq 1, \; a_p(-p) = a_p(p) = 1, a_p(s + 2p) = a_p(s),$$

$$|a_p(s_1) - a_p(s_2)| \leq \frac{L}{p}|s_1 - s_2|. \tag{5}$$

In particular, setting $s_1 = p$ and $s_2 = p + s$, we see that for $|s| < \frac{p\varepsilon}{L}$,

$$1 - \varepsilon < a_p(p + s) \leq 1. \tag{6}$$

Now choose a strictly increasing sequence of positive integers k_i such that $8|k_i$. (This latter condition will only be used for our final example.) Define, inductively, $p_0 = 1$ and

$p_{i+1} = k_i p_i$, and let

$$a_\infty(s) = \sup\,[a_{p_0}(s), a_{p_1}(s), a_{p_2}(s), \ldots].$$

Since $1 \geq a_\infty \geq a_{p_i}$ from (6) we immediately obtain:

$$1 - \varepsilon < a_\infty(p_i + s) \leq 1 \text{ if } |s| < \frac{p_i \varepsilon}{L}. \tag{7}$$

If m is an odd integer $a_p(mp + s) = a_p(p + s)$, so we may replace p by mp in (6) and p_i by mp_i in (7) to get:

$$1 - \varepsilon < a_p(mp + s) \leq 1 \text{ if } |s| < \frac{p\varepsilon}{L} \quad \text{and } m \text{ is odd} \tag{8}$$

and

$$1 - \varepsilon < a_\infty(mp_i + s) \leq 1 \text{ if } |s| < \frac{p_i \varepsilon}{L} \quad \text{and } m \text{ is odd.} \tag{9}$$

Note that (since $2p_j | 2p_i$ whenever $j < i$) we have

$$a_\infty(s + 2p_i) = \sup\,[a_{p_0}(s), \ldots, a_{p_i}(s), a_{p_{i+1}}(s + 2p_i), a_{p_{i+2}}(s + 2p_i), \ldots].$$

Using the elementary result that

$$|\sup\{x_i\} - \sup\{y_i\}| \leq \sup\{|x_i - y_i|\}$$

we obtain

$$|a_\infty(s + 2p_i) - a_\infty(s)| \leq \sup_{j>i}\{|a_{p_j}(s + 2p_i) - a_{p_j}(s)|\} \leq \sup_{j>i}\{\frac{2Lp_i}{p_j}\} = \frac{2L}{k_i}. \tag{10}$$

Applying the same elementary result to (5), since $\frac{L}{p_i} \leq L$, we also obtain a Lipschitz estimate for a_∞:

$$|a_\infty(s + r) - a_\infty(s)| \leq L|r|$$

for all $s, r \in \mathbb{R}$.

Now we are ready to define our uniformly rigid system. For a fixed choice of the sequence $\{k_i\}$, we define the sequence $\alpha \in I^{\mathbb{N}}$ by $\alpha(n) = \{a_\infty(n)\}$. We call this the WK sequence associated with the function a_0. If we begin with the constant function $a_0 = 1$, then $a_\infty = 1$ and the associated WK sequence, which we denote by θ, is the constant sequence $a_\infty(n) = 1$ for all n. Of course θ is a fixed point of S: $S(\theta) = \theta$.

Theorem 4.1. *Let α be the sequence defined in the previous paragraph and let X be the orbit closure of α in the Bebutov system $(I^{\mathbb{N}}, S)$. Then the subsystem (X, S) is a uniformly rigid system which has the fixed point $\{\theta\}$ as its unique minimal set. If $a_0(0) \neq 1$ then (X, S) is not minimal.*

Proof. We first show that for all $\beta \in X$ and all i we have

$$D(S^{2p_i}(\beta), \beta) \leq \frac{2L}{k_i}.$$

For $\beta = \alpha$ this inequality holds because of the corresponding inequality (10) for a_∞. It follows from the definition of D that the same inequality holds for any translate $\beta = S^k(\alpha)$. It is now easy to see that it holds for pointwise limits of such translates. That is, it holds for all $\beta \in X$.

It follows that S^{2p_i} tends uniformly to the identity 1_X so the system (X, S) is uniformly rigid. (In particular α is a recurrent point so $X = \omega S(\alpha)$.

Next we show that the fixed point θ is in X. This follows from (9) which implies that $|S^{mp_i}(\alpha)(n) - 1| < \varepsilon$ if $n < \frac{p_i \varepsilon}{L}$ and if m is an odd integer. This in turn implies

$$d(S^{mp_i}(\alpha), \theta) < \varepsilon \tag{11}$$

for m an odd integer and p_i sufficiently large. Therefore the minimal set $\{\theta\}$ is in X and in fact, using Lemma 1.3 (2), we see that $\{\theta\}$ is the unique minimal set in X.

Note that this implies that (X, S) is not itself minimal unless $X = \{\theta\}$ which can only happen if $\alpha = \theta$. This trivial situation is excluded as long as $a_0(0) \neq 1$. □

We now turn our attention to the question of almost equicontinuity for these examples. Our next theorem provides a sufficient condition.

Theorem 4.2. *Suppose that* a_0 *has a strict minimum at* 0 ($a_0(s) > a_0(0)$ *whenever* $0 < |s| \leq 1$). *Let* α *be the associated WK sequence and let* X *be the orbit closure of* α *under the shift map* S. *Then the system* (X, S) *is almost equicontinuous and not minimal.*

We will need the following lemma:

Lemma 4.3. *Suppose that* a_0 *has a strict minimum at* 0. *For every* $\varepsilon > 0$ *there is a* $\delta > 0$ *such that* $a_1(t) < a_1(0) + \delta$ *implies* $|a_1(s+t) - a_1(s)| \leq \varepsilon$ *for all* $s \in \mathbb{R}$

Proof of the Lemma. Let $\varepsilon > 0$. Then there is a $\delta > 0$ such that if $t \in [-1, 1]$ and $a_0(t) < a_0(0) + \delta$ then $|t| < \frac{\varepsilon}{L}$. Now suppose $t \in \mathbb{R}$ such that $a_1(0) + \delta > a_1(t)$. Write $t = u + t_1$ where u is an even integer and $t_1 \in [-1, 1]$. Then $a_0(0) + \delta = a_1(0) + \delta > a_1(t) = a_0(t_1)$ so $|t_1| < \frac{\varepsilon}{L}$. Hence if $s \in \mathbb{R}$ we have $|a_1(s+t) - a_1(s)| = |a_1(s+t_1) - a_1(s)| \leq L|t_1| < \varepsilon$. □

Proof of Theorem 4.2. We show that for every $\varepsilon > 0$ there is a $\delta > 0$ such that $|a_\infty(t) - a_\infty(0)| < \delta$ implies $|a_\infty(s+t) - a_\infty(s)| \leq \varepsilon$ for all $s \in \mathbb{R}$.

If $|a_\infty(t) - a_\infty(0)| < \delta$ then $a_1(0) + \delta = a_\infty(0) + \delta > a_\infty(t) \geq a_{p_i}(t) = a_1(\frac{t}{p_i})$ for all p_i so $|a_{p_i}(s+t) - a_{p_i}(s)| = |a_1(\frac{s}{p_i} + \frac{t}{p_i}) - a_1(\frac{s}{p_i})| < \varepsilon$ by Lemma 4.3. Taking the supremum over all i then establishes the required property for a_∞ (by (10)).

Now we show that the system (X, S) is almost equicontinuous. For, if $\varepsilon > 0$, let $\delta > 0$ be chosen to correspond to ε as above. Then, if $d(S^n(\alpha), \alpha) < \delta$ we have (recalling that α is just a_∞ restricted to $\mathbb{N}$) that $|a_\infty(n) - a_\infty(0)| = |\alpha(n) - \alpha(0)| < \delta$ so $|a_\infty(n+k) - a_\infty(k)| < \varepsilon$ for all $k \in \mathbb{N}$. But this says that $D(S^n(\alpha), \alpha) < \varepsilon$ which implies that $d(S^{n+k}(\alpha), S^k(\alpha)) < \varepsilon$ for all $k \in \mathbb{N}$. Condition (4) in Theorem 3.6 holds and so by condition (1) of that theorem (X, S) is almost equicontinuous. Since

$a_0(0) < a_0(1)$, X properly contains the minimal set $\{\theta\}$ so (X, S) is not itself minimal. □

We complete our study by constructing a uniformly rigid system which is sensitive (not almost equicontinuous). Consider the function a_0 defined by $a_0(t) = 0$ for $|t| \leq \frac{1}{2}$, and $a_0(t) = 2|t| - 1$ for $\frac{1}{2} \leq |t| \leq 1$. Note that a_0 has Lipschitz constant $L = 2$.

Proposition 4.4. *Let α be the WK sequence associated with the a_0 defined above, and let X be the orbit closure of α under the shift map S. Then (X, S) is not almost equicontinuous.*

Proof. We will show for a certain sequence $\{r_i\}$ that $S^{r_i}(\alpha) \to \alpha$ but S^{r_i} does not converge uniformly to the identity. By Theorem 3.6 (2) this shows that (X, S) is not almost equicontinuous.

With k_i and p_i defined as above, we set $r_i = \frac{p_i}{4}$. We will show that $\alpha(n+r_i) = \alpha(n)$ for $r_i \geq n$, so that $S^{r_i}\alpha \to \alpha$. We will also show that $\alpha(2r_i) = 0$ and that $\alpha(3r_i) \geq \frac{1}{2}$ for all i, so that $d(S^{r_i}(S^{2r_i}(\alpha)), S^{2r_i}(\alpha)) \geq \frac{1}{2}$. This will show that $S^{r_i} \not\to 1_X$ uniformly.

We first show:

$$\text{If } j < i \text{ and } s \in \mathbb{R}, a_{p_j}(s + r_i) = a_{p_j}(s) \tag{12}$$

and

$$\text{If } j \geq i \text{ and } |s| \leq r_i, a_{p_j}(s + r_i) = a_{p_j}(s) = 0. \tag{13}$$

If $j < i$, $a_{p_j}(s + r_i) = a_1(\frac{s}{p_j} + \frac{p_i}{4p_j}) = a_1(\frac{s}{p_j}) = a_{p_j}(s)$. (Recall that in this case $\frac{p_i}{4p_j}$ is an even integer and a_1 is periodic of period 2.) If $j \geq i$ and $|s| \leq r_i$ then $|\frac{s}{p_j}| \leq \frac{r_i}{p_j} \leq \frac{1}{4}$. Therefore (from the definition of a_0) $a_{p_j}(s + r_i) = a_{p_j}(s) = 0$. Thus (12) and (13) are proved, and thus we obtain

$$a_\infty(s + r_i) = a_\infty(s) \text{ for } |s| \leq r_i \tag{14}$$

which in turn gives $\alpha(n + r_i) = \alpha(n)$ for $n \leq r_i$, the first of our required properties.

Now recall $a_p(s) = a_1(\frac{s}{p})$. Observe that $\frac{2r_i}{p_j}$ is an even integer if $j < i$, and at most $\frac{1}{2}$ if $j \geq i$. In either case $a_{p_j}(2r_i) = 0$ and so $\alpha(2r_i) = a_\infty(2r_i) = 0$.

Finally we observe that $a_{p_i}(3r_i) = a_1(\frac{3}{4}) = \frac{1}{2}$ so $\alpha(3r_i) = a_\infty(3r_i) \geq \frac{1}{2}$. In fact $\alpha(3r_i) = \frac{1}{2}$, but the inequality suffices to complete the proof. □

References

[1] E. Akin, The General Topology of Dynamical Systems, Grad. Stud. Math. 1, American Mathematical Society, Providence 1993.

[2] J. Auslander and J. Yorke, Interval maps, factors of maps, and chaos, Tôhuko J. Math. (2) 32 (1980), 177–188.

[3] J. Banks, J. Brooks, G. Cairns, G. Davis, and P. Stacey, On Devaney's Definition of Chaos, Amer. Math. Monthly 99 (1992), 332–33.

[4] S. Glasner and D. Maon, Rigidity in topological dynamics, Ergodic Theory Dynam. Systems 9 (1989), 309–320.

[5] E. Glasner and B. Weiss, Sensitive dependence on initial conditions, Nonlinearity 6 (1993), 1067–1075.

[6] Y. Katznelson and B. Weiss, When all points are recurrent/generic, in: Ergodic Theory and Dynamical Systems, I (College Park, 1979–80), ed. A. Katok, pp. 195–210, Progr. Math. 10, Birkhäuser Verlag, Boston 1981.

[7] V. V. Nemitsky, Topological Problems of the Theory of Dynamical Systems, Uspekhi Mat. Nauk (4) 6(34) (1949), 91–153 (Russian), English Translation in: Translations (1) 5, Stability and Dynamical Systems, pp. 414–497, Amer. Math. Soc., Providence 1962.

Department of Mathematics
City College of New York (CUNY)
New York, NY 10031-9100, U.S.A.

Department of Mathematics
University of Maryland
College Park, MD 20742-0001, U.S.A.

Department of Mathematics
University of Maryland
College Park, MD 20742-0001, U.S.A.

Une Nouvelle Loi Forte des Grands Nombres

Mohamed Atlagh et Michel Weber

English Title: New Law of Large Numbers

Résumé. On obtient des théorèmes central limite presque sûrs, dans lesquels figure la densité usuelle sur les entiers, en lieu et place de la densité logarithmique. Nous montrons l'équivalence entre l'ASCLT et le CLT dans le cas i.i.d.

Abstract. We obtain an almost sure central limit theorem (ASCLT) based on an argument of which is purely ergodic and Brownian. Our result is expressed in terms of the usual density on integers, instead of the logarithmic density. We also prove in the i.i.d. case that the ASCLT and the CLT are equivalent.

I Introduction. Résultats principaux

Depuis sa première version, le théorème central limite (C.L.T) a suscité l'intérêt de nombreux chercheurs. Il décrit le comportement asymptotique en loi des sommes de variables aléatoires indépendantes.

Considéré comme étant l'un des outils fondamentaux en théorie des Probabilités et en Statistique, depuis 1988, diverses versions du type Glivenko–Cantelli ont été établies pour le C.L.T; c'est ce qu' on appelle le *théorème central limite presque sûr (A.S.C.L.T, almost sure central limit theorem)*.

Le présent travail constitue une contribution à l'étude de l'$A.S.C.L.T.$ et ce que l'on pourrait appeler le principe d'invariance en densité. Nous allons tout d'abord rappeler succinctement l'essentiel des résultats connus. Cela va nous permettre de définir le point de vue que nous adopterons et les problèmes que nous allons traiter.

Découvert indépendamment par Brosamler **[Br]** et Schatte **[Sc]**, (voir aussi le travail initial de Fischer **[Fi]**), l'$A.S.C.L.T.$, concernant les sommes de variables aléatoires indépendantes et identiquement distribuées et relativement à la densité logarithmique, a trouvé dans la note de Lacey–Philipp **[LP]** une bonne formulation (même conditions que le C.L.T). Nous en rappelons l'énoncé:

Théorème [LP]. *Soit* $\{X, X_i,\ i \geq 1\}$ *une suite de variables aléatoires réelles indépendantes et équidistribuées, vérifiant:*

$$\mathbb{E}X = 0 \ \text{ et } \ \mathbb{E}(X^2) = 1.$$

Soit

$$S_n = X_1 + X_2 + \cdots + X_n, \quad n \geq 1.$$

Alors

$$\text{p.s.} \qquad \lim_{N\to\infty} \frac{1}{\log N} \sum_{j=1}^{N} \frac{1}{j}\, \delta_{\{S_j/\sqrt{j}\}} \overset{\mathcal{D}}{=} N(0,1),$$

où δ_z *désigne la mesure de Dirac au point* z.

Concernant l'$A.S.C.L.T$ sur les sous-suites, Schatte a établi les résultats suivants: Supposons que la suite $(X_i,\ i \geq 1)$ a un moment d'ordre 3 ($\mathbb{E}|X|^3 < \infty$). Soient a et b deux réels tels que $a < b$. Considérons les intervalles d'entiers n pour lesquels:

$$a \leq S_n(\omega)/\sqrt{n} \leq b. \tag{1.1.1}$$

Alors

Théorème [Sc1, Th. 3]. *Soient* $\eta > 1$, ψ *et* φ *deux fonctions définies dans* $\mathbb{N}$ *à valeurs entières, telles que*

$$\psi(N) \geq N \quad \text{et} \quad \lim_{N\to\infty} \frac{\varphi(N)}{\log N} = \infty.$$

Alors pour presque tout $\omega \in \Omega$, *il existe un* $N_0 = N_0(\omega) < \infty$ *tel que tout intervalle d'entiers défini par:*

$$\psi(N) \leq \log n \leq \psi(N) + \varphi(N)\big(\log\log\psi(N)\big)^{\eta}, \qquad N \geq N_0$$

contienne au moins un n *vérifiant* (1.1.1).

Et

Théorème [Sc1, Th. 4]. *Si la fonction* $\psi(N)$ *est telle que:* $\psi(N+1) \geq \psi(N) + 1$ *et*

$$\lim_{N\to\infty} \frac{\big(\log\log\psi(N)\big)^2 + \varphi(N)}{\log N} = 0,$$

alors pour presque tout $\omega \in \Omega$, *il existe une infinité d'entiers* N, *tels que pour tout* n *vérifiant*

$$\psi(N) \leq \log n \leq \psi(N) + \frac{\varphi(N)}{\log\log\psi(N)},$$

(1.1.1) *n'ait jamais lieu.*

Soit $\{K_j,\ j \geq 1\}$ une suite croissante de réels strictement positifs telle que:

$$\forall j \geq 1,\ K_{j+1} - K_j \geq 1.$$

On définit pour toute partie A de $\mathbb{R}$, une sous-suite croissante d'entiers $\mathcal{N}(A)$ par:

$$\mathcal{N}(A) = \{j \geq 1 : \exists k \in [K_j, K_{j+1}[\ \cap\ \mathbb{N} : \frac{S_k}{\sqrt{k}} \in A\}$$

et

$$d_N(\mathcal{N}(A)) = \frac{\operatorname{card}\{\mathcal{N}(A) \cap [1, N]\}}{N}.$$

On note $\underline{d}(\mathcal{N}(A))$ (resp. $\overline{d}(\mathcal{N}(A))$), la limite inférieure (resp. la limite supérieure), lorsque N tend vers l'infini de $d_N(\mathcal{N}(A))$.

On désigne enfin $d(\mathcal{N}(A))$ la valeur commune de ces expressions lorsqu'elles coîncident; et l'on a pour tout couple (A, B) d'ouverts de $\mathbb{R}$:

$$d\big(\mathcal{N}(A \cup B)\big) \le d\big(\mathcal{N}(A)\big) + d\big(\mathcal{N}(B)\big). \tag{1.1.2}$$

A la lumière des résultats précédents, on peut se demander si (1.1.1) peut s'énoncer à l'aide de la densité naturelle sur les entiers; et si l'on peut caractériser la propriété suivante:

$$\text{p.s.} \quad d\big(\mathcal{N}(A)\big) = \gamma(A) \tag{1.1.3}$$

pour tout ouvert A de $\mathbb{R}$ tel que $\gamma(\partial A) = 0$, où ∂A désigne la frontière de A pour la topologie usuelle et γ une fonction d'ensemble (non nécessairement additive) définie sur les ouverts de $\mathbb{R}$.

Le théorème ci-dessous donne une première solution à ce problème, qui s'apparente à une nouvelle loi forte des grands nombres.

Théorème 1.1.1. *Si $K_j = M^j$, $j \ge 1$ où $M > 1$, alors sous les conditions de moments*

$$\mathbb{E}X = 0 \text{ et } \mathbb{E}(X^2) = 1,$$

pour tout ouvert A de $\mathbb{R}$ tel que $\gamma(\partial A) = 0$, (1.1.3) est réalisée avec

$$\gamma(A) = \gamma_1(A) = \mathbb{P}\Big\{\exists t \in [1, M] : \frac{W(t)}{\sqrt{t}} \in A\Big\},$$

où W est le processus de Wiener standard issu de 0.

L'énoncé suivant s'intéresse au cas où $\lim_{j\to\infty} K_{j+1}/K_j = 1$. La condition $\gamma(\partial A) = 0$ devient $\lambda(\partial A) = 0$, où λ désigne la mesure de Lebesgue.

Théorème 1.1.2. *Si $K_j = e^{j^b}$, $j \ge 1$, avec $0 < b < 1$, ou si $K_j = e^{j/\log j}$, $j \ge 3$, alors sous les mêmes conditions, pour tout ouvert A de $\mathbb{R}$ tel que $\gamma_2(\partial A) = 0$, (1.1.3) est réalisée avec*

$$\gamma_2(A) = N(0, 1)(A).$$

En fait nous établissons le résultat suivant, plus général que celui du Théorème 1.1.2:

Proposition 1.1.2′. *Soit $(j_k,\ k \ge 1)$ une suite de variables aléatoires définies sur $(\Omega, \mathfrak{A}, \mathbb{P})$ et à valeurs dans $\mathbb{N}$. S'il existe une suite $(a_k,\ k \ge 1)$ de réels strictement positifs avec $\lim_{k\to\infty} a_k = +\infty$, telle que:*

a) $$\forall \varepsilon > 0, \quad \lim_{k\to\infty} \mathbb{P}\Big\{\omega \in \Omega : \Big|\frac{j_k(\omega)}{a_k} - 1\Big| < \varepsilon\Big\} = 1$$

et

b) $$\sum_{n\geq 1}\frac{1}{n^4}\sum_{1\leq k<l\leq n^2}\sqrt{\frac{a_k}{a_l}}<\infty;$$

alors, sous les conditions de moments

$$\mathbb{E}X=0 \text{ et } \mathbb{E}(X^2)=1,$$

presque sûrement pour tout A *borélien de* $\mathbb{R}$ *telle que* $\lambda(\partial A)=0$, *on a:*

$$\lim_{N\to\infty}\frac{1}{N}\sum_{k=1}^{N}\mathbf{1}_{\mathrm{A}}\Big(S_{j_k(\omega)}/\sqrt{j_k(\omega)}\Big)=N(0,1)(A).$$

Ceci généralise et améliore le résultat de Schatte; la condition de moment, $\mathbb{E}|X|^3$ finie, n'étant plus nécessaire.

II Démonstration des résultats

II.1 Démonstration du Théorème 1.1.1. Soit A un ouvert de $\mathbb{R}$ dont la frontière ne charge pas γ_1. Soit $\{X, X_i,\ i\geq 1\}$ une suite de variables aléatoires indépendantes et équidistibuées, vérifiant:

$$\mathbb{E}X=0 \text{ et } \mathbb{E}(X^2)=1.$$

Le théorème d'approximation de Skorohod (**[Sk]**, p. 163) établit l'existence d'une suite i.i.d. de temps d'arrêts $T_i\geq 0,\ i\geq 1$ tels que:

$$\mathbb{E}T_i=1 \text{ et } \Big(W(\sum_{i=1}^{n}T_i)\Big)_{n\geq 1}\overset{\mathcal{D}}{=}\big(S_n(X)\big)_{n\geq 1}.$$

Soient $M>1,\ h>0,\ V_h=]1-h,1+h[$; et $I_{M,k}=[M^k,M^{k+1}[,\ \ k\geq 1$. En vertu de la loi forte des grands nombres, il existe un ensemble Ω_1 de probabilité 1 tel que: $\forall\omega\in\Omega_1,\ \exists j_0=j_0(\omega)\in\mathbb{N}:\ \forall i\geq j_0,$

$$n\in I_{M,i}\cap\mathbb{N}\Longrightarrow\exists\theta\in V_h,\ \exists s\in[1,M[\ :\sum_{k=1}^{n}T_k=M^i\theta s. \tag{2.1.1}$$

2.1. Majoration. Pour tout ouvert A, considérons:

$$\begin{cases} C_i=\left\{\exists\theta\in V_h\ ,\ \exists s\in[1,M[\ :\frac{W(M^i\theta s)}{\sqrt{M^i s}}\in A\right\}\\ C=\limsup_{j\to\infty}\frac{1}{j}\sum_{i=1}^{j}\mathbf{1}_{C_i}\\ L(M)=L=\limsup_{j\to\infty}\dfrac{\#\left\{i\leq j:\exists n\in I_{M,i}\cap\mathbb{N}:\frac{W(\sum_{i=1}^{n}T_i)}{\sqrt{n}}\in A\right\}}{j}\end{cases} \tag{2.1.2}$$

Nous avons

$$\text{(2.1.3)} \qquad \text{p.s.} \qquad L \leq C.$$

Comme $(U_M(t),\ t \geq 0$ est un processus gaussien stationnaire, nous savons que le système dynamique $\big(\mathcal{C}(\mathbb{R}^+, \mathbb{R}), \mathcal{L}(U_M), T\big)$, où $\mathcal{L}(U_M)$ est la loi de U_M et $Tf = f(.+1)$, est fortement mélangeant en vertu du théorème de G. Maruyama (voir aussi **[W3]**, Théorème A, p. 477). Il en résulte que, presque sûrement,

$$\text{(2.1.4)} \qquad C = \mathbb{P}\{C_1\} = \mathbb{P}\Big\{\exists\theta \in V_h\ ,\ \exists s \in [1, M[\ : \frac{W(M\theta s)}{\sqrt{Ms}} \in A\Big\}$$

$$= \mathbb{P}\Big\{\exists\theta \in V_h\ ,\ \exists s \in [1, M[\ : \frac{W(\theta s)}{\sqrt{s}} \in A\Big\}.$$

Par conséquent en mettant ensemble (2.1.3) et (2.1.4), puis en faisant tendre h vers 0, nous obtenons

$$\text{(2.1.5)} \qquad \text{p.s.} \qquad L \leq \mathbb{P}\Big\{\exists t \in [1, M[\ : \frac{W(t)}{\sqrt{t}} \in \overline{A}\Big\},$$

d'où

$$\text{(2.1.6)} \qquad \text{p.s.} \qquad \overline{d}\big(\mathcal{N}(A)\big) \leq \gamma_1(\overline{A}),$$

pour tout ouvert A de $\mathbb{R}$. Donc:

$$\text{(2.1.7)} \qquad \text{p.s.} \qquad \overline{d}\big(\mathcal{N}(A)\big) \leq \gamma_1(A)$$

pour tout ouvert A tel que $\gamma_1(\partial A) = 0$. D'où la majoration.

2.2. Minoration. Soient A un ouvert de $\mathbb{R}$ dont la frontière ne charge pas γ_1 et $I = [-1, +1]$. Pour tout $\varepsilon > 0$ assez petit, posons

$$\text{(2.1.8)} \qquad A^{\varepsilon} = A + \varepsilon I = \{x : d(x, A) < \varepsilon\}.$$

Puisque $\gamma(\partial A) = 0$, $\lim\limits_{\varepsilon \to 0} \gamma(A^{\varepsilon}) = \gamma(A)$.

Soient μ un entier positif et C_μ une partie de $\{0, 1, 2, \ldots, \mu - 1\}$ que l'on précisera plus tard, $g(\mu) = \operatorname{card} C_\mu$ et $Y_t = \frac{W(t)}{\sqrt{t}}$, $t > 0$. Soit $\varepsilon > 0$. Introduisons pour tout $i \in \mathbb{N}$ les quantités:

$$\text{(2.1.9)} \qquad B_i = \bigcup_{l \in C_\mu} \bigcup_{v \in I_{m,\mu i + l}} \big\{Y_v \in A\big\}$$

$$D_i = \bigcap_{l \in C_\mu} \Big\{ \bigcap_{\{u,t\} \subset I_{m,\mu i + l}} \bigcap_{\theta \in V_h} \big|Y_u - Y_{t\theta}\sqrt{\theta}\big| \leq \varepsilon \Big\}.$$

$(B_i,\ i \in \mathbb{N})$ et $(D_i,\ i \in \mathbb{N})$ sont bien définis car A est un ouvert de $\mathbb{R}$ et $(Y_t,\ t > 0)$ est à trajectoires continues.

Étape 1. Soit $m = M^{\frac{1}{\mu}}$, soit i_μ un entier positif tel que $M^{i_\mu}(m-1)$ soit strictement supérieur à 1. Alors, $\forall \omega \in \Omega_1$, $\forall i \geq \sup(j_0(\omega), i_\mu)$:

$$\omega \in B_i \cap D_i \Longrightarrow \omega \in \left\{\exists n \in I_{M,i} \cap \mathbb{N} : \frac{W(\sum_{k=1}^n T_k)}{\sqrt{n}} \in A^\varepsilon\right\} \tag{2.1.10}$$

et donc

$$\lim_{j\to\infty} \frac{\sum_{i=1}^{j} \mathbf{1}_{B_i \cap D_i}}{j} \leq \lim_{j\to\infty} \frac{\sum_{i=1}^{j} \mathbf{1}_{\left\{1\leq i\leq j:\ \exists n\in I_{M,i}\,\cap\,\mathbb{N}:\ \frac{W(\sum_{k=1}^n T_k)}{\sqrt{n}}\in A^\varepsilon\right\}}}{j}. \tag{2.1.10$'$}$$

En effet, soit $\omega \in B_i \cap D_i \cap \Omega_1$ avec $i \geq \sup(j_0(\omega), i_\mu)$. Alors, puisque C_μ est fini:

$$\omega \in B_i \Longrightarrow \exists l_\omega \in C_\mu,\ \exists u_\omega \in I_{m,\mu i+l_\omega} : Y_{u_\omega}(\omega) \in A;$$

donc, pour tout θ élément de V_h et pour tout t élément de $I_{m,\mu i+l_\omega}$ on a:

$$\frac{W(t\theta)}{\sqrt{t}}(\omega) = Y_{t\theta}(\omega)\sqrt{\theta} - Y_{u_\omega}(\omega) + Y_{u_\omega}(\omega) \in A + \varepsilon I = A^\varepsilon.$$

Soit n un entier dans $I_{m,\mu i+l_\omega}$ (il en existe car i est supérieur à i_μ), donc n est un élément de $I_{M,i}$ et, par (2.1.1),

$$\frac{W(\sum_{k=1}^n T_k)(\omega)}{\sqrt{n}} = \frac{W(n\theta)}{\sqrt{n}}(\omega),$$

avec $\theta \in V_h$ et donc

$$\omega \in \left\{\exists n \in I_{M,i} \cap \mathbb{N} : \frac{W(\sum_{k=1}^n T_k)}{\sqrt{n}} \in A^\varepsilon\right\},$$

d'où (2.1.10).

Étape 2. Posons $U_M(t) = W(M^t)\, M^{-t/2}$, $t \geq 0$. L'ergodicité du processus gaussien stationnaire $U = \{W(M^t)M^{-\frac{t}{2}},\ t \geq 0\}$, car fortement mélangeant, implique que presque sûrement,

$$b_\mu = \lim_{j\to\infty} \frac{1}{j} \sum_{i=1}^{j} \mathbf{1}_{B_i} = \int_{\Omega_1} \mathbf{1}_{B_1} d\mathbb{P}$$

et

$$d_\mu = \lim_{j\to\infty} \frac{1}{j} \sum_{i=1}^{j} \mathbf{1}_{D_i} = \int_{\Omega_1} \mathbf{1}_{D_1} d\mathbb{P}.$$

Donc

$$\begin{aligned} b_\mu &= \mathbb{P}\{\exists l \in C_\mu,\ \exists v \in I_{m,\mu+l} : Y_v \in A\} \\ &= \mathbb{P}\{\exists l \in C_\mu,\ \exists v \in I_{m,l} : Y_v \in A\} \\ &= \mathbb{P}(B_0) \end{aligned} \tag{2.1.11}$$

et

$$(2.1.12)\qquad d_\mu = \mathbb{P}\{\forall l \in C_\mu,\ \sup_{(u,t)\subset I_{m,\mu+l}} \sup_{\theta\in V_h} |Y_u - Y_{t\theta}\sqrt{\theta}| \le \varepsilon\}$$

$$= \mathbb{P}\{\forall l \in C_\mu,\ \sup_{(u,t)\subset I_{m,l}} \sup_{\theta\in V_h} |Y_u - Y_{t\theta}\sqrt{\theta}| \le \varepsilon\}$$

$$= \mathbb{P}(D_0).$$

Alors, par $(2.1.10)'$, il existe un ensemble Ω_2 de probabilité 1 et contenu dans Ω_1 tel que sur Ω_2:

$$\liminf_{j\to\infty} \frac{\mathrm{card}\{1 \le i \le j : \exists n \in I_{M,i} \cap \mathbb{N} :\ \frac{W(\sum_{k=1}^n T_k)}{\sqrt{n}} \in A^\varepsilon\}}{j} \ge b_\mu + d_\mu - 1.$$

Nous allons commencer par évaluer d_μ. Soient l un élément de C_μ et θ un élément de V_h. Pour tout t et u dans $I_{m,l}$, on a:

$$(2.1.13)\qquad |Y_u - Y_{t\theta}\sqrt{\theta}| \le |Y_u - Y_{m^l}| + |Y_{m^l} - Y_t| + |Y_t - Y_{t\theta}\sqrt{\theta}|$$

$$\le 2 \sup_{v\in I_{m,l}} |Y_v - Y_{m^l}| + \sup_{t'\in[1,M]} \sup_{\theta\in V_h} |Y_{t'} - Y_{t'\theta}\sqrt{\theta}|;$$

et pour tout v élément de $I_{m,l}$,

$$(2.1.14)\qquad |Y_v - Y_{m^l}| = \left| \frac{W(v) - W(m^l)}{\sqrt{m^l}} + W(v)\Big(\frac{1}{\sqrt{v}} - \frac{1}{\sqrt{m^l}}\Big)\right|$$

$$\le \left|\frac{W(v) - W(m^l)}{\sqrt{m^l}}\right| + (m-1)\left|\frac{W(v)}{\sqrt{v}}\right|.$$

Alors, en posant

$$\Omega_h = \{\omega : \sup_{t\in[1,M]} \sup_{\theta\in V_h} |Y_t - Y_{t\theta}\sqrt{\theta}| < \frac{\varepsilon}{2}\},$$

et

$$\Omega'_\mu = \{\omega : \sup_{v\in[1,M]} |Y_v| \le \frac{\varepsilon}{8(m-1)}\},$$

On a:

$$\Omega_h \cap \Omega'_\mu \cap \bigcap_{l\in C_\mu} \{\omega : \sup_{v\in[m^l,m^{l+1}]} \left|\frac{W(v) - W(m^l)}{\sqrt{m^l}}\right| \le \frac{\varepsilon}{8}\} \subset D_0.$$

Soit alors k arbitraire dans $\mathbb{N}^*$. A cause de la continuité des trajectoires de $(Y_t,\ t > 0)$, pour μ assez grand et h assez petit, on a:

$$\mathbb{P}(\Omega_h^c) \le \frac{1}{2k} \qquad \text{et} \qquad \mathbb{P}({\Omega'_\mu}^c) \le \frac{1}{2k}$$

et donc

$$(2.1.15)\qquad \begin{aligned} d_\mu &= \mathbb{P}(D_0) \\ &\geq \mathbb{P}\Big(\bigcap_{l\in C_\mu}\Big\{\sup_{v\in[m^l,m^{l+1}]}\Big|\frac{W(v)-W(m^l)}{\sqrt{m^l}}\Big|\leq\frac{\varepsilon}{8}\Big\}\Big)-\frac{1}{k} \\ &= \prod_{l\in C_\mu}\mathbb{P}\Big\{\sup_{v\in[m^l,m^{l+1}]}\Big|\frac{W(v)-W(m^l)}{\sqrt{m^l}}\Big|\leq\frac{\varepsilon}{8}\Big\}-\frac{1}{k} \\ &\geq \Big(\mathbb{P}\Big\{\sup_{0<\theta<M^{\frac{1}{\mu}}-1}|W(\theta)|\leq\frac{\varepsilon}{8}\Big\}\Big)^{g(\mu)}-\frac{1}{k}. \end{aligned}$$

En appliquant le lemme 2.2, p. 121 de **[JJS]**, il existe une constante C telle que:

$$\mathbb{P}\Big\{\sup_{0<\theta<M^{\frac{1}{\mu}}-1}|W(\theta)|>\frac{\varepsilon}{8}\Big\}\leq\frac{8C}{\varepsilon}\sqrt{M^{\frac{1}{\mu}}-1}\exp\Big(\frac{-\varepsilon^2}{128\,(M^{\frac{1}{\mu}}-1)}\Big),$$

donc

$$d_\mu\geq\Big(1-\frac{8C}{\varepsilon}\sqrt{M^{\frac{1}{\mu}}-1}\exp\Big(\frac{-\varepsilon^2}{128\,(M^{\frac{1}{\mu}}-1)}\Big)\Big)^{g(\mu)}-\frac{1}{k}.$$

Soient $\beta_\mu=\frac{8C}{\varepsilon}\sqrt{M^{\frac{1}{\mu}}-1}\exp\Big(\frac{-\varepsilon^2}{128\,(M^{\frac{1}{\mu}}-1)}\Big)$ et $u_\mu=\frac{\varepsilon}{8\,(\sqrt{M^{\frac{1}{\mu}}-1})}$. Si $u_\mu\geq 1$, on a:

$$\beta_\mu=\frac{C}{u_\mu}\exp\Big(\frac{-u_\mu^2}{2}\Big)\leq\frac{2C}{u_\mu^2}.$$

Ainsi, si $\mu\geq\frac{\log M}{\log(1+\varepsilon^2/256C)}$ alors $u_\mu^2\geq 4C$ et l'on a:

$$(1-\beta_\mu)^{g(\mu)}\geq\Big(1-\frac{2C}{u_\mu^2}\Big)^{g(\mu)}\geq\exp\Big(g(\mu)\log\Big(\frac{2C}{u_\mu^2}\Big)\Big).$$

Or $\lim_{\mu\to\infty}u_\mu=\infty$ donc, pour μ assez grand,

$$(1-\beta_\mu)^{g(\mu)}\geq\exp\Big(\frac{-128C\log M}{\varepsilon^2}\frac{g(\mu)}{\mu}\Big).$$

D'où, si $\frac{g(\mu)}{\mu}$ tend vers zéro quand μ tend vers l'infini,

$$\liminf_{\mu\to\infty}(1-\beta_\mu)^{g(\mu)}\geq 1$$

et donc

$$\liminf_{\mu\to\infty} d_\mu \geq 1 - \frac{1}{k} \quad \text{si} \quad \lim_{\mu\to\infty} \frac{g(\mu)}{\mu} = 0.$$

En faisant tendre h vers zéro et k vers l'infini, on obtient:

$$\lim_{\mu\to\infty} d_\mu \geq 1 \quad \text{si} \quad \lim_{\mu\to\infty} \frac{g(\mu)}{\mu} = 0.$$

Choisissons

$$C_\mu = \bigcup_{k=2}^{[\sqrt{\mu}]} \{ [\mu j \frac{\log k}{k}],\ j < k/\log k \}$$

qui est une partie finie de $[0, \mu - 1[$ vérifiant $\lim_{\mu\to\infty} \frac{g(\mu)}{\mu} = 0$, et posons

$$J_\mu = \bigcup_{k=2}^{[\sqrt{\mu}]} \{ M^{\frac{j \log k}{k}},\ j < k/\log k \} \subset [1, M[\,.$$

Nous avons $J_\mu \subset \cup_{l\in C_\mu} I_{m,l}$ et donc $b(\mu) \geq b'(\mu)$ où:

$$b'(\mu) = \mathbb{P}\{ \exists t \in J_\mu : Y_t \in A \}.$$

Comme $\cup_\mu J_\mu$ est dense dans $[1, M[$ et que $(J_\mu, \mu \geq 2)$ est une suite croissante de parties de $[1, M[$, on a:

$$\lim_{\mu\to\infty} b'(\mu) = \mathbb{P}\{ \exists t \in [1, M[: Y_t \in A \} = \gamma_1(A).$$

Par conséquent, sur Ω_2:

$$\underline{d}\big(\mathcal{N}(A^\varepsilon)\big) \geq \gamma_1(A) - \frac{1}{k}; \tag{2.1.16}$$

or

$$\underline{d}\big(\mathcal{N}(A)\big) \geq \underline{d}\big(\mathcal{N}(A^\varepsilon)\big) - \overline{d}\big(\mathcal{N}(A^\varepsilon \cap A^c)\big) \geq \gamma_1(A) - \frac{1}{k} - \overline{d}\big(\mathcal{N}(A^\varepsilon \cap A^c)\big).$$

Puisque $A^\varepsilon \cap A^c$ est un fermé, par (2.1.6),

$$\underline{d}\big(\mathcal{N}(A)\big) \geq \gamma_1(A) - \frac{1}{k} - \gamma_1\big(A^\varepsilon \cap A^c\big).$$

En faisant tendre k vers l'infini, puis ε vers 0, nous obtenons:

$$\underline{d}\big(\mathcal{N}(A)\big) \geq \gamma_1(A). \tag{2.1.17}$$

D'où la minoration. Finalement nous avons:

$$\text{p.s.} \qquad d\big(\mathcal{N}(A)\big) = \gamma_1(A) \tag{2.1.18}$$

pour tout ouvert A tel que $\gamma_1(\partial A) = 0$. D'où le Théorème 1.1.1. □

II.2 Démonstration du Théorème 1.1.2. Soit $(X, X_i,\ i \geq 1)$ une suite de variables aléatoires réelles (v.a.r.) indépendantes et identiquement distribuées (i.i.d.) définie sur un espace probabilisé $(\Omega, \mathfrak{A}, \mathbb{P})$ et telle que:

$$\mathbb{E}X = 0 \text{ et } \mathbb{E}X^2 = 1.$$

Soit A un ouvert de $\mathbb{R}$ dont la mesure de Lebesgue du bord, notée $\lambda(\partial A)$, est nulle. Considérons la variable aléatoire réelle:

$$\begin{aligned} Y_N(\omega) &= \frac{1}{N}\sum_{k=1}^{N} \mathbf{1}_{\{\exists i\in I_k:\ S_i/\sqrt{i}\ \in A\}}(\omega), \qquad \omega \in \Omega \\ &= \frac{1}{N}\sum_{k=1}^{N} \sup_{i\in I_k} \mathbf{1}_{\{S_i/\sqrt{i}\ \in A\}}(\omega), \end{aligned} \tag{2.2.1}$$

où $I_k = [M_k, M_{k+1}[$ et $(M_k,\ k \geq 1)$ est une suite réelle positive croissante telle que $M_k/M_{k+1} = 1$ tende vers 1 quand k tend vers l'infini et que $M_{k+1} - M_k$ soit supérieur à 1. La suite $(M_k,\ k \geq 1)$ n'étant pas géométrique, le théorème de G. Maruyama **[W3]** ne peut être appliqué. Dans cette démonstration, on utilise un principe d'invariance (Lemme 2.1) assurant la convergence, dans l'espace $\mathcal{C}[0, 1]$, des suites de v.a.r. indexées par des sous-suites aléatoires à valeurs entières et la loi forte des grands nombres pour des v.a.r. faiblement orthogonales.

Soit $j_k\colon (\Omega, \mathfrak{A}, \mathbb{P}) \to \mathbb{N}$ la variable aléatoire associant à tout $\omega \in \Omega$ le plus petit entier $j_k(\omega)$ élément de I_k tel que:

$$\forall j \in I_k,\ \mathbf{1}_{\mathrm{A}}\big(S_j(\omega)/\sqrt{j}\big) \leq \mathbf{1}_{\mathrm{A}}\big(S_{j_k(\omega)}/\sqrt{j_k(\omega)}\big). \tag{2.2.2}$$

Pour tout $\omega \in \Omega$, $j_k(\omega)$ existe; en effet

si pour tout j : $\mathbf{1}_{\mathrm{A}}(S_j(\omega)/\sqrt{j}) = 0$, l'entier $j_k(\omega) = \inf\{\mathbb{N} \cap I_k\}$ vérifie (2.2.2);

s'il existe $j \in I_k$ tel que $\mathbf{1}_{\mathrm{A}}(S_j(\omega)/\sqrt{j}) = 1$, l'entier $j_k(\omega) = \inf\{\mathbb{N} \cap I_k : S_j(\omega)/\sqrt{j} \in A\}$ vérifie (2.2.2).

(2.2.1) s'écrit sous la forme

$$Y_N(\omega) = \frac{1}{N}\sum_{k=1}^{N} \mathbf{1}_{\mathrm{A}}\big(S_{j_k(\omega)}/\sqrt{j_k(\omega)}\big)(\omega), \qquad \omega \in \Omega\,. \tag{2.2.3}$$

On va montrer une version fonctionnelle de

$$\text{p.s} \qquad \lim_{N\to\infty} Y_N(\omega) = \gamma_2(A) = N(0, 1)\,(A). \tag{2.2.4}$$

Pour tout $\omega \in \Omega$ et pour tout $k \geq 1$, soit $s_k(., \omega)$ la fonction définie pour tout $t \in [0, 1]$ par:

$$s_k(t, \omega) = \begin{cases} S_i(\omega)/\sqrt{j_k(\omega)} & \text{si } t = i/j_k(\omega), \quad i = 0, 1, \ldots, j_k(\omega), \\ \text{linéaire dans } \left[\frac{i}{j_k(\omega)}, \frac{i+1}{j_k(\omega)}\right], & i = 0, 1, \ldots, j_k(\omega) - 1. \end{cases} \tag{2.2.5}$$

Donc pour tout $\omega \in \Omega$, $s_k(., \omega) \in \mathcal{C}[0, 1]$.

Soit $BL = BL\big(\mathcal{C}[0,1],\ \|\cdot\|_{BL}\big)$ l'espace des fonctions $f\colon \mathcal{C}[0,1] \to \mathbb{R}$ telles que $\| f \|_{BL} = \| f \|_L + \| f \|_\infty < \infty$, avec

$$\| f \|_L = \sup\left\{ \frac{|f(x) - f(y)|}{\| x - y \|} : x,\ y \in \mathcal{C}[0,1],\ x \neq y \right\}.$$

Puisque $BL\big(\mathcal{C}[0,1]\big)$ est séparable pour $\|\cdot\|_\infty$ (**[Du]**, Corollaire 11.2.5, p. 307), en vertu d'un théorème de R. M. Dudley (**[Du]**, Théorème 11.3.3, p. 310), la convergence étroite dans $\mathcal{C}[0,1]$ est déterminée par un nombre dénombrable de fonctions continues bornées lipschitziennes; donc pour avoir (2.2.4) il suffit de démontrer le

Théorème 2.2.1. *Supposons vérifiées les conditions de moments:*

$$\mathbb{E}X = 0 \text{ et } \mathbb{E}(X^2) = 1,$$

alors, pour tout f élément de $BL\big(\mathcal{C}[0,1]\big)$, presque sûrement, on a

$$\lim_{N\to\infty} \frac{1}{N} \sum_{k=1}^{N} f\big(s_k(\cdot)\big) = \mathbb{E}f(W). \tag{2.2.6}$$

Donc l'ensemble négligeable en dehors duquel (2.2.6) est vérifiée ne dépend pas de f; de plus on aura:

$$\mathbb{P}\left\{ \lim_{N\to\infty} \sup\Big\{\Big|\frac{1}{N} \sum_{k=1}^{N} f(s_k) - \mathbb{E}f(W)\Big| : \| f \|_{BL} \leq 1\Big\} = 0 \right\} = 1. \tag{2.2.6$'$}$$

Démonstration du Théorème 2.2.1. On a le

Lemme 2.2.2 (Th. 17.1, p. 146, [Bi]). *Soient θ une constante positive et $(a_k, k \geq 1)$ une suite réelle telle que $\lim\limits_{k\to\infty} a_k = +\infty$. Si*

$$\forall \varepsilon > 0,\ \lim_{k\to\infty} \mathbb{P}\Big\{\omega \in \Omega : \Big|\frac{j_k(\omega)}{a_k} - \theta\Big| < \varepsilon\Big\} = 1, \tag{2.2.7}$$

alors, en posant $\tilde{s}_k(t,\omega) = S_{[t j_k(\omega)]}(\omega)/\sqrt{j_k(\omega)}$:

$$\lim_{k\to\infty} \tilde{s}_k \overset{\mathcal{D}}{=} W$$

où W désigne le mouvement brownien standard issu de 0.

Or,

$$\sup_{0\leq t\leq 1} |s_k(t) - \tilde{s}_k(t)| \leq \sup_{1\leq i\leq j_k} \frac{|X_{i+1}|}{\sqrt{j_k}} \overset{P}{\to} 0$$

par l'inégalité de Tchebychev et en appliquant le Théorème 4.1 (**[Bi]**, p. 25), nous obtenons

$$\lim_{k\to\infty} s_k \overset{\mathcal{D}}{=} W.$$

Prenons $a_k = M_k$; alors la suite $(j_k(\omega),\ k \geq 1)$ vérifie (2.2.7) avec $\theta = 1$, donc

$$\lim_{k\to\infty} s_k \stackrel{\mathcal{D}}{=} W$$

dans $(\mathcal{C}[0,1])$ ou, ce qui est équivalent: pour tout $f \in BL(\mathcal{C}[0,1])$

$$\lim_{k\to\infty} \mathbb{E}f(s_k) = \mathbb{E}f(W)\cdot \tag{2.2.8}$$

Pour tout $\omega \in \Omega$, posons $Z_k(\omega) = f(s_k(\cdot,\omega)) - \mathbb{E}f(s_k(\cdot))$. (2.2.8) entraîne que

$$\lim_{k\to\infty} \frac{1}{N}\sum_{k=1}^{N} \mathbb{E}f(s_k) = \mathbb{E}f(W)\cdot$$

Donc, pour avoir (2.2.6), il suffit de montrer

$$\text{p.s.} \quad \lim_{k\to\infty} \frac{1}{N}\sum_{k=1}^{N} Z_k(\omega) = 0, \tag{2.2.9}$$

qui est une loi forte des grands nombres pour la suite $\big(Z_k(\omega), k \geq 1\big)$ de v.a faiblement orthogonales.

Lemme 2.2.3. *Il existe une constante positive C dépendant de $\| f \|_{BL}$ telle que:*

$$\forall k < l : |\mathbb{E}(Z_k \cdot Z_l)| \leq C \cdot \sqrt{\frac{M_k}{M_l}}. \tag{2.2.10}$$

Démonstration du Lemme 2.2.3. on va adapter la démonstration de **[LP]**, (Lemme 1) à ce cas. Soit $k < l$ (donc $j_k < j_l$) fixés, où $j_k = j_k(\omega)$. Pour $0 \leq t \leq 1$, on définit:

$$r(t) = r(t,\omega) = \begin{cases} 0 & \text{si } 0 \leq t \leq \frac{j_k(\omega)}{j_l(\omega)}, \\ s_l(t,\omega) - \dfrac{S_{j_k(\omega)}}{\sqrt{j_l(\omega)}} & \text{si } \frac{j_k(\omega)}{j_l(\omega)} \leq t \leq 1. \end{cases} \tag{2.2.11}$$

$r(\cdot) \in \mathcal{C}[0,1]$ et l'on a

$$\begin{aligned} \mathbb{E}\| s_l - r \|_\infty &= \mathbb{E}\max_{0\leq t\leq 1} |s_l(t) - r(t)| \\ &\leq 2\,\mathbb{E}\max_{i<M_{k+1}} \left|\frac{S_i}{\sqrt{M_l}}\right| \\ &\leq C_1 \cdot \sqrt{\frac{M_k}{M_l}}\cdot \end{aligned} \tag{2.2.12}$$

Et, puisque $\omega \to s_k(\cdot,\omega)$ et $\omega \to r(\cdot,\omega)$ sont indépendants

$$\begin{aligned} \big|\mathbb{E}(Z_k \cdot Z_l)\big| &= \big|\mathbb{E}\big[f(s_k)\big(f(s_l) - f(r)\big)\big] + \mathbb{E}\big(f(s_k)\ f(r)\big) - \mathbb{E}f(s_k)\cdot\mathbb{E}f(s_l)\big| \\ &\leq 2\,\| f \|_\infty \cdot \mathbb{E}|f(s_l) - f(r)| \end{aligned}$$

$$\leq 2C_1 \parallel f \parallel_{BL}^2 \cdot \sqrt{\frac{M_k}{M_l}}.$$

En posant $C = 2C_1 \parallel f \parallel_{BL}^2$, on obtient (2.2.10). D'où le Lemme 2.2.3. □

Proposition 2.2.4. *Soit* $(M_k,\ k \geq 1)$ *une suite réelle positive croissante vérifiant:*

$$\text{(2.2.13)} \qquad \begin{cases} \lim\limits_{k\to\infty} \dfrac{M_k}{M_{k+1}} = 1 \\ M_{k+1} - M_k \geq 1 \\ \sum_{N\geq 1} \frac{1}{N^4} \sum_{1\leq k<l\leq N^2} \sqrt{\frac{M_k}{M_l}} < \infty. \end{cases}$$

Alors, (2.2.9) *est réalisée.*

Démonstration. On a

$$\mathbb{P}\{\omega : \frac{1}{N^2} | \sum_{1\leq k\leq N^2} Z_k(\omega)| > \varepsilon\} \leq \frac{1}{\varepsilon^2 N^4}\{\sum_{k=1}^{N^2} \mathbb{E}Z_k^2 + 2 \sum_{1\leq k<l\leq N^2} \mathbb{E}(Z_k Z_l)\}$$

$$\leq \frac{C'}{\varepsilon^2 N^2} + \frac{2C}{\varepsilon^2 N^4} \sum_{1\leq k<l\leq N^2} \sqrt{\frac{M_k}{M_l}}.$$

Donc, par (2.2.13),

$$\text{(2.2.14)} \qquad \sum_{N\geq 1} \mathbb{P}\{\omega : \frac{1}{N^2} | \sum_{1\leq k\leq N^2} Z_k(\omega)| > \varepsilon\} < \infty.$$

Par Borel–Cantelli, pour tout $\varepsilon > 0$:

$$\mathbb{P}\{\omega : \limsup_{N\to\infty} \frac{1}{N^2} | \sum_{1\leq k\leq N^2} Z_k(\omega)| > \varepsilon\} = 0;$$

d'où

$$\text{(2.2.15)} \qquad \text{p.s.} \quad \lim_{N\to\infty} \frac{1}{N^2} \sum_{1\leq k\leq N^2} Z_k(\omega) = 0.$$

Reste à voir que la moyenne $\frac{1}{N} \sum_{1\leq k\leq N} Z_k(\omega)$ ne diffère pas beaucoup en limite de $\frac{1}{m^2} \sum_{1\leq k\leq m^2} Z_k(\omega)$, avec $m^2 < N \leq (m+1)^2$. On a:

Lemme 2.2.5. *Soit* $S_n(Z) = \sum_{j=1}^{n} Z_k(\omega)$. *Posons*

$$D_m = \max_{m^2<N\leq(m+1)^2} |S_N(Z) - S_{m^2}(Z)|, \quad m \geq 1.$$

Alors, pour tout $m \geq 1$*:*

$$\mathbb{E}D_m^2 \leq C_1 \cdot m^2 + C_2 \cdot m \sum_{m^2+1 \leq p < q \leq (m+1)^2} \sqrt{\frac{M_p}{M_q}}$$

avec $0 < C_1, C_2 < \infty$.

Démonstration. On va utiliser le lemme suivant dû à N. Kôno **[Ko]**:

Lemme [Ko]. *Soit* $(X_i,\ i \geq 1)$ *une suite de variables alétoires telle que la suite des sommes partielles*

$$S(c,n) = \sum_{j=c+1}^{c+n} X_j\,, \quad c \geq 0,\ n \geq 1,$$

satisfasse les conditions de moments suivantes:

(i) *pour un réel* $r \geq 1$, *il existe une suite* $\{g_{c,n} : c \geq 0,\ n \geq 1\}$ *telle que*

$$\forall c \geq 0,\ \forall n \geq 1 : \ \mathbb{E}|S(c,n)|^r \leq g_{c,n}.$$

et

(ii)
$$\forall c \geq 0,\ \forall k, j \geq 1 : \ g_{c,j} + g_{c+j,k} \leq g_{c,j+k}.$$

Alors

$$\mathbb{E} \max_{1 \leq k \leq n} |S(c,k)|^r \leq (\log 2n)^r g_{c,n}.$$

On a:

$$\begin{aligned} D_m &= \max_{m^2 < N \leq (m+1)^2} |S_N(Z) - S_{m^2}(Z)| \\ &= \max_{m^2 < N \leq (m+1)^2} \Big| \sum_{j=m^2+1}^{N} Z_j \Big| \\ &= \max_{1 \leq k \leq (2m+1)} \Big| \sum_{j=m^2+1}^{m^2+k} Z_j \Big|. \end{aligned}$$

En utilisant (2.2.10), nous obtenons pour tout $c \in \mathbb{N}$:

$$\begin{aligned} \mathbb{E}\Big(\sum_{j=c+1}^{c+k} Z_j\Big)^2 &= \sum_{j=c+1}^{c+k} \mathbb{E}Z_j^2 + 2 \sum_{c+1 \leq p < q \leq c+k} \mathbb{E}Z_p Z_q \\ &\leq C_1 \cdot k + 2C_2 \cdot \sum_{c+1 \leq p < q \leq c+k} \sqrt{\frac{M_p}{M_q}} = g_{c,k}; \end{aligned}$$

où $0 < C_1, C_2 < \infty$. Par ailleurs,

$$\forall j \geq 1,\ \forall k \geq 1\,,\ g_{c,j} + g_{c+j,k} \leq g_{c,j+k}.$$

Le Lemme 2.2.5 s'obtient en appliquant le lemme de N. Kôno et en utilisant le fait que pour tout $m \geq 1$: $(\log m)^2 \leq m$. □

Soit $B = \sum_{n\geq 1} \frac{1}{n^4} \sum_{1\leq k<l\leq n^2} \sqrt{\frac{M_k}{M_l}} < \infty$. Puisque

$$\begin{aligned} B &\geq \sum_{n\geq 1} \frac{1}{n^4} \sum_{i=1}^{n-1} \sum_{i^2<k<l\leq(i+1)^2} \sqrt{\frac{M_k}{M_l}} \\ &= \sum_{i\geq 1} \sum_{i^2<k<l\leq(i+1)^2} \sqrt{\frac{M_k}{M_l}} \sum_{n=i+1}^{\infty} \frac{1}{n^4} \\ &\geq C \sum_{i\geq 1} \frac{1}{i^3} \sum_{i^2<k<l\leq(i+1)^2} \sqrt{\frac{M_k}{M_l}}\,; \end{aligned}$$

où C est constante numérique; et en utilisant (2.2.13) et le Lemme 2.2.5, on a pour tout $\varepsilon > 0$:

$$\begin{aligned} \sum_{m\geq 1} \mathbb{P}\Big\{\frac{1}{m^2} D_m > \varepsilon\Big\} &\leq \frac{1}{\varepsilon^2} \sum_{m\geq 1} \frac{1}{m^4} \mathbb{E} D_m^2 \\ &\leq \frac{C_1}{\varepsilon^2} \cdot \sum_{m\geq 1} \frac{m^2}{m^4} + \frac{C_2}{\varepsilon^2} \cdot \sum_{m\geq 1} \frac{m}{m^4} \sum_{m^2+1\leq p<q\leq(m+1)^2} \sqrt{\frac{M_p}{M_q}} \\ &\leq \frac{C_1}{\varepsilon^2} \cdot \sum_{m\geq 1} \frac{1}{m^2} + \frac{C_2}{\varepsilon^2} \sum_{m\geq 1} \frac{1}{m^3} \sum_{m^2+1\leq p<q\leq(m+1)^2} \sqrt{\frac{M_p}{M_q}} \\ &< \infty. \end{aligned}$$

Par Borel–Cantelli:

$$\text{(2.2.16)} \qquad \text{p.s.} \quad \lim_{m\to\infty} \frac{1}{m^2} D_m = 0.$$

Pour $m^2 < N \leq (m+1)^2$, on a: (2.2.17)

$$\frac{1}{N} \Big| \sum_{k=1}^{N} Z_k(\omega) \Big| \leq \frac{\Big| \sum_{k=1}^{N} Z_k(\omega) - \sum_{k=1}^{m^2} Z_k(\omega) \Big|}{N} + \frac{\Big| \sum_{k=1}^{m^2} Z_k(\omega) \Big|}{N}$$

$$\le \max_{m^2<N\le(m+1)^2}\left(\frac{\left|\sum_{k=1}^{N} Z_k(\omega)-\sum_{k=1}^{m^2} Z_k(\omega)\right|}{N}+\frac{\left|\sum_{k=1}^{m^2} Z_k(\omega)\right|}{N}\right)$$

$$\le \frac{D_m(\omega)}{m^2}+\frac{1}{m^2}\left|\sum_{k=1}^{m^2} Z_k(\omega)\right|.$$

Finalement, en combinant (2.2.15) et (2.2.16), on obtient (2.2.9)

$$\text{p.s.} \qquad \lim_{k\to\infty}\frac{1}{N}\sum_{k=1}^{N} Z_k(\omega)=0.$$

D'où la Proposition 2.2.4. □

La suite $(M_k=e^{k/\log k},\ k\ge 3)$ vérifie (2.2.13). En effet, puisque

$$\begin{aligned}\sum_{k=3}^{l-1} e^{k/2\log k} &\le \int_3^l e^{t/2\log t}\,dt\\ &= \int_3^l e^{(t/2\log t)}\,\frac{\log t-1}{(\log t)^2}\,\frac{(\log t)^2}{\log t-1}dt\\ &\le 2\frac{(\log l)^2}{\log l-1}\int_3^l d(e^{t/2\log t})\\ &\le 2e^{(l/2\log l)}\frac{(\log l)^2}{\log l-1};\end{aligned}$$

on a:

$$\begin{aligned}\sum_{n\ge 3}\frac{1}{n^4}\sum_{3\le k<l\le n^2}\sqrt{\frac{M_k}{M_l}} &= \sum_{n\ge 3}\frac{1}{n^4}\sum_{l=4}^{n^2}\sum_{k=3}^{l-1}\frac{e^{k/2\log k}}{e^{l/2\log l}}\\ &\le C\sum_{n\ge 1}\frac{1}{n^2}\log n\\ &< \infty\end{aligned}$$

et

$$\lim_{k\to\infty}\frac{M_k}{M_{k+1}}=1.$$

Pour établir (2.2.9) pour la suite $(M_k=e^{k^b},\ 0<b<1)$, on va utiliser (2.2.10) et la loi forte des grands nombres de Gaal–Koksma. Comme $0<b<1$, on a:

$$\mathbb{E}\Big(\sum_{k=m+1}^{m+N} Z_k\Big)^2 \le C\cdot N+2\sum_{m+1\le k<l\le m+N}\sqrt{\frac{M_k}{M_l}}$$

$$\leq C \cdot N + 2 \sum_{m+1 \leq k < l \leq m+N} e^{\frac{k^b}{2}} e^{\frac{-l^b}{2}}$$

$$\leq C \cdot N + \frac{4}{b}(N+m)^{1-b} \sum_{m+1 \leq k < m+N} e^{\frac{k^b}{2}} (e^{-\frac{k^b}{2}} - e^{-\frac{(m+N)^b}{2}})$$

$$\leq C'\big(N + N(N+m)^{1-b}\big)$$

$$\leq 2C'\big((N+m)^{2-b} - m^{2-b}\big).$$

Donc par Gaal–Koksma, pour tout $\delta > 0$, il existe une constante positive C'' telle que:

$$\text{p.s.} \qquad \sum_{k=1}^{N} Z_k(\omega) \leq C'' N^{1-\frac{b}{2}} (\log N)^{2+\delta};$$

et donc (2.2.9) est vérifiée. D'où le Théorème 1.1.2. □

Nous venons de démontrer un résultat plus général que le théorème de Schatte, on a la

Proposition 2.2.6. *Soit $(X_i,\ i \geq 1)$ une suite de variables alétoires réelles indépendantes et équidistribuées vérifiant:*

$$\mathbb{E}X = 0 \text{ et } \mathbb{E}(X^2) = 1.$$

Soit $M > 1$. Alors, presque sûrement, pour tout borélien A de $\mathbb{R}$ tel que $\lambda(\partial A) = 0$, on a

$$\lim_{N \to \infty} \frac{1}{N} \sum_{j=1}^{N} \mathbf{1}_{\{S_{[M^j]}/\sqrt{[M^j]} \in A\}} \stackrel{\mathcal{D}}{=} \gamma_1(A).$$

Démonstration. En effet, en posant $a_k = M^k$, la suite $(M^k)_{k \geq 1}$ vérifie la Proposition 1.1.2′. □

II.3 Relation entre le CLT et l'ASCLT. Dans ce paragraphe, nous allons présenter une approche générale concernant l'étude des deux propriétés: le théorème central limite (CLT) et le théorème central limite presque sûr (ASCLT). Ainsi, nous obtenons des résultats qui aideront à clarifier la relation entre le CLT et l'ASCLT et leurs extensions. Nous commencerons d'abord par rappeler quelques définitions et propriétés des méthodes de sommation matricielles.

Soit $\mathcal{A} = \{a_n,\ n \geq 1\}$ où $a_n = \{a_{n,k},\ k \geq 1\}$ une matrice infinie de nombres réels. Elle est dite de Silverman–Toeplitz, si elle vérifie les hypothèses suivantes de régularité:

$$\text{i) } ||\mathcal{A}|| = \sup_{n \geq 1} ||a_n||_1 = \sup_{n \geq 1} \sum_{k=1}^{\infty} |a_{n,k}| < \infty,$$

$$\text{(2.3.1)} \qquad \text{ii) } \lim_{n \to \infty} \sum_{k=1}^{\infty} a_{n,k} = 1, \quad \text{et} \quad a_{n,k} \geq 0,\ n, k \geq 1$$

$$\text{iii) pour tout } K \geq 1,\ \lim_{n \to \infty} \sum_{k \leq K} |a_{n,k}| = 0.$$

Définition 2.3.1. Soit $\mathcal{X} = \{X, X_i,\ i \geq 1\}$ une suite de v.a.r. définies sur un espace probabilisé $(\Omega, \mathfrak{A}, \mathbb{P})$. Posons $S_0 = 0$ et pour tout $n \geq 1$, $S_n(X) = X_1 + \cdots + X_n$. Soit $\mathcal{A}$ une matrice de Silverman–Toeplitz. On dit que $\mathcal{X}$ vérifie la propriété centrale limite en densité par rapport à $\mathcal{A}$, et l'on écrit $\mathcal{X} \in \mathrm{CLT}(\mathcal{A})$ s'il existe un réel $\sigma_{\mathcal{X}} > 0$ tel que pour tout borélien B de $\mathbb{R}$ vérifiant $N(0,1)(\partial B) = 0$,

$$\lim_{n\to\infty} \sum_{k=1}^{\infty} a_{n,k}\, \mathbb{P}\{\, S_k(X)/\sqrt{k} \in B \,\} = N(0, \sigma_{\mathcal{X}})\,(B), \tag{2.3.2}$$

où ∂B désigne le bord de B

On dit que $\mathcal{X}$ vérifie la propriété de la limite centrale presque sûre par rapport à $\mathcal{A}$, et l'on écrit $\mathcal{X} \in \mathrm{ASCLT}(\mathcal{A})$ si

$$\text{p.s.} \qquad \lim_{n\to\infty} \sum_{k=1}^{\infty} a_{n,k}\, \delta_{\{S_k(X)/\sqrt{k}\}} \stackrel{\mathfrak{D}}{=} N(0,1). \tag{2.3.3}$$

Il y a deux méthodes de sommation matricielles qui jouent un rôle pour ces deux propriétés. Dans la suite, nous les noterons $\mathcal{L} = \{l_{n,k},\ n, k \geq 1\}$ et $\mathcal{C} = \{c_{n,k},\ n, k \geq 1\}$.

La première détermine la densité logarithmique et est définie comme suit:

$$\forall n \geq 2,\ k \geq 1, \quad l_{n,k} = \begin{cases} \frac{1}{k \log n} & \text{si } 1 \leq k \leq n \\ 0 & \text{si } k > n. \\ l_{1,1} = 1,\ l_{1,k} = 0 & \text{si } k > 1. \end{cases} \tag{2.3.4}$$

La seconde méthode de sommation matricielle détermine la methode de sommation au sens de Cesarò et est définie par

$$\forall n \geq 1,\ k \geq 1, \quad c_{n,k} = \begin{cases} \frac{1}{n} & \text{si } 1 \leq k \leq n, \\ 0 & \text{si } k > n. \end{cases} \tag{2.3.5}$$

Nous avons le

Théoreme 2.3.2. *Pour toute suite $\mathcal{X} = \{X, X_i,\ i \geq 1\}$ de variables aléatoires réelles indépendantes et équidistribuées, les assertions suivantes sont équivalentes:*

(a) $\mathcal{X} \in \mathrm{CLT}$.
(b) $\mathcal{X} \in \mathrm{CLT}(\mathcal{L})$.
(c) $\mathcal{X} \in \mathrm{ASCLT}$.
(c) $\mathbb{E}(X) = 0$ *et* $\mathbb{E}(X^2) < \infty$.

Il est bien connu que (a) $\Longleftrightarrow$ (d). Par Lacey–Philipp [LP], on a l'implication ((d) $\Longrightarrow$ (c). Les propriétés de la convergence étroite impliquent (c) $\Longrightarrow$ (b). Enfin (b) $\Longrightarrow$ (d) par la Proposition 2.3.3. ci-dessous. Le fait nouveau et intéressant est l'implication (c) $\Longrightarrow$ (d).

Remarque. Nous remercions M. Lacey de nous avoir fait remarqué que I. Berkes et H. Dehling ont obtenus, indépendamment de ce travail, une équivalence entre le CLT et l'ASCLT.

On a

Proposition 2.3.3. *Soit* $\mathcal{X} = \{X, X_i,\ i \in \mathbb{N}\}$ *une suite de variables aléatoires réelles i.i.d. et* $\mathcal{A}$ *une matrice de Silverman–Toeplitz. Alors,*

$$\mathcal{X} \in \mathrm{CLT}(\mathcal{A}) \quad \Longrightarrow \quad \mathbb{E}(X) = 0 \quad et \quad \mathbb{E}(X^2) < \infty.$$

Note. L'implication partielle

$$\mathcal{X} \in \mathrm{CLT}(\mathcal{A}) \quad \Longrightarrow \quad \mathbb{E}(X^2) < \infty,$$

reste vraie quand la matrice $\mathcal{A}$ satisfait seulement les hypothèses i) et ii) de (2.3.1). L'hypothèse iii) sert à montrer que X est centrée.

Démonstration.

Étape 1. $\mathcal{X}$ est symétrique.

Puisque $\mathcal{X} \in \mathrm{CLT}(\mathcal{A})$, alors il existe un réel $\sigma_{\mathcal{X}} > 0$ tel que pour tout $x > 0$

$$\lim_{n\to\infty} \sum_{k=1}^{\infty} a_{n,k}\, \mathbb{P}\{\, |S_k(X)/\sqrt{k}| > x\,\} = 2\, N(0, \sigma_{\mathcal{X}})\,(\,]x, \infty[\,).$$

Comme $\mathcal{X}$ est symétrique, pour tout réel $M > 0$,

$$X = X\mathbf{1}_{\{\,|X|\le M\,\}} + X\mathbf{1}_{\{\,|X|>M\,\}} \quad \text{et} \quad X' = X\mathbf{1}_{\{\,|X|\le M\,\}} - X\mathbf{1}_{\{\,|X|>M\,\}},$$

ont la même loi que X. Par l'inégalité triangulaire, on peut écrire:

$$\limsup_{n\to\infty} \sum_{k=1}^{\infty} a_{n,k}\, \mathbb{P}\{\, |S_k(X\mathbf{1}_{\{|X|\le M\}})/\sqrt{k}| > x\,\} \le 4\, N(0, \sigma_{\mathcal{X}})\,(\,]x, \infty[\,).$$

Posons pour tout $n \ge 1$ et pour tout $x \ge 0$,

$$G_n(x) = 2x \sum_{k=1}^{\infty} a_{n,k}\, \mathbb{P}\{\, |S_k(X\mathbf{1}_{\{|X|\le M\}})/\sqrt{k}| > x\,\}.$$

Par l'inégalité de Hoeffding (Ann. Stat. Assoc. 58 (1963), p. 13–29), il existe une constante universelle K telle que

$$G_n(x) \le Kx \sup_{n\ge 1} \sum_{k=1}^{\infty} a_{n,k} \exp\Big(-\frac{x^2}{2M^2}\Big) = G(x) \in L^1(\mathbb{R}^+, dx).$$

Ecrivons pour tout $k \ge 1$, $\mathcal{G}_k = \sup_{l\ge k} G_l$. Alors,

$$\lim_{n\to\infty} \mathcal{G}_n(x) = \mathcal{G}(x) \le 8x\, N(0, \sigma_{\mathcal{X}})\,(\,]x, \infty[\,) \quad \text{et} \quad \mathcal{G}_n(x) \le G(x).$$

Le théorème de convergence dominée donne

$$\lim_{n\to\infty} \int_0^{\infty} \mathcal{G}_n(x)\, dx = \int_0^{\infty} \mathcal{G}(x)\, dx \le 2\sigma_{\mathcal{X}}^2.$$

Comme $G(x) \leq \mathcal{G}_n = \sup_{l \geq n} G_l$, on a:

$$\limsup_{n \to \infty} \int_0^\infty G_n(x)\, dx \leq 2\sigma_{\mathcal{X}}^2.$$

Mais,

$$\begin{aligned}\int_0^\infty G_n(x)dx &= \sum_{k=1}^\infty a_{n,k}\, \mathbb{E}(\,[S_k(X\mathbf{1}_{\{|X| \leq M\}})/\sqrt{k}]^2) \\ &= \sum_{k=1}^\infty a_{n,k}\, \mathbb{E}([X\mathbf{1}_{\{|X| \leq M\}}]^2) \\ &\to \mathbb{E}(\,[X\mathbf{1}_{\{|X| \leq M\}}]^2),\end{aligned}$$

quand n tend vers l'infini. Donc, il suit que $\mathbb{E}([X\mathbf{1}_{\{|X| \leq M\}}]^2) \leq 2\sigma_{\mathcal{X}}^2$, et comme M est arbitraire, nous avons montré que

$$\mathbb{E}(X^2) \leq 2\sigma_{\mathcal{X}}^2.$$

Étape 2. $\mathcal{X}$ n'est pas symétrique.

Soit $\mathcal{X}'$ une copie indépendante de $\mathcal{X}$ et soit $\mathcal{Y} = \mathcal{X} - \mathcal{X}' = \{Y, Y_i,\ i \in \mathbb{N}\}$. Alors, $\mathcal{Y}$ est symétrique et puisque $\mathcal{X} \in \mathrm{CLT}(\mathcal{A})$; il existe un réel $\sigma_{\mathcal{X}} > 0$ tel que pour tout $x > 0$, en utilisant l'inégalité triangulaire,

$$\limsup_{n \to \infty} \sum_{k=1}^\infty a_{n,k}\, \mathbb{P}\{\, |S_k(Y)/\sqrt{k}| > x\,\} \leq 4\, N(0, \sigma_{\mathcal{X}})\, (\,]\frac{x}{2}, \infty[\,).$$

Une même argumentation permet la majoration

$$\limsup_{n \to \infty} \sum_{k=1}^\infty a_{n,k}\, \mathbb{P}\{\, |S_k(X\mathbf{1}_{\{|X| \leq M\}})/\sqrt{k}| > x\,\} \leq 4\, N(0, \sigma_{\mathcal{X}})\, (\,]\frac{x}{2}, \infty[\,).$$

En mimant le cas symétrique, on établit que

$$\mathbb{E}(Y^2) \leq 4\sigma_{\mathcal{X}}^2.$$

Donc, par le Corollaire 2, p. 246 de **[Lo]**,

$$\mathbb{E}(\,[X - 5X]^2\,) \leq 8\sigma_{\mathcal{X}}^2,$$

μX est la médiane de X. Ceci implique que $\mathbb{E}(X^2)$ est finie.

Étape 3. Centrage.

Nous avons déjà $\mathbb{E}(X^2) < \infty$, et ceci implique une loi forte des grands nombres. Si $\mathbb{E}X = \mu > 0$, alors, pour tout $x > 0$

$$\lim_{k \to \infty} \mathbb{P}\{S_k(X)/\sqrt{k} > x\} = 1.$$

Donc pour tout $\epsilon > 0$ fixé, il existe $K_\epsilon \in \mathbb{N}$ tel que: pour tout $k \geq K_\epsilon$,

$$|\mathbb{P}\{S_k(X)/\sqrt{k} > x\} - 1| \leq \epsilon.$$

Comme

$$\Big|\sum_{k=1}^{\infty} a_{n,k}\, \mathbb{P}\{S_k(X)/\sqrt{k} > x\} - 1\Big| \leq \Big|\sum_{k=1}^{\infty} a_{n,k}\, [\mathbb{P}\{S_k(X)/\sqrt{k} > x\} - 1]\Big| + \Big|\sum_{k=1}^{\infty} a_{n,k} - 1\Big|,$$

et

$$\begin{aligned}\Big|\sum_{k=1}^{\infty} a_{n,k}\, [\mathbb{P}\{S_k(X)/\sqrt{k} > x\} - 1]\Big| \\ &\leq 2 \sum_{k \leq K_\epsilon} a_{n,k} + \sum_{k > K_\epsilon} a_{n,k}\, \big|\mathbb{P}\{S_k(X)/\sqrt{k} > x\} - 1\big| \\ &\leq 2 \sum_{k \leq K_\epsilon} a_{n,k} + \epsilon \sum_{k=1}^{\infty} a_{n,k}\,,\end{aligned}$$

des hypothèses (2.3.1), nous déduisons

$$\limsup_{n\to\infty} \Big|\sum_{k=1}^{\infty} a_{n,k}\, \mathbb{P}\{S_k(X)/\sqrt{k} > x\} - 1\Big| \leq \epsilon,$$

et donc,

$$\lim_{n\to\infty} \Big|\sum_{k=1}^{\infty} a_{n,k}\, \mathbb{P}\{S_k(X)/\sqrt{k} > x\} - 1\Big| = 0.$$

Ce qui donne: pour tout $x > 0$, $N(0, \sigma_X)\,(\,]x, \infty[\,) = 1$; ce qui est impossible. D'où, $\mathbb{E}X = \mu \leq 0$. De même si $\mathbb{E}X = \mu < 0$, on obtient

$$\forall x > 0\,, \quad \lim_{n\to\infty} \sum_{k=1}^{\infty} a_{n,k}\, \mathbb{P}\{S_k(X)/\sqrt{k} < x\} = 1 = N(0, \sigma_X)(\,]x, \infty[\,),$$

ce qui est aussi impossible. Finalement nous avons montré que $\mathbb{E}X = 0$. □

Références

[AW] Atlagh, M., Weber, M., Un théorème central limite presque sûr relatif à des sous-suites, C. R. Acad. Sci. Paris Sér. I Math. 315 (1992), 203–206.

[A1] Atlagh, M., Théorème central limite presque sûr et loi du logarithme itéré pour des sommes de variables alétoires indépendantes, C. R. Acad. Sci. Paris Sér. I Math. 316 (1993), 929–933.

[A2] Atlagh, M., Théorème central limite presque sûr et loi du logarithme itéré. Thèse. Prépublication I.R.M.A, 1993.

[Bi] Billingsley, P., Convergence of Probability Measures, Wiley, New York 1968.

[Br] Brosamler, G. A., An almost everywhere central limit theorem, Math. Proc. Cambridge Philos. Soc. 104 (1988), 561–574.

[Du] Dudley, R. M., Real Analysis and Probability, Wadsworth, Belmont, CA, 1989.

[Fi] Fischer, A., Convex invariant measure and pathwise central limit theorem, Adv. Math. 63, (1987), 213–246.

[JJS] Jain, N. C., Jogdeo, K., Stout, W. F., Upper and lower functions for martingales and mixing processes, Ann. Probab. 3 (1975), 119–145.

[Ko] Kôno, N., Classical limit theorems for dependent sequences having moment conditions, in: Probability Theory and Math. Statistics (Tbilisi, 1982), Lecture Notes in Math. 1021, pp. 315–319, Springer-Verlag, Berlin-Heidelberg-New York 1983.

[LP] Lacey, M., Philipp, W., A note on the almost sure central limit theorem, Statist. Probab. Lett. 9 (1990), 201–205.

[Lo] Loève, M., Probability theory I, 4th. ed, Springer-Verlag, New York 1977.

[Sc1] Schatte, P., On the value distribution of sums of random variables, Teor. Veroyatnost. i Primenen. 33 (1988), 800–804. English Transl. in Theory Probab. Appl. 33 (1988), 743–747.

[Sc2] Schatte, P., Two remarks on the almost sure central limit theorem, Math. Nachr. 154 (1991), 225–229.

[Sc3] Schatte, P., On strong versions of the almost sure central limit theorem, Math. Nachr. 137, (1988), 249–256.

[Sk] Skorohod, A., Studies in the theory of random processes, Addison-Wesley, Reading 1965.

[W1] Weber, M., Laws of the iterated logarithm on subsequences-characterizations, Nagoya Math. J. 118 (1990), 65–97.

[W2] Weber, M., Une version fonctionnelle du théorème ergodique ponctuel, C. R. Acad. Sci. Paris Sér. I Math. 311 (1990), 131–133.

[W3] Weber, M., Sur un théorème de Maruyama, in: Séminaire de Probabilités XIV, Lecture Notes in Math. 784, pp. 475–488, Springer-Verlag, Berlin-Heidelberg-New York 1980.

Institut de Recherche Mathématique Avancée
Université Louis Pasteur et CNRS
7, rue René Descartes
67084 Strasbourg Cedex
France

An Approximation Condition for Large Deviations and Some Applications

John R. Baxter and Naresh C. Jain**

Abstract. For each $k \geq 1$, let $\{\mu_n^{(k)}\}$ be a sequence of probability measures on a metric space S. Let $\{\mu_n\}$ be another sequence of probability measures on S. We show that if $\{\mu_n^{(k)}\}$ approximates $\{\mu_n\}$ in a certain sense as $k \to \infty$, and the large deviation principle holds for each sequence $\{\mu_n^{(k)}\}$ with rate function I_k, then the large deviation principle holds for the sequence $\{\mu_n\}$ with a rate function I which is obtained in terms of the I_k's. This result covers many situations in the large deviation theory where approximations are used. As an application, we obtain some new results for d-dimensional stationary Gaussian and certain non-Gaussian sequences.

1. Introduction

Let (S, d_0) be a metric space with metric d_0 and let $\mathcal{S}$ denote the σ-algebra of Borel subsets of S. $\mathcal{M}_1(S)$ will denote the set of probability measures on $(S, \mathcal{S})$. In the theory of large deviations, one often encounters the following sort of situation: A sequence $\{\mu_n,\ n \geq 1\}$ in $\mathcal{M}_1(S)$ is given for which one wants to establish the large deviation principle. For each $k \geq 1$, the sequence $\{\mu_n^{(k)},\ n \geq 1\}$ is known to satisfy the large deviation principle with a certain rate function I_k, and as $k \to \infty$, the sequence $\{\mu_n^{(k)}\}$ approximates $\{\mu_n\}$ in some sense. Then one infers the large deviation principle for $\{\mu_n\}$ with a certain rate function I. What we do here is to formulate a general approximation condition involving $\{\mu_n^{(k)}\}$ and $\{\mu_n\}$ such that if this condition holds, and for each k the sequence $\{\mu_n^{(k)}\}$ satisfies the large deviation principle with rate function I_k, then $\{\mu_n\}$ satisfies the large deviation principle with a rate function I which is obtained in terms of the I_k's. Many of the desirable properties of the I_k's carry over to I. We would like to observe that for these results the index sets for k and n need not be discrete.

Our general formulation covers a number of known situations, some of which will be discussed in Section 6. As far as new applications are concerned, our main object here is the large deviation principle for Gaussian and non-Gaussian stationary sequences. For a real stationary Gaussian sequence Donsker and Varadhan **[6]** obtain a large deviation principle when the associated spectral density f is continuous on $[0, 2\pi]$ and satisfies

*This work was partially supported by an NSF Grant.

the condition $\int_0^{2\pi} \log f(\theta)d\theta > -\infty$. They obtain a closed form expression for the rate function I. Using our general result, we have been able to extend the result when f is assumed to be only continuous. The Donsker–Varadhan formula for I does not make sense if their additional condition is violated, and we have no comparable formula for I. We are able to show that I still has compact level sets and is strictly positive and finite at some points. We prove our results for d-dimensional sequences. The result for the Gaussian sequences is then extended to a class of non-Gaussian sequences via an exponential inequality (established in Section 5) which may be of some independent interest.

Dembo and Zeitouni **[4]** formulate an approximation condition similar to ours (cf. Theorem 4.2.16 **[4]**). Their result implies a special case of our Theorem 3.1. Also, we are indebted to W. Bryc **[2]** for reference **[9]** where Steinberg and Zeitouni generalize the Donsker–Varadhan **[6]** result to real-valued Gaussian random fields. He pointed out to us that the proof in **[9]** would yield our Theorem 4.25 when $d = 1$, but would not answer the question whether the rate function is nontrivial (cf. Definition 2.8 below).

The paper is organized as follows. General notation and definitions are given in Section 2. Section 3 contains the approximation theorems. Applications to stationary sequences are presented in Section 4. An inequality comparing exponential moments of stationary sequences is given in Section 5. Applications to some known situations are discussed in Section 6. Some open problems are discussed in Section 7.

We would like to remark that the techniques of **[6]** play a significant role in Section 4.

2. General notation and definitions

S will always denote a metric space with metric d_0, and $\mathcal{M}_1(S)$ will denote the set of probability measures on Borel σ-algebra $\mathcal{S}$. The Lévy–Prohorov metric ρ on $\mathcal{M}_1(S)$ is defined by

$$\rho(\mu, \nu) = \inf\{\epsilon > 0 : \mu(A) \leq \nu(A^{\epsilon}) + \epsilon \quad \text{for all} \quad A \in \mathcal{S}\}, \tag{2.1}$$

where

$$A^{\epsilon} = \cup_{x \in A} B_{\epsilon}(x), \tag{2.2}$$

and

$$B_{\epsilon}(x) = \{y \in S : d_0(x, y) < \epsilon\}. \tag{2.3}$$

For $\delta > 0$, we define the pseudo-metric $\rho_{\delta}(\mu, \nu)$ on $\mathcal{M}_1(S)$ by

$$\rho_{\delta}(\mu, \nu) = \inf\{\epsilon > 0 : \mu(A) \leq \nu(A^{\delta}) + \epsilon \quad \text{for all} \quad A \in \mathcal{S}\}. \tag{2.4}$$

If S is Polish (complete separable metric space) then $\mathcal{M}_1(S)$ is Polish under ρ.

Definition 2.5. (Large deviation principle) Let $\{\mu_n\} \subset \mathcal{M}_1(S)$, $\{r(n)\}$, a sequence of positive numbers tending to $+\infty$, and a lower semicontinuous (lsc) function $I\colon S \to [0, \infty]$ be given. Then $\{\mu_n\}$ is said to satisfy the large deviation principle with the normalizing sequence $\{r(n)\}$ and rate function I, if for any F closed and G open,

$$\limsup_{n\to\infty} \frac{1}{r(n)} \log \mu_n(F) \le - \inf_{x\in F} I(x), \tag{2.6}$$

and

$$\liminf_{n\to\infty} \frac{1}{r(n)} \log \mu_n(G) \ge - \inf_{x\in G} I(x), \tag{2.7}$$

hold.

Definition 2.8. A lsc function $I\colon S \to [0, \infty]$ is called *tight* if for each $a > 0$, the level set $\{x : I(x) \le a\}$ is compact. It is called *nontrivial* if for some x, $0 < I(x) < \infty$.

Remark 2.9. A tight rate function is often referred to as "good" in the literature. Perhaps a tight and nontrivial rate function should be called "good".

Definition 2.10. (Local large deviation principle) The local large deviation principle is satisfied by $\{\mu_n\}$ with normalizing sequence $\{r(n)\}$ and lsc rate function I, if (2.7) holds and (2.6) is modified as follows:

Given $x \in S$, given $a < I(x)$, there exists a neighborhood V of x such that

$$\limsup_{n\to\infty} \frac{1}{r(n)} \log \mu_n(V) \le -a. \tag{2.6$'$}$$

Definition 2.11. (Exponential tightness) A sequence $\{\mu_n\} \subset \mathcal{M}_1(S)$ is called exponentially tight with respect to $\{r(n)\}$ if given $a > 0$, there exists a compact K such that

$$\limsup_{n\to\infty} \frac{1}{r(n)} \log \mu_n(K^c) \le -a.$$

Remark 2.12. It is easily seen (cf. **[5]**, where the terminology was introduced) that exponential tightness and the local large deviation principle imply that the large deviation principle holds and I must be tight.

Definition 2.13. (Local upper and lower rate functions) Define

$$\overline{\kappa}(x) = \inf\{\limsup_{n} \frac{1}{r(n)} \log \mu_n(V) : V \text{ open},\ x \in V\},$$

$$\underline{\kappa}(x) = \inf\{\liminf_{n} \frac{1}{r(n)} \log \mu_n(V) : V \text{ open},\ x \in V\}.$$

Remark 2.14. $-\overline{\kappa}$ and $-\underline{\kappa}$ are lsc, and if $\overline{\kappa} = \underline{\kappa}$, then the local large deviation principle holds for $\{\mu_n\}$ with normalizing sequence $\{r(n)\}$ and rate function $I = -\overline{\kappa} = -\underline{\kappa}$.

We will abbreviate "(local) large deviation principle" by "(local) LDP". We will also use the notation

$$\overline{\psi}(A) = \limsup_{n\to\infty} \frac{1}{r(n)} \log \mu_n(A), \tag{2.15}$$

and

$$\underline{\psi}(A) = \liminf_{n\to\infty} \frac{1}{r(n)} \log \mu_n(A). \tag{2.16}$$

3. Approximation theorems

The approximation condition for $\{\mu_n^{(k)}\}$ and $\{\mu_n\}$ is condition (3.2). Theorem 3.1 (for local LDP) and Theorem 3.11 are the main results. Proposition 3.16 is useful for identifying the rate function; if the I_k's are available in a nice form and converge pointwise to some function J, then one would like to identify I with J. This does occur in applications.

Theorem 3.1. *Let S be a metric space. For each $k \geq 1$, let $\{\mu_n^{(k)},\ n \geq 1\}$ be a sequence in $\mathcal{M}_1(S)$ which satisfies the local LDP with a normalizing sequence $\{r(n)\}$ and a lsc rate function I_k. Let $\{\mu_n,\ n \geq 1\}$ be a sequence in $\mathcal{M}_1(S)$ such that for every $\delta > 0$,*

$$\lim_{k\to\infty} \limsup_{n\to\infty} \frac{1}{r(n)} \log \rho_\delta(\mu_n^{(k)},\ \mu_n) = -\infty. \tag{3.2}$$

Then $\{\mu_n\}$ satisfies the local LDP with the same normalizing sequence $\{r(n)\}$ and with the lsc rate function I given by

$$I(x) = \underline{I}(x) = \overline{I}(x), \tag{3.3}$$

where

$$\underline{I}(x) = \lim_{\delta\downarrow 0} \liminf_{k\to\infty} \inf_{y\in B_\delta(x)} I_k(y), \tag{3.4}$$

and

$$\overline{I}(x) = \lim_{\delta\downarrow 0} \limsup_{k\to\infty} \inf_{y\in B_\delta(x)} I_k(y). \tag{3.5}$$

Proof. For convenience we will write

$$\varphi_\delta(k) = \limsup_{n\to\infty} \frac{1}{r(n)} \log \rho_\delta(\mu_n^{(k)}, \mu_n). \tag{3.6}$$

For $\delta > 0$ and $A \in \mathcal{S}$, we have

$$\mu_n(A) \leq \mu_n^{(k)}(A^\delta) + \rho_\delta(\mu_n^{(k)}, \mu_n). \tag{3.7}$$

Therefore, for any $k \geq 1$, using the local LDP for $\{\mu_n^{(k)}\}$, we get

$$\limsup_{n\to\infty} \frac{1}{r(n)} \log \mu_n(B_\delta(x)) \leq \max\{-\inf_{y\in B_{3\delta}(x)} I_k(y), \varphi_\delta(k)\}.$$

Letting $k \to \infty$ and using (3.2) this gives

$$\overline{\psi}(B_\delta(x)) \leq -\limsup_{k\to\infty} \inf_{y\in B_{3\delta}(x)} I_k(y). \tag{3.8}$$

Letting $\delta \downarrow 0$, we conclude that

$$\overline{\kappa}(x) \le -\overline{I}(x), \quad x \in S. \tag{3.9}$$

Switching the roles of μ_n and $\mu_n^{(k)}$ in (3.7) and letting $n \to \infty$ (with $A = B_\delta(x)$), we get

$$\liminf_{n\to\infty} \frac{1}{r(n)} \log \mu_n^{(k)}(B_\delta(x)) \le \max\{\underline{\psi}(B_{2\delta}(x), \varphi_\delta(k)\}.$$

By the local LDP for $\{\mu_n^{(k)}\}$, this gives

$$-\inf_{y\in B_\delta(x)} I_k(y) \le \max\{\underline{\psi}(B_{2\delta}(x)),\ \varphi_\delta(k)\},$$

and now letting $k \to \infty$, then letting $\delta \downarrow 0$, we get

$$-\underline{I}(x) \le \underline{\kappa}(x), \quad x \in S. \tag{3.10}$$

From (3.9) and (3.10) we see that $\underline{I} \doteq \overline{I} = -\overline{\kappa} = -\underline{\kappa}$. Since $-\overline{\kappa}$ and $-\underline{\kappa}$ are lsc, the theorem is proved. □

Theorem 3.11. *Assume that S is Polish and that for each $k \ge 1$, $\{\mu_n^{(k)}\}$ satisfies the LDP with normalizing sequence $\{r(n)\}$ and tight rate function I_k. If $\{\mu_n\} \subset \mathcal{M}_1(S)$ and* (3.2) *holds, then $\{\mu_n\}$ satisfies the LDP with normalizing sequence $\{r(n)\}$ and rate function I given by* (3.3)*; I is tight. Furthermore, if each I_k has a unique zero, then I has a unique zero; if S has a linear structure compatible with the metric and each I_k is convex, then I is convex.*

Proof. Since for each k, $\{\mu_n^{(k)}\}$ satisfies the LDP with tight rate function I_k, $\{\mu_n^{(k)}\}$ is exponentially tight (cf. **[8]**). Let $\delta > 0$ and $a > 0$ be given. By (3.2) we can pick a k_0 such that

$$\varphi_\delta(k_0) < -a, \tag{3.12}$$

when φ_δ is defined in (3.6).

By the exponential tightness of $\{\mu_n^{(k_0)}\}$ we can find a compact set K such that

$$\limsup_{n\to\infty} \frac{1}{r(n)} \log \mu_n^{(k_0)}(K^c) \le -a. \tag{3.13}$$

Since $((K^\delta)^c)^\delta \subset K^c$, by (3.2) and (3.7) for any $\delta > 0$ we get

$$\begin{aligned}\overline{\psi}((K^\delta)^c) &\le \max\{\limsup_{n\to\infty} \frac{1}{r(n)} \log \mu_n^{(k_0)}(K^c),\ \varphi_\delta(k_0)\}\\ &\le -a.\end{aligned}$$

We have shown that for every $a > 0$, given $\delta > 0$ there exists a set V, which is the finite union of balls of radius less than or equal to δ, such that

$$\overline{\psi}(V^c) \le -a.$$

In Lemma 3.14 to follow we will show that this implies the exponential tightness of $\{\mu_n\}$. Since Theorem 3.1 already implies that $\{\mu_n\}$ satisfies the local LDP with rate function I, the LDP for $\{\mu_n\}$ follows. The rate function I must be tight as a consequence of the LDP and exponential tightness by Remark 2.12.

Assume now that $I_k(x_k) = 0$ for a unique point x_k for each k. First we will show that I must vanish at some point. Let $\delta > 0$, and by (3.2) pick k_1 such that $\varphi_\delta(k_1) < 0$. Let V be any neighborhood of x_{k_1}. Then

$$\mu_n^{(k_1)}(V) \leq \mu_n(V^\delta) + \rho_\delta(\mu_n^{(k_1)}, \mu_n)$$

and by the LDP's for $\{\mu_n^{(k_1)}\}$ and $\{\mu_n\}$ with their respective rate functions, we get

$$-\inf_{x \in V} I_{k_1}(x) \leq \max\{-\inf_{x \in C} I(x), \varphi_\delta(k_1)\},$$

where we write $\overline{V^\delta} = C$. Since the left-side is zero and $\varphi_\delta(k_1) < 0$, this implies that $\inf_{x \in C} I(x) = 0$. This fact and the tightness of I imply that $I(z) = 0$ for some z.

Now let z be such that $I(z) = 0$. For any $k \geq 1$, $\delta > 0$, we have

$$\mu_n(B_\delta(z)) \leq \mu_n^{(k)}(B_{2\delta}(z)) + \rho_\delta(\mu_n^{(k)}, \mu_n),$$

so, as above,

$$-\inf_{x \in B_\delta(z)} I(x) \leq \max\{-\inf_{x \in \overline{B_{2\delta}(z)}} I_k(x), \varphi_\delta(k)\},$$

hence for all k sufficiently large depending on δ, we get

$$\inf_{x \in \overline{B_{2\delta}(z)}} I_k(x) = 0.$$

Since I_k is tight and $I_k(x_k) = 0$, we must have that $x_k \in \overline{B_{2\delta}(z)}$, for all k sufficiently large, depending on δ. But this means that $x_k \to z$, and $I(z) = 0$ at this unique point.

Now suppose S has a linear structure compatible with the metric. Let $0 \leq \alpha \leq 1$ and $\beta = 1 - \alpha$. Then

$$I(\alpha x + \beta y) = \lim_{\delta \downarrow 0} \limsup_{k \to \infty} \inf_{z \in B_\delta(\alpha x + \beta y)} I_k(z).$$

For any $\delta > 0$, there exists $\eta > 0$ such that

$$\alpha B_\eta(x) + \beta B_\eta(y) \subset B_\delta(\alpha x + \beta y),$$

therefore

$$I(\alpha x + \beta y) \leq \lim_{\eta \downarrow 0} \limsup_{k \to \infty} \inf_{u \in B_\eta(x), v \in B_\eta(y)} I_k(\alpha u + \beta v),$$

and by the convexity of I_k the result follows from this at once. This proves the theorem. □

It remains to prove the lemma that was used to prove the theorem.

Lemma 3.14. *Let $\{\mu_n\} \subset S$, and assume S to be Polish. If for any $a > 0$ and $\delta > 0$, there exists a set V, which is the union of finite number of balls of radius less than or equal to δ, such that*

$$\overline{\psi}(V^c) \le -a,$$

then $\{\mu_n\}$ is exponentially tight with respect to $\{r(n)\}$.

Proof. Let $a > 0$ be given. For each $j \ge 1$, there exists a set V_j, a finite union of balls of radius $\le 2^{-j}$ such that $\overline{\psi}(V_j^c) \le -2a$. This means that there exists an integer n_j such that

$$\mu_n(V_j^c) \le 2^{-j} e^{-ar(n)}, \quad n \ge n_j. \tag{3.15}$$

Since each μ_n is tight, without any loss of generality we may enlarge each V_j so that it is still a finite union of balls of radius $\le 2^{-j}$ and (3.15) holds for all $n \ge 1$. If we now take $K = \cap_{j=1}^{\infty} \overline{V}_j$, then clearly K is compact and for $n \ge 1$,

$$\begin{aligned}\mu_n(K^c) \le \mu_n(\cup_{j=1}^{\infty} V_j^c) &\le \sum_{j=1}^{\infty} 2^{-j} e^{-ar(n)} \\ &= e^{-ar(n)},\end{aligned}$$

which shows that $\{\mu_n\}$ is exponentially tight, and the lemma is proved. □

The following proposition gives a better understanding of the relationship between the rate functions I_k and I. Sometimes it is useful in identifying I.

Proposition 3.16. *Let $\underline{I}$ and $\overline{I}$ be defined as in* (3.4) *and* (3.5) *in terms of $\{I_k\}$, and if $\underline{I} = \overline{I}$, let I denote their common value. Then the following hold:*

(i) *$x_k \to x$ implies* $\liminf_{k\to\infty} I_k(x_k) \ge \underline{I}(x)$.

(ii) *For every x, there exists a sequence $x_k \to x$ such that* $\lim_{x\to\infty} I_k(x_k) \le \overline{I}(x)$.

(iii) *If $I = \underline{I} = \overline{I}$, then for every x, there exists a sequence $x_k \to x$, such that* $\lim_{k\to\infty} I_k(x_k) = I(x)$. *If $I_k \to$ some function J, pointwise, as $k \to \infty$, and $x_k \to x$ implies* $\liminf_{k\to\infty} I_k(x_k) \ge J(x)$, *then $J = I$.*

Proof. Suppose $x_k \to x$. Given $\delta > 0$, there exists $k_0(\delta)$ such that for $k \ge k_0(\delta)$,

$$I_k(x_k) \ge \inf_{y\in B_\delta(x)} I_k(y),$$

and letting $k \to \infty$ and then $\delta \downarrow 0$ we conclude (i) immediately from this.

To prove (ii), suppose x is such that $\overline{I}(x) = a < \infty$. It suffices to show that $\limsup_{k\to\infty} I_k(x_k) \le a$. Since $\overline{I}(x) = a$, we can pick $\epsilon_j \downarrow 0$, $\delta_j \downarrow 0$ and $k(j) \uparrow \infty$ such that

$$\inf_{y\in B_{\delta_j}(x)} I_k(y) < a + \epsilon_j, \; k \ge k(j).$$

Therefore, given $k \geq k(1)$, there exist $\delta(k) \searrow 0$, $\epsilon(k) \searrow 0$ such that

$$\inf_{y \in B_{\delta(k)}(x)} I_k(y) < a + \epsilon(k);$$

hence there exist $x_k \in B_{\delta(k)}(x)$ such that $I_k(x_k) < a + \epsilon(k)$ and (ii) is proved.

(iii) is an immediate consequence of (i) and (ii), and the proposition is proved. □

The assumption in Theorem 3.11 that S be Polish is important; the following example makes this clear and also sheds some light on condition (3.2).

Example 3.17. $S = \{2^{-k},\ k = 1, 2, \ldots\}$ with the usual metric. Let $\mu_n = \frac{1}{2}(\delta_{2^{-1}} + \delta_{2^{-n}})$, $n \geq 1$, and let $\mu_n^{(k)} = \mu_k$ for all n. Then it is easily verified that for any $A \subset S$,

$$\mu_n(A) \leq \mu_n^{(k)}(A^{2^{-k}}), \quad n \geq k.$$

Thus

$$\rho_{2^{-k}}(\mu_n, \mu_n^{(k)}) = 0, \quad n \geq k,$$

and condition (3.2) is satisfied for the sequences $\{\mu_n^{(k)}\}$ and $\{\mu_n\}$ with any normalizing sequence $\{r(n)\}$. The sequence $\{\mu_n\}$ satisfies the local LDP with any normalizing sequence $\{r(n)\}$ and tight rate function I defined by

$$\begin{aligned} I(2^{-1}) &= 0, \\ I(2^{-k}) &= \infty, \quad k \geq 2. \end{aligned}$$

There is no LDP for $\{\mu_n\}$.

As an easy corollary of Theorem 3.11 we note the following theorem.

Theorem 3.18. *Suppose S is Polish, and $\{\mu_{y,\epsilon}^{(\eta)};\ y \in B_a(x),\ 0 < \epsilon < \epsilon_0,\ 0 < \eta < \eta_0\} \subset \mathcal{M}_1(S)$, $\{\mu_{y,\epsilon};\ y \in B_a(x),\ 0 < \epsilon < \epsilon_0\} \subset \mathcal{M}_1(S)$ are given for some $a > 0$, $\epsilon_0 > 0$, and $\eta_0 > 0$. Suppose $r(\epsilon) \to \infty$ as $\epsilon \to 0$ and for every $\delta > 0$,*

$$\lim_{\eta \to 0} \limsup_{\substack{\epsilon \to 0 \\ y \to x}} \frac{1}{r(\epsilon)} \log \rho_\delta(\mu_{y,\epsilon}^{(\eta)}, \mu_{y,\epsilon}) = -\infty.$$

Also assume that for each η, $0 < \eta < \eta_0$, there exists a tight function $I_\eta : S \to [0, \infty]$ such that

$$\limsup_{\substack{\epsilon \to 0 \\ y \to x}} \frac{1}{r(\epsilon)} \log \mu_{y,\epsilon}^{(\eta)}(C) \leq -\inf_{z \in C} I_\eta(z), \quad C \text{ closed}, \tag{3.19}$$

and

$$\liminf_{\substack{\epsilon \to 0 \\ y \to x}} \frac{1}{r(\epsilon)} \log \mu_{y,\epsilon}^{(\eta)}(G) \geq -\inf_{z \in G} I_\eta(z), \quad G \text{ open}, \tag{3.20}$$

then (3.19) *and* (3.20) *hold with* $\mu_{y,\epsilon}^{(\eta)}$ *replaced by* $\mu_{x,\epsilon}$ *and* I_η *replaced by* I, *where*

$$\begin{aligned} I(z) &= \lim_{\alpha\downarrow 0} \liminf_{\eta\to 0} \inf_{z'\in B_\alpha(z)} I_\eta(z') \\ &= \lim_{\alpha\downarrow 0} \limsup_{\eta\to 0} \inf_{z'\in B_\alpha(z)} I_\eta(z'). \end{aligned} \tag{3.21}$$

Proof. Let $\epsilon_n \to 0$, $y_n \to x$, $\eta_k \to 0$ be any sequences, and apply Theorem 3.11 to $\mu_n^{(k)} = \mu_{y_n,\epsilon_n}^{(\eta_k)}$, $\mu_n = \mu_{y_n,\epsilon_n}$. We get the desired conclusion with the two expressions for I equal along $\{\eta_k\}$. Suppose z is fixed, then there exist sequences $\{\eta_n'\}$ and $\{\eta_n''\}$ such that the first expression for $I(z)$ is achieved along $\{\eta_n'\}$, the second along $\{\eta_n''\}$. If we put these two sequences together and call it $\{\eta_k\}$, then with this $\{\eta_k\}$ used in the initial argument we conclude that (3.21) holds for any z. Once we know that (3.21) holds, the conclusion of the corollary follows immediately. □

The following converse sort of implication is also useful.

Theorem 3.22. *Suppose S is Polish and* $\{\mu_n^{(k)}\} \subset \mathcal{M}_1(S)$, $k \geq 1$, $\{\mu_n\} \subset \mathcal{M}_1(S)$ *satisfy: for any* $\delta > 0$,

$$\limsup_{\substack{n\to\infty \\ k\to\infty}} \frac{1}{r(n)} \log \rho_\delta(\mu_n^{(k)}, \mu_n) = -\infty. \tag{3.23}$$

Suppose $\{\mu_n\}$ *satisfies the LDP with normalizing sequence* $\{r(n)\}$ *and tight rate function* I. *Then if C is closed and G open in S, we have*

$$\limsup_{\substack{n\to\infty \\ k\to\infty}} \frac{1}{r(n)} \log \mu_n^{(k)}(C) \leq -\inf_{x\in C} I(x), \tag{3.24}$$

and

$$\liminf_{\substack{n\to\infty \\ k\to\infty}} \frac{1}{r(n)} \log \mu_n^{(k)}(G) \geq -\inf_{x\in G} I(x). \tag{3.25}$$

If (3.23) *is replaced by* (3.2), *then simultaneous limsup and liminf in* (3.24) *and* (3.25) *will have to be replaced by the appropriate iterated limsup and liminf.*

Proof. Note that

$$\mu_n^{(k)}(C) \leq \mu_n(C^\delta) + \rho_\delta(\mu_n^{(k)}, \mu_n),$$

hence by (3.23), the left-side in (3.24) is dominated by $-\inf_{x\in\overline{C^\delta}} I(x)$ for all $\delta > 0$. It is known and easily seen that for any tight I, $\lim_{\delta\downarrow 0} \inf_{x\in\overline{C^\delta}} I(x) = \inf_{x\in C} I(x)$, and (3.24) follows.

For (3.25), let $x \in G$ and $B_{2\delta}(x) \subset G$. Then

$$\mu_n(B_\delta(x)) \leq \mu_n^{(k)}(B_{2\delta}(x)) + \rho_\delta(\mu_n^{(k)}, \mu_n),$$

and again by (3.23) we get

$$-I(x) \leq \limsup_{\substack{n\to\infty \\ k\to\infty}} \frac{1}{r(n)} \log \mu_n^{(k)}(B_{2\delta}(x)),$$

which implies (3.25) immediately. The final assertion about iterated limits has a similar proof. □

4. Application to stationary sequences

Theorem 4.9 of this section is a fairly general result for stationary sequences, but its applicability depends on whether condition (4.10) can be checked. This condition implies condition (3.2) and Theorem 3.11 is used to prove Theorem 4.9. If the stationary sequence is a d-dimensional Gaussian sequence with a continuous spectral density matrix F, then in Theorem 4.25 we show that condition (4.10) is satisfied. Finally, if the stationary sequence is not Gaussian, but the i.i.d. sequence $\{\eta_n\}$ of d-dimensional vectors that is used in the definition of the given sequence $\{X_j\}$ is such that the tail of $|\eta_1|$ is "smaller" than Gaussian, then we can again check (4.10). This result is proved in Theorem 4.40.

Let $\boldsymbol{B}$ be a separable Banach space, and on some probability space $(\Omega, \mathcal{F}, P)$, let $\{X_n,\ -\infty < n < \infty\}$ be a sequence of $\boldsymbol{B}$-valued random variables. Let $\tilde{\Omega}$ denote the space of doubly infinite $\boldsymbol{B}$-valued sequences, i.e. $x \in \tilde{\Omega}$ means $x = \{x_n,\ -\infty < n < \infty\}$ and $x_n \in \boldsymbol{B}$ for each n. On $\tilde{\Omega}$ we define the metric

$$\tilde{d}(x, y) = \frac{1}{3} \sum_{j=-\infty}^{\infty} 2^{-|j|} \min(\|x_j - y_j\|, 1),$$

where $\|\cdot\|$ denotes the $\boldsymbol{B}$-norm. With this metric the space $\tilde{\Omega}$ is Polish. For convenience, if $x \in \tilde{\Omega}$, we will write

$$|x| = \frac{1}{3} \sum_{j=-\infty}^{\infty} 2^{-|j|} \min(\|x_j\|, 1).$$

$T: \tilde{\Omega} \to \tilde{\Omega}$ denotes the usual backward shift. Writing $X = \{X_n,\ -\infty < x < \infty\}$, define

$$R_n(\omega) = \frac{1}{2n+1} \sum_{|j|\leq n} \delta_{T^j X(\omega)}, \qquad \omega \in \Omega, \tag{4.1}$$

where δ_x denotes the probability measure on $\tilde{\Omega}$ which assigns unit mass to x.

We give the Polish space $\mathcal{M}_1(\tilde{\Omega})$ the Lévy–Prohorov metric ρ. The distribution of R_n is a probability measure μ_n on the Polish space $S = \mathcal{M}_1(\tilde{\Omega})$ and we are interested in the large deviation principle for the sequence $\{\mu_n\}$. For convenience, as usual we will also refer to it as the large deviation principle for $\{R_n\}$. If $\{X_n\}$ is stationary and ergodic with distribution Q on $\tilde{\Omega}$, then as $n \to \infty$, $R_n \to Q$ a.s. with respect to the metric ρ. If V_1 is an open set containing Q, and V_2 is another set disjoint from V_1, then

$P\{R_n \in V_2\} = \mu_n(V_2) \to 0$, and the question is to determine this rate, if possible. More specifically, we want a normalizing sequence $\{r(n)\}$ and a rate function I, preferably tight and nontrivial, such that if $C \subset \mathcal{M}_1(\tilde{\Omega})$ is closed and $G \subset \mathcal{M}_1(\tilde{\Omega})$ is open, then

$$\limsup_{n\to\infty} \frac{1}{r(n)} \log P\{R_n \in C\} \leq - \inf_{Q' \in C} I(Q'), \tag{4.2}$$

and

$$\liminf_{n\to\infty} \frac{1}{r(n)} \log P\{R_n \in G\} \geq - \inf_{Q' \in G} I(Q'). \tag{4.3}$$

Now, let

$$L_n(\omega) = \frac{1}{2n+1} \sum_{|j|\leq n} \delta_{X_j(\omega)}, \ \omega \in \Omega. \tag{4.4}$$

If μ is the distribution of X_0, then $L_n \to \mu$ a.s. in the sense of the Lévy–Prohorov metric on $\mathcal{M}_1(\boldsymbol{B})$. If $\pi(x) = x_0$ is the projection map on $\tilde{\Omega}$, then the map $\psi(Q') = Q' \circ \pi^{-1}$ is a continuous map from $\mathcal{M}_1(\tilde{\Omega})$ to $\mathcal{M}_1(\boldsymbol{B})$. By Varadhan's contraction principle **[10]**,

$$\begin{aligned} \psi\Big(\frac{1}{2n+1} \sum_{|j|\leq n} \delta_{T^j x}\Big) &= \frac{1}{2n+1} \sum_{|j|\leq n} \psi(\delta_{T^j x}) \\ &= \frac{1}{2n+1} \sum_{|j|\leq n} \delta_{X_j} = L_n \end{aligned}$$

satisfies the LDP with rate function $J : \mathcal{M}_1(\boldsymbol{B}) \to [0, \infty]$ given by

$$J(\nu) = \inf\{I(Q') : Q' \in \mathcal{M}_1(\tilde{\Omega}), \ \psi(Q') = \nu\}. \tag{4.5}$$

Next, if $h : \boldsymbol{B} \to \mathbb{R}$ (real line) is a bounded continuous function, then $\psi_1(\mu) = \int h d\mu$ is a continuous map from $\mathcal{M}_1(\boldsymbol{B})$ into $\mathbb{R}$, and

$$\psi_1(L_n) = \int h dL_n = \frac{1}{2n+1} \sum_{|j|\leq n} h(X_j) \to \int h d\mu \tag{4.6}$$

a.s. as $n \to \infty$, and by the contraction principle $\psi_1(L_n)$ satisfies the LDP with rate function

$$J_h(x) = \inf\{J(\nu) : \nu \in \mathcal{M}_1(\boldsymbol{B}), \ \int h d\nu = x\}. \tag{4.7}$$

It is therefore quite useful to obtain the LDP for $\{R_n\}$, whenever possible. It may be remarked that even if I is available in closed form, we don't generally get such an expression for J or J_h.

To obtain the LDP for $\{R_n\}$, we will follow the approximation scheme of Donsker and Varadhan **[6]** for the Gaussian and non-Gaussian cases. First we formulate a general result.

Let $\{X_n, \ -\infty < n < \infty\}$ be a stationary, ergodic sequence of $\boldsymbol{B}$-valued random variables. Suppose for every $k \geq 1$, we have stationary $\{Y_n^{(k)}\}$, $\{Z_n^{(k)}\}$ such that

$$X_n = Y_n^{(k)} + Z_n^{(k)}.$$

Denoting

$$Y^{(k)} = \{Y_n^{(k)}, \ -\infty < n < \infty\},$$

let

$$R_n^{(k)} = \frac{1}{2n+1} \sum_{|j| \leq n} \delta_{T^j Y^{(k)}}. \tag{4.8}$$

The distributions of R_n and $R_n^{(k)}$ in the metric space $\mathcal{M}_1(\tilde{\Omega})$ will be denoted by $\nu_n^{(k)}$ and ν_n, respectively.

Theorem 4.9. *Suppose for each $k \geq 1$, $\{\nu_n^{(k)}\}$ satisfies the LDP with normalizing sequence $\{2n+1\}$ and tight rate function I_k. Suppose for every $\lambda > 0$, there exists $k(\lambda)$ such that for all $k \geq k(\lambda)$, all $n \geq 1$,*

$$E\big\{\exp\big[\lambda \sum_{|j| \leq n} \min(\|Z_j^{(k)}\|^2, 1)\big]\big\} \leq e^n, \tag{4.10}$$

then $\{\nu_n\}$ satisfies the LDP with normalizing sequences $\{2n+1\}$ and rate function I given by (3.3) *in terms of the sequence $\{I_k\}$; I is tight.*

Proof. We will apply Theorem 3.11 with $S = \mathcal{M}_1(\tilde{\Omega})$, $\mu_n^{(k)} = \nu_n^{(k)}$, $\mu_n = \nu_n$. We only need to show that (4.10) implies the condition (3.2).

Let $A \subset \mathcal{M}_1(\tilde{\Omega})$, then for $k \geq 1$, denoting by ρ the Lévy–Prohorov metric in $\mathcal{M}_1(\tilde{\Omega})$, we get

$$\begin{aligned}\nu_n(A) &= P\{R_n \in A\} \\ &= P\{R_n \in A, \ \rho(R_n, R_n^{(k)}) \leq \delta\} \\ &\qquad + P\{R_n \in A, \ \rho(R_n, R_n^{(k)}) > \delta\} \\ &\leq P\{R_n^{(k)} \in A^\delta\} + P\{\rho(R_n, R_n^{(k)}) > \delta\} \\ &= \nu_n^{(k)}(A^\delta) + P\{\rho(R_n, R_n^{(k)}) > \delta\}.\end{aligned}$$

Therefore

$$\rho_\delta(\nu_n^{(k)}, \nu_n) \leq P\{\rho(R_n, R_n^{(k)}) > \delta\}$$

and to check (3.2), it suffices to check that given $\delta > 0$,

$$\lim_{k\to\infty} \limsup_{n\to\infty} \frac{1}{2n+1} \log P\{\rho(R_n, R_n^{(k)}) > \delta\} = -\infty. \tag{4.11}$$

If B is a Borel subset of $\tilde{\Omega}$, then

$$\begin{aligned}
R_n(B) &= \frac{1}{2n+1}\sum_{|j|\le n} 1_B(T^jY^{(k)} + T^jZ^{(k)}) \\
&\le \frac{1}{2n+1}\sum_{|j|\le n}\{1_B(T^jY^{(k)} + T^jZ^{(k)})1_{B_\delta(0)}(T^jZ^{(k)}) + 1_{B_\delta^c(0)}(T^jZ^{(k)})\} \\
&\le \frac{1}{2n+1}\sum_{|j|\le n}\{1_{B^\delta}(T^jY^{(k)})\} + \frac{1}{2n+1}\#\{-n \le j \le n : |T^jZ^{(k)}| > \delta\} \\
&= R_n^{(k)}(B^\delta) + \frac{1}{2n+1}\#\{-n \le j \le n : |T^jZ^{(k)}| > \delta\}.
\end{aligned}$$

Therefore $\rho(R_n, R_n^{(k)}) > \delta$ implies that

$$\frac{1}{2n+1}\#\{-n \le j \le n : |T^jZ^{(k)}| > \delta\} > \delta.$$

This in turn implies

$$\sum_{|j|\le n} |T^jZ^{(k)}|^2 > (2n+1)\delta^3.$$

Thus by Chebyshev's inequality, for any $\lambda > 0$

$$P\{\rho(R_n, R_n^{(k)}) > \delta\} \le \exp(-\lambda(2n+1)\delta^3)E\big\{\exp(\lambda\sum_{|j|\le n}|T^jZ^{(k)}|^2)\big\}. \tag{4.12}$$

Now, by Jensen's inequality

$$\begin{aligned}
&E\big\{\exp(\lambda\sum_{|j|\le n}|T^jZ^{(k)}|^2)\big\} \\
&= E\big\{\exp(\lambda\sum_{|j|\le n}[\tfrac{1}{3}\sum_r 2^{-|r|}\min(\|Z_{r+j}^{(k)}\|, 1)]^2)\big\} \\
&\le E\big\{\exp(\lambda\sum_{|j|\le n}\tfrac{1}{3}\sum_r 2^{-|r|}\min(\|Z_{r+j}^{(k)}\|^2, 1))\big\} \\
&\le \frac{1}{3}\sum_r 2^{-|r|}E\big\{\exp(\lambda\sum_{|j|\le n}\min(\|Z_{r+j}^{(k)}\|^2, 1))\big\},
\end{aligned}$$

and by the stationarity of $\{Z_j^{(k)}\}$, the last expectation is independent of r, and the last expression equals

$$= E\big\{\exp(\lambda\sum_{|j|\le n}\min(\|Z_j^{(k)}\|^2, 1))\big\}.$$

By (4.10) and (4.12) we conclude that given $\lambda > 0$, there exists $k(\lambda)$ such that for $k \ge k(\lambda)$ and for all n

$$P\{\rho(R_n, R_n^{(k)}) > \delta\} \le \exp\{-\lambda(2n+1)\delta^3\}e^n.$$

Therefore for all $\lambda > 0$,

$$\limsup_{k\to\infty} \limsup_{n\to\infty} \frac{1}{2n+1} \log P\{\rho(R_n, R_n^{(k)}) > \delta\} \le -\lambda\delta^3 + \frac{1}{2},$$

which implies (4.11) and the theorem is proved. □

Remark 4.13. In the hypothesis (4.10) of the theorem, it is clear from the proof that $\|Z_j^{(k)}\|^2$ can be replaced by $\|Z_j^{(k)}\|^p$, where $p \ge 1$.

We will now apply the above theorem to a class of d-dimensional stationary ergodic Gaussian sequences.

Let $\{X_j, -\infty < j < \infty\}$ be a d-dimensional, zero mean, stationary Gaussian sequence with real components. Let

$$\Gamma(n) = E(X_0 X_n^T), \quad -\infty < n < \infty, \tag{4.14}$$

denote the $d \times d$ covariance matrix. Here B^T denotes the transpose of the matrix B. By stationarity, $E(X_m X_n^T) = \Gamma(n-m)$, and it follows that

$$\Gamma(k)^T = \Gamma(-k), \quad -\infty < k < \infty. \tag{4.15}$$

It is well-known (cf. **[3]**, Chapter 8) that there exists a unique $d \times d$ matrix $\mu = (\mu^{p,q})$, $1 \le p, q \le d$, such that each $\mu^{p,q}$ is a complex measure on $[0, 2\pi]$ of finite total variation such that

$$\Gamma(j) = \int_{[0,2\pi]} e^{-ij\theta} \mu(d\theta), \quad -\infty < j < \infty, \tag{4.16}$$

where integration is performed componentwise.

We now assume that μ has a continuous density $F/2\pi$, where F is a $d \times d$ matrix with continuous complex-valued components $F^{p,q}$. Then (4.16) is the same as

$$\Gamma(j) = \frac{1}{2\pi} \int_0^{2\pi} e^{-ij\theta} F(\theta) d\theta, \quad -\infty < j < \infty. \tag{4.17}$$

It follows from (4.15), and the fact that Γ is real, that for each $\theta \in [0, 2\pi]$, the matrix $F(\theta)$ is Hermitian and satisfies

$$\overline{F}(\theta) = F(2\pi - \theta), \tag{4.18}$$

where $\overline{A}$ denotes the complex-conjugate of the complex matrix A.

If b is a complex $d \times 1$ matrix, then $b^T \Gamma(j) \overline{b}$, $-\infty < j < \infty$, is the covariance function of the complex-valued stationary process $\{b^T X_j, -\infty < j < \infty\}$, and its spectral density $b^T F(\theta) \overline{b}$ is continuous and nonnegative for each θ. Therefore, for each $\theta \in [0, 2\pi]$, the matrix $F(\theta)$ is nonnegative definite.

We make one more assumption on the matrix F.

Assumption 4.19. *Assume that $F(0) = F(2\pi)$.*

Remark 4.20. If F is real, then (4.18) implies that Assumption 4.19 automatically holds. In particular, if $d = 1$, then this assumption imposes no additional restriction. In general, a Hermitian, nonnegative definite, continuous F need not satisfy the assumption.

Let G be the Hermitian nonnegative definite square-root of F. Then we have the Fourier expansion for G (in $L^2[0, 2\pi]$)

$$G(\theta) = \sum_{n=-\infty}^{\infty} A_n e^{in\theta}, \quad 0 \leq \theta \leq 2\pi, \tag{4.21}$$

where

$$A_n = \frac{1}{2\pi} \int_0^{2\pi} e^{-in\theta} G(\theta) d\theta, \quad -\infty < n < \infty. \tag{4.22}$$

Since G must satisfy (4.18) in place of F, it follows that each A_n is a real matrix, and the Hermitian property of G implies that $A_n^T = A_{-n}$. Note also that

$$G(0) = G(2\pi). \tag{4.23}$$

Let $\{\xi_n, \ -\infty < n < \infty\}$ be a sequence of i.i.d., d-dimensional, Gaussian vectors such that the d components of each vector are themselves i.i.d. $N(0, 1)$ random variables. Define

$$X_j = \sum_{n=-\infty}^{\infty} A_n \xi_{n+j}, \quad -\infty < j < \infty. \tag{4.24}$$

Then $\{X_j\}$ is a stationary ergodic sequence of d-dimensional Gaussian random variables with spectral density $G^2 = F$. The spectral density of the process $\{X_{j,p}, \ -\infty < j < \infty\}$ is $F^{p,p}$, the (p, p) entry of the matrix F, where $X_{j,p}$ denotes the p-th component of X_j.

Our aim now is to prove the following theorem. The notation is that of Theorem 4.9 with $\boldsymbol{B} = \mathbb{R}^d$.

Theorem 4.25. *Let $\{X_j, \ -\infty < j < \infty\}$ be a stationary ergodic d-dimensional Gaussian sequence on the probability space $(\Omega, \mathcal{F}, P)$ with a continuous spectral density $F/2\pi$ such that $F(0) = F(2\pi)$, and F is not identically zero on $[0, 2\pi]$. Let R_n, $n \geq 1$, be defined by (4.1) in terms of the given $\{X_j\}$. Then there exists a nontrivial, tight rate function $I: \mathcal{M}_1(\tilde{\Omega}) \to [0, \infty]$ such that if $C \subset \mathcal{M}_1(\tilde{\Omega})$ is closed and $G \subset \mathcal{M}_1(\tilde{\Omega})$ is open, then*

$$\limsup_{n\to\infty} \frac{1}{2n+1} \log P\{R_n \in C\} \leq - \inf_{Q' \in C} I(Q'), \tag{4.26}$$

and

$$\liminf_{n\to\infty} \frac{1}{2n+1} \log P\{R_n \in G\} \geq - \inf_{Q' \in G} I(Q'). \tag{4.27}$$

Proof. As observed earlier, we use the approximation procedure of Donsker and Varadhan **[4]**. For each positive integer k, define

$$Y_j^{(k)} = \sum_{|n|\le k} A_n\Big(1 - \frac{|n|}{k+1}\Big)\xi_{n+j}, \quad -\infty < j < \infty, \tag{4.28}$$

where A_n is given by (4.22) and X_j is represented by (4.24). For each $k \ge 1$, $\{Y_j^{(k)}, -\infty < j < \infty\}$ is a stationary ergodic Gaussian sequence with spectral density matrix $F_k = G_k^2$, where

$$G_k(\theta) = \sum_{|n|\le k} A_n\Big(1 - \frac{|n|}{k+1}\Big)e^{in\theta}, \tag{4.29}$$

and by Fejér's theorem $G_k \to G$ uniformly on $[0, 2\pi]$ as $k \to \infty$.

In the framework of Theorem 4.9, we define $Z_j^{(k)}$ by

$$X_j = Y_j^{(k)} + Z_j^{(k)}, \quad -\infty < j < \infty. \tag{4.30}$$

The spectral density of $\{Z_j^{(k)}, -\infty < j < \infty\}$ is equal to $(G_k - G)^2$ and $(G_k - G)^2 \to 0$ uniformly on $[0, 2\pi]$ as $k \to \infty$.

We will first check condition (4.10) for $\{Z_j^{(k)}\}$. Here $\|\cdot\|$ denotes the Euclidean norm on $\mathbb{R}^d$. For $\lambda > 0$,

$$E\Big\{\exp\Big(\lambda \sum_{j=-n}^{n} \|Z_j^{(k)}\|^2\Big)\Big\} = E\Big\{\exp\Big(\lambda \sum_{j=-n}^{n}\sum_{p=1}^{d} |Z_{j,p}^{(k)}|^2\Big)\Big\}.$$

By Jensen's inequality, the last expression is dominated by

$$\frac{1}{d}\sum_{p=1}^{d} E\Big\{\exp\Big(d\lambda \sum_{j=-n}^{n} |Z_{j,p}^{(k)}|^2\Big)\Big\}.$$

Since the spectral density of $\{Z_{j,p}^{(k)}, -\infty < j < \infty\}$, which is $((G_k - G)^2)^{p,p}$, converges to 0 uniformly on $[0, 2\pi]$ for any p, $1 \le p \le d$, the argument of Lemma 2.4 **[6]** applies. We include it here for the sake of completeness. The process $\{Z_{j,p}^{(k)}, -n \le j \le n\}$ is real Gaussian with zero mean. Therefore (for given p), if $\lambda\alpha_{n,p}^{(k)} < 1/4$, then

$$E\Big\{\exp\Big(d\lambda \sum_{j=-n}^{n} |Z_{j,p}^{(k)}|^2\Big)\Big\} \le e^{2d\lambda\alpha_{n,p}^{(k)}(2n+1)},$$

where $\alpha_{n,p}^{(k)}$ denotes the maximum eigenvalue of the $(2n+1)\times(2n+1)$ covariance matrix of $\{Z_{j,p}^{(k)},\ -n \le j \le n\}$. For any real $\{\eta_j,\ -n \le j \le n\}$, we have

$$\begin{aligned}\sum_{j,r=-n}^{n} \eta_j \eta_r E(Z_{j,p}^{(k)} Z_{r,p}^{(k)}) &= \frac{1}{2\pi}\int_0^{2\pi} \sum_{j,r=-n}^{n} \eta_j \eta_r e^{i(j-r)(\theta)}((G_k - G)^2)^{p,p}(\theta)d\theta \\ &= \frac{1}{2\pi}\int_0^{2\pi} |\sum_{j=-n}^{n} \eta_j e^{ij\theta}|^2((G_k - G)^2)^{p,p}(\theta)d\theta \\ &\le \sup_{0\le\theta\le 2\pi} ((G_k - G)^2)^{p,p}(\theta) \sum_{j=-n}^{n} \eta_j^2.\end{aligned}$$

Therefore, for any $p,\ 1 \le p \le d$,

$$\sup_{n\ge 1} \alpha_{n,p}^{(k)} \le \sup_{0\le\theta\le 2\pi} ((G_k - G)^2)^{p,p}(\theta) \to 0$$

as $k \to \infty$. Hence, given $\lambda > 0$, there exists $k(\lambda) > 0$ such that for all $k \ge k(\lambda)$, all $n \ge 1$,

$$E\{\exp(\lambda \sum_{j=-n}^{n} \|Z_j^{(k)}\|^2)\} \le e^n,$$

and condition (4.10) is satisfied. □

For each $k \ge 1$, let $\{R_n^{(k)}\}$ be defined in terms of the process $\{Y_n^{(k)}\}$ as in (4.8). The distribution of $R_n^{(k)}$ satisfies the LDP with a tight rate function I_k for the same reason as given in Donsker and Varadhan **[6]**. One may also observe that the process $\{Y_n^{(k)}\}$ is $(2k+1)$-independent, so hypermixing, and then a theorem of Chiyonobu and Kusuoka (**[5]**, Ch. 5) also leads to the same conclusion. Therefore the distribution of R_n satisfies the LDP with a tight rate I given by (3.3) in terms of the I_k's.

We would like to show now that I is in fact nontrivial. Let Q denote the stationary Gaussian distribution of the process on $\tilde{\Omega}$. First, we will show that $I(Q') < \infty$ for some $Q' \ne Q$. Then in Lemma 4.33 we will show that for a hypermixing process the rate function vanishes at a unique point which is the stationary distribution of the process. This would mean that each $I^{(k)}$, the rate function corresponding to $\{Y_j^{(k)}\}$, has a unique zero. By Theorem 3.11, I must have a unique zero which (by Lemma 4.33) must be Q. This will end the proof. We now proceed to show that $I(Q') < \infty$ for some $Q' \ne Q$.

To see this, let J be the rate function defined by (4.5) for L_n. It clearly suffices to show that $J(\mu') < \infty$ for some $\mu' \ne \mu$ in $\mathcal{M}_1(\mathbb{R}^d)$, where μ is the marginal of Q. Let

$$K = \{\mu' : \mu'([-1,1]^d) = 1\}.$$

This is a compact set, and

$$\{\|X_j\|_\infty \le 1,\ -n \le j \le n\} \subset \{L_n \in K\},$$

where $\|x\|_\infty = \max_{1\le p\le d} |x_p|$ for $x = (x_1, \ldots, x_d)$. By the LDP we have

$$\begin{aligned}\limsup_{n\to\infty} \frac{1}{2n+1} \log P\{\|X_j\|_\infty \le 1,\ -n \le j \le n\} \\ \le \limsup_{n\to\infty} \frac{1}{2n+1} \log P\{L_n \in K\} \le -\inf_{\mu'\in K} J(\mu').\end{aligned} \tag{4.31}$$

Let $I_{d\times d}$ denote the $d \times d$ identity matrix. Then we can find $\alpha > 0$ sufficiently large such that $\alpha I_{d\times d} - F(\theta)$ is positive definite for all $\theta \in [0, 2\pi]$. We can define a Gaussian process $\{X_n'\}$ with this spectral density, which is independent of the given process $\{X_n\}$. The underlying probability space is irrelevant for this part of the argument. Then the process $X_j'' = X_j + X_j'$, $-\infty < j < \infty$, is a Gaussian process with spectral density αI. By an inequality of T. W. Anderson **[1]**, we have

$$P\{\|X_j\|_\infty \le 1,\ -n \le j \le n\} \ge P\{\|X_j + X_j'\|_\infty \le 1,\ -n \le j \le n\},$$

and since the process $\{X_j''\}$ is an i.i.d. sequence of d-dimensional Gaussian vectors with vector components also independent $N(0, \alpha)$, we get

$$P\{\|X_j\|_\infty \le 1,\ -n \le j \le n\} \ge c^{(2n+1)d} \tag{4.32}$$

for some $c \in (0, 1)$. By (4.31) and (4.32) we then have

$$\inf_{\mu'\in K} J(\mu') \le -\log c < \infty.$$

Since K is compact and J is lsc, the inf is attained at some $\mu_0 \in K$. The marginal μ of Q is Gaussian, so cannot belong to K. We have thus shown that there exists $\mu_0 \ne \mu$ in $\mathcal{M}_1(\mathbb{R}^d)$ such that $J(\mu_0) < \infty$. The theorem is proved except for the lemma.

We will use the notation of Theorem 4.9 except that $\boldsymbol{B}$ is allowed to be Polish.

Lemma 4.33. *Let $\{X_j,\ -\infty < j < \infty\}$ be a stationary ergodic process with values in $\boldsymbol{B}$. Let Q denote the distribution of X in $\mathcal{M}_1(\tilde{\Omega})$. Suppose the LDP holds for R_n, i.e.* (4.2) *and* (4.3) *hold with a tight rate I. Then $I(Q) = 0$. If the process is hypermixing, then $I(Q') > 0$ if $Q' \ne Q$.*

Proof. By the ergodic theorem, $R_n \to Q$ a.e. in the sense of Lévy–Prohorov metric. Therefore, if V is a neighborhood of Q in $\mathcal{M}_1(\tilde{\Omega})$, then by the LDP,

$$\inf_{Q'\in\bar{V}} I(Q') = 0,$$

and this implies that $I(Q) = 0$.

We now prove the second assertion. Suppose $Q' \ne Q$. Then there exists a $k \ge 0$ and real constants b_j, $-k \le j \le k$, such that the distribution ν of the random variable $S_0 = \sum_{j=-k}^{k} b_j X_j$ under Q is different from its distribution ν' under Q'. Let

$$S_r = \sum_{j=-k}^{k} b_j X_{j+r}, \quad -\infty < r < \infty.$$

Then $\{S_r\}$ is a stationary hypermixing process (assuming that $\{X_j\}$ is) with marginal distribution ν. Let

$$\hat{L}_n = \frac{1}{2n+1} \sum_{j=-n}^{n} \delta_{S_j},$$

and let $\hat{J}$ denote the rate function in the LDP for $\hat{L}_n$. If $I(Q) = I(Q')$, then the contraction principle of Varadhan would imply that $\hat{J}(\nu) = \hat{J}(\nu')$. It thus suffices to show that $\hat{J}(\nu) \neq \hat{J}(\nu')$ if $\nu \neq \nu'$ for *any* hypermixing process. Thinking of $\{X_j\}$ as a *generic* stationary hypermixing process, it thus suffices to show that if J is the rate function for L_n, μ is the marginal of Q, and U is a neighborhood of μ in $\mathcal{M}_1(\boldsymbol{B})$, then $\inf\{J(\mu') : \mu' \notin U\} > 0$. Let f_j, $1 \leq j \leq r$, be bounded continuous functions on $\boldsymbol{B}$. Let $\bar{f}_j = f_j - \int f_j d\mu$. Then $E\bar{f}_j(X_1) = 0$, $1 \leq j \leq r$. Let $\epsilon > 0$ and let

$$V = \{\nu : |\int \bar{f}_j d\nu| < \epsilon, 1 \leq j \leq r\}$$

be a neighborhood of μ. Then

$$\begin{aligned} &\limsup_{n\to\infty} \frac{1}{2n+1} \log P\{L_n \in V^c\} \\ &\qquad \leq \max_{1\leq j\leq r} \limsup_{n\to\infty} \frac{1}{2n+1} \log P\{|\int \bar{f}_j dL_n| \geq \epsilon\}. \end{aligned} \tag{4.34}$$

Now, for any bounded continuous f on $\boldsymbol{B}$, by Chebyshev,

$$\begin{aligned} P\{\int f dL_n \geq \epsilon\} &= P\{\sum_{j=-n}^{n} f(X_j) \geq (2n+1)\epsilon\} \\ &\leq e^{-\lambda(2n+1)\epsilon} E\{\exp(\lambda \sum_{j=-n}^{n} f(X_j))\}. \end{aligned}$$

By hypermixing we can find $\ell > 0$ such that

$$E\{\exp(\lambda \sum_{j=-n}^{n} f(X_j))\} \leq E\{\exp(2\lambda\ell f(X_1))\}^{\frac{(2n+1)}{2\ell}}.$$

Therefore by Chebyshev for every $\lambda > 0$

$$\limsup_{n\to\infty} \frac{1}{2n+1} \log P\{\int \bar{f}_j dL_n \geq \epsilon\} \leq \{-\lambda\epsilon + \frac{1}{2\ell} \log \varphi_j(2\lambda\ell)\}, \tag{4.35}$$

where $\varphi_j(u) = E\{\exp(u\bar{f}_j(X_1))\}$. Since (4.35) is true for all $\lambda > 0$, we may replace the right-side with its inf over all $\lambda > 0$. Also, $E\{\bar{f}_j(X_1)\} = 0$ implies that $\log \varphi_j(u) = O(u^2)$ as $u \to 0$, therefore

$$\inf_{\lambda>0} \{-\lambda\epsilon + \frac{1}{2\ell} \log \varphi_j(2\lambda\ell)\} = \eta_j < 0.$$

Applying the same reasoning with $\overline{f}_j$ replaced by $-\overline{f}_j$, we get the corresponding quantity $\tilde{\eta}_j < 0$, and

$$\limsup_{n\to\infty} \frac{1}{2n+1} \log P\{L_n \in V^c\} \le \max_{1\le j\le r} \max(\eta_j, \tilde{\eta}_j) = \eta < 0. \tag{4.36}$$

Let $V_1 = \{\nu : |\int \bar{f}_j d\nu| < 2\epsilon,\ 1 \le j \le r\}$ and $U = \{\nu : |\int \bar{f}_j d\nu| < 3\epsilon,\ 1 \le j \le r\}$. Then $U^c \subset \bar{V}_1^c \subset V^c$, and by (4.35) and the LDP we get

$$-\inf_{\nu\in U^c} J(\nu) \le -\inf_{\nu\in \bar{V}_1^c} J(\nu) \le \eta < 0.$$

Since every neighborhood of μ contains a neighborhood of the form U, the lemma is proved. □

In the next theorem we will obtain an analogue of Theorem 4.25 for a class of non-Gaussian stationary, ergodic, sequences.

Let $\eta = (\eta^{(1)}, \dots, \eta^{(d)})$ be a d-dimensional random vector such that

$$E\{\eta^{(j)}\} = 0, \quad 1 \le j \le d, \tag{4.37}$$

and there exist $0 < \alpha < 1$ and $x_0 > 0$ such that for $x \ge x_0$,

$$\alpha P\{|\eta^{(j)}| > x\} \le \frac{1}{(2\pi)^{1/2}} \int_{[u:|u|>x]} e^{-u^2/2} du, \tag{4.38}$$

for $1 \le j \le d$.

Let F be a $d \times d$ continuous Hermitian nonnegative definite spectral density matrix on $[0, 2\pi]$ such that $F(0) = F(2\pi)$, let G be its square root, and let A_n be defined by (4.22). Let $\{\eta_n\}$ be a sequence of i.i.d. d-dimensional random vectors, such that each component of η_1 satisfies (4.37) and (4.38). Define

$$X_j = \sum_{n=-\infty}^{\infty} A_n \eta_{n+j}, \quad -\infty < j < \infty. \tag{4.39}$$

Theorem 4.40. *Let the stationary ergodic sequence $\{X_j, -\infty < j < \infty\}$ of d-dimensional random variables be defined by* (4.39). *Let $\{R_n\}$ be defined by* (4.1) *in terms of $\{X_j\}$. Then* (4.26) *and* (4.27) *hold with a tight rate function I, which vanishes at exactly one point, namely the stationary distribution $Q \in \mathcal{M}_1(\tilde{\Omega})$ of the process.*

Proof. We will use an exponential inequality proved in Theorem 5.1 in the next section to check the condition (4.10) of Theorem 4.9. As we did in the proof of Theorem 4.25, we define

$$Y_j^{(k)} = \sum_{|n|\le k} A_n \left(1 - \frac{|n|}{k+1}\right) \eta_{n+j}, \quad -\infty < j < \infty,$$

and define $\{R_n^{(k)}\}$ in terms of $\{Y_j^{(k)}\}$. Since $\{Y_j^{(k)}\}$ is hypermixing, the LDP holds for $\{R_n^{(k)}\}$ with a tight rate function I_k. By Lemma 4.33, I_k vanishes at a unique point. Let

$Z_j^{(k)}$ be defined by

$$X_j = Y_j^{(k)} + Z_j^{(k)}, \quad -\infty < j < \infty.$$

Denoting by $\|\cdot\|$ the Euclidean norm on $\mathbb{R}^d$, we have for any $\lambda > 0$, $k \geq 1$

$$E\{\exp(\lambda \sum_{j=-n}^{n} \|Z_j^{(k)}\|^2)\} \leq \prod_{p=1}^{d} E^{1/d}\{\exp(\lambda d \sum_{j=-n}^{n} |Z_{j,p}^{(k)}|^2)\}, \tag{4.41}$$

where $Z_{j,p}^{(k)}$ is the p-th component of $Z_j^{(k)}$. By Theorem 5.1, there exists a constant C independent of λ and n such that for $1 \leq p \leq d$,

$$E\{\exp(\lambda d \sum_{j=-n}^{n} |Z_{j,p}^{(k)}|^2)\} \leq E\{\exp(C\lambda d \sum_{j=-n}^{n} |\hat{Z}_{j,p}^{(k)}|^2)\},$$

where for each $k \geq 1$, the d-dimensional sequence $\{\hat{Z}_j^{(k)}\}$ is defined by replacing the i.i.d. sequence $\{\eta_r\}$ of d-dimensional vectors by an i.i.d. sequence $\{\xi_n\}$ of d-dimensional Gaussian vectors with i.i.d. $N(0, 1)$ components. Since the spectral density of the Gaussian process $\{\hat{Z}_{j,p}^{(k)}\}$ tends to zero uniformly as $k \to \infty$, $1 \leq p \leq d$, the proof of Theorem 4.25 shows that there exists $k(\lambda)$ such that for $k \geq k(\lambda)$

$$E\{\exp(C\lambda d \sum_{j=-n}^{n} |\hat{Z}_{j,p}^{(k)}|^2)\} \leq e^n, \quad 1 \leq p \leq d.$$

Therefore (4.41) shows that condition (4.10) of Theorem 4.9 is satisfied and the LDP for $\{R_n\}$ holds with a tight rate function I which is defined in terms of the I_k's by (3.3). As we already observed, each I_k has a unique zero, hence by Theorem 3.11, I has a unique zero which is the stationary distribution of the process $\{X_j\}$. This proves the theorem. □

5. An exponential inequality

Theorem 5.1. *Let $\{\eta_n\}$ and $\{\xi_n\}$ be sequences of i.i.d. real-valued random variables with zero mean and variance* 1. *Suppose there exist $0 < \alpha \leq 1$ and $x_0 > 0$ such that*

$$\alpha P\{|\eta_1| > x\} \leq P\{|\xi_1| > x\}, \quad x \geq x_0. \tag{5.2}$$

Let $\{c_r\}$ be an ℓ^2 sequence, and let

$$U_j = \sum_{r=-\infty}^{\infty} c_r \eta_{r+j}, \; V_j = \sum_{r=-\infty}^{\infty} c_r \xi_{r+j}; \quad -\infty < j < \infty. \tag{5.3}$$

Then for $p \geq 1$,

$$E\{\exp(\sum_{j=1}^{n} |U_j|^p)\} \leq E\{\exp(C \sum_{j=1}^{n} |V_j|^p)\}, \tag{5.4}$$

for all $n \geq 1$, where $C = \max\{(64x_1/\beta)^p, (64/\alpha^2)^p\}$, $\beta = E|\xi_1 - \xi_1'|$, and $x_1 \geq x_0$ is such that $P\{|\xi_1| \leq x_1\} \geq \alpha/2$; here ξ_1' is an independent copy of ξ_1.

The proof of the theorem is based on ideas used in **[7]**.

We introduce a sequence $\{\mathcal{E}_n\}$ of i.i.d. Bernoulli random variables such that

$$P\{\mathcal{E}_1 = 1\} = P\{\mathcal{E}_1 = -1\} = 1/2,$$

and we will take $\{\mathcal{E}_n\}$ independent of $\{\eta_n\}$ and $\{\xi_n\}$. Then the symmetric sequences $\{\mathcal{E}_n|\eta_n|\}$ and $\{\mathcal{E}_n|\xi_n|\}$ satisfy the conditions of the theorem. We will denote

$$W_j = \sum_{r=-\infty}^{\infty} c_r\mathcal{E}_{r+j}, \quad U_j' = \sum_{r=-\infty}^{\infty} c_r\mathcal{E}_{r+j}|\eta_{r+j}|, \quad V_j' = \sum c_r\mathcal{E}_{r+j}|\xi_{r+j}| \tag{5.5}$$

for $-\infty < r < \infty$. The first step in the proof of the theorem will be to prove (5.4) with U_j' and V_j' in place of U_j and V_j, respectively. Then this symmetry condition will be removed. First we prove some lemmas.

Lemma 5.6. *For $p \geq 1$, we have*

$$E\{\exp(\sum_{j=1}^{n} |W_j|^p)\} \leq E\{\exp(\theta^{-p}\sum_{j=1}^{n} |V_j'|^p)\},$$

where $\theta = E|\xi_1|$.

Proof. By Jensen's inequality,

$$E\{\exp(\sum_{j=1}^{n}|\sum_{r} c_r\mathcal{E}_{r+j}|\xi_{r+j}||^p)|\mathcal{E}_r, \ -\infty < r < \infty\} \geq \{\exp(\theta^p \sum_{j=1}^{n} |W_j|^p)\}, \quad \text{a.s.,}$$

if we now replace c_r by $c_r\theta^{-1}$ and take expectations, we get the desired inequality. □

Lemma 5.7. *Suppose $\{\delta_r\}$ is a nonrandom sequence of 0's and 1's. Let $\{a_r\}$ be a real sequence such that for every j, $\{c_r a_{r+j}\} \in \ell^2$. Then for $p \geq 1$,*

$$E\{\exp(\sum_{j=1}^{n}|\sum_{r} c_r\delta_{r+j}a_{r+j}\mathcal{E}_{r+j}|^p)\} \leq E\{\exp(\sum_{j=1}^{n}|\sum_{r} c_r a_{r+j}\mathcal{E}_{r+j}|^p)\}.$$

Proof. Since $\{\delta_r\mathcal{E}_r\}$ and $\{(1-\delta_r)\mathcal{E}_r\}$ are independent sequences, by Jensen's inequality,

$$E\{\exp(\sum_{j=1}^{n}|\sum_{r} c_r a_{r+j}\mathcal{E}_{r+j}|^p)|\delta_r\mathcal{E}_r, \ -\infty < r < \infty\}$$

$$\geq \exp(\sum_{j=1}^{n}|\sum_{r} c_r a_{r+j}\delta_{r+j}\mathcal{E}_{r+j}|^p), \quad \text{a.s.}$$

The conclusion of the lemma follows by taking expectations in the above inequality. □

Lemma 5.8. *Suppose $\{\zeta_r\}$ is a sequence of random variables such that $|\eta_r| \le |\zeta_r|$, a.s. for all r, and $\{\mathcal{E}_r\}$ is independent of $\{\zeta_r\}$. Then for all $p \ge 1$,*

$$E\{\exp(\sum_{j=1}^{n} |U_j'|^p)\} \le E\{\exp(\sum_{j=1}^{n} |\sum_r c_r \mathcal{E}_{r+j} |\zeta_{r+j}||^p)\}. \tag{5.9}$$

Proof. We write

$$|\eta_r| = (|\eta_r|/|\zeta_r|)|\zeta_r|,$$

interpreting the right side to be zero whenever $\zeta_r = 0$. Thus we can write

$$|\eta_r| = \sum_{k=1}^{\infty} 2^{-k} \delta_{r,k} |\zeta_r|,$$

where $\delta_{r,k}$ are random variables taking values 0 and 1. Then

$$\exp(\sum_{j=1}^{n} |\sum_r c_r \mathcal{E}_{r+j} |\eta_{r+j}||^p) = \exp(\sum_{j=1}^{n} |\sum_r c_r \mathcal{E}_{r+j} (\sum_{k=1}^{\infty} 2^{-k} \delta_{r+j,k}) |\zeta_{r+j}||^p)$$

and by Jensen's inequality, the last expression is

$$\le \sum_{k=1}^{\infty} 2^{-k} \exp(\sum_{j=1}^{r} |\sum_r c_r \mathcal{E}_{r+j} \delta_{r+j,k} |\zeta_{r+j}||^p).$$

The sequence $\{\mathcal{E}_r\}$ is independent of $\{\delta_{r,k}, \zeta_r\}$, $k \ge 1$, thus, first taking conditional expectation given $\{\delta_{r,k}, \zeta_r, -\infty < r < \infty, k \ge 1\}$ and applying Lemma 5.7, the conditional expectation of the left side in (5.9) is less than or equal to the conditonal expectation of the right side. The lemma follows if we now take expectations. □

Proof of Theorem 5.1. Let $\tau_r = 1_{[|\eta_r| \le x_0]}$ and $\tau_r' = 1 - \tau_r$. Then first using the inequality $|x+y|^p \le 2^{p-1}(|x|^p + |y|^p)$ for real x, y and then using the Cauchy–Schwartz inequality we get

$$E\{\exp(\sum_{j=1}^{n} |\sum_r c_r \mathcal{E}_{r+j} |\eta_{r+j}||^p)\} \le$$

$$E^{1/2}\{\exp(2^p \sum_{j=1}^{n} |\sum_r c_r \mathcal{E}_{r+j} |\eta_{r+j} \tau_{r+j}||^p)\} \tag{5.10}$$

$$\times E^{1/2}\{\exp(2^p \sum_{j=1}^{n} |\sum_r c_r \mathcal{E}_{r+j} |\eta_{r+j} \tau_{r+j}'||^p)\}.$$

By Lemma 5.8, with $|\zeta_r| \equiv x_0$, the first term on the right in (5.10) is

$$\le E^{1/2}\{\exp((2x_0)^p \sum_{j=1}^{n} |W_j|^p)\}.$$

We now deal with the second term. Let $\{\theta_r\}$ be a sequence of i.i.d. random variables, independent of $\{\eta_r\}$ such that $P\{\theta_r = 1\} = \alpha,\ P\{\theta_r = 0\} = 1 - \alpha$. Then we have

$$P\{|\theta_1 \eta_1 \tau_1'| > x\} \leq P\{|\xi_1| > x\}$$

for all $x \geq 0$. Therefore, we can define sequences of i.i.d. random variables $\{\eta_r'\}$ and $\{\xi_r\}$ on the same probability space such that $|\eta_r'| \leq |\xi_r|$, a.s., for all r, where $|\eta_r'|$ has the same distribution as $|\theta_r \eta_r \tau_r'|$. By Lemma 5.8, with $\zeta_r = \xi_r$, $-\infty < r < \infty$, we get

$$E\{\exp(2^p \sum_{j=1}^{n} |\sum_r c_r \mathcal{E}_{r+j} |\theta_{r+j} \eta_{r+j} \tau_{r+j}'||^p)\}$$
$$\leq E\{\exp(2^p \sum_{j=1}^{n} |\sum_r c_r \mathcal{E}_{r+j} |\xi_{r+j}||^p)\}.$$

We apply Jensen's inequality to the left side conditioning on $\{\eta_r\}$ and $\{\mathcal{E}_r\}$ to get (taking expectation again and noting that $E\{\theta_r\} = \alpha$)

$$E\{\exp((2\alpha)^p \sum_{j=1}^{n} |\sum_r c_r \mathcal{E}_{r+j} |\eta_{r+j} \tau_{r+j}'||^p)\}$$
$$\leq E\{\exp(2^p \sum_{j=1}^{n} |\sum_r c_r \mathcal{E}_{r+j} |\xi_{r+j}||^p)\}.$$

Finally, using Lemma 5.6 and the above estimates in (5.10) we conclude that

$$E\{\exp(\sum_{j=1}^{n} |U_j'|^p)\} \leq E\{\exp(C_1 \sum_{j=1}^{n} |V_j'|^p)\}, \tag{5.11}$$

where we may take ($\theta = E|\xi_1|$)

$$C_1 = \max((2x_0/\theta)^p,\ (2/\alpha)^p). \tag{5.12}$$

It remains to remove the symmetry condition. Since $\mathcal{E}|\eta_1|$ and η_1 have the same distribution if η_1 is symmetric, we have proved the theorem as long as η_1 and ξ_1 are symmetric random variables. For the general, zero-mean case, let $\{\tilde{\eta}_r\}$ and $\{\tilde{\xi}_r\}$ be independent copies of $\{\eta_r\}$ and $\{\xi_r\}$, respectively. Then

$$P\{|\eta_1 - \tilde{\eta}_1| > x\} \leq 2P\{|\eta_1| > \frac{x}{2}\}$$
$$\leq (2/\alpha) P\{|\xi_1| > \frac{x}{2}\},$$

by (5.2), if $x \geq 2x_0$. Let $x_1 \geq x_0$ be such that $P\{|\xi_1| \leq x_1\} \geq \alpha/2$. Then

$$P\{|\xi_1| > \frac{x}{2}\} \leq \frac{2}{\alpha} P\{|\xi_1| > \frac{x}{2},\ |\tilde{\xi}_1| \leq x_1\}$$
$$\leq \frac{2}{\alpha} P\{|\xi_1 - \tilde{\xi}_1| > \frac{x}{2} - x_1\}.$$

Therefore, if $x \geq 2x_1$, then

$$P\{|\eta_1 - \tilde{\eta}_1| > x\} \leq \frac{4}{\alpha^2} P\{|\xi_1 - \tilde{\xi}_1| > \frac{x}{2} - x_1\}.$$

This implies that if $x \geq 4x_1$, then

$$\frac{\alpha^2}{4} P\{\frac{1}{4}|\eta_1 - \tilde{\eta}_1| > x\} \leq P\{|\xi_1 - \tilde{\xi}_1| > x\}.$$

We now apply (5.11) with $\mathcal{E}_r|\eta_r| = \frac{1}{4}(\eta_r - \tilde{\eta}_r)$ and $\mathcal{E}_r|\xi_r| = (\xi_1 - \tilde{\xi}_1)$, to get

$$E\{\exp(4^{-p} \sum_{j=1}^{n} |\sum_r c_r(\eta_r - \tilde{\eta}_r)|^p)\} \leq E\{\exp(C_2 \sum_{j=1}^{n} |\sum_r c_r(\xi_r - \tilde{\xi}_r)|^p)\}, \tag{5.13}$$

where $C_2 = \max((8x_1/\beta)^p, (8/\alpha^2)^p)$, and $\beta = E|\xi_1 - \tilde{\xi}_1|$. By Jensen's inequality (conditioning on $\{\tilde{\eta}_r\}$), the left side in (5.13) dominates

$$E\{\exp(4^{-p} \sum_{j=1}^{n} |U_j|^p)\}.$$

The right side in (5.13) is easily seen to be dominated by

$$E\{\exp(2^p C_2 \sum_{j=1}^{n} |V_j|^p)\}.$$

therefore, (5.13) implies the conclusion of the theorem with

$$C = \max\{(64x_1/\beta)^p,\ (64/\alpha^2)^p\}.$$

□

6. Applications to some known situations

The main purpose here is to indicate how easily our condition (3.2) translates into some key conditions that one checks in many known situations; the point is that once such a condition has been checked, no further work needs to be done if one appeals to our Theorem 3.11.

As a specific case, we briefly recall the Freidlin–Wentzell theory (cf. **[5]**, **[10]**) of small random perturbations of a dynamical system. Here one considers the stochastic differential equation

$$\begin{cases} dX_\epsilon^x(t) &= \epsilon^{1/2}\sigma(X_\epsilon^x(t))d\beta(t) + b(X_\epsilon^x(t))dt,\ 0 < t \leq T, \\ X_\epsilon^x(0) &= x, \end{cases} \tag{6.1}$$

where $\sigma\colon \mathbb{R}^d \to d \times d$ matrices, and $b\colon \mathbb{R}^d \to \mathbb{R}^d$ are maps satisfying certain conditions and β is a d-dimensional standard Brownian motion. Suppose b and σ are such that (6.1) has at least a unique weak solution with distribution $v_{\epsilon,x}$ in the space of continuous functions $C([0,T];\mathbb{R}^d)$. To prove the large deviation principle holds for $\{v_\epsilon, x\}$ using

Theorem 3.11, it suffices to check a single condition, (6.3) below, which can be verified under rather general assumptions. The problem of identifying the rate function can be dealt with as a separate problem, and is handled by Proposition 3.16.

For $\eta > 0$, let

$$X^x_{\epsilon,\eta}(t) = x + \epsilon^{1/2}\int_0^t \sigma(X^x_{\epsilon,\eta}(\psi_\eta(s)))d\beta(s) + \int_0^t b(X^x_{\epsilon,\eta}(s))ds, \ 0 \le t \le T, \quad (6.2)$$

where $\psi_\eta(t) = [t/\eta]\eta$ is a step function; $[t]$ denotes the largest integer $\le t$. Let $v^\eta_{\epsilon,x}$ denote the distribution of $X^x_{\epsilon,\eta}$ in $C([0,T];\mathbb{R}^d)$. If A is a Borel subset of $C([0,T];\mathbb{R}^d)$, then

$$\begin{aligned} v_{\epsilon,x}(A) &= P\{X^x_\epsilon(\cdot) \in A\} \\ &\le P\{X^x_\epsilon(\cdot) \in A,\ \|X^x_\epsilon(\cdot) - X^x_{\epsilon,\eta}(\cdot)\|_\infty \le \delta\} \\ &\quad + P\{\|X^x_\epsilon(\cdot) - X^x_{\epsilon,\eta}(\cdot)\|_\infty > \delta\} \\ &\le P\{X^x_\epsilon(\cdot) \in A^\delta\} + P\{\|X^x_\epsilon(\cdot) - X^x_{\epsilon,\eta}(\cdot)\|_\infty > \delta\}, \end{aligned}$$

where $\|\cdot\|_\infty$ denotes the sup norm on $C([0,T];\mathbb{R}^d)$ and $A^\delta = \{y \in C([0,T];\mathbb{R}^d) : \|y - x\|_\infty < \delta \text{ for some } x \in A\}$. It follows that

$$\rho_\delta(v_{\epsilon,x}, v^\eta_{\epsilon,x}) \le P\{\|X^x_\epsilon(\cdot) - X^x_{\epsilon,\eta}(\cdot)\|_\infty > \delta\}.$$

Hence to check condition (3.2) between $v_{\epsilon,x}$ and the approximating measure $v^\eta_{\epsilon,x}$, $\eta \to 0$, it suffices to check the usual condition: For any $\delta > 0$,

$$\lim_{\eta\to 0}\limsup_{\epsilon\to 0} \epsilon \log P\{\|X^x_\epsilon(\cdot) - X^x_{\epsilon,\eta}(\cdot)\| > \delta\} = -\infty. \quad (6.3)$$

For each $\eta > 0$, the map φ_η in (6.2) from $\beta(\cdot) \in C([0,T];\mathbb{R}^d)$ to $X^x_{\epsilon,\eta}(\cdot) \in C([0,T];\mathbb{R}^d)$ is continuous and the large deviation principle for $\{v^x_{\epsilon,\eta},\ \epsilon > 0\}$ holds by Varadhan's contraction principle with the tight rate function

$$J_\eta(f) = \inf\{I(g) : \varphi_\eta(g) = f\}, \quad (6.4)$$

where, writing $\dot{g}$ for the derivative of the absolutely continuous function g,

$$\begin{aligned} I(g) &= \frac{1}{2}\int_0^T |\dot{g}(t)|^2 dt, \text{ if } g \text{ is absolutely continuous} \\ &= \infty \quad \text{otherwise.} \end{aligned} \quad (6.5)$$

If one can check condition (6.3), which is possible even under conditions weaker than uniform ellipticity of $\sigma\sigma^*$ and the usual Lipschitz conditions on σ and b, then it follows immediately from Theorem 3.11 that the measures $\{v_{\epsilon,x},\ \epsilon > 0\}$ satisfy the large deviation principle with some tight rate function J. The identification of J with $\lim_{\eta\to 0} J_\eta$ follows immediately from Proposition 3.16 (via Fatou's lemma) under the uniform ellipticity and Lipschitz conditions; under weaker conditions additional work would be necessary to identify J with an expression in computable form.

We close the section by mentioning one more application. In **[5]**, Lemma 2.1.4 was invented to deal with approximations. Without going into details, we would like to remark that this lemma follows as an easy corollary of Theorem 3.11. Consequently, the theorem would apply in any situation where this lemma is applicable.

7. Some open questions

(a) Even though the rate functions I_k, at least in the real ($d = 1$) Gaussian case, are available by the Donsker–Varadhan formula, and I is determined by the I_k's via (3.3), we do not have a good expression for I. When $d \geq 1$, we do not have any idea how I or even the I_k look like. Of course, one can use the hypermixing property of $\{Y_j^{(k)}\}$ and obtain an expression for I_k as in **[5]**.

(b) For $d = 1$, Gaussian case, if the spectral density f is defined by

$$f(\theta) = 1_{[0,\frac{\pi}{2}]\cup[\frac{3\pi}{2},2\pi]}(\theta),$$

we do not know if the LDP holds. Thus when the spectral density is discontinuous (even if $\int \log f d\theta > -\infty$) we don't know the answer. W. Bryc **[2]** has shown us an example of a stationary Gaussian sequence where the spectral density is bounded but not continuous and the large deviation principle fails. This example does not cover the simple situation mentioned here.

(c) The rate function J for $\{L_n\}$ (see Section 4) is determined by I. Is there a good expression for J?

(d) Going from sequences to continuous time stationary processes presents new difficulties.

References

[1] T. W. Anderson, The integral of a symmetric unimodal function over a symmetric convex set and some probability inequalities, Proc. Amer. Math. Soc. 6 (1955), 170–176.

[2] W. Bryc, Private communication.

[3] H. Cramér and M. R. Leadbetter, Stationary and Related Stochastic Processes, John Wiley & Sons, New York 1967.

[4] A. Dembo and O. Zeitouni, Large Deviations Techniques and Applications, Jones and Bartlett, Boston 1993.

[5] J.-D. Deuschel and D. W. Stroock, Large Deviations. Academic Press, New York 1989.

[6] M. D. Donsker and S. R. S. Varadhan, Large deviations for stationary Gaussian processes, Comm. Math. Phys. 97 (1985), 187–210.

[7] N. C. Jain and M. B. Marcus, Integrability of infinite sums of independent vector-valued random variables, Trans. Amer. Math. Soc. 212 (1975), 1–36.

[8] A. A. Pukhalski, On functional principle of large deviations, in: New Trends in Probability and Statistics, Vol. 1 (Bakuriani, 1990), pp. 198–218, V. Sazonov and T. Shervashidze (Editors), VSP/Mokslas, Utrecht 1991.

[9] Y. Steinberg and O. Zeitouni, On tests for normality, IEEE Trans. Inform. Theory 38 (1992), 1779–1787.

[10] S. R. S. Varadhan, Large Deviations and Applications, SIAM, Philadelpha 1984.

School of Mathematics
University of Minnesota
Minneapolis, MN 55455, U.S.A.
e-mail: baxter@math.umn.edu
jain@math.umn.edu

Sets of Recurrence and Generalized Polynomials

Vitaly Bergelson and Inger Johanne Håland

1. Introduction

A set $S \subset \mathbb{Z}$ is called a *set of recurrence* if for any invertible measure preserving system $(X, \mathcal{B}, \mu, T)$ and any $A \in \mathcal{B}$ with $\mu(A) > 0$ there exists $n \in S$, $n \neq 0$, such that $\mu(A \cap T^{-n}A) > 0$. For example, for any infinite $E \subset \mathbb{N}$, the set of differences $E - E = \{x - y \mid x, y \in E, x > y\}$ is a set of recurrence. Another simple example of a set of recurrence is given by any set containing arbitrarily long arithmetic progressions of the form $\{a, 2a, \ldots, ka\}$. Less trivial example is the following. Let $p(t)$ be a polynomial taking integer values on integers and satisfying $p(0) = 0$. Then $\{p(n) \mid n \in \mathbb{Z}\}$ is a set of recurrence. This result is due, independently, to H. Furstenberg **[7]** and A. Sárközy **[13]**. One can show that if a polynomial $p(t)$ takes integer values on integers then the set $\{p(n) \mid n \in \mathbb{Z}\}$ is a set of recurrence if and only if it contains a multiple of a for any $a \in \mathbb{N}$. (See, for example, **[12]**.)

The notion of a set of recurrence was introduced by H. Furstenberg in connection with number theoretical applications. (See, for example **[7]** and **[6]**.) The following brief remarks give some necessary background related to this connection.

For a subset $E \subset \mathbb{Z}$ its *upper Banach density* is defined by

$$d^*(E) = \limsup_{|I| \to \infty} \frac{|E \cap I|}{|I|},$$

where I ranges over all intervals of $\mathbb{Z}$.

The *upper density* of E, $\bar{d}(E)$ is defined by

$$\bar{d}(E) = \limsup_{n \to \infty} \frac{|E \cap [-n, n]|}{2n + 1}.$$

(If $E \subset \mathbb{N}$ one puts $\bar{d}(E) = \limsup_{n \to \infty} \frac{|E \cap [1,n]|}{n}$.)

If the limit in question exists, we say that E has *density* and denote it by $d(E)$.

Furstenberg's correspondence principle (cf. [5], [7], see also [2]). *Let $E \subset \mathbb{Z}$ with $d^*(E) > 0$. Then there exist a measure preserving system $(X, \mathcal{B}, \mu, T)$ and a set $A_E \in \mathcal{B}$ with $\mu(A_E) = d^*(E)$ such that for any $k \in \mathbb{N}$ and for any $n_1, n_2, \ldots, n_k \in \mathbb{Z}$ one has:*

$$\mu(A_E \cap T^{-n_1} A_E \cap \cdots \cap T^{-n_k} A_E) \leq d^*(E \cap (E - n_1) \cap \cdots \cap (E - n_k)).$$

The following corollary of Furstenberg's correspondence principle makes the connection between sets of recurrence and number theory quite apparent (cf. **[3]**, see also **[2**, Theorem 4.4**]** for a stronger result.)

Proposition 1.1. *A set $R \subset \mathbb{Z}$ is a set of recurrence if and only if for any $E \subset \mathbb{Z}$ with $d^*(E) > 0$ one has*

$$(E - E) \cap R \neq \emptyset.$$

In other words, R is a set of recurrence if and only if one can always solve the diophantine equation $x - y = z$ with $x, y \in E,\ z \in R$. Before giving the proof of Proposition 1.1 we formulate a convenient lemma (which will also be utilized in the proof of Proposition 1.3 (the *uniformity* of recurrence) below.

Lemma 1.2. [1]. *Let $(X, \mathcal{B}, \mu, T)$ be a measure preserving system and let $A \in \mathcal{B}$, $\mu(A) = a > 0$. Then there exists an increasing sequence $\{n_m\}_{m=1}^{\infty}$ with $d(\{n_m\}) \geq a$ such that for any $m \in \mathbb{N}$ one has*

$$\mu(A \cap T^{-n_1}A \cap \cdots \cap T^{-n_m}A) > 0.$$

Proof of Proposition 1.1. In one direction the result follows immediately from Furstenberg's correspondence principle. Indeed, if $E \subset \mathbb{Z},\ d^*(E) > 0$, then for any n, and in particular for any $n \in R$, one has $d^*(E \cap (E-n)) \geq \mu(A_E \cap T^{-n}A_E)$. If R is a set of recurrence one can find $n \in R$ such that $\mu(A_E \cap T^{-n}A_E) > 0$. Then $d^*(E \cap (E-n)) > 0$ and, in particular, $E \cap (E - n) \neq \emptyset$ which is equivalent to $(E - E) \cap R \neq \emptyset$.

Assume now that for any $E \subset \mathbb{Z}$ with $d^*(E) > 0$ one has $(E - E) \cap R \neq \emptyset$. We shall show that R is a set of recurrence. Let $(X, \mathcal{B}, \mu, T)$ be a measure preserving system and $A \in \mathcal{B}$ with $\mu(A) > 0$. By Lemma 1.2 there exists a set $E = \{n_m\}_{m=1}^{\infty}$ with $d(E) > 0$ such that for any $n_1, n_2 \in E,\ \mu(T^{-n_1}A \cap T^{-n_2}A) > 0$ which implies $\mu(A \cap T^{-(n_2-n_1)}A) > 0$. Since $(E - E) \cap R \neq \emptyset$, there exist distinct $n_1, n_2 \in E$, $n \in R$ with $n_1 - n_2 = n$. This gives $\mu(A \cap T^{-n}A) > 0$. We are done. □

Proposition 1.3. (Uniformity of recurrence). *If $R \subset \mathbb{Z}$ is a set of recurrence, then for any $a > 0$ there exists $K = K(a)$, such that for any invertible measure preserving system $(X, \mathcal{B}, \mu, T)$ and $A \in \mathcal{B}$ with $\mu(A) > a$ there exists $n \in R$ with $|n| < K$ satisfying $\mu(A \cap T^{-n}A) > 0$.*

Proof. Let us show first that for any $a > 0$ there exists $L = L(a)$ such that if $I = [N + 1, \ldots, M]$ is any interval in $\mathbb{Z}$ with $|I| = M - N > L$ and $F \subset I$ is any subset in I satisfying $\frac{|F \cap I|}{|I|} > a$, then there exist $x, y \in F$ and $k \in R$ with $x - y = k$ (which is the same as $(F - F) \cap R \neq \emptyset$). Indeed, if this was not the case, one could find arbitrarily long intervals I_n and subsets $F_n \subset I_n$ satisfying $\frac{|F_n \cap I_n|}{|I_n|} > a$ but such that $(F_n - F_n) \cap R = \emptyset$. Let $|I_n| = N_n$ and let

$$\mathcal{F}_n(b) = \{F \subset I \mid I = \{m + 1, \ldots, m + N_n\},\ m \in \mathbb{Z},\ \frac{|F \cap I|}{|I|} > b,\ (F - F) \cap R = \emptyset\}.$$

Note that $\mathcal{F}_n(b)$ is a shift-invariant family, i.e. if $F \in \mathcal{F}_n(b)$ then also $F - m = \{a - m \mid a \in F\} \in \mathcal{F}_n(b)$ for any $m \in \mathbb{Z}$. Note also that $\mathcal{F}_n(a) \subset \mathcal{F}_n(b)$ for all $0 < b < a$ and that $\mathcal{F}_n(a) \neq \emptyset$ for all n. Let $0 < a_0 < a$, and let $N_{n+1} = l_n N_n + r_n$, where $0 \leq r_n < N_n$. By going to a subsequence of $N_n, n = 1, 2, \dots$, if necessary, one may assume that l_n are sufficiently large so that $\sum_{n=1}^{\infty} 1/l_n < a - a_0$. Let $\varepsilon_n = \sum_{i=1}^{n} 1/l_i$, and let $F \in \mathcal{F}_{n+1}(a_0 + \varepsilon_n)$, $F \subset I_{n+1}$. There exists a subinterval $J \subset I_{n+1}$, $|J| = N_n$, such that $\frac{|F \cap J|}{|J|} > a_0 + \varepsilon_n - 1/l_n = a_0 + \varepsilon_{n-1}$. Hence, $F \cap J \in \mathcal{F}_n(a_0 + \varepsilon_{n-1}) \subset \mathcal{F}_n(a_0)$. So for any $k \in \mathbb{N}$, there exist sets $G_n \in \mathcal{F}_n(a_0 + \varepsilon_{n-1}) \subset \mathcal{F}_n(a_0)$, $n = 1, \dots, k$, such that $G_1 \subset \cdots \subset G_n \subset G_{n+1} \subset \cdots \subset G_k$. If $G_1 \subset \{m + 1, \dots, m + N_1\}$, then $G_1 - m \subset \{1, \dots, N_1\}$, $G_n - m \in \mathcal{F}_n(a_0)$, $n = 1, \dots, k$, and $(G_1 - m) \subset \cdots \subset (G_n - m) \subset (G_{n+1} - m) \subset \cdots \subset (G_k - m)$. Since there are only finitely many sets $F \subset \{1, \dots, N_1\}$, there exists an infinite sequence $G_1 \subset \cdots \subset G_n \subset G_{n+1} \subset \cdots$, where $G_n \in \mathcal{F}_n(a_0)$ for each n. By taking $E = \bigcup_{n=1}^{\infty} G_n$, one gets a set E with $\bar{d}(E) \geq a_0$ such that $(E - E) \cap R = \emptyset$ which contradicts Proposition 1.1.

Now let $(X, \mathcal{B}, \mu, T)$ be a measure preserving system and let $A \in \mathcal{B}$ with $\mu(A) > a > 0$. By Lemma 1.2 there exists a set $S = \{n_i\}_{i=1}^{\infty}$ with $d(S) \geq \mu(A) > a$ and such that $\mu(A \cap T^{-n_1} A \cap T^{-n_2} A \cap \cdots \cap T^{-n_k} A) > 0$ for any $k \in \mathbb{N}$. Let $I = [N+1, \dots, M]$ be an interval in $\mathbb{Z}$ with $2L > |I| > L$ and such that $\frac{|S \cap I|}{|I|} > a$ (since $d(S) > a$, this is always possible). Then for some $n_1, n_2 \in S \cap I$, $n \in R$ one has $n_1 - n_2 = n$. This gives $\mu(T^{-n_1} A \cap T^{-n_2} A) = \mu(A \cap T^{-n} A) > 0$ with $|n| < 2L$. This shows that one can put $K(a) = 2L(a)$. We are done. □

Remarks. (i) One can find another proof of Proposition 1.3 in **[4]**.

(ii) So far we have only treated subsets of $\mathbb{Z}$ and $\mathbb{Z}$-actions. However, everything extends to the $\mathbb{Z}^k$-case. Let us say that a set $E \subset \mathbb{Z}^k$ has *positive upper density* $\bar{d}(E)$ if

$$\bar{d}(E) = \limsup_{N \to \infty} \frac{|E \cap \{-N, \dots, N\}^k|}{(2N+1)^k} > 0.$$

A set $R \subset \mathbb{Z}^k$ is a *set of recurrence* if for any probability space $(X, \mathcal{B}, \mu)$, any commuting invertible measure preserving transformations $T_1, T_2, \dots, T_k$ and any $A \in \mathcal{B}$ with $\mu(A) > 0$, there exists $r = (r_1, \dots, r_k) \in R$, $r \neq (0, \dots, 0)$, such that $\mu(A \cap T_r^{-1} A) > 0$, where $T_r = T_1^{r_1} T_2^{r_2} \dots T_k^{r_k}$.

Note that the $\mathbb{Z}^k$-version of Proposition 1.3 (uniformity of recurrence) is proved completely analogously to the case $k = 1$.

(iii) See Proposition 2.1 below.

Corollary 1.4. *For any fixed $k \in \mathbb{N}$, any sequence $\{b_i\}_{i=1}^{\infty} \subset \mathbb{Z}$ and any increasing sequence of integers $\{N_i\}_{i=1}^{\infty}$ the set $R = \bigcup_{i=1}^{\infty} \{b_i, 2^k b_i, 3^k b_i, \dots, N_i^k b_i\}$ is a set of recurrence.*

Proof. Given an arbitrary measure preserving system $(X, \mathcal{B}, \mu, T)$ and $A \in \mathcal{B}$ with $\mu(A) > 0$ we have to show that there exists $n \in R$ with $\mu(A \cap T^{-n} A) > 0$. For the set of recurrence $\{n^k\}_{n=1}^{\infty}$ and for any a satisfying $\mu(A) > a > 0$ let $K = K(a)$ as defined in the formulation of Proposition 1.3. Let i be such that $N_i > K$. Applying the uniformity of recurrence to the system $(X, \mathcal{B}, \mu, S)$ where $S = T^{b_i}$ and to $\{n^k\}_{n=1}^{\infty}$, we get n, $1 \leq n \leq N_i$ such that $\mu(A \cap T^{-b_i n^k} A) = \mu(A \cap S^{-n^k} A) > 0$. □

From this corollary it follows that, for example, $\{[\alpha n^2] \mid n \in \mathbb{N}\}$ is a set of recurrence. To see this, notice that for any $x \in \mathbb{R}$ and $h \in \mathbb{N}$, we have the following identity

$$[hx] = h[x] + \sum_{i=0}^{h-1} i \, 1_{[\frac{i}{h}, \frac{i+1}{h})}(\{x\}), \tag{1}$$

so that $[hx] = h[x]$ if and only if $h < \frac{1}{\{x\}}$. Let $\{m_i\}_{i=1}^{\infty} \subset \mathbb{N}$ be a sequence such that $\{\alpha m_i^2\} \to 0$, and let $b_i = [\alpha m_i^2]$ and $N_i = [1/\{\alpha m_i^2\}^{1/2}] - 1$. Then

$$[\alpha(nm_i)^2] = [\alpha m_i^2] n^2 = b_i n^2 \quad \text{for} \quad n = 1, 2, \ldots, N_i.$$

Hence, $\{[\alpha n^2] \mid n \in \mathbb{N}\}$ is a set of recurrence by Corollary 1.4. To treat more complicated expressions like $[\alpha n^2] + [\beta n]$ one needs a little more elaborated technique, which will be developed in Section 2.

In this paper we shall exploit the so called *generalized polynomials* to give some new examples of and answer some questions about sets of recurrence.

Generalized polynomials form a natural family of functions which are obtained from polynomials by the use of the greatest integer function $[\cdot]$, addition and multiplication. For example, $[\alpha x]\beta x^2$ or $[\alpha x][\beta x] + [[\delta x][\epsilon x^3]^2 \eta]$ are generalized polynomials. A few words of caution are in place here. First of all, a generalized polynomial can have different symbolic representations. For example, $[(1 + \alpha)n]\beta n = \beta n^2 + [\alpha n]\beta n$. This will not cause any confusion since we are going to be interested in the values taken by generalized polynomials. Second, on many occasions we shall not distinguish between the set of values of a generalized polynomial, $\{q(n) \mid n \in \mathbb{N}\}$, and the naturally formed sequence $(q(n))_{n=1}^{\infty}$. The "sequential" vision of generalized polynomials will be especially natural when we shall be checking whether the set $\{q(n) \mid n \in \mathbb{N}\}$ is an *averaging* set of recurrence.

Definition 1.5. An ordered set of recurrence $\{x_n \mid n \in \mathbb{N}\} \subset \mathbb{Z}$ is an *averaging set of recurrence* if for any invertible measure preserving system $(X, \mathcal{B}, \mu, T)$ and any $A \in \mathcal{B}$ with $\mu(A) > 0$ one has

$$\lim_{N\to\infty} \frac{1}{N} \sum_{n=1}^{N} \mu(A \cap T^{-x_n} A) > 0.$$

For example, any set of density one in $\mathbb{N}$ is an averaging set of recurrence. On the other hand, one can easily construct an infinite set $E \subset \mathbb{N}$ such that the naturally ordered set of recurrence $\{x - y \mid x, y \in E, \ x > y\}$ is not an averaging set of recurrence.

Given a polynomial $p(n)$ which takes integer values on integers one can show that $p(\mathbb{N})$ is a set of recurrence if and only if it is an averaging set of recurrence (and if and only if $p(\mathbb{N}) \cap a\mathbb{Z} \neq \emptyset$ for all $a \in \mathbb{N}$). The criteria for when a set $q(\mathbb{N})$ is a set of recurrence, where $q(n)$ is a generalized polynomial, are more complicated. Here are a few illustrative examples. Notice that a necessary condition for a set R to be a set of recurrence is that $R \cap a\mathbb{Z} \neq \emptyset$ for any $a \in \mathbb{N}$. If $\alpha > 1$ is irrational, then $\left[[2\alpha n]\frac{1}{\alpha}\right] = \left[2n - \{2\alpha n\}\frac{1}{\alpha}\right] = 2n - 1$, so that $\{\left[[2\alpha n]\frac{1}{\alpha}\right] \mid n \in \mathbb{N}\}$ is not a set of recurrence. However, $\{\left[[\sqrt{2}n]\sqrt{2}\right] \mid n \in \mathbb{N}\} \cap a\mathbb{Z} \neq \emptyset$ for any $a \in \mathbb{N}$ but is not a set of recurrence. On the other hand, $\{\left[[\sqrt{2}n]\sqrt{2}\right]^2 \mid n \in \mathbb{N}\}$ is an averaging set of recurrence. Curiously enough, both the sets $\{\left[[\sqrt{3}n]\sqrt{3}\right] \mid n \in \mathbb{N}\}$ and $\{\left[[\sqrt{3}n]\sqrt{3}\right]^2 \mid n \in \mathbb{N}\}$ are not sets of recurrence. (See Prop. 4.1, (1) and (2)).

Conjecture. Let $q(n)$ be a generalized polynomial, taking integer values on integers. Then $q(\mathbb{N})$ is a set of recurrence if and only if it is an averaging set of recurrence.

Note that in order to prove that x_n, $n = 1, 2, \ldots$, is an averaging set of recurrence, it is enough to show two things:

(i) $\limsup_{N\to\infty} \frac{1}{N} \sum_{n=1}^{N} \mu(A \cap T^{-x_n} A) > 0$ for any $(X, \mathcal{B}, \mu, T)$ and $A \in \mathcal{B}$ with $\mu(A) > 0$,

(ii) $\lim_{N\to\infty} \frac{1}{N} \sum_{n=1}^{N} e^{2\pi i x_n \lambda}$ exists for any $\lambda \in \mathbb{R}$.

(To see that (ii) implies the existence of $\lim_{N\to\infty} \frac{1}{N} \sum_{n=1}^{N} \mu(A \cap T^{-x_n} A)$ use the spectral theorem and Lebesgue convergence theorem:

$$\lim_{N\to\infty} \frac{1}{N} \sum_{n=1}^{N} \mu(A \cap T^{-x_n} A) = \lim_{N\to\infty} \frac{1}{N} \sum_{n=1}^{N} \langle 1_A, T^{x_n} 1_A \rangle$$
$$= \lim_{N\to\infty} \frac{1}{N} \sum_{n=1}^{N} \int_0^1 e^{2\pi i x_n \lambda} d\nu_A(\lambda).)$$

In light of this it would be nice to know the answer to the following:

Question. Is it true that the limit $\lim_{N\to\infty} \frac{1}{N} \sum_{n=1}^{N} e^{2\pi i q(n)\lambda}$ exists for any generalized polynomial $q(n)$ and any $\lambda \in \mathbb{R}$?

In Sections 2 and 3 two different methods of studying sets of recurrence of the form $q(\mathbb{N})$ where q is a generalized polynomial are presented. In Section 4 we give conditions for some concrete families of generalized polynomials to generate averaging sets of recurrence. A sample of results to be found in Section 4 was brought before Conjecture above.

2. Generalized polynomials whose values contain polynomial progressions

By a polynomial progression we mean a finite sequence of the form $p(n)$, $n = 1, 2, \ldots, N$, where $p(t)$ is a polynomial. In what follows we shall need a generalization of Corollary 1.4. We shall obtain this generalization as a corollary of the following.

Proposition 2.1. *Let $p_1(n), p_2(n), \ldots, p_k(n)$ be any polynomials taking integer values on integers and satisfying $p_i(0) = 0$, $i = 1, 2, \ldots, k$. For any $a > 0$ there exist positive constants $C(a)$ and $c(a)$ such that for any probability space $(X, \mathcal{B}, \mu)$, any commuting invertible measure preserving transformations $T_1, T_2, \ldots, T_k$ of $(X, \mathcal{B}, \mu)$ and any set $A \in \mathcal{B}$ with $\mu(A) > a$ there exists $n \in \mathbb{N}$, $n < C(a)$, such that*

$$\mu(A \cap T_n A) > c(a),$$

where $T_n = T_1^{p_1(n)} T_2^{p_2(n)} \ldots T_k^{p_k(n)}$.

Proof. Let $(X, \mathcal{B}, \mu)$ be a probability space, $T_1, T_2, \ldots, T_k$ measure preserving transformations and let $A \in \mathcal{B}$ with $\mu(A) > a > 0$. By a $\mathbb{Z}^k$-version of Furstenberg's correspondence principle it follows ([**1**, Cor. 4.2.1]) that for any $E \subset \mathbb{Z}^k$ with $\bar{d}(E) > 0$ one has

$$(E - E) \cap P \neq \emptyset$$

where $P = \{(p_1(n), p_2(n), \ldots, p_k(n)) \mid n \in \mathbb{Z}\}$. By Proposition 1.3 and the Remark after it, one sees that the following form of this statement is true: For any $a > 0$ there exists $L = L(a)$ such that if $I = \prod_{i=1}^k [a_i, b_i] \subset \mathbb{Z}^k$ with $\min_i (b_i - a_i) \geq L$, then for any $F \subset I$ with $\frac{|F \cap I|}{|I|} \geq a/2$ one has $(F - F) \cap P \neq \emptyset$. Consider now

$$f(x) = \frac{1}{L^k} \sum_{1 \leq n_1, n_2, \ldots, n_k \leq L} 1_A(T_1^{n_1} T_2^{n_2} \ldots T_k^{n_k} x).$$

Let

$$B = \{x \mid f(x) \geq a/2\}.$$

Since $0 \leq f(x) \leq 1$ and $\int f(x) d\mu = \mu(A) > a$, it follows that $\mu(B) \geq a/2$. But for $x \in B$, $f(x) \geq a/2$ implies that the set

$$F = \{(n_1, n_2, \ldots, n_k) \in \mathbb{Z}^k \mid 1 \leq n_i \leq L,\ i = 1, 2, \ldots, k,\ x \in T_1^{-n_1} T_2^{-n_2} \ldots T_k^{-n_k} A\}$$

satisfies $\frac{|F \cap I|}{|I|} \geq a/2$, where $I = [1, L]^k$, and hence $(F - F) \cap P \neq \emptyset$. So

$$B \subset \bigcup (T_1^{-n_1} T_2^{-n_2} \ldots T_k^{-n_k} A \cap T_1^{-(n_1 + p_1(n))} T_2^{-(n_2 + p_2(n))} \ldots T_k^{-(n_k + p_k(n))} A)$$

where the union is taken over all $(n_1, n_2, \ldots, n_k) \in \mathbb{Z}^k$ and $n \in \mathbb{Z}$ such that both vectors $(n_1, n_2, \ldots, n_k)$ and $(n_1 + p_1(n), n_2 + p_2(n), \ldots, n_k + p_k(n))$ belong to I. But the number of such pairs does not exceed L^{k+1}. So one of the intersections

$$(T_1^{-n_1} T_2^{-n_2} \ldots T_k^{-n_k} A \cap T_1^{-(n_1 + p_1(n))} T_2^{-(n_2 + p_2(n))} \ldots T_k^{-(n_k + p_k(n))} A)$$

has measure at least $\frac{a}{2L^{k+1}}$. □

Analogously to Corollary 1.4 we get

Corollary 2.2. *Let* $p_1(n), p_2(n), \ldots, p_k(n)$ *be any linearly independent polynomials taking integer values on integers, and satisfying* $p_i(0) = 0$, $i = 1, 2, \ldots, k$. *For any integer-valued sequences* $\{b_{1i}\}_{i=1}^{\infty}, \ldots, \{b_{ki}\}_{i=1}^{\infty}$ *and any increasing sequence of integers* $\{N_i\}_{i=1}^{\infty}$, *the set*

$$R = \bigcup_{i=1}^{\infty} \{ \sum_{j=1}^{k} b_{ji} p_j(n) \mid n = 1, 2, \ldots, N_i \}$$

is a set of recurrence.

As in the example after Corollary 1.4 it now follows that $\{ [\alpha n^2] + [\beta n] \mid n \in \mathbb{Z} \}$ is a set of recurrence. By the following theorem it is also an averaging set of recurrence.

We call $R = \{x_n \mid n \in \mathbb{N}\}$ an *almost* averaging set of recurrence if for any invertible measure preserving system $(X, \mathcal{B}, \mu, T)$ and $A \in \mathcal{B}$ with $\mu(A) > 0$, $\limsup_{N\to\infty} \frac{1}{N} \sum_{n=1}^{N} \mu(A \cap T^{-x_n} A) > 0$.

Note that there are generalized polynomials $q(n)$ for which $\{q(n) \mid n \in \mathbb{Z}\}$ does not contain arbitrarily long polynomial progressions. For example, $\{ [[\sqrt{2}n]\sqrt{2}] \mid n \in \mathbb{Z} \}$ is shown not to be a set of recurrence in Section 4.

Theorem 2.3. *Let* $q(n)$ *be an integer-valued generalized polynomial. If* $\{ q(n) \mid n \in \mathbb{N} \}$ *contains arbitrarily long polynomial progressions* $p_m(n)$, $n = 1, \ldots, N$, *where*

$$p_m(x) = \sum_{i=1}^{l} a_i(m) x^{r_i}$$

and where $r_1 < r_2 < \cdots < r_l$ *are natural numbers and* $a_1(n), \ldots, a_l(n)$ *are generalized polynomials, then* $\{ q(n) \mid n \in \mathbb{N} \}$ *is a set of recurrence. If there exists an increasing sequence* $\{N_i\}_{i=1}^{\infty}$ *such that the set*

$$S_k = S_k(q) = \{m \in \mathbb{N} \mid q(mn) = p_m(n), \ n = 1, 2, \ldots, N_k\} \tag{2}$$

has positive upper density for any $k \in \mathbb{N}$ *then* $\{ q(n) \mid n \in \mathbb{N} \}$ *is an almost averaging set of recurrence. If in addition, the limit* $\lim_{N\to\infty} \frac{1}{N} \sum_{n=1}^{N} e^{2\pi i q(n)\lambda}$ *exists for each* $\lambda \in \mathbb{R}$, *then* $\{q(n) \mid n \in \mathbb{N}\}$ *is an averaging set of recurrence.*

Proof. The first part follows from Corollary 2.2. Let $(X, \mathcal{B}, \mu, T)$ be a measure preserving system and $A \in \mathcal{B}$ with $\mu(A) > 0$. Let $R = \{(n^{r_1}, \ldots, n^{r_l}) \mid n \in \mathbb{N}\}$. By Proposition 2.1 there exist some $C(a) \in \mathbb{N}$ and $c(a) > 0$ such that for each $m \in S_k$, where k is chosen so that $N_k > C(a)$, there is some $n \in \{1, 2, \ldots, N_k\}$ such that with

$T_i = T^{a_i(m)}$, we have $\mu(A \cap T^{-p_m(n)}A) = \mu(A \cap T_1^{-n^{r_1}} \dots T_l^{-n^{r_l}} A) > c(a)$. Clearly,

$$\limsup_{N\to\infty} \frac{1}{N}\sum_{n=1}^{N} \mu(A \cap T^{-q(n)}A) \geq \limsup_{N\to\infty} \frac{1}{N}\sum_{m=1}^{[N/n]} \mu(A \cap T^{-q(mn)}A)$$

for all $n = 1, 2, \dots$, so that

$$\begin{aligned}\limsup_{N\to\infty} \frac{1}{N}\sum_{n=1}^{N} \mu(A \cap T^{-q(n)}A) &\geq \frac{1}{N_k}\sum_{n=1}^{N_k} \limsup_{N\to\infty} \frac{1}{N}\sum_{m=1}^{[N/n]} \mu(A \cap T^{-q(mn)}A)\\ &\geq \frac{1}{N_k} \limsup_{N\to\infty} \frac{1}{N}\sum_{m=1}^{[N/N_k]}\sum_{n=1}^{N_k} \mu(A \cap T^{-q(mn)}A)\\ &\geq \frac{1}{N_k} \limsup_{N\to\infty} \frac{1}{N}\sum_{m\in S_k([N/N_k])}\sum_{n=1}^{N_k} \mu(A \cap T^{-p_m(n)}A)\\ &> \frac{c(a)}{N_k^2}\,\bar{d}(S_k) > 0,\end{aligned}$$

where $S_k([N/N_k]) = S_k \cap \{1, 2, \dots, [N/N_k]\}$. By the observation made after Conjecture at the end of Section 1, we are done. □

If

$$q_1(n) = \left[\left[[\alpha n]\beta n^2\right]^3 [\lambda n + \delta]\gamma n\right]$$

is (a representation) of an integer-valued generalized polynomial, then α, β, λ, δ, γ are called the *coefficients* of (the representation of) the generalized polynomial $q_1(n)$. Denote by $C(q)$ all the coefficients of an integer-valued generalized polynomial $q(n)$, and let $P(q) = \bigcup C(v)$, where the union is taken over all generalized polynomials $v(n)$ which are obtained from $q(n)$ by removing any number of *nested* brackets in $q(n)$. Thus $P(q_1) = C(q_1) \cup \{\alpha\beta, \alpha\beta\gamma, \beta\gamma, \lambda\gamma, \delta\gamma\}$.

Definition 2.4. A (representation of an) integer-valued generalized polynomial $q(n)$ has *independent coefficients* if $P(q) \cup \{1\}$ is rationally independent.

Proposition 2.5. *If $q(n)$ is an integer-valued generalized polynomial with independent coefficients such that $q(0) = 0$, then $\{q(n) \mid n \in \mathbb{N}\}$ is an almost averaging set of recurrence.*

Proof. First, consider *simple* generalized polynomials $q(n)$ by which we mean generalized polynomials whose representations do not contain any sums, as for example

$$q_2(n) = \left[[\alpha n]\beta n^2\right]^3 [\lambda n]n.$$

Note that $q_1(n)$ above is not simple. Since $q_2(n)$ has independent coefficients, then $(\alpha n, [\alpha n]\beta n^2, \lambda n)$, $n = 1, 2, \dots$, is uniformly distributed (mod 1) in $\mathbb{R}^3$, **[10,** Prop. 3.4].

By (1), $[xm] = m[x]$ iff $\{x\} < 1/m$. So if m is in the set

$$\{m \in \mathbb{N} \mid \{\alpha m\} < \frac{1}{k},\ \{[\alpha m]\beta m^2\} < \frac{1}{k^3},\ \{\lambda m\} < \frac{1}{k}\} \tag{3}$$

which has positive density, then

$$\big[[\alpha mn]\beta(mn)^2\big]^3[\lambda mn]mn = \big[[\alpha m]\beta m^2n^3\big]^3[\lambda m]mn^2 = \big[[\alpha m]\beta m^2\big]^3[\lambda m]m\, n^{11}$$

for $n = 1, 2, \ldots, k$. Hence, $S_k(q_2)$ defined by (2) (with $N_k = k$) is contained in the set (3), and it follows from Theorem 2.3 that $\{q_2(n) \mid n \in \mathbb{N}\}$ is an almost averaging set of recurrence. We call αn, $[\alpha n]\beta n^2$, λn the (inner) subpolynomials of $q_2(n)$. The proof for any simple generalized polynomial $q(n)$ is done the same way. For if $v_1(n), \ldots, v_l(n)$ are all the subpolynomials of $q(n)$, then by **[10**, Prop. 3.4**]**, $(v_1(n), \ldots, v_l(n))$, $n = 1, 2, \ldots,$ is uniformly distributed (mod 1) in $\mathbb{R}^l$ so that the fractional part of each $v_i(n)$ can be chosen arbitrary small.

Now if $q(n)$ is not simple, then by **[10**, Lemma 2.7**]**, we have

$$q(n) = \sum_{i=1}^{L} q_i(n) + \sum_i t_i(n) 1_A(r_{1i}(n), r_{2i}(n)),$$

where $1_A(x, y) = \begin{cases} 1 & \text{if } \{x\} + \{y\} \geq 1 \\ 0 & \text{if } \{x\} + \{y\} < 1 \end{cases}$ and where each $q_i(n)$ is a simple non-constant generalized polynomial and $r_{ji}(n)$ is a sum of simple subpolynomials of $q_i(n)$ (possibly with some indicator functions involved). Let N_k be given and let $v_1(n), \ldots, v_l(n)$ be all the subpolynomials involved in the $q_i(n)$'s and $r_{ji}(n)$'s. Then by **[10**, Prop. 3.4**]**, $(v_1(n), \ldots, v_l(n))$, $n = 1, 2, \ldots,$ is uniformly distributed (mod 1) so that the fractional part of each $v_i(m)$ can be chosen small enough to make each $1_A(r_{1i}(mn), r_{2i}(mn)) = 0$ for $n = 1, \ldots, N_k$. For these m's, we have $q(mn) = \sum_{i=1}^{L} q_i(mn)$, and similarly as in the above simple case, $q(mn) = \sum_{i=1}^{L} q_i(m)\, n^{\deg(q_i)}$ for $n = 1, \ldots, N_k$. □

This method of finding arbitrarily long polynomial progressions in $\{[q(n)] \mid n \in \mathbb{N}\}$, works for many other generalized polynomials as well. For example,

$$\{[\alpha_1 n][\alpha_2 n] \ldots [\alpha_k n] \mid n \in \mathbb{N}\}$$

is an averaging set of recurrence for any $k \in \mathbb{N}$, $\alpha_i \in \mathbb{R}$. Indeed, if $N \in \mathbb{N}$, there exists $k_1 \in \mathbb{N}$, $1 < k_1 \leq k$, such that by possibly reordering the α_i's the set M of m's for which $\{\alpha_i m\} < 1/N$, $i = 1, \ldots, k_1$, and $\{\alpha_i m\} > 1 - 1/N$, $i = k_1 + 1, \ldots, k$, has positive density. Then for $m \in M$ and $n = 1, 2, \ldots, N$ we have $[\alpha_i mn] = n[\alpha_i m]$, $i = 1, \ldots, k_1$, and $[\alpha_i mn] = n[\alpha_i m] + n - 1$, $i = k_1 + 1, \ldots, k$, so that $[\alpha_1 mn][\alpha_2 mn] \ldots [\alpha_k mn] = p_m(n)$, $n = 1, 2, \ldots, N$, where $p_m(n)$ is a polynomial with $p_m(0) = 0$. Note that if $k_1 = k$ then $p_m(n) = n^k[\alpha_1 m][\alpha_2 m] \ldots [\alpha_k m]$. Furthermore,

$$\lim_{N\to\infty}\frac{1}{N}\sum_{n=1}^{N}e^{2\pi i[\alpha_1 n][\alpha_2 n]\dots[\alpha_k n]\lambda}$$

exists for any $\lambda\in\mathbb{R}$. For by **[11]**, if $[\alpha_1 n][\alpha_2 n]\dots[\alpha_k n]\lambda$, $n=1,2,\dots$, is not uniformly distributed (mod 1), then either λ is rational or $k<3$ and λ depends rationally on $1, 1/\alpha_1$ $(k=1)$ or on $1,\alpha_1/\alpha_2=\sqrt{c}$ for some $c\in\mathbb{Q}^+$ $(k=2)$, in which cases we have identities like $[\alpha n]\frac{1}{\alpha}=n-\{\alpha n\}\frac{1}{\alpha}$ and

$$[\alpha\sqrt{c}n][\alpha n]\sqrt{c}\equiv-\frac{1}{2}\big(\{\alpha\sqrt{c}n\}-\sqrt{c}\{\alpha n\}\big)^2 \pmod 1.$$

If $1,\alpha,\alpha\beta$ are rationally independent, then the set $\{[[\alpha n]\beta]\mid n\in\mathbb{N}\}$ contains arbitrarily long progressions starting at 0 since $(\alpha n,[\alpha n]\beta)$, $n=1,2,\dots$, is uniformly distributed (mod 1). However, if $\alpha\beta$ is rationally dependent on $1,\alpha$, then it is not obvious if one can find such progressions. We need therefore a different method. In Section 4 we show that $\{[[\sqrt{2}n]\sqrt{2}]^2\mid n\in\mathbb{N}\}$ is a set of recurrence while $\{[[\sqrt{2}n]\sqrt{2}]\mid n\in\mathbb{N}\}$ is not. Therefore, $\{[[\sqrt{2}n]\sqrt{2}]^2\mid n\in\mathbb{N}\}$ cannot contain any quadratic polynomial progressions, and possibly no polynomial progressions at all.

3. The spectral method

Furstenberg used spectral theory in **[7]** to show that polynomials $p(t)$ taking integer values on the integers and with $p(0)=0$, form sets of recurrence. We shall use a similar method.

Recall that a real-valued sequence y_n, $n=1,2,\dots$, is uniformly distributed (mod 1) if and only if for any Riemann-integrable periodic mod 1 function g, $\frac{1}{N}\sum_{n=1}^{N}g(y_n)\to\int_0^1 g\,dx$, and that this is equivalent to Weyl's criterion:

$$\frac{1}{N}\sum_{n=1}^{N}e^{2\pi i r\,y_n}\to 0\quad\text{for all}\quad r\in\mathbb{Z}\setminus\{0\}.$$

Let $\{x_n\mid n\in\mathbb{N}\}$ be a subset of $\mathbb{Z}$, and let

$$\Lambda=\Lambda(x_n)=\{\lambda\in\mathbb{R}\mid x_n\lambda, n=1,2,\dots,\ \text{is not uniformly distributed}\ \pmod 1\}.$$

For many integer-valued generalized polynomials $q(n)$ we can show that the sequence $q(n)\lambda$, $n=1,2,\dots$, is uniformly distributed (mod 1) when $\lambda\in\mathbb{R}$ is outside a certain countably set Λ_e which only depends on the coefficients of $q(n)$. For example,

$$[[\alpha n]\beta]\lambda=\alpha\beta\lambda n-\{\alpha n\}\beta\lambda-\{\alpha\beta n-\{\alpha n\}\beta\}\lambda,\ n=1,2,\dots,$$

is uniformly distributed (mod 1) if $\alpha\beta\lambda$ is rationally independent of $1,\ \alpha,\ \alpha\beta$, i.e. if $\lambda\notin\Lambda_e=\{a+\frac{b}{\beta}+\frac{c}{\alpha\beta}\mid a,b,c\in\mathbb{Q}\}$. Here it depends on α and β if Λ is equal to Λ_e.

Often it is hard to determine precisely what $\Lambda(x_n)$ is. Therefore, to avoid unnecessary problems of finding $\Lambda(x_n)$, we shall sometimes let $\Lambda_e = \Lambda_e(x_n)$ to be a subset of $\mathbb{R}$ containing $\Lambda(x_n)$. Note that $\mathbb{Q} \subseteq \Lambda(x_n) \subseteq \Lambda_e(x_n)$, so that $\Lambda(x_n)$ is always infinite. However, it is not known whether $\Lambda(q(n))$ is always countable when $q(n)$ is an integer-valued generalized polynomial.

Definition 3.1. An integer valued sequence x_n, $n = 1, 2, \ldots$, has *property P* if there exists a countable set Λ_e containing $\Lambda(x_n)$ and such that for any $\lambda_1, \ldots, \lambda_l \in \Lambda_e$ there exist real-valued sequences $v_1(n), \ldots, v_k(n)$, $n = 1, 2, \ldots$, and Riemann-integrable periodic mod 1 functions g_i on $\mathbb{R}^k$ with $\{x_n\lambda_i\} = g_i(v_1(n), \ldots, v_k(n))$, $i = 1, 2, \ldots, l$, and such that

$$(x_n\theta, v_1(n), \ldots, v_k(n)), \quad n = 1, 2, \ldots,$$

is uniformly distributed (mod 1) in $\mathbb{R}^{k+1}$ for any $\theta \notin \Lambda_e$.

Question. Is it true that for any integer-valued generalized polynomial $q(n)$, the sequence $q(n)$, $n = 1, 2, \ldots$ has property P?

Let $\{x_n \mid n \in \mathbb{N}\} \subset \mathbb{Z}$ be a sequence having the property P and let $\Lambda_e(x_n) \supseteq \Lambda(x_n)$ be a countable set. If $(X, \mathcal{B}, \mu, T)$ is a measure preserving system, then denote by $\mathcal{H}_{\Lambda_e}$ the subspace of $\mathcal{H} = L^2(X, \mathcal{B}, \mu)$ spanned by all the eigenfunctions f of T for which $Tf = e^{2\pi i\lambda} f$ for some $\lambda \in \Lambda_e(x_n)$. If $\delta > 0$ and $\Lambda_0 \subset \Lambda_e(x_n)$ is a finite subset, let

$$G = G(\Lambda_0, \delta) = \{n \in \mathbb{N} \mid |e^{2\pi i x_n \lambda} - 1| < \delta \text{ for all } \lambda \in \Lambda_0\} \tag{4}$$

and let $G_N = G \cap \{1, 2, \ldots, N\}$.

Lemma 3.2. *If x_n, $n = 1, 2, \ldots$, is an integer-valued sequence with the property P, let $\Lambda_e = \Lambda_e(x_n) \supseteq \Lambda(x_n)$ be a countable set, let Λ_0 be some finite subset of Λ_e, let $\delta > 0$ and let $G = G(\Lambda_0, \delta)$. Then $d(G) = \lim_{N\to\infty} \frac{|G_N|}{N}$ exists, and for any $\lambda \in \mathbb{R}$, the limits*

$$\lim_{N\to\infty} \frac{1}{|G_N|} \sum_{n\in G_N} e^{2\pi i x_n \lambda} \quad \text{and} \quad \lim_{N\to\infty} \frac{1}{N} \sum_{n=1}^{N} e^{2\pi i x_n \lambda} \tag{5}$$

exist. If $\lambda \notin \Lambda_e$, then both limits in (5) *equal* 0.

Proof. Let $\Lambda_0 = \{\lambda_1, \ldots, \lambda_l\}$. Since x_n, $n = 1, 2, \ldots$, has the property P, there exist Riemann-integrable periodic mod 1 functions g_i, $i = 1, \ldots, l$, and sequences $v_1(n), \ldots, v_k(n)$, $n = 1, 2, \ldots$, such that $(v_1(n), \ldots, v_k(n))$ is uniformly distributed (mod 1) and $\{x_n\lambda_i\} = g_i(v_1(n), \ldots, v_k(n))$ for all $\lambda_i \in \Lambda_0$. Let $\varepsilon > 0$ be such that if $\{y\} = y - [y] \in I(\varepsilon) = [0, \varepsilon) \cup (1 - \varepsilon, 1]$ then $|e^{2\pi i y} - 1| < \delta$. Define

$$g_0(y_1, \ldots, y_k) = \prod_{i=1}^{l} 1_{I(\varepsilon)}(g_i(y_1, \ldots, y_k)).$$

Then $n \in G$ if and only if $g_0(v_1(n), \ldots, v_k(n)) = \prod_{i=1}^{l} 1_{I(\varepsilon)}(\{x_n\lambda_i\}) = 1$. Therefore,

$$d(G) = \lim_{N\to\infty} \frac{|G_N|}{N} = \lim_{N\to\infty} \frac{1}{N} \sum_{n=1}^{N} g_0(v_1(n), \ldots, v_k(n)) = \int_0^1 \cdots \int_0^1 g_0 \, dy_1 \ldots dy_k$$

exists.

If $\lambda \notin \Lambda_e$, then by the definition of Λ_e, $\lim_{N\to\infty} \frac{1}{N} \sum_{n=1}^{N} e^{2\pi i x_n \lambda} = 0$, and

$$\begin{aligned} \frac{1}{|G_N|} \sum_{n\in G_N} e^{2\pi i x_n \lambda} &= \frac{1}{|G_N|} \sum_{n=1}^{N} e^{2\pi i x_n \lambda} g_0(v_1(n), \ldots, v_k(n)) \\ &= \frac{N}{|G_N|} \frac{1}{N} \sum_{n=1}^{N} g(x_n\lambda, v_1(n), \ldots, v_k(n)) \end{aligned} \tag{6}$$

where $g(y_0, y_1, \ldots, y_k) = e^{2\pi i y_0} g_0(y_1, \ldots, y_k)$ is a Riemann-integrable periodic mod 1 function. Since $(x_n\lambda, v_1(n), \ldots, v_k(n))$, $n = 1, 2, \ldots$, is uniformly distributed (mod 1), (6) converges to $\frac{1}{d(G)} \int_0^1 \cdots \int_0^1 g \, dy_0 \ldots dy_k = 0$.

If $\lambda \in \Lambda_e$, then for $\lambda, \lambda_1, \ldots, \lambda_l$ there exist Riemann-integrable periodic mod 1 functions f and g_i, $i = 1, \ldots, l$, and real-valued sequences $v_1(n), \ldots, v_k(n)$, $n = 1, 2, \ldots$, so that $\{x_n\lambda\} = f(v_1(n), \ldots, v_k(n))$ and $\{x_n\lambda_i\} = g_i(v_1(n), \ldots, v_k(n))$ for all $\lambda_i \in \Lambda_0$. So

$$\begin{aligned} \lim_{N\to\infty} \frac{1}{N} \sum_{n=1}^{N} e^{2\pi i x_n \lambda} &= \lim_{N\to\infty} \frac{1}{N} \sum_{n=1}^{N} e^{2\pi i f(v_1(n), \ldots, v_k(n))} \\ &= \int_0^1 \cdots \int_0^1 e^{2\pi i f(y_1, \ldots, y_k)} \, dy_1 \ldots dy_k \end{aligned}$$

and

$$\begin{aligned} \lim_{N\to\infty} \frac{1}{|G_N|} \sum_{n\in G_N} e^{2\pi i x_n \lambda} &= \lim_{N\to\infty} \frac{1}{|G_N|} \sum_{n=1}^{N} e^{2\pi i f(v_1(n), \ldots, v_k(n))} g_0(v_1(n), \ldots, v_k(n)) \\ &= \frac{1}{d(G)} \int_0^1 \cdots \int_0^1 e^{2\pi i f(y_1, \ldots, y_k)} g_0(y_1, \ldots, y_k) \, dy_1 \ldots dy_k. \end{aligned}$$

□

Lemma 3.3. *Let x_n, $n = 1, 2, \ldots$, be an integer-valued sequence with property P, let $\Lambda_e = \Lambda_e(x_n) \supseteq \Lambda(x_n)$ be a countable set, let Λ_0 be some finite subset of Λ_e, let $\delta > 0$ and let $G = G(\Lambda_0, \delta)$ be defined by (4). If f is orthogonal to any function in $\mathcal{H}_{\Lambda_e}$, then*

$$\lim_{N\to\infty} \frac{1}{|G_N|} \sum_{n\in G_N} T^{x_n} f = 0.$$

Proof. By the spectral theorem,

$$\left\| \frac{1}{|G_N|} \sum_{n \in G_N} T^{x_n} f \right\|^2 = \frac{1}{|G_N|^2} \sum_{n,m \in G_N} \langle T^{x_n - x_m} f, f \rangle$$
$$= \frac{1}{|G_N|^2} \sum_{n,m \in G_N} \int_0^1 e^{2\pi i (x_n - x_m)\lambda} d\nu_f(\lambda)$$
$$= \int_0^1 \left| \frac{1}{|G_N|} \sum_{n \in G_N} e^{2\pi i x_n \lambda} \right|^2 d\nu_f(\lambda). \tag{7}$$

By Lemma 3.2, $\frac{1}{|G_N|} \sum_{n \in G_N} e^{2\pi i x_n \lambda} \to 0$ for all $\lambda \notin \Lambda_e$. Since f is orthogonal to $\mathcal{H}_{\Lambda_e}$, the measure ν_f does not have atoms in Λ_e. Therefore, since Λ_e is countable, we have

$$\frac{1}{|G_N|} \sum_{n \in G_N} e^{2\pi i x_n \lambda} \to 0 \qquad \text{a.e. } (\nu_f).$$

Hence, (7) converges to 0. □

Theorem 3.4. *Let x_n, $n = 1, 2, \ldots$, be an integer-valued sequence with property P, and let $\Lambda_e = \Lambda_e(x_n) \supseteq \Lambda(x_n)$ be a countable set. If for any $\delta > 0$ and any finite subset $\Lambda_0 \subset \Lambda_e$, the set $G(\Lambda_0, \delta)$ defined by (4) has positive density, then the set $\{x_n \mid n \in \mathbb{N}\}$ is an averaging set of recurrence. On the other hand, if for some $\Lambda_0 \subset \Lambda_e$ and $\varepsilon > 0$, the set $G(\Lambda_0, \varepsilon)$ is of zero density, then $\{x_n \mid n \in \mathbb{N}\}$ cannot be an averaging set of recurrence, and if for some $\Lambda_0 \subset \Lambda_e$ and $\varepsilon > 0$, the set $G(\Lambda_0, \varepsilon)$ is finite, then $\{x_n \mid n \in \mathbb{N}\}$ is not a set of recurrence.*

Question. Is there an example of an integer-valued sequence x_n, $n = 1, 2, \ldots$, for which $\{x_n \mid n \in \mathbb{N}\}$ is a set recurrence and such that there exist $\Lambda_0 \subset \Lambda_e(x_n)$ and $\varepsilon > 0$ so that the set $G(\Lambda_0, \varepsilon)$ is of zero density?

Proof. Let $(X, \mathcal{B}, \mu, T)$ be a measure preserving system and let $A \in \mathcal{B}$ with $\mu(A) > 0$. We will show that if all $G(\Lambda_0, \delta)$ has positive density, then

$$\lim_{N \to \infty} \frac{1}{N} \sum_{n=1}^{N} \mu(A \cap T^{-x_n} A) = \lim_{N \to \infty} \frac{1}{N} \sum_{n=1}^{N} \langle 1_A, T^{x_n} 1_A \rangle > 0.$$

As before, denote by $\mathcal{H}_{\Lambda_e}$ the subspace of $\mathcal{H} = L^2(X, \mathcal{B}, \mu)$ spanned by the eigenfunctions of T corresponding to the eigenvalues $e^{2\pi i \lambda}$, $\lambda \in \Lambda_e(x_n)$. Let $f \in \mathcal{H}$ be such that

$$\lim_{N \to \infty} \frac{1}{|G_N|} \sum_{n \in G_N} \langle f, T^{x_n} f \rangle = 0$$

for any $G = G(\Lambda_0, \delta)$. We have $f = f_1 + f_2$, where $f_1 \in \mathcal{H}_{\Lambda_e}$ and f_2 is orthogonal to $\mathcal{H}_{\Lambda_e}$. Since $f_1 \in \mathcal{H}_{\Lambda_e}$, we have $f_1 = \sum_{\lambda \in \Lambda_e} a_\lambda g_\lambda$, where $T g_\lambda = e^{2\pi i \lambda} g_\lambda$ and

$|g_\lambda| = 1$. Let $\varepsilon > 0$, and choose a finite subset $\Lambda_0 \subset \Lambda_e$ so that $\sum_{\lambda \notin \Lambda_0} |a_\lambda|^2 < \varepsilon$. Then

$$\|f_1 - \sum_{\lambda \in \Lambda_0} a_\lambda g_\lambda\| = \|\sum_{\lambda \notin \Lambda_0} a_\lambda g_\lambda\| < \varepsilon.$$

Let $0 < \delta < \frac{\varepsilon}{\|f_1\|}$ and $G = G(\Lambda_0, \delta)$. By assumption, $d(G) > 0$. If $n \in G$, then

$$\begin{aligned}
&\|f_1 - T^{x_n} f_1\| \\
&\le \|f_1 - \sum_{\lambda \in \Lambda_0} a_\lambda g_\lambda\| + \|\sum_{\lambda \in \Lambda_0} a_\lambda g_\lambda - T^{x_n} \sum_{\lambda \in \Lambda_0} a_\lambda g_\lambda\| + \|T^{x_n} \sum_{\lambda \in \Lambda_0} a_\lambda g_\lambda - T^{x_n} f_1\| \\
&\le 2\varepsilon + \|\sum_{\lambda \in \Lambda_0} a_\lambda (1 - e^{2\pi i x_n \lambda}) g_\lambda\| \le 2\varepsilon + \delta \|f_1\| < 3\varepsilon,
\end{aligned}$$

so that

$$\left\| \frac{1}{|G_N|} \sum_{n \in G_N} T^{x_n} f_1 - f_1 \right\| < 3\varepsilon.$$

By Lemma 3.3, $\frac{1}{|G_N|} \sum_{n \in G_N} T^{x_n} f_2 \to 0$. So for large N,

$$\left\| \frac{1}{|G_N|} \sum_{n \in G_N} T^{x_n} f - f_1 \right\| < 3\,\varepsilon.$$

Since $\left|\left\langle \frac{1}{|G_N|} \sum_{n \in G_N} T^{x_n} f - f_1, f \right\rangle\right| \le 3\,\varepsilon\,\|f\|$ and

$$\left|\left\langle \frac{1}{|G_N|} \sum_{n \in G_N} T^{x_n} f - f_1, f \right\rangle\right| = \left| \frac{1}{|G_N|} \sum_{n \in G_N} \langle T^{x_n} f, f \rangle - \langle f_1, f \rangle \right| \to |\langle f_1, f \rangle|$$

for any $\varepsilon > 0$, it follows that $\langle f_1, f \rangle = 0$, i.e. f is orthogonal to $\mathcal{H}_{\Lambda_e}$.

Consider now $f = 1_A$. Since $\langle 1_A, 1 \rangle = \mu(A) > 0$ and $1 \in \mathcal{H}_{\Lambda_e}$, 1_A is not orthogonal to $\mathcal{H}_{\Lambda_e}$. Hence,

$$\limsup_{N \to \infty} \frac{1}{|G_N|} \sum_{n \in G_N} \mu(A \cap T^{-x_n} A) = a > 0$$

for some $G = G(\Lambda_0, \delta)$ of positive density. By Lemma 3.2, $\lim_{N \to \infty} \frac{1}{|G_N|} \sum_{n \in G_N} e^{2\pi i x_n \lambda}$ exists for any $\lambda \in \mathbb{R}$. So by the comments at the end of Section 1

$$\lim_{N \to \infty} \frac{1}{|G_N|} \sum_{n \in G_N} \mu(A \cap T^{-x_n} A)$$

exists. We therefore get

$$\lim_{N \to \infty} \frac{1}{N} \sum_{n=1}^{N} \mu(A \cap T^{-x_n} A) \ge \lim_{N \to \infty} \frac{|G_N|}{N} \frac{1}{|G_N|} \sum_{n \in G_N} \mu(A \cap T^{-x_n} A) = d(G)\,a > 0$$

which shows that $\{x_n \mid n \in \mathbb{N}\}$ is an averaging set of recurrence.

Now, let $\Lambda_0 = \{\lambda_1, \ldots, \lambda_l\}$ and $\delta > 0$, and consider the measure preserving system $(K^l, \mathcal{B}, \mu, T)$, where $K^l \simeq \mathbb{R}^l/\mathbb{Z}^l$ is the l-dimensional torus and

$$T(x_1, x_2, \ldots, x_l) = (x_1 + \lambda_1, x_2 + \lambda_2, \ldots, x_l + \lambda_l).$$

Let

$$A = \{y \in K^l \mid |y_i - 1| < \delta/2,\ i = 1, 2, \ldots, l\}.$$

For each n for which $\mu(A \cap T^{-x_n}A) > 0$, there exists $(e^{2\pi i z_1}, \ldots, e^{2\pi i z_l}) \in A \cap T^{-x_n}A$ so that $|e^{2\pi i z_i} - 1| < \delta/2$ and $|e^{2\pi i (z_i + x_n\lambda_i)} - 1| < \delta/2$ for $i = 1, 2, \ldots, l$. This implies that $|e^{2\pi i x_n \lambda_i} - 1| < \delta$ for $i = 1, 2, \ldots, l$. Hence,

$$\{n \mid \mu(A \cap T^{-x_n}A) > 0\} \subset G(\Lambda_0, \delta) \tag{8}$$

which shows that $\{x_n \mid n \in \mathbb{N}\}$ cannot be an averaging set of recurrence if $G(\Lambda_0, \delta)$ has zero density. Furthermore, if $G(\Lambda_0, \delta) = \{n_1, \ldots, n_r\}$ then let

$$\Lambda_1 = \{\lambda_1, \ldots, \lambda_l, \theta_1, \ldots, \theta_r\}$$

where $\theta_i = \theta_0/x_{n_i}$ and θ_0 is chosen so that $|e^{2\pi i \theta_0} - 1| > \delta$. This way $|e^{2\pi i x_{n_i} \theta_i} - 1| > \delta$ for each $i = 1, \ldots, r$, so that $G(\Lambda_1, \delta) = \emptyset$. Hence, by using (8) with Λ_1 instead of Λ_0 and for a new system $(K^{l+r}, \mathcal{B}, \mu, T)$, we see that $\{x_n \mid n \in \mathbb{N}\}$ is not a set of recurrence. □

We shall apply this theorem to what we call *manageable* generalized polynomials. If

$$q(n) = \big[[\alpha n^2 + \beta n]\lambda n\big][\gamma n]$$

then we call αn^2, βn, λn, γn, $[\alpha n^2]\lambda n$, $[\beta n]\lambda n$, $[\alpha n^2 + \beta n]\lambda n$ *subpolynomials* of $q(n)$. We say that $q(n)$ is *manageable* if for any subpolynomials $a_1(n), \ldots, a_k(n)$ of $q(n)$ there exist generalized polynomials $u_1(n), \ldots, u_l(n)$ and Riemann-integrable periodic mod 1 functions g_i on $\mathbb{R}^l$, $i = 1, \ldots, k$, such that $(u_1(n), \ldots, u_l(n))$, $n = 1, 2, \ldots$, is uniformly distributed (mod 1) and such that

$$\{a_i(n)\} = g_i(u_1(n), \ldots, u_l(n)),\ i = 1, \ldots, k.$$

Note that if all generalized polynomials are manageable, then the answer to the question at the end of Section 1 is positive.

Corollary 3.5. *If $q(n)$ is a manageable integer-valued generalized polynomial and $q(n)\lambda$, $n = 1, 2, \ldots$, is uniformly distributed* (mod 1) *for any irrational number λ, then $\{q(n) \mid n \in \mathbb{N}\}$ is an averaging set of recurrence if $d(\{n \in \mathbb{N} \mid q(n) \equiv 0$ (mod s)$\}) > 0$ for each $s \in \mathbb{N}$.*

Proof. Here we let $\Lambda_e = \Lambda = \mathbb{Q}$. So since $q(n)$ is manageable, $q(n)$ has the property P and $d(\{n \in \mathbb{N} \mid q(n) \equiv 0 \pmod{s}\})$ exists for each $s \in \mathbb{N}$. If $\Lambda_0 = \{\frac{r_1}{s_1}, \ldots, \frac{r_l}{s_l}\}$ and $\delta < \min\{\frac{1}{s_i} \mid i = 1, \ldots, l\}$, then

$$G(\Lambda_0, \delta) = \{n \in \mathbb{N} \mid q(n)\frac{r_i}{s_i} \equiv 0 \pmod{1},\ i = 1, \ldots, l\} =$$

$$= \{n \in \mathbb{N} \mid q(n) \equiv 0 \pmod{s},\ s \text{ is least common multiple of } s_1, \dots, s_l\}.$$

□

For example, the generalized polynomials $[\alpha_1 n][\alpha_2 n] \dots [\alpha_k n]$, $k \geq 3$, and $[\alpha n]n^2$ satisfy the conditions of Corollary 3.5 (see **[11]** and **[9**, Thm. 3.2.5]). However, there are many generalized polynomials which this corollary does not apply to. For example, if $q(n) = \big[[q_0(n)]\beta n^l\big]$ then $q(n)\lambda = [q_0(n)]\beta\lambda n^l - \{[q_0(n)]\beta n^l\}\lambda$ is not uniformly distributed (mod 1) if $\lambda = 1/\beta$. The next proposition covers far more generalized polynomials.

We shall use a result from **[8]** on *IP-recurrent* functions. A subset $S \subset \mathbb{N}$ is called an *IP-set* if it consists of a sequence $p_1, p_2, \dots$ together with all finite sums $p_{i_1} + p_{i_2} + \dots + p_{i_k}$ with $i_1 < i_2 < \dots < i_k$. A bounded function $f(n)$ is called *IP-recurrent* if for any $\varepsilon > 0$ the set

$$\{n \in \mathbb{N} \mid |f(n) - f(0)| < \varepsilon\}$$

has non-empty intersection with any IP-set. A set which non-trivially intersects any IP-set is called an IP*-set. It is known that the intersection of finitely many IP*-sets is an IP*-set. Any IP*-set has bounded gaps and hence has positive upper density. (See **[7]** for more details.) So if $f_1(n), \dots, f_k(n)$ are IP-recurrent, then for any $\varepsilon > 0$ the set

$$\{n \in \mathbb{N} \mid |f_i(n) - f_i(0)| < \varepsilon,\ i = 1, 2, \dots, k\}$$

is an IP*-set, and in particular it has positive upper density. A function $g(n)$ is called LIPR if $e^{2\pi i g(n)\lambda}$ is IP-recurrent for any $\lambda \in \mathbb{R}$.

Lemma 3.6. [8]. *Let $q(n)$ be a generalized polynomial. There exists a countable set $\Theta(q)$ such that if $\theta \in \mathbb{R} \setminus \Theta(q)$, and each bracket in $q(n)$ is replaced by the function $I_\theta(x) = [x + \theta]$, then the resulting generalized polynomial $q_\theta(n)$ is LIPR.*

So if $q(n) = \big[[\alpha n]\beta n\big]$, then $q_\theta(n) = \big[[\alpha n + \theta]\beta n + \theta\big]$ for some $\theta \in \mathbb{R}$. For example, $[\alpha n + \theta]^k$ are LIPR if and only if $\theta \neq l\alpha$, $l \in \mathbb{Z}$, **[9**, Prop. 4.2.5].

Remark. It is not necessary to use the same θ for all the brackets even though we have found it convenient to do so here.

Proposition 3.7. *Let $q(n)$ be a manageable integer-valued generalized polynomial which has the property P and is such that $q(0) = 0$. Then there exists a real number θ such that if $q_\theta(n)$ is the generalized polynomial obtained from $q(n)$ by replacing each bracket by I_θ, the set*

$$\{q_\theta(n) \mid n \in \mathbb{N}\}$$

is an averaging set of recurrence.

Proof. By Lemma 3.6, the θ 's may be chosen arbitrary small so that $q_\theta(0) = 0$. Since $q(n)$ is manageable it follows that if $q(n)\lambda$ is uniformly distributed (mod 1), then $q_\theta(n)\lambda$ is also uniformly distributed (mod 1), and $q_\theta(n)$ inherits the property P. The proposition now follows by Lemma 3.6 and Theorem 3.4. □

Note that there are many generalized polynomials $q(n)$ for which $\{q(n) \mid n \in \mathbb{N}\}$ is a set of recurrence even though $q(n)$ is not LIPR. However, in Proposition 4.1 we give examples of generalized polynomials $q(n)$ for which $\{q(n) \mid n \in \mathbb{N}\}$ are not sets of recurrence.

4. Some special results

We give conditions for when the values of certain generalized polynomials of degree one and two form an averaging set of recurrence. It follows from **[9]** that these generalized polynomials $q(n)$ have the property P. For most of these generalized polynomials, Theorem 2.3 is not (easily) applicable.

Proposition 4.1. *The following conditions hold, where ASR means averaging set of recurrence.*

(1) $\{[[\alpha n]\frac{a}{\alpha}] \mid n \in \mathbb{N}\}$, $a \in \mathbb{Q}^+$, $\alpha > 0$ *irrational, is ASR iff* $\alpha < 1$ *or* $a < \frac{\alpha}{[\alpha]}$.
(2) $\{[[\alpha n]\frac{a}{\alpha}]^2 \mid n \in \mathbb{N}\}$, $a \in \mathbb{Q}^+, \alpha > 0$ *irrational, is ASR iff* $\alpha < 1$ *or* $a < \frac{\alpha}{\alpha - 1}$.
(3) $\{[[\alpha n + \theta]\frac{a}{\alpha}] \mid n \in \mathbb{N}\}$, $a \in \mathbb{Q}^+$, $\alpha > 0$ *irrational, is ASR if* $0 < \theta < \alpha/a$.
(4) $\{[[\alpha n]\frac{a}{\alpha} + \theta] \mid n \in \mathbb{N}\}$, $a \in \mathbb{Q}^+$, $\alpha > 0$ *irrational, is ASR if* $\theta < 1$.
(5) $\{[\alpha n + \gamma] \mid n \in \mathbb{N}\}$, α *irrational, is ASR iff there exists* $k \in \mathbb{Z}$ *such that* $\gamma - \alpha k \in [0, 1]$.
(6) $\{[\alpha n + \gamma]^2 \mid n \in \mathbb{N}\}$, α *irrational,* $\gamma \in \mathbb{R}$, *is ASR.*
(7) $\{[[\alpha n]\beta n] \mid n \in \mathbb{N}\}$, $\alpha, \beta \in \mathbb{R}$, *is ASR.*

Remark. (i) The generalized polynomials of (1) and (2) are ASR for any rational value of α, while some conditions are needed for those in (3)–(6) if α is rational.

(ii) Note that if $c > 1$ is a rational number such that $\sqrt{c}$ is irrational, then the values of $[[\sqrt{c}n]\sqrt{c}]$ do not form a set of recurrence. However, if $c < \frac{3+\sqrt{5}}{2}$, then $\{[[\sqrt{c}n]\sqrt{c}]^2 \mid n \in \mathbb{N}\}$ is an averaging set of recurrence. Hence, $\{[[\sqrt{2}n]\sqrt{2}]^2 \mid n \in \mathbb{N}\}$ is, but $\{[[\sqrt{3}n]\sqrt{3}]^2 \mid n \in \mathbb{N}\}$ is not an averaging set of recurrence.

Proof. In the proof of (1)–(4) let $a = l/k$.

(1) If $\alpha < 1$ then $\{[\alpha n] \mid n \in \mathbb{N}\} = \mathbb{N}$ so that $\{[[\alpha n]\frac{a}{\alpha}] \mid n \in \mathbb{N}\} = \{[n\frac{a}{\alpha}] \mid n \in \mathbb{N}\}$ is an averaging set of recurrence by Proposition 2.5.

Let $\alpha > 1$. First we show that the condition $\frac{\alpha}{[\alpha]} > \frac{l}{k}$ which is equivalent to $\{\alpha\} > (1 - \frac{k}{l})\alpha$, is sufficient by using Theorem 3.4. Since the sequence

$$[[\alpha n]\frac{a}{\alpha}]\beta = a\beta n - \{\alpha n\}\frac{a}{\alpha}\beta - \{an - \{\alpha n\}\frac{a}{\alpha}\}\beta$$

is uniformly distributed (mod 1) if $1, \alpha, \beta$ are rationally independent, we let $\Lambda_e = \mathbb{Q} + \alpha\mathbb{Q}$. Let $\Lambda_0 = \{\frac{l_1}{k_1}, \ldots, \frac{l_{i_0}}{k_{i_0}}, \beta_1, \ldots, \beta_{j_0}\}$, where $\beta_i = \frac{r_i \alpha}{s_i}$. Let $0 < \varepsilon < 1 - \{\alpha\}$, $k_0 = \prod_i k_i s_i$, $r = \prod_i r_i$, $n = kk_0 m + 1$ and $S = \{n = kk_0 m + 1 \mid \{\alpha m\} < \frac{\varepsilon}{kk_0^2 rl}\}$.

We have $d(S) > 0$. We will show that $S \subset G(\Lambda_0, \varepsilon)$. Now, if $n \in S$, then $\{\alpha n\} = \{\alpha mkk_0 + \alpha\} > \{\alpha\} > (1 - \frac{k}{l})\alpha$ so that $0 \le \frac{l}{k}(1 - \{\alpha n\}\frac{1}{\alpha}) < \frac{l}{k}(1 - (1 - \frac{k}{l})\alpha\frac{1}{\alpha}) = 1$ and therefore

$$\left[[\alpha n]\frac{l}{\alpha k}\right] = \left[n\frac{l}{k} - \{\alpha n\}\frac{l}{\alpha k}\right] = lk_0m + \left[\frac{l}{k}(1 - \{\alpha n\}\frac{1}{\alpha})\right] = lk_0m.$$

Hence, $\left[[\alpha n]\frac{a}{\alpha}\right]\frac{l_i}{k_i} = lk_0m\frac{l_i}{k_i} \equiv 0 \pmod 1$ and $\left[[\alpha n]\frac{a}{\alpha}\right]\beta_i = lk_0m\frac{r_i\alpha}{s_i}$ is ε-close to 0 (mod 1) for $n \in S$. Hence, $S \subset G(\Lambda_0, \varepsilon)$ and therefore $d(G) > 0$. This completes the proof of the sufficiency.

To show that $\frac{\alpha}{[\alpha]} > \frac{l}{k}$ is necessary, let $l > 1$ and $n = mk + r$, $0 \le r \le k-1$. Then

$$\left[[\alpha n]\frac{l}{k\alpha}\right] = \left[\frac{l}{k}n + \frac{rl}{k} - \{\alpha n\}\frac{l}{k\alpha}\right] = lm + \left[\frac{l}{k}(r - \{\alpha n\}\frac{1}{\alpha})\right]$$

is divisible by l if and only if $\left[\frac{l}{k}(r - \{\alpha n\}\frac{1}{\alpha})\right] \in \{-l, 0\}$, since $\frac{l}{k}(r - \{\alpha n\}\frac{1}{\alpha}) < \frac{l}{k}(k-1) < l$ for any n. Now, if $\left[\frac{l}{k}(r - \{\alpha n\}\frac{1}{\alpha})\right] = -l$ then $r = 0$ and $\frac{l-1}{l}k\alpha < \{\alpha n\} < 1 < k\alpha$ which implies that $\alpha k < \frac{l}{l-1} < 2$ so that $k = 1$ and hence $n = m$ and $1 < \alpha < 1 + \frac{1}{l-1}$. Therefore, $0 < 1 - \frac{\alpha}{l} = \frac{l-1}{l}\alpha - \{\alpha\} < \{\alpha n\} - \{\alpha\} < 1 - \{\alpha\} < 1$ which means that while $\left[[\alpha n]\frac{l}{\alpha k}\right]\frac{1}{l} \equiv 0 \pmod 1$, $\{\left[[\alpha n]\frac{l}{\alpha k}\right]\frac{\alpha k}{l}\} = \{\alpha n - \alpha\}$ is bounded away from 0 and 1. Hence, if $\{\left[[\alpha n]\frac{l}{k\alpha}\right] \mid n \in \mathbb{N}\}$ is a set recurrence, then $\left[\frac{l}{k}(r - \{\alpha n\}\frac{1}{\alpha})\right] = 0$ so that $\left[[\alpha n]\frac{l}{\alpha k}\right] = lm$. Therefore $r = 1$ and

$$1 > \{\alpha n\} > (1 - \frac{k}{l})\alpha. \tag{9}$$

Now, $\left[[\alpha n]\frac{l}{\alpha k}\right]\frac{\alpha k}{l} = mk\alpha = \alpha n - \alpha$ comes close to 0 (mod 1) if and only if $\{\alpha n\}$ is close to $\{\alpha\}$. So by (9) we need $\{\alpha\} > (1 - \frac{k}{l})\alpha$. Hence, the condition $\frac{\alpha}{[\alpha]} > \frac{l}{k}$ is necessary.

(2) If β is irrational, then $\left[[\alpha n]\frac{a}{\alpha}\right]^2\beta$ is uniformly distributed (mod 1). Hence, by Theorem 3.4, we need only to check for which a and α the set $\{\left[[\alpha n]\frac{a}{\alpha}\right]^2 \mid n \in \mathbb{N}\}$ is divisible by any $s \in \mathbb{N}$. It follows from the case $n = mk+1$ above that $\left[[\alpha n]\frac{l}{k\alpha}\right]^2 = l^2m^2$ when $\{\alpha n\} > (1 - \frac{k}{l})\alpha$, which is true for a set of n of positive density if and only if $(1 - \frac{k}{l})\alpha < 1$, i.e., if and only if $\frac{l}{k} < \frac{\alpha}{\alpha - 1}$.

(3) Let $s \in \mathbb{N}$ and $0 < \varepsilon < \frac{\theta}{ks}$. We show that for those $n = ksm$ for which $\{\alpha m\} > 1 - \varepsilon/l$, both $\left[[\alpha n + \theta]\frac{l}{\alpha k}\right]$ is divisible by s and $\left[[\alpha n + \theta]\frac{l}{\alpha k}\right]\frac{\alpha}{s}$ is ε-close to 0 (mod 1). For under these conditions $\{\alpha n + \theta\} < \theta$ such that $\left[[\alpha n + \theta]\frac{l}{\alpha k}\right] = \left[lsm + \frac{l}{\alpha k}(\theta - \{\alpha n + \theta\})\right] = lsm$ and $\left[[\alpha n + \theta]\frac{l}{\alpha k}\right]\frac{\alpha}{s} = lm\alpha$.

(4) Let $s \in \mathbb{N}$ and $n = ksm$. If $\{\alpha n\}$ and $\{\alpha m\}$ are sufficiently small, then $\left[[\alpha n]\frac{l}{\alpha k} + \theta\right]\frac{1}{s} = ml + \left[\theta - \{\alpha n\}\frac{l}{\alpha k}\right]\frac{1}{s} = ml \equiv 0 \pmod 1$ and $\left[[\alpha n]\frac{l}{\alpha k} + \theta\right]\frac{\alpha}{s} = \alpha ml$ is close to 0 (mod 1).

(5) $[\alpha n + \gamma]\beta$ is uniformly distributed (mod 1) unless $1, \frac{1}{\alpha}, \beta$ are rationally dependent. Observe that $[\alpha n + \gamma]\frac{1}{s} \equiv 0 \pmod 1$ if and only if $\{(\alpha n + \gamma)/s\} < 1/s$ and is so for a

set of n of positive density. Now, writing $n = sm + r,\ 0 \le r \le s-1$, we have

$$[\alpha n+\gamma]\frac{1}{\alpha s} = n/s + \frac{\gamma}{\alpha s} - \{\alpha n+\gamma\}\frac{1}{\alpha s} = m + r/s + (\gamma - \{\alpha n+\gamma\})\frac{1}{\alpha s}$$

which is close to $0 \pmod 1$ for a set of n of positive density if and only if $\gamma-(ks-r)\alpha \in [0,1]$ for some integer k. When $s=1$ this reduces to $\gamma - k\alpha \in [0,1]$ for some $k \in \mathbb{Z}$. Observe that this is also a sufficient condition in this linear case.

(6) Since $[\alpha n+\gamma]^2\beta$ is uniformly distributed (mod 1) for any irrational β and $\{[\alpha n+\gamma] \mid n \in \mathbb{N}\}$ is divisible by any $l \in \mathbb{N}$, $\{[\alpha n+\gamma]^2 \mid n \in \mathbb{N}\}$ is an averaging set of recurrence.

(7) If $\alpha^2 \notin \mathbb{Q}$ and β is irrational or $\alpha^2 \in \mathbb{Q}$ and β is rationally independent of $1, \alpha$, then $(\alpha n, [\alpha n]\beta n)$ is uniformly distributed (mod 1) **[10**, Prop. 5.3], so that $\{\big[[\alpha n]\beta n\big] \mid n \in \mathbb{N}\}$ is an averaging set of recurrence by Theorem 2.3. This follows also easily if $\beta \in \mathbb{Q}$. Suppose $\alpha = \sqrt{c}$, $c \in \mathbb{Q}^+$, and $\beta = \frac{r}{s}\sqrt{c}+\frac{a}{b}$, $r, s, a, b \in \mathbb{N}$. We will show that for any $l \in \mathbb{N}$ such that $\big[[\sqrt{c}n](\frac{r}{s}\sqrt{c}+\frac{a}{b})n\big]$ is divisible by l, $\big[[\sqrt{c}n](\frac{r}{s}\sqrt{c}+\frac{a}{b})n\big]\frac{\sqrt{c}}{l}$ comes arbitrary close to $0 \pmod 1$ for a set of n of positive density. Let $\varepsilon > 0$, and let $n = 2lsbm$ and $\frac{2-\varepsilon}{2lsb} < \{\sqrt{c}m\} < \frac{2}{2lsb}$. Then $1-\varepsilon < \{\sqrt{c}n\} < 1$ and $[\sqrt{c}n] = 2lsb[\sqrt{c}m]+1$, so that by the identity $a[b]+b[a] = ab+[a][b]-\{a\}\{b\}$,

$$\begin{aligned}\big[[\sqrt{c}n](\frac{r}{s}\sqrt{c}+\frac{a}{b})n\big] &= \big[\frac{rc}{2s}n^2 + \frac{r}{2s}[\sqrt{c}n]^2 - \frac{r}{2s}\{\sqrt{c}n\}^2 + [\sqrt{c}n]\frac{a}{b}n\big] \\ &\equiv \big[\frac{r}{2s}(1-\{\sqrt{c}n\}^2)\big] \pmod l \\ &= 0\end{aligned}$$

whenever $1-\{\sqrt{c}n\}^2 < \frac{2s}{r}$. Furthermore,

$$\begin{aligned}&\big[[\sqrt{c}n](\frac{r}{s}\sqrt{c}+\frac{a}{b})n\big]\frac{\sqrt{c}}{l} \\ &= [\sqrt{c}n]\frac{cr}{sl}n + [\sqrt{c}n]\frac{a}{bl}\sqrt{c}n - \big\{[\sqrt{c}n](\frac{r}{s}\sqrt{c}+\frac{a}{b})n\big\}\frac{\sqrt{c}}{l} \\ &\equiv \frac{a}{2bl}(cn^2+[\sqrt{c}n]^2-\{\sqrt{c}n\}^2) - \big\{\frac{r}{2s}(cn^2+[\sqrt{c}n]^2-\{\sqrt{c}n\}^2)\big\}\frac{\sqrt{c}}{l} \pmod 1 \\ &\equiv \frac{a}{2bl}(1-\{\sqrt{c}n\}^2) - \big\{\frac{r}{2s}(1-\{\sqrt{c}n\}^2)\big\}\frac{\sqrt{c}}{l} \pmod 1\end{aligned}$$

which, depending on ε, can be chosen arbitrary small. □

Acknowledgement. The authors wish to thank the referee whose penetrating remarks helped us to eliminate some flaws in the original version of the paper.

References

[1] V. Bergelson, Sets of Recurrence of $\mathbb{Z}^m$-actions and Properties of Sets of Differences in $\mathbb{Z}^m$, J. London Math. Soc. (2) 31 (1985), 295–304.

[2] V. Bergelson, Ergodic Ramsey Theory, in: Logic and Combinatorics, ed. S. Simpson, Contemp. Math. 65, pp. 63–87, Amer. Math. Soc., Providence 1987.

[3] A. Bertrand-Mathis, Ensembles Intersectifs et Recurrence de Poincare, Israel J. Math. 55 (1986), 184–198.

[4] A. H. Forrest, Recurrence in Dynamical Systems: A Combinatorial Approach, Ph.D. Thesis, The Ohio State University, 1990.

[5] H. Furstenberg, Ergodic behavior of diagonal measures and a theorem of Szemerédi on arithmetic progressions, J. Analyse Math. 31 (1977), 204–256.

[6] H. Furstenberg, Poincaré Recurrence and Number Theory, Bull. Amer. Math. Soc. 5 (1981), 211–234.

[7] H. Furstenberg, Recurrence in Ergodic Theory and Combinatorial Number Theory, Princeton Univ. Press, Princeton, N.J., 1981.

[8] H. Furstenberg and B. Weiss, Simultaneous Diophantine Approximation and IP-sets, Acta Arith. 49 (1988), 413–426.

[9] I. J. Haaland, Uniform distribution of generalized polynomials, Ph.D. Thesis, The Ohio State University, 1992.

[10] I. J. Håland, Uniform distribution of generalized polynomials, J. Number Theory 45 (1993), 327–366.

[11] I. J. Håland, Uniform distribution of generalized polynomials of the product type, Acta Arith. 67 (1994), 13–27.

[12] T. Kamae and M. Mendès France, Van der Corput's difference theorem, Israel J. Math. 31 (1978), 335–342.

[13] A. Sárkösy, On difference sets of sequences of integers I, Acta Math. Hungar. 31 (1978), 125–149.

Department of Mathematics
The Ohio State University
Columbus, OH 43210, U.S.A.
vitaly@math.ohio-state.edu

Faculty of Engineering
Agder College
Grooseveien 36
N-4890 Grimstad, Norway
Inger.Haland@hia.no

The Strong Arc-Sine Law in Higher Dimensions

N. H. Bingham

1. Introduction

We begin with the two one-dimensional results which motivate this note. Write $BM(\mathbb{R}^d)$ for standard Brownian motion in d dimensions, AS for the arc-sine law (density $1/(\pi x^{\frac{1}{2}}(1-x)^{\frac{1}{2}})$ on $[0, 1]$).

Theorem A. (Lévy's arc-sine law). *With* $B = (B_t)$ $BM(\mathbb{R})$,

$$\frac{1}{t}\int_0^t I(B_u > 0)du = AS \qquad \textit{in distribution.}$$

Theorem B. (Lévy's strong arc-sine law).

$$\frac{1}{\log t}\int_1^t I(B_u > 0)du/u \to \frac{1}{2} \qquad (t \to \infty) \qquad a.s.$$

Theorem A dates back to Lévy (1939); for a modern proof see e.g. Rogers and Williams (1987), VI.53. The logarithmic average in Theorem B dates from Lévy (1937); cf. Brosamler (1973), Th. 1, Brosamler (1988), Th. 1.6. For a comparison between the Cesàro and logarithmic averages here, and for generalizations and further references, see Bingham and Rogers (1991).

Theorem A is intimately connected with the classical fluctuation theory of one-dimensional random walks due to Sparre Andersen, Spitzer, Baxter and others; see e.g. Bingham, Goldie and Teugels (1989), §8.9 and references therein. Some, but by no means all, of fluctuation theory generalises to higher dimensions. For a study of Theorem A in higher dimensions, and for background and references here, we refer to Bingham and Doney (1988).

We consider Theorem B in higher dimensions in §2 and its partial-sum version in §3. We turn to Theorem A in higher dimensions in §4, and close with some remarks in §5.

2. Theorem B in higher dimensions

By a cone C with vertex the origin in d dimensions, we shall understand a set closed under vector addition and multiplication by positive scalars.

Theorem 1. *With $B = (B_t)$ $BM(\mathbb{R}^d)$ and C a cone with vertex the origin,*

$$\frac{1}{\log t}\int_1^t I(B_u \in C)du/u \to p \qquad (t \to \infty) \qquad a.s.,$$

where p is the fraction of the total angle at the origin subtended by the cone.

Proof. We have to show

$$\frac{1}{t}\int_0^t I(B(e^v) \in C)dv \to p \qquad \text{a.s.},$$

or writing $Y_t := \exp(-\frac{1}{2}t)B(e^t)$,

$$\frac{1}{t}\int_0^t I(Y_v \in C)dv \to p \qquad \text{a.s.}$$

Now Y is an Ornstein–Uhlenbeck process in $\mathbb{R}^d$, $OU(\mathbb{R}^d)$, with covariance matrix

$$\Sigma(s,t) = e^{-\frac{1}{2}|t-s|}I.$$

For $OU(\mathbb{R})$, the limit (stationary) distribution exists, and is centred Gaussian (see e.g. Karlin and Taylor (1981), 219, 221, or Rogers and Williams (1987), V.5.2(ii), V.52.1–2). By independence of the coordinate processes, for $OU(\mathbb{R}^d)$ the stationary distribution exists and is centred Gaussian, with covariance matrix of the form $\sigma^2 I$.

Now take two independent Ornstein–Uhlenbeck processes Y and Y', one started at the origin (say), one with the stationary distribution. By the reflection coupling construction of Lindvall and Rogers (1986), §4 (cf. Lindvall (1992), VI.2), Y and Y' may be coupled with probability one. Then the required limit for Y is the same as that with Y replaced by Y'. But this is p, by rotational symmetry of the multivariate normal distribution $N(0, \sigma^2 I)$: project onto the unit sphere, and use rotational symmetry of d-dimensional Lebesgue measure, the Haar measure of $\mathbb{R}^d$. □

The most important case is, of course, that of an orthant:

Corollary. *With B $BM(\mathbb{R}^d)$ and Q the positive orthant of $\mathbb{R}^d$,*

$$\frac{1}{\log t}\int_1^t I(B_u \in Q)du/u \to 1/2^d \qquad (t \to \infty) \qquad a.s.$$

3. Partial sums

Theorem 2. *Let* $X, X_1, X_2, \ldots$ *be i.i.d. and* $\mathbb{R}^d$*-valued, with mean* 0 *and covariance matrix* Σ. *If* $S_n := \sum_{k=1}^n X_k$ *and* C *is a cone with vertex the origin,*

$$\frac{1}{\log n}\sum_{k=1}^{n} I(S_k \in C)/k \to p \qquad (n \to \infty) \qquad a.s.,$$

where if Y *is multivariate normal* $N(0, \Sigma)$,

$$p := P(Y \in C).$$

Proof. This is essentially a special case of a result of Lacey and Philipp (1990); see their Theorem 2 and Remark c. First, their (4) of Theorem 2 (weak convergence of logarithmic averages of Dirac masses to Wiener measure) is equivalent to their (6) (convergence of expectations of bounded Lipschitz-continuous functions thereof). Then, steps (a), (b) of their proof proceed as before and steps (c), (d) on using Donsker's theorem in d dimensions. The crux of the proof – their Lemma – follows as before: one uses a blocking technique and the Rademacher–Menshov theorem (maximal inequality for orthogonal random variables: see e.g. Doob (1953), 156-7), extended to bounded quasi-orthogonal variables, concluding by a Chebychev and Borel–Cantelli argument, as in Chung (1974), proof of Theorem 5.1.2.

To finish the proof, consider the intersection C' of the cone C with a $(d-1)$-dimensional hyperplane not containing the origin. Approximate the indicator function of C' above and below by bounded Lipschitz-continuous functions, and use these to bound the indicator I_C above and below by such functions in d dimensions. The Lacey–Philipp result applies to each approximant, hence for I_C, giving the result. □

A simpler argument suffices if we assume finiteness of moments beyond the second. With $(2+\delta)$th moments finite for some $\delta > 0$, the almost-sure invariance principle of Philipp (1979) (Theorem 2) shows that we may redefine (X_n) on a new probability space – without change of distribution – on which is defined a Brownian motion (B_t) with

$$S_{[t]} - B_t = o(t^{1/(2+\delta)} \log t) \qquad \text{a.s.}$$

Since the error term is $o(t^{\frac{1}{2}})$ a.s., one may argue as in Theorem 1 of Lacey and Philipp (1990) to reduce Theorem 2 to Theorem 1 above. Higher moment conditions yield smaller error terms; see Einmahl (1989).

While the significance of the a.s. limit p above is clear, its calculation (in terms of C and Σ) is not. Even in the case $C = Q$ of an orthant, there is in general no closed form for multivariate normal orthant probabilities. However, multinormal orthant probabilities have been extensively studied; see e.g. Johnson and Kotz (1972), Gupta (1963a), (1963b). By projection onto the unit ball, one may consider equivalently the volume cut off on the d-ball by hyperplanes; in this form, the problem goes back to Schläfli in 1858 (see Schläfli (1953), 219–270).

This lack of an explicit expression for normal orthant probabilities is a recurring theme; it is the source of complications in §4 below.

4. Theorem A in higher dimensions

For $d = 1$, the analogue of Theorem A for partial sums may be reduced to the Brownian case by the central limit theorem and the continuous mapping theorem, as in Billingsley (1968). The central limit theorem in $\mathbb{R}^d$ allows us a similar reduction of the Cesàro average

$$\frac{1}{n}\sum_{k=1}^{n} I(S_k \in C)$$

to

$$\frac{1}{t}\int_0^t I(B_u \in C)du,$$

which by Brownian scaling is the same in law as

$$\int_0^1 I(B_u \in C)du.$$

Even in the simplest case with $d > 1$, namely $d = 2$ and C the first quadrant, Q, no closed form for this law is known. Some information on moments in this case was obtained in §2 of Bingham and Doney (1988). We note here that the same method extends to d dimensions. Proceeding as there,

$$\mu_{1,d} := E\int_0^1 I(B_u \in Q)du = P(B_1 \in Q) = 1/2^d,$$

$$\mu_{2,d} := E\int_0^1\int_0^1 I(B(u_1) \in Q, B(u_2) \in Q)\,du_1du_2 = 2 \iint_{v_i>0,\ v_1+v_2<1} P^d\,dv_1dv_2,$$

where

$$P := P(X(v_1) > 0, X(v_1 + v_2) > 0)$$

with $X\,BM(\mathbb{R})$ (by independence of coordinate processes). By Sheppard's formula,

$$P = \frac{1}{4} + \frac{\arcsin\rho}{2\pi}, \qquad \rho := v_1/(v_1 + v_2)^{\frac{1}{2}},$$

whence

$$\mu_{2,d} = 2\int_0^1 \rho\Big[\frac{1}{4} + \frac{\arcsin\rho}{2\pi}\Big]^d d\rho = \frac{1}{2}\int_0^\pi \sin\phi\Big[\frac{1}{4} + \frac{\phi}{4\pi}\Big]^d d\phi$$

($\rho = \sin\theta, \theta = \frac{1}{2}\phi$). This yields the reduction formula

$$\mu_{2,d} = \frac{2^d + 1}{4^d} - \frac{d(d-1)\mu_{2,d-2}}{16\pi^2},$$

whence

$$\mu_{2,1} = \frac{3}{8},\quad \mu_{2,2} = \frac{5}{32} - \frac{1}{8\pi^2},\quad \mu_{2,3} = \frac{9}{64}\Big(1 - \frac{1}{\pi^2}\Big),\quad \mu_{2,4} = \frac{17}{256} - \frac{15}{128\pi^2} + \frac{3}{32\pi^4},$$

etc. The same method for third moments yields

$$\mu_{3,d} = 8\int_0^1\int_0^1 uv^3\Big[\frac{1}{8} + \frac{1}{4\pi}\big(\arcsin u + \arcsin v + \arcsin uv\big)\Big]^d du dv.$$

5. Remarks

Theorem B is closely related to the almost-sure (or pathwise) central limit theorem of Brosamler (1988) and Schatte (1988):

$$\frac{1}{\log n}\sum_1^n I_A(S_k/\sqrt{k})/k \to \Phi(A) \qquad (n \to \infty) \qquad \text{a.s.}$$

(see Lacey and Philipp (1990) or Bingham and Rogers (1991) for further references). This important and suggestive result has inspired recent work in several directions. In addition to the higher-dimensional setting considered here, we mention: (a) most general possible sets A (Bingham and Rogers (1991)), (b) other limit theorems involving logarithmic density (Berkes and Dehling (1993)), (c) counter-examples (Berkes, Dehling and Mori (1991)), (d) stable limits, log-log analogues etc. (Peligrad and Révész (1991)).

References

Berkes, I and Dehling, H. (1993), Some limit theorems in log density, Ann. Probab. 21, 1640–1970.

Berkes, I., Dehling, H. and Mori, T. F. (1991), Counter-examples related to the a.s. central limit theorem, Studia Sci. Math. Hungar. 26, 153–164.

Billingsley, P. (1968), Convergence of probability measures, Wiley.

Bingham, N. H. and Doney, R. A. (1988), On higher-dimensional analogues of the arc-sine law, J. Appl. Probab. 25, 120–131.

Bingham, N. H., Goldie, C. M. and Teugels, J. L. (1989), Regular variation (2nd ed.), Encyclopedia Math. Appl. 27, Cambridge Univ. Press (1st ed. 1987).

Bingham, N. H. and Rogers, L. C. G. (1991), Summability methods and almost-sure convergence, in: Almost everywhere convergence II (ed. Bellow, A. and Jones, R. L.), pp. 69–83, Academic Press.

Brosamler, G. A. (1973), The asymptotic behaviour of certain functionals of Brownian motion, Invent. Math. 20, 87–96.

Brosamler, G. A. (1988), An almost-everywhere central limit theorem. Math. Proc. Cambridge Phil. Soc. 104, 561–574.

Chung, K. L. (1974), A course in probability theory (2nd ed.), Academic Press.

Doob, J. L. 1953), Stochastic processes, Wiley.

Einmahl, U. (1989), Extension of results of Komlós, Major and Tusnády to the multivariate case, J. Multivariate Anal. 28, 20–68.

Gupta, S. S. (1963a), Probability integrals of multivariate normal and multivariate t, Ann. Math. Statist. 34, 792–828.

Gupta, S. S. (1963b), Bibliography on the multivariate normal integrals and related topics, Ann. Math. Statist. 34, 829–838.

Johnson, N. L. and Kotz, S. (1972), Distributions in statistics. Continuous multivariate distributions, Wiley.

Karlin, S. and Taylor, H. M. (1981), A second course in stochastic processes, Academic Press.

Lacey, M. T. and Philipp, W. (1990), A note on the almost sure central limit theorem, Statist. Probab. Lett. 9, 201–205.

Lévy, P. (1937), Théorie de l'addition des variables aléatoires, Gauthier-Villars, Paris.

Lévy, P. (1939), Sur certaines processus stochastiques homogènes, Compositio Math. 7, 283–339.

Lindvall, T. (1992), Lectures on the coupling method, Wiley.

Lindvall, T. and Rogers, L. C. G. (1986), Coupling of multidimensional diffusions by reflection, Ann. Probab. 14, 860–872.

Peligrad, M. and Révész, P. (1991), On the almost-sure central limit theorem, in: Almost everywhere convergence II (ed. Bellow, A. and Jones, R. L.), pp. 209–225, Academic Press.

Philipp, W. (1979), Almost sure invariance principles for sums of B-valued random variables, in: Probability in Banach spaces II, pp. 171–193, Lecture Notes in Math. 709, Springer-Verlag.

Rogers, L. C. G. and Williams, D. (1987), Diffusions, Markov processes and martingales, Volume II: Itô calculus, Wiley.

Schatte, P. (1988), On strong versions of the central limit theorem, Math. Nachr. 137, 249–256.

Schläfli, L. (1953), Gesammelte mathematische Abhandlungen, Band II, Birkhäuser Verlag, Basel.

Department of Statistics
Birkbeck College
University of London
Malet Street
London WC1E 7HX
England

Integer and Fractional Parts of Good Averaging Sequences in Ergodic Theory

Michael Boshernitzan[1], *Roger L. Jones*[2], *and Máté Wierdl*[3]

Abstract. Let $p \geq 1$, and let (a_k) be a sequence of real numbers which is a universally good averaging sequence in the sense that for any measure-preserving flow $(U_t)_{t\in\mathbb{R}}$ of a probability space the averages

$$\frac{1}{n}\sum_{k=1}^{n} f(U_{a_k}x)$$

converge almost everywhere for $f \in L^p$. We show that certain transformations of the sequence (a_k) preserve its good averaging property. In particular, we show that the sequences of integer parts $(\llbracket a_k \rrbracket)$ and fractional parts $(\langle\!\langle a_k \rangle\!\rangle)$ are also good averaging sequences for L^p. These results extend work of Bourgain on integer parts of polynomial sequences.

Let (X, Σ, m, U) denote a dynamical system, where (X, Σ, m) is a Lebesgue space, and U_t is a measure preserving flow. Let $(a_k)_{k=1}^{\infty}$ denote a sequence of real numbers. We will say that the sequence $(a_k)_{k=1}^{\infty}$ is good for a.e. Cesàro convergence for $L^p(X)$ if the Cesàro averages $\frac{1}{N}\sum_{k=1}^{N} f(U_{a_k}x)$ converge a.e. for all $f \in L^p(X)$ and all measurable measure preserving flows $U_t\colon X \to X$. In this paper we show that certain transformations preserve the class of sequences which are good for a.e. Cesàro convergence for $L^p(X)$. In particular, we will prove the following.

Theorem 1. *If the sequence $(a_k)_{k=1}^{\infty}$ is good for a.e. Cesàro convergence for $L^p(X)$ then so is the sequence of integer parts $(\llbracket a_k \rrbracket)_{k=1}^{\infty}$ and the sequence of fractional parts $(\langle\!\langle a_k \rangle\!\rangle)_{k=1}^{\infty}$ of the a_k's.*

The question of the size of the set of sequences which are good for a.e. Cesàro convergence for $L^p(X)$ is also addressed. This set turns out to be small: a nowhere dense closed set in the box topology. (See Theorem 12.) Let $(a_k)_{k=1}^{\infty}$ be a sequence of real numbers, and consider perturbations $(a_k + \delta_k)_{k=1}^{\infty}$ where $(\delta_k)_{k=1}^{\infty}$ is a bounded sequence. It is interesting to note that while for integer sequences, $(a_k)_{k=1}^{\infty}$ and $(\delta_k)_{k=1}^{\infty}$, bounded perturbations preserve the property that the Cesàro averages satisfy a maximal inequality, it is not the case that bounded perturbations of a sequence of integers which

1 Partially supported by NSF Grant
2 Partially supported by NSF Grant 9302012
3 Partiallysupported by NSF Grant 9224667

is good for a.e. Cesàro convergence for L^p will necessarily be good for a.e. Cesàro convergence even for bounded f. Moreover, in the case of sequences of real numbers, bounded perturbations may even fail to preserve a maximal inequality. On the other hand, the class of good sequences for mean convergence (in any $L^p(X)$, $1 \le p < \infty$) is not that small. Such sequences form a non-empty open set in the box topology. In fact adding a sequence which converges to 0 does not change the L^p convergence of the Cesàro averages, although it may change the pointwise convergence properties.

In Theorem 10 we show that for real polynomials, $p(x) \in \mathbb{R}[x]$, the sequence $(p(k))_{k=1}^{\infty}$ is good for a.e. Cesàro convergence for $L^2(X)$, and consequently the sequence obtained by taking the integer parts of the sequence $(p(k))_{k=1}^{\infty}$ is also good for a.e. Cesàro convergence for L^2. The fact that the integer parts of a polynomial sequence form a sequence which is good for a.e. Cesàro convergence for L^2 is due to Bourgain **[B3]**, but our proof is quite different from his. We also show sequences such as $([\![\alpha k^2]\!] + [\![\beta k^3]\!])_{k=1}^{\infty}$ are good for a.e. Cesàro convergence for L^2. Several other examples are also given.

Taking integer parts can produce good sequences for a.e. Cesàro convergence even in cases when the original sequence does not have that property. In particular, the sequence $(\sqrt{k})_{k=1}^{\infty}$ is known not to be good for a.e. Cesàro convergence, (this was first proved in **[BBB]**, but see also **[JW1]**), even though the associated Cesàro averages converge in L^p norm, $1 \le p < \infty$. Further, for any real α, the sequence obtained by taking the integer parts of the sequence $(\alpha\sqrt{k})_{k=1}^{\infty}$ is good for a.e. Cesàro convergence for L^p, $1 \le p < \infty$. (See **[JW2]** for the case $\alpha = 1$; for general α the same argument works.)

In fact we prove several more general versions of Theorem 1. (See Theorems 2, 7, 8, 10, and associated corollaries below.) To state and prove these more general versions we need some additional notation.

Let $\mathcal{M}$ denote the family of Borel probability measures on $\mathbb{R}$. Each such measure defines an operator on $L^p(\mathbb{R})$ for each p, $p \ge 1$, by

$$\mu(f)(x) = \int_{\mathbb{R}} f(x+t)\, d\mu(t).$$

The measure μ also induces an operator on the dynamical system by

$$\mu(f)(x) = \int_{\mathbb{R}} f(U_t x)\, d\mu(t).$$

For a real number x we will denote by $[\![x]\!]$ the greatest integer less than or equal to x, and by $\langle\!\langle x \rangle\!\rangle = x - [\![x]\!]$ the fractional part of x. We will write $x \oplus t$ to denote $x + t$ (mod 1).

For $\mu \in \mathcal{M}$ we consider two associated measures, $[\![\mu]\!]$ and $\langle\!\langle \mu \rangle\!\rangle$ defined by

$$[\![\mu]\!](I) = \mu(\{x \in \mathbb{R} : [\![x]\!] \in I\}) \text{ and } \langle\!\langle \mu \rangle\!\rangle(I) = \mu(\{x \in \mathbb{R} : \langle\!\langle x \rangle\!\rangle \in I\})$$

for every interval $I \subset \mathbb{R}$. We will refer to these measures respectively as the integer part and fractional part of μ. We can think of $\langle\!\langle \mu \rangle\!\rangle$ as a measure on the torus, $[0, 1) = \mathbb{R}/\mathbb{Z}$, while of course, $[\![\mu]\!]$ is always supported on $\mathbb{Z}$.

Definition. Let $p \geq 1$. A sequence of measures $(\mu_n)_{n=1}^{\infty}$ in $\mathcal{M}$ is said to be *good for pointwise convergence in* L^p if given any non-atomic Lebesgue space (X, Σ, m), and any measure preserving flow $U_t: X \to X$, we have for each $f \in L^p(X)$, $\int_X f(U_t x) d\mu_n(t)$ converges pointwise for a.e. $x \in X$. A sequence $(\mu_n)_{n=1}^{\infty}$ in $\mathcal{M}$ is *good for norm convergence in* L^p if given any non-atomic Lebesgue space (X, Σ, m), and any measure preserving flow $U_t: X \to X$, the sequence of averages $\int_X f(U_t x) d\mu_n(t)$ converges in L^p norm for each $f \in L^p(X)$.

Remark. In the above definition we defined a sequence of measures to be "good for pointwise convergence in L^p" based on what happens with measure preserving flows. However, if we have an integer sequence it is not difficult to see that there is no difference between the definition above, and a definition where the flows are replaced by invertible measure preserving point transformations.

In this paper we show the relationship between good behavior of $(\mu_n)_{n=1}^{\infty}$ and good behavior of $(\llbracket \mu_n \rrbracket)_{n=1}^{\infty}$ and $(\langle\!\langle \mu_n \rangle\!\rangle)_{n=1}^{\infty}$. In particular, we will prove the following:

Theorem 2. *Let $(\mu_n)_{n=1}^{\infty}$ denote a sequence of measures in $\mathcal{M}$, each of which has discrete support, and which is good for pointwise convergence in L^p for some p, $1 \leq p \leq \infty$. Then both sequences of measures, $(\llbracket \mu_n \rrbracket)_{n=1}^{\infty}$ and $(\langle\!\langle \mu_n \rangle\!\rangle)_{n=1}^{\infty}$, are good for pointwise convergence in L^p.*

Theorem 2 will follow from Lemma 5 and Proposition 6 below, and of course Theorem 1 is a special case of Theorem 2. Before stating Lemma 5 and Proposition 6 we introduce some additional notation, and give some related results.

We will also be interested in sequences of measures which are badly behaved. We will use the following terminology (see Rosenblatt **[R]**) and show that bad sequences are typical, and that small perturbations of a bad sequence results in a bad sequence.

Definitions. A sequence of measures, $(\mu_n)_{n=1}^{\infty}$ is L^p *universally bad* if for all dynamical systems (X, Σ, m, U_t), with X a non-atomic Lebesgue probability space, and $U_t: X \to X$ is an aperiodic measure preserving flow, the associated operators $\mu_n f$ have the property that there exists $f \in L^p(X)$ such that $\lim_{n\to\infty} \mu_n f(x)$ fails to exist for all x in a set of positive measure.

A sequence of measures $(\mu_n)_{n=1}^{\infty}$ has the *strong sweeping out property* if for all dynamical systems (X, Σ, m, U_t), with X a non-atomic Lebesgue probability space, and $U_t: X \to X$ an ergodic measure preserving flow, the associated operators $\mu_n f$ have the property that given $\epsilon > 0$, there is a measurable set E, with $m(E) < \epsilon$, such that $\limsup_{n\to\infty} \mu_n \chi_E(x) = 1$ a.e. and $\liminf_{n\to\infty} \mu_n \chi_E(x) = 0$ a.e.

A sequence of measures $(\mu_n)_{n=1}^{\infty}$ supported on a discrete set $\mathcal{A}$ is *dissipative* if for each fixed $a \in \mathcal{A}$, $\mu_n(a) \to 0$.

A sequence of distinct real numbers $(a_k)_{k=1}^{\infty}$ is L^p *Cesàro bad* if the associated sequence of Cesàro averages, $\frac{1}{n}\sum_{k=1}^{n} \delta_{a_k}$, is L^p universally bad. The sequence $(a_k)_{k=1}^{\infty}$ is *universally bad* if for any dissipative sequence of measures, $(\mu_n)_{n=1}^{\infty}$, supported on the elements of the sequence, the associated sequence of operators $(\mu_n)_{n=1}^{\infty}$ defined by $\mu_n f(x) = \sum_{k=1}^{\infty} \mu_n(a_k) f(U_{a_k} x)$ is universally bad, and has the *strong sweeping out property* if each such sequence $(\mu_n)_{n=1}^{\infty}$ has the strong sweeping out property. Examples of sequences with this strong sweeping out property include lacunary sequences. (See **[Aetal]**.)

Let $(\mu_n)_{n=1}^{\infty}$ denote a sequence of measures in $\mathcal{M}$. The sequence $(\mu_n)_{n=1}^{\infty}$ is said to *converge* mod 1 if the sequence of measures $(\langle\!\langle \mu_n \rangle\!\rangle)_{n=1}^{\infty}$ converges weakly to some measure on $[0, 1) = \mathbb{R}/\mathbb{Z}$. That is, there is a measure μ on $[0, 1)$ such that

$$\lim_{n\to\infty} \int_0^1 f(x)\, d\langle\!\langle \mu_n \rangle\!\rangle(x) = \int_0^1 f(x)\, d\mu(x)$$

for all continuous periodic functions on $\mathbb{R}$ with period 1.

The sequence $(\mu_n)_{n=1}^{\infty}$ is said to be *regular* if the sequence of measures $(\langle\!\langle \mu_n \rangle\!\rangle)_{n=1}^{\infty}$ converges weakly to some measure on $[0, 1]$, that is: there is a measure μ on $[0, 1]$ such that

$$\lim_{n\to\infty} \int_0^1 f(x)\, d\langle\!\langle \mu_n \rangle\!\rangle(x) = \int_0^1 f(x)\, d\mu(x)$$

for all continuous functions on $[0, 1]$.

It is clear that a sequence of measures which is regular also converges mod 1. However the opposite implication is false. To see this take $\mu_n = \delta_{\left\{\frac{(-1)^n}{n}\right\}}$ where $\delta_{\{x\}}$ denotes the Dirac measure at the point x. Then take any function which is continuous on $[0, 1]$ such that $f(0) \neq f(1)$. For such functions, convergence fails.

Proposition 3. *If for some* $p \geq 1$, $(\mu_n)_{n=1}^{\infty}$ *is good for norm convergence in* L^p, *then* $(\mu_n)_{n=1}^{\infty}$ *converges* mod 1. *(The converse is also true by the spectral theorem.)*

Proof. Take $X = [0, 1)$, and consider the flow $U_t x = x \oplus t$. Consider the functions $e_k(x) = e^{2\pi i k x}$ for $k \in \mathbb{Z}$. We have that $\int e_k(U_t x) d\mu_n(t) = \hat{\mu}_n(k)$. The hypothesis implies that for each fixed k, $(\hat{\mu}_n(k))_{n=1}^{\infty}$ is a convergent sequence of numbers. The theorem now follows by observing that finite linear combinations of the functions $(e_k)_{k=-\infty}^{\infty}$ are dense in the space of continuous functions of period 1 on $\mathbb{R}$. □

Corollary 4. *If for some* $p \geq 1$, $(\mu_n)_{n=1}^{\infty}$ *is good for pointwise convergence in* L^p, *then* $(\mu_n)_{n=1}^{\infty}$ *converges* mod 1.

Lemma 5. *If a sequence of measures* $(\mu_n)_{n=1}^{\infty}$ *converges* mod 1 *then it is regular if and only if*

$$\lim_{\epsilon\to 0} \limsup_{n\to\infty} \langle\!\langle \mu_n \rangle\!\rangle([0, \epsilon)) = \lim_{\epsilon\to 0} \liminf_{n\to\infty} \langle\!\langle \mu_n \rangle\!\rangle([0, \epsilon)). \tag{1}$$

Proof. For each ϵ, $0 < \epsilon < 1/2$, let f_ϵ be a continuous "characteristic function" of the interval $[0, \epsilon]$, that is let $f_\epsilon \colon [0, 1] \to \mathbb{R}$ be continuous and satisfy the following properties

(i) $0 \leq f_\epsilon \leq 1$;

(ii) $f_\epsilon(x) = 1$ for $0 \le x \le \epsilon$;
(iii) $f_\epsilon(x) = 0$ for $2\epsilon \le x \le 1$.

Then the condition in (1) is equivalent with

$$\lim_{\epsilon \to 0^+} \limsup_{n \to \infty} \int f_\epsilon \, d \langle\!\langle \mu_n \rangle\!\rangle = \lim_{\epsilon \to 0^+} \liminf_{n \to \infty} \int f_\epsilon \, d \langle\!\langle \mu_n \rangle\!\rangle . \tag{2}$$

This follows immediately from the inequalities

$$\langle\!\langle \mu \rangle\!\rangle([0, \epsilon]) \le \int f_\epsilon \, d \langle\!\langle \mu \rangle\!\rangle \le \langle\!\langle \mu \rangle\!\rangle([0, 2\epsilon]) \le \int f_{2\epsilon} \, d \langle\!\langle \mu \rangle\!\rangle,$$

which is valid for every $\mu \in \mathcal{M}$ and $0 < \epsilon < 1/4$.

Now, if (μ_n) is regular then (2) clearly holds, so we just have to prove that (2) implies the regularity of the sequence (μ_n). Let $g\colon [0, 1] \to \mathbb{R}$ be continuous; we want to show that $\int g d \langle\!\langle \mu_n \rangle\!\rangle$ converges. Since we assumed that $(\mu_n)_{n=1}^{\infty}$ converges mod 1, we can assume that $g(0) \ne g(1)$. Applying a linear transformation ($ag + b$ with some constants a, b) to g we can assume that $g(0) = 1$ and $g(1) = 0$. For $\epsilon > 0$ let us define $h_\epsilon = g - f_\epsilon$. Then $h_\epsilon(0) = h_\epsilon(1) = 0$, hence $\lim_{n \to \infty} \int h_\epsilon d \langle\!\langle \mu_n \rangle\!\rangle$ exists for (μ_n) converges mod 1. But then

$$\limsup_{n \to \infty} \int g \, d \langle\!\langle \mu_n \rangle\!\rangle - \liminf_{n \to \infty} \int g \, d \langle\!\langle \mu_n \rangle\!\rangle = \limsup_{n \to \infty} \int f_\epsilon \, d \langle\!\langle \mu_n \rangle\!\rangle - \liminf_{n \to \infty} \int f_\epsilon \, d \langle\!\langle \mu_n \rangle\!\rangle,$$

and, letting $\epsilon \to 0^+$, we see, by (2), that indeed $\int g d \langle\!\langle \mu_n \rangle\!\rangle$ converges. □

Note that under the conditions of Proposition 3 the sequence of measures may not be regular. However in the case when all measures are discrete we can say more.

Proposition 6. *Assume that the sequence of measures $(\mu_n)_{n=1}^{\infty} \subset \mathcal{M}$ is good for pointwise convergence in L^∞. If all μ_n are discrete then $(\mu_n)_{n=1}^{\infty}$ is regular and*

$$\lim_{\epsilon \to 0} \lim_{n \to \infty} \langle\!\langle \mu_n \rangle\!\rangle([-\epsilon, 0)) = 0.$$

Proof. By Theorem 3 we know that the sequence $(\mu_n)_{n=1}^{\infty}$ converges mod 1. To show it is regular, by Lemma 5 above, it will be enough to show that condition (1) is satisfied. Passing to a subsequence, we can assume $\lim_{n \to \infty} \langle\!\langle \mu_n \rangle\!\rangle\{0\} = \alpha$ for some $0 \le \alpha < 1$. In this case we can assume that $\langle\!\langle \mu_n \rangle\!\rangle\{0\} = 0$ for every n, since if not define a new sequence of measures, $(\nu_n)_{n=1}^{\infty}$ by $\nu_n(I) = \frac{\langle\!\langle \mu_n \rangle\!\rangle(I)}{1 - \langle\!\langle \mu_n \rangle\!\rangle(0)}$ for $I \subset (0, 1)$, and $\nu_n(0) = 0$. The new sequence will be regular if and only if the original sequence was regular. Now we show that $(\nu_n)_{n=1}^{\infty}$ is regular. To do this, we will first show that the left side (and hence the right side) of equation (1) is zero. Assume not. Then there is a $\delta > 0$ such that $\lim_{\epsilon \to 0} \limsup_{n \to \infty} \langle\!\langle \nu_n \rangle\!\rangle([0, \epsilon)) = \delta$. Consequently we can find a sequence $(n_k)_{k=1}^{\infty}$ and $(\epsilon_k)_{k=1}^{\infty}$ such that $\langle\!\langle \nu_{n_k} \rangle\!\rangle([0, \epsilon_k)) > \frac{\delta}{2}$. Now define new sequence of measures

$(\tilde{\nu}_k)_{k=1}^{\infty}$ such that for $I \subset [0,1)$ we have

$$\tilde{\nu}_k(I) = \frac{\langle\!\langle \nu_{n_k} \rangle\!\rangle (I \cap [0, \epsilon_k))}{\langle\!\langle \nu_{n_k} \rangle\!\rangle ([0, \epsilon_k))}.$$

This sequence of measures is a dissipative sequence, with $\lim_{k\to\infty} \tilde{\nu}_k([0,\epsilon)) = 1$, and supported on a discrete set. For any dissipative sequence with this property, we know that we have the strong sweeping out property. (See **[Aetal]** for this and related results, also see **[B1]** and **[AdJL]**.) Let $X = [0,1)$. On X define a flow by $U_t x = x \oplus t$. Since we have the strong sweeping out property, we can find a set $E \subset [0,1)$ such that $|E| < \frac{\delta}{100}$ but such that

$$\limsup_{k\to\infty} \int \chi_E(U_t x)\, d\tilde{\nu}_k = 1 \text{ and } \liminf_{k\to\infty} \int \chi_E(U_t x)\, d\tilde{\nu}_k = 0 \text{ a.e.}$$

By hypothesis, we know that $(\nu_n \chi_E(x))_{n=1}^{\infty}$ converges for a.e. x. However

$$\begin{aligned}
\limsup_{k\to\infty} \int \chi_E(U_t x)\, d\nu_{n_k}(t) &= \limsup_{k\to\infty} \int \chi_E(U_t x)\, d \langle\!\langle \nu_{n_k} \rangle\!\rangle \\
&= \limsup_{k\to\infty} \int \chi_E(U_t x) \langle\!\langle \nu_{n_k} \rangle\!\rangle ([0,\epsilon_k))\, d \frac{\langle\!\langle \nu_{n_k} \rangle\!\rangle (I \cap [0, \epsilon_k))}{\langle\!\langle \nu_{n_k} \rangle\!\rangle ([0, \epsilon_k))} \\
&\geq \limsup_{k\to\infty} \int \nu_{n_k}([0,\epsilon_k)) \chi_E(U_t x)\, d\tilde{\nu}_k \\
&\geq \frac{\delta}{2} \limsup_{k\to\infty} \int \chi_E(U_t x)\, d\tilde{\nu}_k \geq \frac{\delta}{2}.
\end{aligned}$$

However since $|E| < \delta/100$ we see that we have a contradiction. This contradiction proves that (1) is satisfied, and we see the sequence is regular.

To see that $\lim_{\epsilon\to 0} \lim_{n\to\infty} \langle\!\langle \mu_n \rangle\!\rangle ([-\epsilon, 0)) = 0$ we assume this is false, and select a sequence $(\epsilon_k)_{k=1}^{\infty}$ such that $\lim_{n\to\infty} \langle\!\langle \mu_n \rangle\!\rangle ([-\epsilon_k, 0)) > \delta/2$ in the same way as above. Now define a dissipative sequence of measures $(\tilde{\nu}_k)_{k=1}^{\infty}$ by

$$\tilde{\nu}_k(I) = \frac{\langle\!\langle \nu_{n_k} \rangle\!\rangle (I \cap [-\epsilon_k, 1))}{\langle\!\langle \nu_{n_k} \rangle\!\rangle ([-\epsilon_k, 1))},$$

and argue as before. □

Theorem 7. *Let* $(\mu_n)_{n=1}^{\infty}$ *denote a sequence of measures in* $\mathcal{M}$, *which is good for pointwise convergence in* L^p. *If* $(\mu_n)_{n=1}^{\infty}$ *is regular, then the sequence* $(\langle\!\langle \mu_n \rangle\!\rangle)_{n=1}^{\infty}$ *is good for pointwise convergence in* L^p. *If* $(\mu_n)_{n=1}^{\infty}$ *is regular, and one of*

a) $\lim_{\epsilon\to 0} \lim_{n\to\infty} \langle\!\langle \mu_n \rangle\!\rangle ([0,\epsilon)) = 0$,
b) $\lim_{\epsilon\to 0} \lim_{n\to\infty} \langle\!\langle \mu_n \rangle\!\rangle ((-\epsilon, 0)) = 0$,

holds, then $([\![\mu_n]\!])_{n=1}^{\infty}$ *is good for pointwise convergence in* L^p.

Proof. First we will consider the fractional part. Consider the flow on the circle defined by $U_t x = x \oplus t$. For this flow we have convergence of the averages

$$\lim_{n\to\infty}\int_{\mathbb{R}} f(U_t x)\,d\mu_n(t) = \lim_{n\to\infty}\int_{\mathbb{R}} f(x\oplus t)\,d\mu_n(t)$$

$$= \lim_{n\to\infty}\int_{\mathbb{R}} f(x\oplus t)\,d\langle\!\langle \mu_n\rangle\!\rangle(t) = \lim_{n\to\infty}\int_{\mathbb{R}} f(U_t x)\,d\langle\!\langle \mu_n\rangle\!\rangle(t),$$

and we are done in this special case.

We now consider another special case, the case of a flow on the real line $\mathbb{R}$ defined by $U_t x = x + t$. This case will then be used to prove the result on a general dynamical system. For a non-negative function $f \in L^p(\mathbb{R})$ we define for each integer m, a function

$$g_m(x) = f(m + x) + f(m + 1 + x),$$

for $0 \le x < 1$. Clearly $x \oplus \alpha = x + \alpha$ if $x + \alpha < 1$ and $x \oplus \alpha = x + \alpha - 1$ for $1 \le x + \alpha < 2$. Thus we have $f(m + x + \alpha) = f(m + x \oplus \alpha)$ or $f(m + x + \alpha) = f(m + 1 + x \oplus \alpha)$. In either case we have $f(m + x + \alpha) \le |g_m(x \oplus \alpha)|$. Considering our operators, we see that

$$|\langle\!\langle \mu_n f(m+x)\rangle\!\rangle| = \Big|\int_{\mathbb{R}} f(m+x+t)\,d\,\langle\!\langle \mu_n\rangle\!\rangle(t)\Big|$$
$$\le \int |g_m(x\oplus t)|\,d\,\langle\!\langle \mu_n\rangle\!\rangle(t).$$

The right hand side is the periodic case considered above, hence converges. Thus we have a weak type (p, p) maximal inequality. Hence

$$m\{x \in [0,1) : \sup_n |\langle\!\langle \mu_n\rangle\!\rangle f(m+x)| > \lambda\} < \frac{c}{\lambda^p}\int_m^{m+2} |f(x)|^p\,dx.$$

Adding these inequalities for each m, we see that

$$m\{x \in \mathbb{R} : \sup_n |\langle\!\langle \mu_n\rangle\!\rangle f(x)| > \lambda\} < \frac{2c}{\lambda^p}\|f\|_p^p.$$

To prove a.e. convergence it is enough to have a maximal inequality and convergence on a dense class. We have just established the maximal inequality, and the dense class is just the continuous functions. To see this, let f be a continuous function on $\mathbb{R}$. Fix an x and define a function $g_x(y) = f(x + y)$ for $0 \le y < 1$, and extend g to be periodic with period 1. Since our sequence is regular, and g_x is bounded and continuous on $[0, 1)$, we know

$$\langle\!\langle \mu_n\rangle\!\rangle f(x) = \langle\!\langle \mu_n\rangle\!\rangle g_x(0) = \int_{\mathbb{R}} g_x(t)\,d\,\langle\!\langle \mu_n\rangle\!\rangle(t)$$

converges. Hence we have convergence for all $f \in L^p(\mathbb{R})$.

Consider a general dynamical system. If the system is aperiodic, we proceed as follows. Assume that for some function $f \in L^p(X)$, divergence occurs for $x \in E$,

$m(E) > 0$. Form a tall Rohlin tower of height N and base B so that $m\{[0, N-1)\times B\} > 1 - m(E)/2$. Thus the region $[0, N-1) \times B$ contains a set of measure at least $m(E)/2$ for which divergence occurs. Note that on each "fiber" over $x \in B$ for which $\int_0^N |f(U_t x)|^p dt$ is finite, we have convergence, for a.e. t, $0 < t < N-1$, at the point $U_t x$. However the dB measure of the set of fibers for which we do not have an L^p function is zero. Hence we have convergence a.e. in $[0, N-1) \times B$, a contradiction.

If the system is periodic, with period ρ, consider

$$g(t) = f(U_t x)\chi_{[0,\rho)}(t) + f(U_t x)\chi_{[\rho,2\rho)}(t) + \cdots + f(U_t x)\chi_{[(N-1)\rho, N\rho)}(t).$$

For each N, we are in $L^p(0, N\rho)$, and hence we can apply the above result on the real line with translation to get convergence of $\langle\!\langle \mu_n \rangle\!\rangle\, g(t)$ for a.e. t. However, since these operators are "local", for N large enough, this will also imply convergence of

$$\int_{\mathbb{R}} f(U_{t+s}x)\, d\langle\!\langle \mu_n \rangle\!\rangle(s) = \int_{\mathbb{R}} f(U_s(U_t x))\, d\langle\!\langle \mu_n \rangle\!\rangle(s) = \langle\!\langle \mu_n \rangle\!\rangle\, f(U_t x)$$

for a.e. t, and hence for a.e. x, since $U_t x$ ranges over all of X.

For the integer part we proceed as follows. Fix a transformation T. Consider the space $Y = X \times [0, 1) = \{(x, s) : x \in X,\ s \in [0, 1)\}$. On Y define a flow by $U_t(x, s) = (T^{[s+t]}x, s \oplus t)$. Fix $f \in L^p(X)$. Define $\tilde{f}$ on Y by $\tilde{f}(x, s) = f(x)$. By hypothesis, we know

$$\mu_n \tilde{f}(x, s) = \int \tilde{f}(U_t(x, s))d\mu_n(t)$$

converges for a.e. (x, s). If we knew we had convergence of

$$\begin{aligned}\mu_n \tilde{f}(x, 0) &= \int \tilde{f}(U_t(x, 0))\, d\mu_n(t) = \mu_n \tilde{f}(x, s)\\ &= \int \tilde{f}(U_t(x, s))\, d[\![\mu_n]\!](t) = [\![\mu_n]\!]\, f(x)\end{aligned}$$

we would be done. Unfortunately we only know convergence for a.e. (x, s) and $X \times \{0\}$ is a set of measure zero. However by Fubini, we know that for a.e. fixed $\delta \geq 0$, we have convergence of

$$\mu_n \tilde{f}(x, \delta) = \int \tilde{f}(U_t(x, \delta))\, d\mu_n(t).$$

To see that we have convergence on a dense class, let $f \in L^\infty(X)$. Assume first that condition b) holds. Select a sequence $(\delta_j) \to 0$ with δ_j having the property that for a.e.

x we have convergence of $\mu_n \tilde{f}(x, \delta_j)$. Consider

$$\begin{aligned}
&\limsup_{n\to\infty} \Big| \int \tilde{f}(U_t(x,0))d\mu_n(t) - \int \tilde{f}(U_t(x,\delta_j))d\mu_n(t) \Big| \\
&= \limsup_{n\to\infty} \Big| \int_{\langle\langle t \rangle\rangle < 1-\delta_j} \tilde{f}(U_t(x,0))d\mu_n(t) - \int_{\langle\langle t \rangle\rangle < 1-\delta_j} \tilde{f}(U_t(x,\delta_j))d\mu_n(t) \Big| \\
&\quad + \limsup_{n\to\infty} \Big| \int_{\langle\langle t \rangle\rangle \geq 1-\delta_j} \tilde{f}(U_t(x,0))d\mu_n(t) - \int_{\langle\langle t \rangle\rangle > 1-\delta_j} \tilde{f}(U_t(x,\delta_j))d\mu_n(t) \Big| \\
&= \limsup_{n\to\infty} \Big| \int_{\langle\langle t \rangle\rangle < 1-\delta_j} \tilde{f}(U_t(x,0)) - \tilde{f}(U_t(x,\delta_j))d\mu_n(t) \Big| \\
&\quad + \limsup_{n\to\infty} \Big| \int_{\langle\langle t \rangle\rangle \geq 1-\delta_j} \tilde{f}(U_t(x,0)) - \tilde{f}(U_t(x,\delta_j))d\mu_n(t) \Big| \\
&\leq \limsup_{n\to\infty} \Big| \int_{\langle\langle t \rangle\rangle \geq 1-\delta_j} \tilde{f}(U_t(x,0)) - \tilde{f}(U_t(x,\delta_j))d\mu_n(t) \Big| \\
&\leq 2\|f\|_\infty \limsup_{n\to\infty} \langle\langle \mu_n \rangle\rangle([1-\delta_j, 1)),
\end{aligned}$$

and by assumption b), we have $\lim_{j\to\infty} \lim_{n\to\infty} \langle\langle \mu_n \rangle\rangle([1-\delta_j, 1)) = 0$.

If we assume condition a) holds instead of condition b), the argument is almost the same. Now we take a sequence $(\delta_j) \to 1$ from below, and such that for each δ_j, we have $\lim_{n\to\infty} \mu_n \tilde{f}(x, \delta_j)$ exists for a.e. x. Break the integral into the same two regions as above. However now we compare $\mu_n \tilde{f}(x, \delta_j)$ with $\mu_n \tilde{f}(\tau x, 0)$. We see that the difference will dominated by $2\|f\|_\infty \limsup_{n\to\infty} \langle\langle \mu_n \rangle\rangle([0, 1-\delta_j))$ and this converges to 0 by assumption.

To see that we have a maximal inequality, note that if δ is such that $\langle\langle t \rangle\rangle + \delta \geq 1$ then $f(T^{[\![n_k]\!]}(Tx)) = \tilde{f}(U_t(x,\delta))$, and if $\langle\langle t \rangle\rangle + \delta < 1$ then $f(T^{[\![t]\!]}(Tx)) = \tilde{f}(U_t(Tx,\delta))$. In either case we have $f(T^{[\![t]\!]}(Tx)) \leq |\tilde{f}(U_t(x,\delta))| + |\tilde{f}(U_t(Tx,\delta))$. Pick a δ for which we have convergence, then we have a maximal inequality of the form

$$m\{x : \sup_n [\![\mu_n]\!] f(Tx) > \lambda\} \leq \frac{c}{\lambda^p} \int_X |\tilde{f}(x,\delta)|^p dm(x).$$

Since we have convergence on a dense class, and a maximal inequality, we have pointwise convergence for all $f \in L^p$. □

We are now in a position to prove Theorem 2.

Proof of Theorem 2. By Proposition 6, we see that part b) of the hypothesis of Theorem 7 is satisfied. We now apply Theorem 7.

Example 1. The converse of Theorem 1 is false. To see this consider the Cesàro averages along a sequence $(a_k)_{k=1}^{\infty}$. The sequence will be defined in blocks, with each block much longer than the union of the prior blocks. In the first block, if k is even we let $a_k = k + \frac{1}{2}$ and if k is odd, we let $a_k = k$. Thus the sequence starts out with

$a_1 = 1$, $a_2 = 2\frac{1}{2}$, $a_3 = 3$, $a_4 = 4\frac{1}{2}$, etc. In the second block if k is even we let $a_k = k$ and if k is odd we let $a_k = k + \frac{1}{2}$. We alternate between these two types of blocks. This sequence can be given by the formula

$$a_k = k + \frac{1}{4}\left(1 + (-1)^{k+[\![\log_{10}\log_{10}(k+10)]\!]}\right).$$

Then for $\mu_n = \frac{1}{n}\sum_{k=0}^{n-1}\delta_{a_k}$, clearly both $([\![\mu_n]\!])_{n=1}^{\infty}$ and $(\langle\!\langle \mu_n \rangle\!\rangle)_{n=1}^{\infty}$ are good for pointwise convergence. But $(\mu_n)_{n=1}^{\infty}$ is L^p universally bad, that is, $(a_k)_{k=1}^{\infty}$ is L^p Cesàro bad for $1 \le p \le \infty$, and in fact is bad for norm convergence. Consider $X = [0, 2)$ and the flow $U_t x = x + t$, mod 2. Then on the blocks where $a_k = k + \frac{1}{2}$ for k even, $U_{a_k}x$ alternates between x and $x + \frac{1}{2}$, and on the blocks where $a_k = k$ for k even, $U_{a_k}x$ alternates between x and $x + \frac{3}{2}$. Since these blocks alternate, and each is longer than the prior blocks, no convergence is possible.

To see that we need an extra assumption such as a) or b) in Theorem 7, we now construct an example of a sequence of measures for which we have a.e. convergence for all $f \in L^p$ for each $p > 1$, and which is regular, but such that the integer parts of the sequence fail to be good.

Example 2. There is a sequence of measures $(c_n)_{n=1}^{\infty}$ which is regular, and which is good for pointwise convergence for all $f \in L^p$ for each $p > 1$, but such that $([\![c_n]\!])_{n=1}^{\infty}$ fails to be good even for $f \in L^{\infty}$. (We note that for a sequence of measures which is good for pointwise convergence for $f \in L^p$, and which is regular, the fractional parts are always good for pointwise convergence for $f \in L^p$.)

The idea of the example is to construct two good sequences of measures, both of which have the same limit, but whose integer parts have disjoint supports. We then get the desired sequence $(c_n)_{n=1}^{\infty}$ by alternating between these two good sequences of measures.

Fix a sequence $(\epsilon_k)_{k=1}^{\infty}$ such that $\lim_{k\to\infty}\epsilon_k \to 0$. Define the sequence of measures $(\nu_k)_{k=1}^{\infty}$ by

$$\nu_k(x) = \begin{cases} \frac{1}{\epsilon_k}\chi_{[0,\epsilon_k)} & \text{k even;} \\ \frac{1}{\epsilon_k}\chi_{(-\epsilon_k,0)} & \text{k odd.} \end{cases}$$

Then by the Lebesgue differentiation theorem, we know that for all $f \in L^p$, $p \ge 1$, $\lim_{k\to\infty}\nu_k f(x) = f(x)$ for a.e. x, and for $p > 1$, $\|\sup_k |\nu_k f|\|_p \le c\|f\|_p$.

Now define the measures $(\mu_n)_{n=1}^{\infty}$ by $\mu_n = \frac{1}{n}\sum_{k=1}^{n}\nu_k \star \delta_k$. For this sequence of measures we have a.e. convergence for all $f \in L^p$ for $p > 1$. To see this we argue as in **[JW2]**, where a more general result is proved.

Note first that

$$\mu_k f(x) = \frac{1}{n}\sum_{k=1}^{n}\nu_k f(U_k x) = \frac{1}{n}\sum_{k=1}^{n} f(U_k x) + \frac{1}{n}\sum_{k=1}^{n}(\nu_k f(U_k x) - f(U_k x)).$$

Since by the ergodic theorem we have a.e. convergence of $\frac{1}{n}\sum_{k=1}^{n} f(U_k x)$, it will be enough to show that for each $\lambda > 0$ we have

$$m\Big\{x \mid \lim_{N\to\infty} \sup_{n\geq N} \Big|\frac{1}{n}\sum_{k=1}^{n} (\nu_k f(U_k x) - f(U_k x))\Big| > \lambda\Big\} = 0.$$

Let $f_k = \nu_k f(x) - f(x)$. Then we know $\lim_{k\to\infty} f_k(x) = 0$ a.e. and

$$\lim_{k_0\to\infty} \|\sup_{k>k_0} |f_k|\|_p = 0. \tag{3}$$

We need to show that

$$m\Big\{x \mid \lim_{N\to\infty} \sup_{n\geq N} \Big|\frac{1}{n}\sum_{k=1}^{n} f_k(U_k x)\Big| > \lambda\Big\} = 0.$$

Since for $p > 1$ the maximal function $Mf(x) = \sup_{n>0} |\frac{1}{n}\sum_{k=1}^{n} f(U_k x)|$ is a bounded operator from L^p to L^p, we see that (3) implies for each $\epsilon > 0$,

$$\lim_{k_0\to\infty} m\{x \mid M \sup_{k>k_0} |f_k| > \epsilon\} = 0. \tag{4}$$

For fixed k_0 we estimate

$$\begin{aligned}
&m\Big\{x \mid \lim_{N\to\infty} \sup_{n\geq N} \Big|\frac{1}{n}\sum_{k=1}^{n} f_k(U_k x)\Big| > \lambda\Big\} \\
&\leq m\Big\{x \mid \lim_{N\to\infty} \sup_{n\geq N} \Big|\frac{1}{n}\sum_{k=1}^{k_0-1} f_k(U_k x)\Big| > \frac{\lambda}{2}\Big\} + m\Big\{x \mid \lim_{N\to\infty} \sup_{n\geq N} \Big|\frac{1}{n}\sum_{k=k_0}^{n} f_k(U_k x)\Big| > \frac{\lambda}{2}\Big\} \\
&\leq \sum_{k=1}^{k_0-1} m\Big\{x \mid \lim_{N\to\infty} \sup_{n\geq N} \Big|\frac{1}{n} f_k(U_k x)\Big| > \frac{\lambda}{2k_0}\Big\} + m\Big\{x \mid \sup_{n} \Big|\frac{1}{n}\sum_{k=k_0}^{n} f_k(U_k x)\Big| > \frac{\lambda}{2}\Big\} \\
&\leq m\Big\{x \mid \sup_{n} \Big|\frac{1}{n}\sum_{k=k_0}^{n} f_k(U_k x)\Big| > \frac{\lambda}{2}\Big\} \\
&\leq m\Big\{x \mid M \sup_{k\geq k_0} |f_k(x)| > \frac{\lambda}{2}\Big\},
\end{aligned}$$

and by (4) we know $m\{x \mid M \sup_{k\geq k_0} |f_k(x)|) > \frac{\lambda}{2}\} \to 0$.

A similar computation shows that for the sequence of measures defined by

$$\tilde{\nu}_k(x) = \begin{cases} \frac{1}{\epsilon_k}\chi_{[0,\epsilon_k)} & k \text{ odd;} \\ \frac{1}{\epsilon_k}\chi_{(-\epsilon_k,0)} & k \text{ even.} \end{cases}$$

the sequence of measures

$$\tilde{\mu}_n = \frac{1}{n}\sum_{k=1}^{n} \tilde{\nu}_k \star \delta_k$$

is good for a.e. convergence for $f \in L^p$ for all $p > 1$. Further, both $(\mu_n f)_{n=1}^{\infty}$ and $(\tilde{\mu}_n f)_{n=1}^{\infty}$ converge a.e. to the same invariant function. We now define the sequence of measures (c_n) by

$$c_n = \begin{cases} \mu_n & \text{if } n \text{ is even} \\ \tilde{\mu}_n & \text{if } n \text{ is odd.} \end{cases}$$

Clearly the $(c_n)_{n=1}^{\infty}$ are good for a.e. convergence for all $f \in L^p$, $p > 1$, and the sequence is regular, since it converges to $\frac{1}{2}(\delta_0 + \delta_1)$. However we see that the integer parts will not converge, since for even n, $[\![c_n]\!]$ is supported only on even integers, and for odd n, $[\![c_n]\!]$ is supported only on odd integers.

We now give some examples to show that there are many examples of good sequences for a.e. Cesàro convergence in L^p.

Theorem 8. *Assume that the sequence $(n_k)_{k=1}^{\infty}$ is good for a.e. Cesàro convergence for L^p. Then for any $\alpha \in \mathbb{R}$ the sequence $([\![\alpha n_k]\!])_{k=1}^{\infty}$ is also good for a.e. Cesàro convergence for L^p.*

Proof. Since the sequence $(n_k)_{k=1}^{\infty}$ is good for a.e. Cesàro convergence for L^p, we know that for every measure preserving flow, U_t, we have convergence of the averages

$$\frac{1}{N}\sum_{k=1}^{N} f(U_{n_k}x)$$

for a.e. x. Define a new flow, V_t by $V_t x = U_{\frac{t}{\alpha}} x$. Hence we have convergence of

$$\frac{1}{N}\sum_{k=1}^{N} f(V_{\alpha n_k}x) = \frac{1}{N}\sum_{k=1}^{N} f(U_{n_k}x)$$

for a.e. x. Now an application of Theorem 1 completes the proof.

Corollary 9. *Let $\alpha_1, \alpha_2, \ldots, \alpha_r$ denote r non-zero real numbers. If the sequence $(n_k)_{k=1}^{\infty}$ is good for a.e. Cesàro convergence for L^p. Then the sequence*

$$([\![\alpha_1 [\![\alpha_2 [\![\ldots [\![\alpha_r n_k]\!] \ldots]\!]]\!]]\!])_{k=1}^{\infty}$$

is also good for a.e. Cesàro convergence for L^p.

Proof. Since the sequence $(n_k)_{k=1}^{\infty}$ is good for a.e. Cesàro convergence for L^p, we just need to apply Theorem 8, r times.

Example. For each positive integer k, let p_k denote the kth prime. Since $(p_k)_{k=1}^{\infty}$ is good for a.e. Cesàro convergence for L^p, $p > 1$, (see **[W2]**) we have that for any non-zero real number α, the sequence $([\![\alpha p_k]\!])_{k=1}^{\infty}$ is good for a.e. Cesàro convergence for L^p for each $p > 1$. Further, if $\alpha_1, \alpha_2, \ldots, \alpha_r$ denote r non-zero real numbers, then the sequence $([\![\alpha_1 [\![\alpha_2 [\![\ldots [\![\alpha_r p_k]\!] \ldots]\!]]\!]]\!])_{k=1}^{\infty}$ is good for a.e. Cesàro convergence for L^p. More generally, if q is any non-constant polynomial with integer coeficients, then the sequence $([\![\alpha_1 [\![\alpha_2 [\![\ldots [\![\alpha_r q(p_k)]\!] \ldots]\!]]\!]]\!])_{k=1}^{\infty}$ is good for a.e. Cesàro convergence for L^p. This follows from the fact that $(q(p_k))_{k=1}^{\infty}$ is good for a.e. Cesàro convergence for L^p. (See **[W1]** or **[N]**.)

Example. Let $\alpha_1, \alpha_2, \ldots, \alpha_r$ denote r non-zero real numbers, then the sequence

$$(\llbracket \alpha_1 \llbracket \alpha_2 \llbracket \ldots \llbracket \alpha_r \llbracket n^\delta \rrbracket \rrbracket \ldots \rrbracket \rrbracket \rrbracket)_{k=1}^{\infty}$$

is good for a.e. Cesàro convergence for L^p. This follows from the fact that $(\llbracket n^\delta \rrbracket)_{n=1}^{\infty}$ is good for a.e. Cesàro convergence for L^p. (See **[W]**.)

Theorem 10. *Let $p(x) \in \mathbb{R}[x]$ denote a polynomial with real coefficients. Then*
a) *the sequence $(p(k))_{k=1}^{\infty}$ is good for a.e. Cesàro convergence for $L^2(X)$,*
b) *the sequence obtained by taking the integer parts of the sequence $(p(k))_{k=1}^{\infty}$ is good for a.e. Cesàro convergence for L^2.*

Proof. Let $p(x) = a_0 + a_1 x + \cdots + a_r x^r$. We can assume $r \geq 1$, and $a_r \neq 0$. Let U_t be a measure preserving flow. Define the measure preserving transformations τ_j by $\tau_j^k x = U_{a_j j^k} x$. Let $T_j f(x) = f(\tau_j x)$. Then the family T_j, $j = 0, 1, \ldots, r$ are commuting transformation, and we can apply Theorem 6 from **[B2]** to conclude that the averages

$$\frac{1}{N} \sum_{k=1}^{N} T_0^{k^0} T_1^{k^1} \ldots T_r^{k^r} f(x)$$

converge a.e. for all $f \in L^2(X)$. However

$$T_0^{k^0} T_1^{k^1} \ldots T_r^{k^r} f(x) = f(U_{a_0} U_{a_1 k^1} \ldots U_{a_r k^r} x) = f(U_{p(k)} x).$$

Consequently

$$\frac{1}{N} \sum_{k=1}^{N} f(U_{p(k)} x)$$

converges a.e. for all $f \in L^2$, proving part a). Part b) is now an immediate consequence of Theorem 1.

We can also consider higher dimensional version of the above theorems. For example the same techniques can be used to prove the following theorem:

Theorem 11. *Let $S = ((a_k, b_k))_{k=1}^{\infty}$ be a sequence in $\mathbb{R}^2$. Assume the probability measures (μ_n) are supported on S, and that $\mu_n f(x) = \sum_{k=1}^{\infty} \mu_n(a_k, b_k) f(U_{a_k} V_{b_k} x)$ converges a.e. for all $f \in L^p$ for some $p \geq 1$ and for all commuting measure preserving flows U and V. Then the associated operators*

$$\langle\!\langle \mu_n \rangle\!\rangle f(x) = \sum_{k=1}^{\infty} \mu_n(a_k, b_k) f(U_{\langle\!\langle a_k \rangle\!\rangle} V_{\langle\!\langle b_k \rangle\!\rangle} x)$$

and

$$[\![\mu_n]\!] f(x) = \sum_{k=1}^{\infty} \mu_n(a_k, b_k) f(U_{[\![a_k]\!]} V_{[\![b_k]\!]} x)$$

all converge a.e. for all $f \in L^p(X)$.

Proof. Since the proof is similar to the proof of Theorem 7, we just sketch the main ideas. First we consider the flows U and V defined on $\mathbb{T}^2 = [0,1) \times [0,1)$ by $U_t V_s(x,y) = (x \oplus t, y \oplus s)$. We then have a.e. convergence of $(\langle\!\langle \mu_n \rangle\!\rangle f)_{n=1}^{\infty}$ for all $f \in L^p(\mathbb{T}^2)$. Now let $f \in L^p(\mathbb{R}^2)$ and for each pair of integers (m,n) define a function $g_{m,n}$ on $\mathbb{T}^2$ by $g_{m,n}(x,y) = f(m+x, n+y) + f(m+1+x, n+y) + f(m+x, n+1+y) + f(m+1+x, n+1+y)$. We have a.e. convergence of $(\langle\!\langle \mu_k \rangle\!\rangle g_{m,n})_{k=1}^{\infty}$ for each pair (m,n). This implies a weak type maximal inequality, and hence a.e. convergence for $(\langle\!\langle \mu_n \rangle\!\rangle f)_{n=1}^{\infty}$ for $f \in L^p(\mathbb{R}^2)$, and hence $f \in L^2(X)$ as in the proof of Theorem 7.

To handle the integer part, consider a space X with two commuting measure preserving transformations τ and σ. We define a space $Y = X \times \mathbb{T}^2$ and flows U and V such that $U_t V_s(x,a,b) = (\tau^{[\![t+a]\!]} \sigma^{[\![s+b]\!]}, t \oplus a, s \oplus b)$. On Y define a new function $\tilde{f}(x,a,b)$ by $\tilde{f}(x,a,b) = f(x)$. We now have convergence of $[\![\mu_n]\!] \tilde{f}(x,a,b)$ for a.e. choice of (x,a,b). From this we obtain a maximal inequality for $[\![\mu_n]\!] f(x)$, so it only remains to obtain convergence on a dense subspace of $L^p(X)$. We will be done if we can show convergence for all $f \in L^\infty(X)$. Let $f \in L^\infty(X)$. For a.e. (a,b) we have a.e. convergence of $[\![\mu_n]\!] \tilde{f}(x,a,b)$. Take a sequence (a_k, b_k) which converges to $(0,0)$, and such that for a.e. x we have convergence of $[\![\mu_n]\!] f(x, a_k, b_k)$ We now want to show that $|[\![\mu_n]\!] f(x,a_k,b_k) - [\![\mu_n]\!] f(\tau\sigma x)| \to 0$. This will follow if we have

$$\lim_{\epsilon \to 0} \lim_{n \to \infty} \langle\!\langle \mu_n \rangle\!\rangle([1-\epsilon, 1) \times [0,1) \cup [0,1) \times [1-\epsilon, 1) = 0.$$

To see this, first consider the case when V_t is the identity. Then by our one-dimensional case we have

$$\lim_{\epsilon \to 0} \lim_{n \to \infty} \langle\!\langle \mu_n \rangle\!\rangle([1-\epsilon, 1) \times [0,1)) = 0.$$

In the same way we see that for U_t the identity, we get

$$\lim_{\epsilon \to 0} \lim_{n \to \infty} \langle\!\langle \mu_n \rangle\!\rangle([0,1) \times [1-\epsilon, 1)) = 0.$$

Thus the proof is complete. □

Example. Let α and β be non-zero real numbers, then the sequence

$$([\![\alpha k^2]\!] + [\![\beta k^3]\!])_{k=1}^{\infty}$$

is good for a.e. Cesàro convergence for L^2. To see this let $(a_k, b_k)_{k=1}^{\infty} = (\alpha k^2, \beta k^3)_{k=1}^{\infty}$. By the techniques in the proof of Theorem 10, we know the sequence $(\alpha k^2, \beta k^3)_{k=1}^{\infty}$ is good for a.e. Cesàro convergence for L^2. Consequently the sequence

$$([\![\alpha k^2]\!], [\![\beta k^3]\!])_{k=1}^{\infty}$$

is good for a.e. Cesàro convergence for L^2. Now consider the two dimensional flow $V_{s,t} = U_s U_t = U_{s+t}$, and the result follows.

Denote by $\mathcal{B}^p \subset \mathbb{R}^{\mathbb{Z}^+}$ the set of real sequences which are L^p Cesàro bad, and by $\mathcal{B} \subset \mathbb{R}^{\mathbb{Z}^+}$ the set of real sequences such that the associated Cesàro averages have the strong sweeping out property. Define the box topology to be the topology on $\mathbb{R}^{\mathbb{Z}^+}$ such that the set $S_1 \times S_2 \times \cdots \subset \mathbb{R}^{\mathbb{Z}^+}$ is open whenever each S_i is open in $\mathbb{R}$.

Then we have the following theorem.

Theorem 12. *The sets $\mathcal{B}^p$ and $\mathcal{B}$ are an open dense subsets of $\mathbb{R}^{\mathbb{Z}^+}$ in the box topology.*

Proof. We will show that $\mathcal{B}$ is open, the argument for $\mathcal{B}^p$ being the same. We have to show that if we have a sequence $r = (r_k)_{k=1}^{\infty} \in \mathcal{B}$ then there is a sequence of positive numbers $(\delta_k)_{k=1}^{\infty}$ such that $t = (t_k)_{k=1}^{\infty} \in \mathcal{B}$ whenever $|t_k - r_k| < \delta_k$ for all $k \geq 1$.

Since $(r_k)_{k=1}^{\infty}$ has the strong sweeping out property, given $\eta > 0$, a non-atomic Lebesgue space, (X, Σ, m), and a measure preserving flow, $U_t \colon X \to X$, we can find a set E with $m(E) < \eta$ such that the sequence of averages $C_n^r \chi_E(x) = \frac{1}{n} \sum_{k=1}^{n} \chi_E(U_{r_k} x)$ satisfy $\limsup C_n^r \chi_E(x) = 1$ a.e. and $\liminf C_n^r \chi_E(x) = 0$ a.e. For every $k \geq 1$ choose δ_k such that $\|\chi_E \circ U_s - \chi_E\|_1 < 2^{-k}$ for all $0 \leq s < \delta_k$. Thus if $|t_k - r_k| < \delta_k$ then $\|\chi_E \circ U_{t_k} - \chi_E \circ U_{r_k}\|_1 < 2^{-k}$. Hence, $\sum_{k=1}^{\infty} |\chi_E(U_{t_k} x) - \chi_E(U_{r_k} x)|$ converges for a.e. x. Thus for a.e. x there is $K(x)$ such that if $k \geq K(x)$ then $\chi_E(U_{t_k} x) = \chi_E(U_{r_k} x)$. It follows that we can estimate, for $n > K(x)$, as

$$\left|\frac{1}{n} \sum_{k=1}^{n} \chi_E(U_{t_k} x) - \frac{1}{n} \sum_{k=1}^{n} \chi_E(U_{r_k} x)\right| \leq \frac{1}{n} \sum_{k=1}^{K(x)} |\chi_E(U_{t_k} x) - \chi_E(U_{r_k} x)| \leq \frac{K(x)}{n}.$$

Since $\frac{K(x)}{n} \to 0$, and by assumption, for a.e. x,

$$\limsup_{n\to\infty} \frac{1}{n} \sum_{k=1}^{n} \chi_E(U_{r_k} x) = 1 \text{ and } \liminf_{n\to\infty} \frac{1}{n} \sum_{k=1}^{n} \chi_E(U_{r_k} x) = 0,$$

we have

$$\limsup_{n\to\infty} \frac{1}{n} \sum_{k=1}^{n} \chi_E(U_{t_k} x) = 1 \text{ and } \liminf_{n\to\infty} \frac{1}{n} \sum_{k=1}^{n} \chi_E(U_{t_k} x) = 0,$$

for a.e. x, which shows $(t_k) \in \mathcal{B}$.

To see that $\mathcal{B}$ is dense, note that any sequence whose terms are linearly independent over the rationals has the strong sweeping out property. (See **[JW1]** and **[Aetal]**, also see **[BBB]**.) Such sequences are dense in the box topology. Since $\mathcal{B} \subset \mathcal{B}^p$, we see that $\mathcal{B}^p$ is dense also. □

References

[Aetal] Akcoglu, M., Bellow, A., Jones, R., Losert, V., Reinholt-Larsson, K., and Wierdl, M., The strong sweeping out property for lacunary sequences, Riemann sums, convolution powers, and related matters, Ergodic Theory Dynam. Systems 16 (1996), 1–46.

[AdJL] Akcoglu, M., del Junco, A., and Lee, F., A solution to a problem of A. Bellow, in: Almost Everywhere Convergence II (Evanston, IL, October 1989), A. Bellow and R. Jones, eds., Academic Press, New York 1991.

[B1] Bourgain, J., Almost sure convergence and bounded entropy, Israel J. Math. 63 (1988), 79–97.

[B2] Bourgain, J., On the maximal ergodic theorem for certain subsets of the integers, Israel J. Math. 61 (1988), 39–72.

[B3] Bourgain, J., Pointwise ergodic theorems for arithmetic sets, Inst. Hautes Études Sci. Publ. Math. 69 (1989), 5–45.

[BBB] Bergelson, V., Boshernitzan, M., and Bourgain, J., Some results on non-linear recurrence, J. Analyse Math. 62 (1994), 29–46.

[JW1] Jones, R. and Wierdl, M., Convergence and divergence of ergodic averages, Ergodic Theory Dynam. Systems 14 (1994), 515–535.

[JW2] Jones, R. and Wierdl, M., Convergence of ergodic averages, in: Convergence in Ergodic Theory and Probability, Ohio State Univ. Math. Res. Inst. Publ. 5 (1996), 229–247.

[N] Nair, R., On polynomials in primes and J. Bourgain's circle method approach to ergodic theorems, Ergodic Theory Dynam. Systems 11 (1991), 485–499.

[R] Rosenblatt, J., Universally bad sequences in ergodic theory, in: Almost Everywhere Convergence II (Evanston, IL, October 1989), A. Bellow and R. Jones, eds., Academic Press, New York 1991.

[W1] Wierdl, M., Almost everywhere convergence and recurrence along subsequences in ergodic theory, Ph.D Thesis, Ohio State University, 1989.

[W2] Wierdl, M., Pointwise ergodic theorems along the prime numbers, Israel J. Math. 64 (1988), 315–336.

Department of Mathematics
Rice University
Houston, Texas 77251, U.S.A.

Department of Mathematics
DePaul University
2219 N. Kenmore
Chicago, IL 60614, U.S.A.
MATRLJ@DePaul.Bitnet

Department of Mathematics
Northwestern University
Evanston, IL 60201, U.S.A.

The Uniqueness of Induced Operators

Robert E. Bradley[1]

Abstract. Suppose that T is a positive linear operator of type (p, q) and that there is a function u such that $\|Tu\| = \|T\| \, \|u\|$. Using u and its image as weight functions, we may define an associated operator of type (r, s) for certain choices of r and s. We show that an operator so defined does not depend on the choice of u.

1. Introduction

Suppose that L_p and L_q are the usual L_p spaces and T is a linear operator from L_p to L_q. When T is bounded with norm $\|T\|$ we say T is of ***type*** $\boldsymbol{(p, q)}$. We call $u \in L_p$ a ***norming function*** for T if $\|Tu\|_q = \|T\| \, \|u\|_p$. We say that T is ***positive*** if $Tf \geq 0$ whenever $f \geq 0$. In **[3]** the following was proved:

Theorem 1.1. *Suppose T is a positive operator of type (p, q), $p, q \in [1, \infty)$, with a norming function u, where $u > 0$ and $v = Tu > 0$ a.e. If $1 \leq s \leq r < \infty$ then the equation*

$$Sf = v^{\frac{q}{s}-1} T\left(u^{1-\frac{p}{r}} f\right)$$

defines a positive operator of type (r, s) with $\|S\| = \|T\|^{\frac{q}{s}} \|u\|_p^{\frac{q}{s}-\frac{p}{r}}$.

When $p = q$ and $r = s$, then the operator so defined is independent of the choice of positive norming function u and we call it the L_r-operator ***induced*** by T, denoted T_r. This was proved in **[1]**, and the induced operators were used to prove an alternating sequence theorem, generalizing a theorem of Stein in **[8]** and Rota's alternating sequence theorem **[7]**; see also **[5]** and **[2]** and the historical survey in **[3]**.

When $qr \neq ps$ then the operator depends on $\|u\|$. Moreover, it was demonstrated in **[3]** that when $p = q$ and $r \neq s$, then the operator S depends upon the choice of u even if we consider only norming vectors of unit length.

The primary purpose of this note is to show that when $p \neq q$ and $qr = ps$ (in which case we necessarily have $q < p$), then the operator S is independent of the choice of norming function. This gives an affirmative answer to an open question in **[3]**.

In **[6]**, Kan characterizes the norming functions of L_p-operators. In particular, he shows that when $q < p$, then T has at most one positive norming function, up to

---

1 AMS subject classification: primary 47B38; secondary 47A30, 47A35.

scalar multiplicity (this was shown for a special case in **[3]**). Thus, the operator S in Theorem 1.1 is unique in the case cited above.

Kan's characterization also allows us to consider non-positive norming functions. As long as the function Tu in the definition of S is replaced by its complex conjugate $\overline{Tu}$, the condition $u > 0$ may be loosened considerably when $p = q$ and dropped entirely when $q < p$. This is proved by means of a pointwise comparison of the operator S given by an arbitrary u with the one given by a positive u, whose existence in guaranteed by Theorem 1.1.

2. The case $p > q$

Suppose that $(X, \mathcal{F}, \mu)$ and $(Y, \mathcal{G}, \nu)$ are σ-finite Lebesgue spaces. Let $p, q \in [1, \infty)$ and $L_p = L_p(X, \mathcal{F}, \mu)$ and $L_q = L_q(Y, \mathcal{G}, \nu)$ be the usual Banach spaces. We consider positive bounded linear operators T from L_p to L_q. Let $\mathcal{N}(T)$ denote the set of all norming functions of T. If $\mathcal{N}(T)$ contains anything other than the zero function $\mathbf{0}$ then we say that T is ***norm-attaining***. Since $|Tu| \leq T|u|$ for positive operators, we see that $|u| \in \mathcal{N}(T)$ whenever u is. The ***support*** of a function f is the set $\operatorname{supp} f = \{x \mid f(x) \neq 0\}$. We say that $E \in \mathcal{F}$ is a ***reducing*** set for T if $(Tf)(Tg) = \mathbf{0}$ whenever $\operatorname{supp} f \subseteq E$ and $\operatorname{supp} g \subseteq X - E$. Note that identities such as $(Tf)(Tg) = \mathbf{0}$ are to be understood as statements concerning the associated measure algebras, which are complete.

Let $\mathcal{F}(T) = \{E \in \mathcal{F} \mid E \text{ is reducing}\}$. The composition operator $Tf = f \circ \tau$ satifies $\mathcal{F}(T) = \mathcal{F}$; any operator satisfying this condition is called ***Lamperti***. 1_A is the characteristic function of the set A and $\mathbf{1} = 1_X$. The natural set mapping $A \mapsto \operatorname{supp} T1_A$ of Lamperti operators generalizes to a set mapping $\Phi\colon (\mathcal{F}(T), \mu) \to (\mathcal{G}, \nu)$ given by

$$\Phi A = \sup\{\operatorname{supp} Tf \mid \operatorname{supp} f \subseteq A\}.$$

The following is immediate.

Lemma 2.1. *If $A, B \in \mathcal{F}(T)$ are disjoint, then so are ΦA and ΦB.*

Define $s(T) = \inf\{\operatorname{supp}(u) \mid Tf = \mathbf{0}$ for any f satisfying $uf = \mathbf{0}\}$. We will assume that $s(T) = X$ throughout this paper. The general case follows from this special case by replacing X with $s(T)$ and Y with ΦX.

The following results are in **[6]** as Theorems 2.1 (i) and 4.1 (a).

Theorem 2.2. *$\mathcal{F}(T)$ is a sub-σ-algebra of $\mathcal{F}$ and when $1 \leq q < p < \infty$ then if $\mathbf{0} \neq u \in \mathcal{N}(T)$, we have*

$$\mathcal{N}(T) = \{c\theta u \mid c \geq 0,\ \theta \text{ is } \mathcal{F}(T)\text{-measurable and } |\theta| = \mathbf{1}\}.$$

Thus there is at most one norming function u for T satisfying $u > 0$, up to scalar multiplicity. Therefore when $s < r$ and $qr = ps$ in Theorem 1.1, the operator S is independent of the choice of positive norming function.

In what follows, we understand f^t to be shorthand for the function whose value at x is $\operatorname{sgn}(f(x))|f(x)|^t$, where $\operatorname{sgn}(z)$ is the complex number of unit modulus having the same argument as z.

Theorem 2.3. *Suppose that T is a positive norm-attaining operator of type (p, q), $1 \leq q < p < \infty$. Let $u \in \mathcal{N}(T)$ and $v = (\overline{Tu})$. Suppose that $1 \leq s < \infty$ and $qr = ps$. Then $T_{r,s}$, given by*

$$T_{r,s} f = v^{\frac{p}{r}-1} T\left(u^{1-\frac{p}{r}} f\right) \tag{1}$$

for $f \in L_r$, is a positive operator of type (r, s) independent of the choice of u and satisfying $\|T_{r,s}\| = \|T\|^{\frac{p}{r}}$.

Proof. Let u be the unique positive function of unit norm in $\mathcal{N}(T)$, and let S be the operator given by equation 1 for this choice of u. By Theorem 1.1, S has the desired properties.

By Theorem 2.2, if u' is any other norming function then $u' = c\theta u$ for some $\mathcal{F}(T)$-measurable θ of unit modulus. We may assume $c = 1$ since the formula for $T_{r,s}$ is clearly homogeneous. Let S' be the operator given by equation 1 for u'. We wish to show that for every $f \in L_r$, $Sf = S'f$.

Let $t = 1 - p/r$ and let $A = \theta^{-1}(\alpha)$, where α is any complex number of unit modulus. By Theorem 2.2, $A \in \mathcal{F}(T)$. Let $B = \Phi A$. By Lemma 2.1 it is enough to consider the restrictions of S and S' to the set B.

$$\begin{aligned}
1_B S' f &= \left(\overline{T(1_A u')}\right)^{-t} T(1_A u'^t f) \\
&= 1_B \bar{\alpha} (\overline{Tu})^{-t} \alpha T(u^t f) \\
&= 1_B S f.
\end{aligned}$$

Since $|\theta| = \mathbf{1}$, this completes the proof. □

3. The case $p = q$

When $p = q$, we call an operator of type (p, q) an L_p-operator. An L_p-operator may have various distinct positive unit norming functions, and these need not have full support. In particular, there are non-trivial L_p-isometries.

The following is a consequence of Theorem 3.4 in **[6]**.

Theorem 3.1. *Suppose $p \in (1, \infty)$ and that T is a positive norm-attaining L_p-operator. Then there is a function $\mathbf{0} \neq u_0 \geq 0$ with $\operatorname{supp} u_0 \in \mathcal{F}(T)$ and*

$$\mathcal{N}(T) = \{\, \theta u_0 \mid \theta \in \mathcal{F}'(T) \,\} \cap L_p,$$

where $\mathcal{F}'(T) = \{\, F \in \mathcal{F}(T) \mid F \subseteq \operatorname{supp} u_0 \,\}$.

Consider the direct sum of an isometry on one L_p space with the operator T on the L_p space of the unit interval, where $Tf(x) = xf(x)$, as an example of a norm-attaining operator for which $\operatorname{supp} u_0 \neq X$.

We say u has ***full support*** when $\operatorname{supp} u = X$ (or, more generally, when $\operatorname{supp} u = s(T)$). The following theorem follows from Theorems 1.1 and 3.1 with a proof similar to that of Theorem 2.3. We note that when $p = 1$ the operator need not be unique, see **[4]**.

Theorem 3.2. *Suppose that $p \in (1, \infty)$ and T is a positive norm-attaining L_p-operator. Suppose $u \in \mathcal{N}(T)$ has full support and let $v = (\overline{Tu})$. Suppose that $r \in [1, \infty)$. Then T_r, given by*

$$T_r f = v^{\frac{p}{r}-1} T\left(u^{1-\frac{p}{r}} f\right)$$

for $f \in L_r$, is a positive L_r-operator independent of the choice of u and satisfying $\|T_r\| = \|T\|^{\frac{p}{r}}$.

References

[1] M. A. Akcoglu, R. E. Bradley, Alternating sequences and induced operators, Trans. Amer. Math. Soc. 325 (1991), 765–791.

[2] M. A. Akcoglu, L. Sucheston, Pointwise convergence of alternating sequences, Canad. J. Math. 40 (1988), 610–632.

[3] R. E. Bradley, On induced operators, Canad. J. Math. 43 (1991), 477–494.

[4] R. E. Bradley, Concerning induced operators and alternating sequences, in: Almost Everywhere Convergence II, A. Bellow, R. L. Jones, eds., Academic Press, Boston 1991.

[5] D. L. Burkholder and Y. S. Chow, Iterates of conditional expectation operators, Proc. Amer. Math. Soc. 12 (1961), 490–495.

[6] C. H. Kan, Norming vectors of linear operators between L_p spaces, Pacific J. Math. 150 (1991), 309–327.

[7] G. C. Rota, An "alternierende Verfahren" for general positive operators, Bull. Amer. Math. Soc. 68 (1962), 95–102.

[8] E. M. Stein, On the maximal ergodic theorem, Proc. Nat. Acad. Sci. USA 47 (1961), 1894–1897.

Department of Mathematics & Computer Science
Adelphi University
Garden City, NY 11530, U.S.A.
e-mail: bradley@panther.adelphi.edu

On Convergence of Partial Sums of Independent Random Variables

Nasrollah Etemadi

Abstract. We will show how the removal of *centering and/or moment assumptions* from Kolmogorov's maximal inequalities will lead to a unified treatment of classical convergence theorems for partial sums of independent random variables (random vectors taking values in separable Banach space).

Let $\{ X_n : n \geq 1 \}$ be a sequence of independent *random variables* taking values in a separable Banach space $\boldsymbol{B}$. Let $S_n = \sum_{i=1}^{n} X_i$. We have shown recently, see Etemadi (1987, 1992), how the maximal inequalities in Etemadi (1987) can be used to obtain necessary and sufficient conditions for L_p-convergence, $p \geq 0$, of sums of independent random vectors in the general set up that was originally formulated for the weak law of large numbers—the case where$p = 0$—by Kolmogorov and Feller. In this work first we will give a slightly weaker version of one of our maximal inequalities. This inequality is much simpler to prove but apparently as effective as the old one. Then we will *complete* our previous work by showing, among other things, that weak convergence of S_n implies its convergence in probability, thus showing how the removal of centering and/or moment assumptions from Kolmogorov's maximal inequalities, see Loève(1963), p. 235, can lead to a unified treatment of convergence results for sums of independent random vectors. The fact that weak convergence of sums of random vectors implies its convergence in probability is well known, see Tortrat (1965), p. 227, or Ito and Nisio (1968), p. 37. The proof for the real-valued random variables can be found in almost any textbook, see e.g. Dudley (1989). Our method has the advantage that it avoids the method of characteristic functions (functionals).

The basic inequalities, in Lemma 1 and Lemma 2, are presented here in a more general fashion than they are needed in the proofs. This is done not for aiming at generalization, but because we simply get them for free.

Let $(\Omega, \mathcal{F}, \mathbb{P})$ be the underlying probability space. Let $(\boldsymbol{S}, +, \mathcal{S})$ be an additive (abelian) group with a σ-field $\mathcal{S}$ such that $+ : \boldsymbol{S} \times \boldsymbol{S} \to \boldsymbol{S}$ is measurable with respect to the product σ-field. Let $\| \, \| : \boldsymbol{S} \to [0, \infty]$ be a measurable function such that for some constant α and all $x, y \in \boldsymbol{S}$,

$$\begin{aligned} &(a) \quad \|0\| = 0, \\ &(b) \quad \|x + y\| \leq \alpha(\|x\| + \|y\|), \qquad (1) \\ &(c) \quad \|x\| = \| - x\|. \end{aligned}$$

We will also assume, to avoid trivialities, that for some $x_0 \in \boldsymbol{S}$, $\|x_0\| \in (0, \infty)$. This assumption together with (b) implies that $\alpha \geq 1$ and for $n \geq 2$,

$$\|\sum_{i=1}^{n} x_i\| \leq \sum_{i=1}^{n-1} \alpha^i \|x_i\| + \alpha^{n-1} \|x_n\|. \tag{2}$$

The proof of part (a) of the following lemma is a slight improvement of the one in Etemadi (1985) or Billingsley (1986), p. 297. Although the inequality in part (b) is weaker than the one given in Lemma 1.4 in Etemadi (1987), but its effectiveness in applications seems to be the same. For instance all the L_p-convergence theorems in Etemadi(1992) are also consequences of the following lemma.

Lemma 1. *Let* $\{X_i : 1 \leq i \leq n\}$, $n \geq 1$, *be independent* $\boldsymbol{S}$*-valued random variables and* $S_i = \sum_{k=1}^{i} X_k$, $1 \leq i \leq n$. *Then for all* $t > 0$,

(a) $\mathbb{P}\{ \max_{1\leq i\leq n} \|S_i\| > 3\alpha^2 t \} \leq 3 \max_{1\leq i\leq n} \mathbb{P}\{ \|S_i\| > t \}$,

(b) $(1 - 2\alpha^2 \mathbb{P}\{ \max_{1\leq i\leq n} \|S_i\| > t \})\mathbb{E} \max_{1\leq i\leq n} \|S_i\| \leq 2\alpha^2 (t + \mathbb{E} \max_{1\leq i\leq n} \|X_i\|)$.

Proof. Let $t > 0$ and define $S_0 = 0$. On the set $\{ 3\alpha^2 t < \|S_j\|, \|S_n\| \leq \alpha t \}$ for $1 \leq j < n$ we have

$$\begin{aligned} 3\alpha^2 t < \|S_j\| = \|S_j - S_n + S_n\| &\leq \alpha(\|S_j - S_n\| + \alpha t) \\ &= \alpha \|S_n - S_j\| + \alpha^2 t. \end{aligned} \tag{3}$$

Therefore

$$\begin{aligned}
&\mathbb{P}\{ \max_{1\leq i\leq n} \|S_i\| > 3\alpha^2 t, \|S_n\| \leq \alpha t \} \\
&\qquad = \sum_{j=1}^{n} \mathbb{P}\{ \max_{0\leq i<j} \|S_i\| \leq 3\alpha^2 t < \|S_j\|, \|S_n\| \leq \alpha t \} \\
(\text{ by (3) }) &\qquad \leq \sum_{j=1}^{n} \mathbb{P}\{ \max_{0\leq i<j} \|S_i\| \leq 3\alpha^2 t < \|S_j\|, \|S_n - S_j\| > 2\alpha t \} \\
(\text{ by independence }) &\qquad = \sum_{j=1}^{n} \mathbb{P}\{ \max_{0\leq i<j} \|S_i\| \leq 3\alpha^2 t < \|S_j\| \}\mathbb{P}\{ \|S_n - S_j\| > 2\alpha t \}
\end{aligned}$$

$$\begin{aligned}
&\leq \max_{1\leq j<n} \mathbb{P}\{ \|S_n - S_j\| > 2\alpha t \}\mathbb{P}\{ \max_{1\leq i\leq n} \|S_i\| > 3\alpha^2 t \} \\
&\leq \max_{1\leq j<n} \mathbb{P}\{ \|S_n - S_j\| > 2\alpha t \} \\
&\leq \max_{1\leq j<n} \mathbb{P}\{ \|S_n\| + \|S_j\| > 2t \} \\
&\leq \mathbb{P}\{ \|S_n\| > t \} + \max_{1\leq j\leq n} \mathbb{P}\{ \|S_j\| > t \}.
\end{aligned} \tag{4}$$

Consequently

$$\begin{aligned}
\mathbb{P}\{ \max_{1\le i\le n} \|S_i\| > 3\alpha^2 t \} &\le \mathbb{P}\{ \max_{1\le i\le n} \|S_i\| > 3\alpha^2 t,\ \|S_n\| \le \alpha t \} + \mathbb{P}\{ \|S_n\| > \alpha t \} \\
(\alpha \ge 1) \qquad &\le \max_{1\le i\le n} \mathbb{P}\{ \|S_i\| > t \} + 2\mathbb{P}\{ \|S_n\| > t \} \\
&\le 3 \max_{1\le i\le n} \mathbb{P}\{ \|S_i\| > t \}.
\end{aligned} \tag{5}$$

To see part (b), first note that if $\mathbb{E} \max_{1\le i\le n} \|X_i\| = \infty$ then the inequality is obvious. Otherwise for $t > 0$ define $A_n = \{ \max_{1\le i\le n} \|S_i\| > t \}$. As usual consider the event $B_j = \{ \max_{0\le i<j} \|S_i\| \le t < \|S_j\| \}$, i.e. the first time the process $\{ S_i : i \ge 1 \}$ reaches beyond t. Then

$$\begin{aligned}
\max_{1\le i\le n} \|S_i\| I_{B_j} &\le [\max_{0\le i<j} \|S_i\| + \max_{j\le i\le n} \|S_i\|] I_{B_j} \\
&\le [t + \max_{j\le i\le n} \|S_i - S_j + S_{j-1} + X_j\|] I_{B_j} \\
&\le [t + \alpha \max_{j\le i\le n} \|S_i - S_j\| + \alpha^2 (t + \|X_j\|)] I_{B_j} \\
&\le [(1+\alpha^2)t + \alpha^2 \max_{1\le i\le n} \|X_i\| + \alpha \max_{j\le i\le n} \|S_i - S_j\|] I_{B_j}.
\end{aligned} \tag{6}$$

Therefore by independence

$$\begin{aligned}
\mathbb{E}(\max_{1\le i\le n} \|S_i\| I_{B_j}) &\le [(1+\alpha^2)t + \alpha \mathbb{E} \max_{j\le i\le n} \|S_i - S_j\|] \mathbb{P}(B_j) + \alpha^2 \mathbb{E}(\max_{1\le i\le n} \|X_i\| I_{B_j}) \\
&\le [(1+\alpha^2)t + 2\alpha^2 \mathbb{E} \max_{1\le i\le n} \|S_i\|] \mathbb{P}(B_j) + \alpha^2 \mathbb{E}(\max_{1\le i\le n} \|X_i\| I_{B_j}).
\end{aligned} \tag{7}$$

Since $A_n = \cup_{j=1}^n B_j$, and B_j 's are disjoint,

$$\begin{aligned}
\mathbb{E} \max_{1\le i\le n} \|S_i\| &= \mathbb{E}(\max_{1\le i\le n} \|S_i\| I_{A_n^c}) + \mathbb{E}(\max_{1\le i\le n} \|S_i\| I_{A_n}) \\
&\le t\mathbb{P}(A_n^c) + (1+\alpha^2)t\mathbb{P}(A_n) + \alpha^2 \mathbb{E} \max_{1\le i\le n} \|X_i\| + 2\alpha^2 \mathbb{P}(A_n) \mathbb{E} \max_{1\le i\le n} \|S_i\| \\
&\le 2\alpha^2 [t + \mathbb{P}(A_n) \mathbb{E} \max_{1\le i\le n} \|S_i\| + \mathbb{E} \max_{1\le i\le n} \|X_i\|],
\end{aligned} \tag{8}$$

and because of (2) $\mathbb{E} \max_{1\le i\le n} \|S_i\| < \infty$ and we are through. □

Note that by Chebyshev's inequality, when it is needed, and the fact that $\max_{1\le i\le n} \mathbb{E}\|S_i\| \le \mathbb{E} \max_{1\le i\le n} \|S_i\|$,

$$\frac{1}{2\alpha^2}\left(1 - 2\alpha^2 \frac{t + \mathbb{E} \max_{1\le i\le n} \|X_i\|}{\mathbb{E} \max_{1\le i\le n} \|S_i\|}\right) \le \mathbb{P}\{ \max_{1\le i\le n} \|S_i\| > t \} \le \frac{\mathbb{E} \max_{1\le i\le n} \|S_i\|}{t}, \tag{9}$$

$$\frac{1}{2\alpha^2}\left(1 - 2\alpha^2 \frac{t + \mathbb{E} \max_{1\le i\le n} \|X_i\|}{\max_{1\le i\le n} \mathbb{E}\|S_i\|}\right) \le \mathbb{P}\{ \max_{1\le i\le n} \|S_i\| > t \} \le 16\alpha^2 \frac{\max_{1\le i\le n} \mathbb{E}\|S_i\|}{t}. \tag{10}$$

We also need upper and lower bounds for the tail probability of the random variable $\max_{1\le i\le n} \|X_i\|$. The following lemma not only provides us with such bounds, but it also has other nice applications, e.g. it gives us the Borel–Cantelli lemma for pairwise independent random variables. For the proof and other applications see Etemadi (1984, 1985).

Lemma 2. *Let* $\{ A_i : 1 \le i \le n \}$, $n \ge 1$, *be a set of pairwise independent events. Then*

$$\frac{\sum_{i=1}^n \mathbb{P}(A_i)}{1 + \sum_{i=1}^n \mathbb{P}(A_i)} \le \mathbb{P}\left(\bigcup_{i=1}^n A_i \right) \le \sum_{i=1}^n \mathbb{P}(A_i).$$

In order to prove that convergence of S_n 's in distribution (weak convergence) implies their convergence in probability, we need to know when $\sup_{n\ge 1} \|S_n\| = \infty$ when the X_n 's are i.i.d., see Theorem 1. The following propositions show us how Lemma 2 and the inequalities in (9, 10) can easily provide us with necessary and sufficient conditions for $\sup_{n\ge 1} \|S_n\| < \infty$ a.e. in general, i.e. where X_n 's are independent but not necessarily identically distributed. Of course, Kolmogorov 0-1 law will then take care of the other case where $\sup_{n\ge 1} \|S_n\| = \infty$ a.e. We resort, as usual, to the method of truncation. Namely, for $c > 0$ and $n \ge 1$ define

$$X_{n,c} = X_n I\{ \|X_n\| \le c \} \ ; \ S_{n,c} = \sum_{i=1}^n X_{i,c}. \tag{11}$$

Proposition 1. *Let* $\{ X_n : n \ge 1 \}$, *be a sequence of independent* **S**-*valued random variables. Let* $S_n = \sum_{i=1}^n X_i$, $S_{n,c}$ *be defined as in* (11) *and* $p > 0$. *Then* $\sup_{n\ge 1} \|S_n\| < \infty$ *a.e. if and only if for some* $c > 0$, *independent of* p,

$$\text{(a)} \ \sum_{n=1}^\infty \mathbb{P}\{ \|X_n\| > c \} < \infty \ ; \ \text{(b)} \ \sup_{n\ge 1} \mathbb{E}\|S_{n,c}\|^p < \infty.$$

Furthermore if (a) *and* (b) *are true for some* $p > 0$, *then they are true for all* $p > 0$.

Proof. Assume $\sup_{n\ge 1} \|S_n\| < \infty$ a.e. Since

$$\sup_{n\ge 1} \|X_n\| = \sup_{n\ge 1} \|S_n - S_{n-1}\| \le 2\alpha \sup_{n\ge 1} \|S_n\|, \tag{12}$$

$\sup_{n\ge 1} \|X_n\| < \infty$ a.e. Now let in Lemma 2, $A_i = \{ \|X_i\| > c \}, i \ge 1$, and let $c > 0$ be large enough so that $\mathbb{P}\{ \sup_{n\ge 1} \|X_n\| > c \} < 1$. Thus not only (a) holds for the same c, but by the Borel–Cantelli lemma we also have

$$\sup_{n\ge 1} \|S_{i,c}\|^p < \infty \ a.e. \tag{13}$$

Therefore by (10), $\sup_{n\ge 1} \mathbb{E}\|S_{i,c}\|^p$ can not be infinite which gives us part (b).

Next suppose (a) and (b) are true. (b), via (10), clearly implies (13) and because of (a) and by Borel–Cantelli again, we have $\sup_{n\geq 1} \|S_n\| < \infty$ a.e. □

For a random variable X taking values in $\boldsymbol{S}$ if the notion of "expectation" is well defined such that $\|\mathbb{E}X\| \leq \mathbb{E}\|X\|$, we may also define the variance of this random variable by $\operatorname{Var} X = \mathbb{E}\|X - \mathbb{E}X\|^2$. We could then express condition (b) in Proposition 1 in terms of expectations and variances by simply replacing (b) by

$$\text{(c)}\ \sup_{n\geq 1} \|\mathbb{E}S_{n,c}\| < \infty\,; \ \text{(d)}\ \sup_{n\geq 1} \operatorname{Var} S_{n,c} < \infty. \tag{14}$$

This can be certainly done when $\boldsymbol{S}$ is a separable Banach space with norm $\|\ \|$ in general or a separable Hilbert space in particular. $\mathcal{S}$ is then taken as the Borel σ-fields and the *expectation* can be defined by the Bochner integral (see e.g. Dunford and Schwartz (1967), Chapter III). For Hilbert space-valued random variables we can rewrite our proposition as follows:

Proposition 2. *Let* $\{X_i : i \geq 1\}$, *be a sequencs of independent random variables taking values in a separable Hilbert space. Let* $S_n = \sum_{k=1}^n X_k$, $n \geq 1$. *Then*

$$\sup_{i\geq 1} \|S_i\| < \infty \ a.e. \quad \text{iff} \quad \begin{array}{ll} \text{(a)} & \sup_{n\geq 1} \sum_{i=1}^n \mathbb{P}\{\,\|X_i\| > c\,\} < \infty, \\ \text{(b)} & \sup_{n\geq 1} \|\sum_{i=1}^n \mathbb{E}(X_i I\{\,\|X_i\| \leq c\,\})\| < \infty, \\ \text{(c)} & \sup_{n\geq 1} \sum_{i=1}^n \operatorname{Var}(X_i I\{\,\|X_i\| \leq c\,\}) < \infty, \end{array}$$

for some $c > 0$. □

Now we are in a position to "complete" our work in the sense we explained in the introduction. Again, let $\{X_i : i \geq 1\}$ be a sequence of independent *random variables* taking values in a separable Banach space $\boldsymbol{S}$. Let $S_n = \sum_{i=1}^n X_i$. It is known that the weak convergence of S_n implies its convergence in probability (see Tortrat (1965), p. 227, Ito and Nisio (1968), p. 37 or Araujo and Giné (1980), p. 105). The proof for real-valued random variables can be found in almost any textbook see, e.g. Dudley (1989). The proof given here is based on our inequalities and thus it avoids the method of characteristic functions (functionals).

Theorem 1. *Let* $\{X_i : i \geq 1\}$ *be a sequence of i.i.d. random variables taking values in a separable Banach space. Let* $S_n = \sum_{i=1}^n X_i$ *be its partial sums. Then either* $X_1 = 0$ *a.e. or* $\sup_{n\geq 1} \|S_n\| = \infty$ *a.e.* .

Proof. The real-valued case can be handled readily by Proposition 2. Namely, if $\sup_{n\geq 1} |S_n| = \infty$ a.e. fails, by the Kolmogorov 0-1 law $\sup_{n\geq 1} |S_n| < \infty$ a.e. and (a) to (c) of Proposition 2 imply that for some $c > 0$, $|X_1| \leq c$ a.e., $\mathbb{E}X_{1,c} = 0$, and $\mathbb{E}X_{1,c}^2 = 0$ respectively.Therefore $X_1 = 0$ a.e. For an alternative proof see Chung (1974), p. 264. For the general case, note that if $f(X_1) = 0$ a.e. for all f in S^*, the dual space of S, then because S is separable, there is a countable number of linear functionals $\{ f_i \}$ in the unit ball of S^* such that for all $x \in S$, $\|x\| = \sup |f_i(x)|$. Thus we get $X = 0$ a.e. □

Theorem 2. *Let X and Y be two independent random variables taking values in a separable Banach space. If $X + Y$ and X have the same distribution, then $Y = 0$ a.e.*

Proof. Let $\{ Y_i : i \geq 1 \}$ be a sequence of independent random variables having the same distribution as Y. Let Y_i's be also independent of X. Let $T_n = \sum_{i=1}^n Y_i$, $n \geq 1$ be its partial sum. Then clearly $X + T_n$ and X have the same distribution. Assume Y is non-degenarate at zero. Then by Lemma 1 and Theorem 1, for all $t > 0$, we have

$$\begin{aligned} 1 = \mathbb{P}\{ \sup_{n \geq 1} \|T_n\| > 6t \} &\leq 3 \sup_{n \geq 1} \mathbb{P}\{ \|T_n\| > 2t \} \\ &\leq 3 \sup_{n \geq 1} \mathbb{P}\{ \|X + T_n\| + \| - X\| > 2t \} \\ &\leq 3(\sup_{n \geq 1} \mathbb{P}\{ \|X + T_n\| > t \} + \mathbb{P}\{ \|X\| > t \}) \\ &= 6\,\mathbb{P}\{ \|X\| > t \}, \end{aligned} \tag{15}$$

which is obviously impossible. □

The following theorem is an obvious consequence of Skorohod's theorem which states that in some sense a weakly convergent sequence of random variables taking values in a complete separable metric space can be replaced by one that converges a.e., see Dudley (1989), p. 325. Although we recommend using Skorohod's theorem for real-valued random variables, where the proof is easy (see Billingsley (1986), p. 343) one can bypass this theorem altogether, thanks to Jim Dai, as follows:

Theorem 3. *Let $\{U_i : i \geq 1\}$ and $\{ V_i : i \geq 1\}$ be two independent sequences of random variables taking values in a separable Banach space. If U_n and V_n converge weakly to U and V respectively, then U and V can be chosen to be independent and $U_n + V_n$ converges weakly to $U + V$.*

Proof. Since μ_{U_n} and μ_{V_n} the measures corresponding to U_n and V_n are tight, given $\epsilon > 0$, we can find a compact set K such that $\mu_{U_n}(K) \geq 1 - \epsilon$ and $\mu_{V_n}(K) \geq 1 - \epsilon$ for all n. Let f be a bounded continuous function on the Banach space. For all $\boldsymbol{x}$ in the Banach space define, $g_n(\boldsymbol{x}) = Ef(U_n + \boldsymbol{x})$ and $g(\boldsymbol{x}) = Ef(U + \boldsymbol{x})$. Since

$$\begin{aligned} |g_n(\boldsymbol{y}) - g_n(\boldsymbol{z})| &\leq \int_K |f(\boldsymbol{x}+\boldsymbol{y}) - f(\boldsymbol{x}+\boldsymbol{z})|\, d\mu_{U_n}(\boldsymbol{x}) + \int_{K^c} |f(\boldsymbol{x}+\boldsymbol{y}) - f(\boldsymbol{x}+\boldsymbol{z})|\, d\mu_{U_n}(\boldsymbol{x}) \\ &\leq \int_K |f(\boldsymbol{x}+\boldsymbol{y}) - f(\boldsymbol{x}+\boldsymbol{z})|\, d\mu_{U_n}(\boldsymbol{x}) + 2\,\|f\|\epsilon, \end{aligned} \tag{16}$$

and f is uniformly continuous on K, g_n is equicontinuous on K and we have

$$\begin{aligned}|Ef(U_n+V_n)-Ef(U+V)| &= \Big|\int g_n(\boldsymbol{x})\,d\mu_{V_n}(\boldsymbol{x})-\int g(\boldsymbol{x})d\,\mu_V(\boldsymbol{x})\Big| \\ &\le \int |g_n(\boldsymbol{x})-g(\boldsymbol{x})|\,d\mu_{V_n}(\boldsymbol{x}) \\ &\quad +\Big|\int g(\boldsymbol{x})\,d\mu_{V_n}(\boldsymbol{x})-\int g(\boldsymbol{x})\,d\mu_V(\boldsymbol{x})\Big| \qquad (17)\\ &\le \int_K |g_n(\boldsymbol{x})-g(\boldsymbol{x})|\,d\mu_{V_n}(\boldsymbol{x})+2\,\|f\|\epsilon \\ &\quad +\Big|\int g(\boldsymbol{x})\,d\mu_{V_n}(\boldsymbol{x})-\int g(\boldsymbol{x})\,d\mu_V(\boldsymbol{x})\Big|,\end{aligned}$$

where μ_V is the measure induced on the Banach space by V. Now because $g_n(\boldsymbol{x})$ clearly converges to $g(\boldsymbol{x})$ pointwise, by equicontinuity it converges uniformly and we have the desired result. □

Theorem 4. *Let* $\{X_i : i \geq 1\}$ *be a sequence of independent random variables taking values in a separable Banach space. Let* $S_n = \sum_{i=1}^{n} X_i$, $n \geq 1$. *Then the weak convergence of* S_n *implies its convergence in probability.*

Proof. Let $\{m_i : i \geq 1\}$ and $\{n_i : i \geq 1\}$ be two arbitrary subsequences of positive integers such that $m_i < n_i$, $i \geq 1$. It suffices to show that $S_{n_i} - S_{m_i}$ converges to zero in probability. For $i \geq 1$, define $U_i = S_{n_i} - S_{m_i}$ and $V_i = S_{m_i}$. Since $U_n + V_n$ and V_n converge weakly and furthermore U_n and V_n are independent, U_n is tight (see Araujo and Giné (1980), p. 26). Next assume U_{n_k} converges weakly to U for a subsequence of the positive integers and S_n converges weakly to V. Then, S_n converges weakly to both V and $U+V$ with U and V independent. Consequntly $U+V$ has the same distribution as V and by Theorem 3, $U = 0$ a.e., which implies that U_n converges to zero weakly. Hence U_n also converges to zero in probability (see Dudley (1989), p. 305) and we are through. □

Note that by Lemma 1 for all $t > 0$,

$$\begin{aligned}\lim_{n\to\infty}\sup_{m\geq n}\mathbb{P}\{\,\|S_m-S_n\|>3t\,\} &\le \lim_{n\to\infty}\mathbb{P}\{\,\sup_{m\geq n}\|S_m-S_n\|>3t\,\}\\ &\le 3\lim_{n\to\infty}\sup_{m\geq n}\mathbb{P}\{\,\|S_m-S_n\|>t\,\}. \qquad (18)\end{aligned}$$

Therefore S_n converges in probabitity iff it converges a.e. Finally we know almost everywhere convergence trivially implies weak convergence, and thus for S_n the notion of convergence a.e., in probability, and in distribution are *equivalent*.

References

[1] Araujo, A and Giné E. (1980), The Central Limit Theorm for Real and Banach Valued Random Variables, John Wiley & Sons, New York.

[2] Billingsley, P. (1986), Probability and Measure, 2d ed., John Wiley & Sons, New York.

[3] Chung, K. L. (1974), A Course in Probability Theory, 2d ed., Academic Press, New York.

[4] Dudley R. M. (1989), Real Analysis and Probability, Wadsworth & Brooks/Cole, Pacific Grove, California.

[5] Dunford, N. and Schwartz, J. T. (1967), Linear Operators, Vol. I, Interscience, New York.

[6] Etemadi, N. (1984), On the maximal inequalities for the averages of pairwise i.i.d. random variables, Comm. Statist. Theory Methods 13, 2749–2756.

[7] Etemadi, N. (1985), On some classical results in probability theory, Sankhyā, Series A, Vol. 47, part 2, 215–221.

[8] Etemadi, N. (1987), On sums of independent random vectors, Commun. Statist. Theory Methods 16, 241-252.

[9] Etemadi, N. (1992), Necessary and sufficient conditions for L^p-convergence of sums of independent random variables, Preprint.

[10] Ito, K. and Nisio, M. (1968), On the convergence of sums of independent Banach space valued random variables, Osaka J. Math. 5, 35–48.

[11] Loève, M. (1963), Probability Theory, 3rd ed., Van Nostrand Company, New York.

[12] Tortrat, A. (1965), Lois de probabilité sur un espace topologique complètement régulier et produits infinis à termes indépendants dans un groupe topologique, Ann. Inst. H. Poincaré Probab. Statist. 1, 217–237.

Dept. of Mathematics, Statistics & Computer Science
University of Illinois at Chicago
8515 Morgan
Chicago, IL 60607-7045, U.S.A.

Harmonic Analysis of Operators Associated with a Multiparameter Group of Dilations

Robert A. Fefferman

Abstract. This article is concerned with one of the simplest problems concerning classes of multiplier operators on $\mathbb{R}^n$ which are invariant under the action of a multiparameter group of dilations. This is, in some sense, the "next step" after the theory of product space operators, whose theory is now relatively well understood, and represents some recent joint work of Jill Pipher and the author.

1. Introduction

At present the properties of operators from harmonic analysis such as singular integrals, maximal functions, or multipliers invariant with respect to the action of the usual dilation group on $\mathbb{R}^n$ are well understood. The sharp estimates for Riesz transforms and their generalizations, Hörmander multipliers, and the Hardy–Littlewood Maximal Function have been known for many years now. If, instead of the usual dilations of $\mathbb{R}^n$, we consider the group of product dilations,

$$\rho_{\delta_1,\delta_2,\ldots,\delta_n} : (x_1, x_2, \ldots, x_n) \to (\delta_1 x_1, \delta_2 x_2, \ldots, \delta_n x_n), \quad \delta_i > 0, \quad i = 1, 2, \ldots, n,$$

then the situation is quite different. Operators invariant under this group, such as the multiple Hilbert transform and its generalizations, Marcinkiewicz multipliers, and the Strong Maximal Function, have been investigated intensely and are just now starting to be understood with a picture approaching the beautiful simplicity and completeness of the classical theory for the usual dilations. The product dilations are the simplest example of a group of dilations indexed by several parameters. The operator theory for the other multiparameter dilation groups remains at a rather primitive level. Here, in this article, we shall consider what is perhaps the next simplest example of dilations, as follows: Consider, in $\mathbb{R}^3$, the dilations $\rho_{\delta_1,\delta_2}(x, y, z) = (\delta_1 x, \delta_2 y, \delta_1\delta_2 z)$, $\delta_1, \delta_2 > 0$.[1] Then rather little is known about the associated operators. As far as the maximal function is

1 Our methods extend easily to some other groups of dilations, for example, in $\mathbb{R}^n$, the group $\rho_{\delta_1,\delta_2,\ldots,\delta_{n-1}}(x_1, x_2, \ldots, x_n) = (\delta_1 x_1, \delta_2 x_2, \ldots, \delta_{n-1}x_{n-1}, (\delta_1^{a_1}\delta_2^{a_2}\ldots\delta_{n-1}^{a_{n-1}})x_n)$, $\delta_1, \delta_2, \ldots, \delta_n > 0$. Here $a_i > 0$ for $1 \le i \le n-1$.

concerned, we are considering

$$\mathrm{M}_{\mathfrak{z}}(f)(x,y,z) = \sup_{\substack{(x,y,z)\in R \\ R\in\mathcal{R}_{\mathfrak{z}}}} \frac{1}{|R|}\int_R |f|$$

where $\mathcal{R}_{\mathfrak{z}}$ denotes the family of all rectangles in $\mathbb{R}^3$ whose sides are parallel to the coordinate axes and with side lengths of the form s (in the x direction), t (in the y direction), and st (in the z direction). Of course it is immediate that

$$\mathrm{M}_{\mathfrak{z}} f \le \mathrm{M_S} f$$

if $\mathrm{M_S}$ is the Strong Maximal operator on $\mathbb{R}^3$, and so $\mathrm{M}_{\mathfrak{z}}$ is a bounded operator on $L^p(R^3)$ for all $1 < p \le \infty$. What is much more interesting is the proof of the so-called "Zygmund Conjecture" by A. Cordoba **[1]**, namely that for f supported on the unit cube, Q, of $\mathbb{R}^3$, we have the sharp estimate

$$m\{(x,y,z)\in Q \,|\, \mathrm{M}_{\mathfrak{z}}(f)(x,y,z) > \alpha\} \le \frac{C}{\alpha}\|f\|_{L\log^+ L(Q)}.$$

This estimate certainly does not follow from the obvious domination of the operator $\mathrm{M}_{\mathfrak{z}}$ by $\mathrm{M_S}$, since it depends crucially upon the two parameter nature of the dilations under which $\mathrm{M}_{\mathfrak{z}}$ is invariant. Here we shall consider the singular integral or more precisely, multipliers invariant under the $\rho_{\delta_1,\delta_2}(x,y,z)$. These are defined as follows: We shall consider a "fundamental domain," $A \subseteq \mathbb{R}^3$ for these dilations. This means that the regions $A_{k,j} = \rho_{2^k,2^j}(A)$ are pairwise disjoint for $j,k \in \mathbb{Z}$ and $\mathbb{R}^3 = \bigcup_{k,j\in\mathbb{Z}} A_{k,j}$. Then we shall consider a smooth cutoff function η which is a smooth version of χ_A and consider the class $\mathcal{M}_{\mathfrak{z}}$ of multipliers on $\mathbb{R}^3$ satisfying

$$|\partial^\alpha(m\eta)(\xi)| \le C_\alpha \qquad \text{for all multi-indices } \alpha \text{ and all } \xi \in \mathbb{R}^3 \tag{$*$}$$

(C_α depends only on α, not ξ). We shall also assume that $(*)$ is satisfied uniformly by the dilates m_{δ_1,δ_2} of m given by $m_{\delta_1,\delta_2} = m \circ \rho_{\delta_1,\delta_2}$. It turns out that one can view operators in $\mathcal{M}_{\mathfrak{z}}$ as three parameter product operators so that the operators corresponding to multipliers in $\mathcal{M}_{\mathfrak{z}}$ are bounded in $L^p(R^3)$ for all $1 < p < \infty$ (R. Fefferman–Stein **[6]**). This is the analogue, for singular integrals, of the domination of $\mathrm{M}_{\mathfrak{z}}$ by $\mathrm{M_S}$. In this article, our aim is to obtain the simplest results possible for the multipliers in $\mathcal{M}_{\mathfrak{z}}$ which cannot be obtained directly by viewing the corresponding operators as product operators. The simplest issue of this type that we know is the weighted estimates for the multipliers. Specifically, suppose that we define a Muckenhoupt class $A^p(\mathfrak{z})$, adjusted to be invariant under the dilations at hand: Say that $w(x,y,z) \in A^p(\mathfrak{z})$ provided

$$\left(\frac{1}{|R|}\int_R w\right)\cdot\left(\frac{1}{|R|}\int_R w^{-1/(p-1)}\right)^{p-1} \le C$$

for all rectangles R in $\mathbb{R}^3$ with sides parallel to the axes and side lengths of the form s, t, and st. Then our main result here, obtained jointly with J. Pipher, is the following:

Theorem. *Suppose $m \in \mathcal{M}_3$ and $1 < p < \infty$. Set*

$$Tf\hat{}(\xi) = m(\xi) \cdot \hat{f}(\xi), \qquad f \in C_0^\infty(\mathbb{R}^3).$$

Then if $w \in A^p(\mathfrak{z})$ we have the a priori estimate

$$\|T(f)\|_{L^p(w)} \leq C\|f\|_{L^p(w)}$$

where C depends only on p and $\|w\|_{A^p(\mathfrak{z})}$.

This is just essentially the appropriate version of the Hunt–Muckenhoupt–Wheeden Theorem **[7]** in the case of our two parameter group of dilations.

The method of proof makes use of the Calderón–Zygmund philosophy in two basic ways: (1) The L^2 theory is somehow considerably more elementary than the L^p $(p \neq 2)$ theory, so that the Hilbert space case is done first. (2) The relevant maximal operator, $M_\mathfrak{z}$, must be understood and then its properties applied to pass from the L^2 to the L^p results for the multiplier. Our method, more specifically, is to introduce the right Littlewood–Paley area integral (adapted to the dilation group we are considering) and to reduce the desired weighted estimates on L^2 for the multiplier to those for the area integral, $S(f)$. The key point which explains why the case $p = 2$ is special is that the weighted L^2 theory for $S(f)$ may be obtained by a simple iteration argument, while the weighted L^p theory cannot be. (This stands in marked contrast to the case of product operators, where the $L^p(w)$ theory is obtained directly by iteration, for any value of p). We next pass from Hilbert space to the L^p spaces by using the theory of the maximal operator. In fact, according to the Extrapolation Theorem (Rubio de Francia **[11]**), it is precisely the weighted L^p theory of the appropriate maximal operator $M_\mathfrak{z}$ which is needed in order to see that the weighted L^2 estimates for the multipliers imply their weighted L^p estimates. Finally, the required analysis of the operator $M_\mathfrak{z}$ is carried out by applying the standard covering lemma for rectangles in $\mathbb{R}^n$ (see A. Cordoba and R. Fefferman **[2]** and R. Fefferman **[5]**). The arguments given here may serve to illustrate possible applications to the more general multiple parameter setting of the familiar tools of extrapolation and covering lemmas.

Before beginning the proof of our main result, let us briefly indicate the organization of this paper. In retrospect, it is the maximal operator $M_\mathfrak{z}$ which is controlling the multipliers we consider here. We therefore begin with a treatment of the necessary results for this maximal function in Section 2. Then, in Section 3, we introduce the appropriate Littlewood–Paley function $S_\mathfrak{z}(f)$ and derive its L^2 theory. We then apply all of this in the final section to complete the proof of the main result on weighted L^p estimates for the multipliers under consideration.

Finally, we should remark that the multipliers in question here are connected to the theory of multiple singular integrals along hypersurfaces of $\mathbb{R}^n$, and some related results (in the unweighted case) can be found in the literature. Of these, we shall be content to refer to three particularly striking ones: The interested reader should consult Nagel and Wainger **[9]**, Duoandikoetxea **[3]**, and the recent very general results of Ricci and Stein **[10]**.

2. The maximal function $\mathrm{M}_{\mathfrak{z}}$

Suppose, as in the Introduction, that $\mathcal{R}_{\mathfrak{z}}$ denotes the family of rectangles in $\mathbb{R}^3$ whose sides are parallel to the axes and have side lengths s, t, and st in the x, y, and z directions respectively (for some values of the parameters $s, t \geq 0$). Let

$$\mathrm{M}_{\mathfrak{z}} f(x, y, z) = \sup_{\substack{(x,y,z)\in R \\ R\in\mathcal{R}_{\mathfrak{z}}}} \frac{1}{|R|} \int_R |f|$$

and, as above we denote by $A^p(\mathfrak{z})$ those weights in $\mathbb{R}^3$ for which the A^p condition

$$\Big(\frac{1}{|R|}\int_R w\Big) \cdot \Big(\frac{1}{|R|}\int_R w^{-1/(p-1)}\Big)^{p-1} \leq C$$

is satisfied uniformly over all rectangles $R \in \mathcal{R}_{\mathfrak{z}}$.

We wish to prove, in this section, that for each $w \in A^p(\mathfrak{z})$ and $1 < p < \infty$, we have ([5])

$$\|\mathrm{M}_{\mathfrak{z}}(f)\|_{L^p(w)} \leq C \|f\|_{L^p(w)}$$

where C depends only on p and $\|w\|_{A^p(\mathfrak{z})}$.

To do this, we require two lemmas:

Lemma 2.1. *Suppose $w \in A^p(\mathfrak{z})$. Then there exists a $\delta > 0$ depending only on $\|w\|_{A^p(\mathfrak{z})}$ for which the reverse Hölder inequality*

$$\Big(\frac{1}{|R|}\int_R w^{1+\delta}\Big)^{\frac{1}{1+\delta}} \leq C\Big(\frac{1}{|R|}\int_R w\Big)$$

holds uniformly over every $R \in \mathcal{R}_{\mathfrak{z}}$.

Proof. Fixing $s > 0$, and letting $t \to 0$ differentiation of the integral shows that the function $w_{y,z}(x) = w(x, y, z)$ is uniformly in the class $A^p(\mathbb{R}^1)$ for each $y, z \in \mathbb{R}^1$. Similarly, fixing an interval on the x axis, I, of length s, we claim that the function $w_I(y, z) = \int_I w(x, y, z)\, dx$ belongs to the class of weights which satisfy the A^p condition in $\mathbb{R}^2$ uniformly over the class of rectangles whose side lengths are of the form t, st, for some $t > 0$. Indeed, let R be such a rectangle in the y-z plane. Then

$$\begin{aligned}
&\Big(\frac{1}{|R|}\int_R w_I\, dy\, dz\Big)\Big(\frac{1}{|R|}\int_R w_I^{-1/(p-1)}\, dy\, dz\Big)^{p-1} \\
&= \frac{1}{|R|}\int_R\Big(\frac{1}{|I|}\int_I w(x, y, z)\, dx\Big)\, dy\, dz \cdot \Big(\frac{1}{|R|}\int_R\Big(\frac{1}{|I|}\int_I w(x, y, z)\, dx\Big)^{\frac{-1}{p-1}} dy\, dz\Big)^{p-1} \\
&\leq \Big(\frac{1}{|R \times I|}\int_{R\times I} w\Big)\Big(\frac{1}{|R \times I|}\int_{R\times I} w^{-1/(p-1)}\Big)^{p-1} \\
&\leq \|w\|_{A^p(\mathfrak{z})}.
\end{aligned}$$

Now, let us fix a rectangle $\tilde{R} \in \mathcal{R}_3$ of the form $\tilde{R} = I \times R$, where I is an interval on the x axis and R is a rectangle in the y-z plane with sides t and $|I|t$. By the classical one parameter theory of A^p weights, there exists $\delta > 0$ so that (in the notation above)

$$\Big(\frac{1}{|I|}\int_I w_{y,z}(x)^{1+\delta}\,dx\Big)^{\frac{1}{1+\delta}} \le C\Big(\frac{1}{|I|}\int_I w_{y,z}(x)\,dx\Big)$$

for all $y, z \in R$ (C depends only on $\|w\|_{A^p(3)}$) and

$$\Big(\frac{1}{|R|}\int_R w_I(y,z)^{1+\delta}\,dy\,dz\Big)^{\frac{1}{1+\delta}} \le C\Big(\frac{1}{|R|}\int_R w_I(y,z)\,dy\,dz\Big).$$

Then combining these estimates,

$$\begin{aligned}
\Big(\frac{1}{|\tilde{R}|}\int_{\tilde{R}} w^{1+\delta}\Big)^{\frac{1}{1+\delta}} &= \Big[\frac{1}{|R|}\int_R\Big\{\Big(\frac{1}{|I|}\int_I w_{y,z}(x)^{1+\delta}\,dx\Big)^{\frac{1}{1+\delta}}\Big\}^{1+\delta} dy\,dz\Big]^{\frac{1}{1+\delta}} \\
&\le C\Big[\frac{1}{|R|}\int_R\Big(\frac{1}{|I|}\int_I w_{y,z}(x)\,dx\Big)^{1+\delta} dy\,dz\Big]^{\frac{1}{1+\delta}} \\
&= \frac{C}{|I|}\Big[\frac{1}{|R|}\int_R w_I(y,z)^{1+\delta}\,dy\,dz\Big]^{\frac{1}{1+\delta}} \\
&\le \frac{C^2}{|I|}\frac{1}{|R|}\int_R w_I(y,z)\,dy\,dz \\
&= C^2\frac{1}{|\tilde{R}|}\int_{\tilde{R}} w.
\end{aligned}$$

This proves Lemma 2.1.

In order to state Lemma 2.2, we need to introduce some notation and terminology. Let w be a positive locally integrable function on $\mathbb{R}^3$. As is familiar practice in proving weighted estimates for maximal functions, we introduce a weighted version of the operator we are dealing with: Define

$$\mathrm{M}_3^w(f)(x,y,z) = \sup_{\substack{(x,y,z)\in R\\ R\in\mathcal{R}_3}} \frac{1}{w(R)}\int_R |f|w.$$

We say that $w(x, y, z)$ is uniformly A^∞ in the x variable if fixing any y and z the function $w_{y,z}(x) = w(x, y, z)$ satisfies a classical A^∞ condition in the x variable, and the A^∞ constant of $w_{y,z}$ can be taken to be independent of y and z. Of course we can make the obvious modifications to define "w is uniformly A^∞ in the y or the z variable." We shall say that w satisfies on $\mathcal{R}$ a doubling condition in the z variable provided whenever $R = S \times J \in \mathcal{R}$ and J is an interval of the z axis with concentric double $\tilde{J}$, it follows that

$$w(\tilde{R}) = \int_{\tilde{R}} w \le Cw(R) \quad \text{for } \tilde{R} = S \times \tilde{J}.$$

With all this terminology, we finally have:

Lemma 2.2. *Suppose $w \geq 0$ is a locally integrable function on $\mathbb{R}^3$, which is uniformly A^∞ in each of the x and y variables, and $\mathcal{R}_3$ doubling in the z variable. Then the operator M_3^w is bounded on all the spaces $L^p(w)$, for all $1 < p \leq \infty$.*

Proof. Since the proof of this lemma can be found in **[5]** (stated in the case of dyadic rectangles) and in a more general setting in **[8]** we shall merely sketch the idea of the proof. The idea is to prove the following covering lemma with $w\,dx\,dy\,dz$ replacing Lebesgue measure: Let $1 < p < \infty$. Given a collection $\{R_i\}$ of rectangles in $\mathcal{R}_3$, there exists a subcollection $\{\tilde{R}_i\}$ so that

$$\Big\| \sum \chi_{\tilde{R}_i} \Big\|_{L^p(w)} \leq C w\Big(\bigcup \tilde{R}_i\Big)^{1/p} \quad \text{and} \quad w\Big(\bigcup R_i\Big) \leq C w\Big(\bigcup \tilde{R}_i\Big).$$

This implies that M_3^w is bounded on $L^p(w)$ by standard arguments.

To prove this covering lemma, assume the R_i are listed so that their z side lengths are decreasing. Having chosen $\tilde{R}_j$, for $j < k$, let $\tilde{R}_k$ be the first rectangle R on the list after $\tilde{R}_{k-1}$ such that

$$\Big|R \cap \Big[\bigcup_{j<k} \tilde{R}_j^d\Big]\Big| \leq \frac{1}{2}|R|$$

where $\tilde{R}_j^d$ denotes the rectangle concentric with $\tilde{R}_j$, but whose z side length has been multiplied by 5 (the x and y sidelengths are the same for $\tilde{R}_j^d$ as for $\tilde{R}_j$). Then, if R is not one of the $\tilde{R}_j$, it is easy to see that

$$\Big|S \cap \Big[\bigcup \tilde{S}_j^d\Big]\Big| > \frac{1}{2}|S|$$

where S is a slice of R by some plane parallel to the x-y plane which passes through R. Because w belongs to A^∞ uniformly in the x and y variables, it follows that

$$w\left(S \cap \Big[\bigcup \tilde{S}_j^d\Big]\right) > \epsilon\, w(S),$$

for some $\epsilon > 0$, hence that $M_2^w(\chi_{\cup \tilde{R}_j^d}) > \epsilon$ on $\bigcup R_j$. Here M_2^w denotes the strong maximal function with respect to w measure in the x-y plane, i.e.

$$M_2^w(f)(p) = \sup_S \frac{1}{w(S)} \int_S |f| w\,dx\,dy$$

where the sup is taken over all two-dimensional rectangles oriented parallel to the x-y plane, and passing through p.

By another covering argument, (see **[4]**) it can be shown that for w an A^∞ measure in the x and y variables, M_2^w is bounded on $L^p(w)$ for all $1 < p < \infty$, and this shows that

$$w\Big(\bigcup R_j\Big) \leq C w\Big(\bigcup \tilde{R}_j^d\Big) \leq C' w\Big(\bigcup \tilde{R}_j\Big),$$

which is (2).

As for (1), let $\varphi \in L^{p'}(w)$, $\|\varphi\|_{L^{p'}(w)} = 1$, and we estimate $\int \sum \chi_{\tilde{R}_j} \varphi\, w\, dx\, dy\, dz$ by slicing $\mathbb{R}^3$ with a plane parallel to the x-y plane, and estimating $\int \sum \chi_{\tilde{S}_j} \varphi\, w\, dx\, dy$, where $\tilde{S}_j$ is the slice that this plane cuts of $\tilde{R}_j$. Then

$$\int \sum \chi_{\tilde{S}_j} \varphi\, w\, dx\, dy \leq \sum w(\tilde{S}_j) \Big[\frac{1}{w(\tilde{S}_j)} \int_{\tilde{S}_j} \varphi\, w\, dx\, dy \Big]$$

and setting $\tilde{E}_j = \tilde{S}_j - \bigcup_{i<j} \tilde{S}_i$ we see that $w(\tilde{E}_j) > \epsilon w(\tilde{S}_j)$ so that

$$\int \sum \chi_{\tilde{S}_j} \varphi\, w\, dx\, dy \leq \frac{1}{\epsilon} \int_{\cup \tilde{S}_j} M_2^w(\varphi) w\, dx\, dy.$$

Integrating this in z, we have

$$\Big\| \sum \chi_{\tilde{R}_j} \Big\|_{L^p(w)} \leq C_\epsilon \big\| M_2^w(\varphi) \big\|_{L^{p'}(w)} \big\| \chi_{\cup \tilde{R}_j} \big\|_{L^p(w)}$$

and this gives (i) since M_2^w is bounded on $L^{p'}(w)$.

Now it is an easy matter to prove:

Theorem 2.1. *If $w \in A^p(\mathfrak{z})$ then $\mathrm{M}_\mathfrak{z}$ is bounded on $L^p(w)$.*

Proof. Applying the definition of $A^p(\mathfrak{z})$ with Hölder's inequality gives the standard estimate

$$\mathrm{M}_\mathfrak{z}(f)(x, y, z) \leq C \mathrm{M}_\mathfrak{z}^w (f^p)^{1/p}(x, y, z)$$

and, since, by Lemma 2.1, $w^{-1/(p-1)}$ satisfies a reverse Hölder inequality, we have $w \in A^{p-\epsilon}(\mathfrak{z})$ for some $\epsilon > 0$ depending only on $\|w\|_{A^p(\mathfrak{z})}$. It follows that $\mathrm{M}_\mathfrak{z}(f)(x, y, z) \leq C_\epsilon \mathrm{M}_\mathfrak{z}^w (f^{p-\epsilon})^{1/(p-\epsilon)}(x, y, z)$. We now observe that any $w \in A^p(\mathfrak{z})$ is automatically uniformly A^∞ in each of the x and y variables, and is $\mathcal{R}_\mathfrak{z}$ doubling in the z direction. Hence, by Lemma 2.2, the operator $\mathrm{M}_\mathfrak{z}^w (f^{p-\epsilon})^{1/(p-\epsilon)}$ is bounded on $L^p(w)$, finishing the proof of Theorem 2.1.

3. Estimates for the Littlewood–Paley function

In this section, we shall introduce the appropriate dilation-invariant $\mathrm{S}_\mathfrak{z}$ function needed to analyze multipliers in $\mathcal{M}_\mathfrak{z}$. First it is convenient to define, for $(x, y, z) \in \mathbb{R}^3$,

$$\Gamma_\mathfrak{z}(x, y, z) = \big\{ (u, v, w, s, t) \;\big|\; u, v, w, \in \mathbb{R}^1 \text{ and } s, t > 0, \\ \text{with } |x - u| < s\,,\ |y - v| < t\,,\ \text{and } |z - w| < st \big\}.$$

This is the cone region we shall work with. Suppose, now, that ψ_1 and ψ_2 belong to $\mathcal{S}(\mathbb{R}^1)$, ψ_i not identically 0, and assume that

$$\int_{\mathbb{R}^1} \psi_1(|x|)\, dx = \int_{\mathbb{R}^2} \psi_2(|(y, z)|)\, dy\, dz = 0.$$

Set

$$\psi(x, y, z) = \psi_1(|x|)\psi_2(|(y, z)|),$$

and for $s, t > 0$, let

$$\psi_{s,t}(x, y, z) = (st)^{-2}\psi(x/s, y/t, z/st).$$

Then we define the area integral S_3 by

$$S_3^2(f)(x, y, z) = \int\cdots\int_{\Gamma_3(x,y,z)} |f * \psi_{s,t}(u, v, w)|^2 \frac{du\,dv\,dw\,ds\,dt}{s^3t^3}.$$

Our theorem of this section is then

Theorem 3.1. *If $w \in A^2(3)$, then there exist positive constants c_1 and c_2, depending only on $\|w\|_{A^2(3)}$, so that*

$$c_1\|f\|_{L^2(w)} \leq \|S(f)\|_{L^2(w)} \leq c_2\|f\|_{L^2(w)}.$$

Proof.

$$\int_{\mathbb{R}^3} S^2(f) w\,dx\,dy\,dz$$

$$= \int_{(x,y,z)\in\mathbb{R}^3} \Big(\int_{\Gamma_3(x,y,z)} |f * \psi_{s,t}(u, v, w)|^2 \frac{du\,dv\,dw\,ds\,dt}{s^3t^3}\Big) w\,dx\,dy\,dz \qquad [0]$$

$$= \int_{\mathbb{R}^3}\int_{s,t>0} |f * \psi_{s,t}(x, y, z)|^2 w(R_{s,t}(x, y, z)) \frac{ds\,dt}{st}\,dx\,dy\,dz \qquad (\dagger)$$

where $R_{s,t}(x, y, z)$ denotes the rectangle in $\mathbb{R}^3$ whose sides are parallel to the axes, of length $2s$, $2t$, and $4st$ respectively, and whose center is (x, y, z).

Now fix $x \in \mathbb{R}^1$ and $s > 0$. Define

$$w_{x,s}(y, z) = \int_{|u-x|<s} w(u, y, z)\,du.$$

Then as we have already observed in Section 2, $\|w_{x,s}\|_{A^2(\mathcal{R}_{2s})} \leq \|w\|_{A^2(3)}$ where $A^2(\mathcal{R}_{2s})$ denotes the class of all weights ω in $\mathbb{R}^2$ so that

$$\Big(\frac{1}{|R|}\int_R \omega\Big)\Big(\frac{1}{|R|}\int_R \omega^{-1}\Big) \leq C \quad \text{for all } R \in \mathcal{R}_{2s}$$

and where $\mathcal{R}_{2s}$ denotes the class of all rectangles in $\mathbb{R}^2$ with sides parallel to the axes and whose lengths are of the form t and $(2s)t$ for some $t > 0$. Then

$$(\dagger) = \int_{\substack{x\in\mathbb{R}^1\\ s>0}} \Big(\int_{\substack{(y,z)\in\mathbb{R}^2\\ t>0}} |f_{x,s} * (\psi_2)_{s,t}(y, z)|^2 w_{x,s}(R_{t,st}(y, z)) \frac{dt}{t}\,dy\,dz\Big)\,dx\,\frac{ds}{s}$$

where

$$(\psi_2)_{s,t}(y,z) = s^{-1}t^{-2}\psi_2(y/t, z/st)$$

and

$$f_{x,s}(y,z) = \int f(x-u, y, z)(\psi_1)_s(u)\,du.$$

The inner integral inside the parentheses above is just the weighted L^2 norm $\|S(f_{x,s})\|_{L^2(w_{x,s})}$ where the S operator is defined by convolution with the (standard) dilates of the bump function $s^{-1}\psi_2(y, z/s)$. By the classical, one parameter theory, and a linear change of variables, the integral in parenthesis is equivalent to

$$\int_{\mathbb{R}^2} |f_{x,s}(y,z)|^2 w_{x,s}(y,z)\,dy\,dz,$$

so that if $A \sim B$ denotes that the ratio A/B is bounded above and below by positive constants depending only on $\|w\|_{A^2(\mathfrak{z})}$, we have

$$(\dagger) \sim \int_{\substack{x\in\mathbb{R}^1 \\ s>0}} \Big(\int_{(y,z)\in\mathbb{R}^2} |f_{x,s}(y,z)|^2 w_{x,s}(y,z)\,dy\,dz\Big)\,dx\,\frac{ds}{s}.$$

Now, fix y and z. Then

$$\int_{\substack{x\in\mathbb{R}^1 \\ s>0}} |f_{x,s}(y,z)|^2 w_{x,s}(y,z)\,dx\,\frac{ds}{s} \sim \int_{x\in\mathbb{R}^1} |f(x,y,z)|^2 w(x,y,z)\,dx$$

because $w(\cdot, y, z)$ is uniformly in $A^2(\mathbb{R}^1)$ in the x variable, and we are simply applying the classical weighted inequalities on L^2 in one dimension for the S operator. Integrating this in y and z, and then using Fubini's Theorem, we see that

$$\int \mathrm{S}^2_{\mathfrak{z}}(f) w\,dx\,dy\,dz \sim \int f^2 w\,dx\,dy\,dz,$$

concluding the proof of Theorem 3.1.

4. The main result: weighted estimates for multipliers

In this section, we combine the machinery from the previous sections in order to give the weighted estimates for the operators invariant under the relevant dilation group. Let us begin by defining the operators precisely.

Let $\zeta : [0,\infty) \to [0,\infty)$ be a C_c^∞ function which is identically one on the interval $[\frac{1}{2}, 2]$ and vanishes in a neighborhood of 0 (and of ∞). We shall consider multipliers $m(\xi)$, $\xi \in \mathbb{R}^3$, which satisfy

$$\Big\| \partial_\xi^\alpha \big[m(\xi)\zeta(|\xi_1|)\zeta\big(|(\xi_2,\xi_3)|\big) \big] \Big\|_\infty \le C_\alpha$$

for each multi-index $\alpha = (\alpha_1, \alpha_2, \alpha_3)$, and also satisfy the same estimate (with the same value of C_α) when $m(\xi)$ is replaced by any one of the functions $m(\delta_1\xi_1, \delta_2\xi_2, \delta_1\delta_2\xi_3)$. We shall call the class of multipliers defined by these estimates $\mathcal{M}_{\mathfrak{z}}$.

Our theorem for multipliers in $\mathcal{M}_{\mathfrak{z}}$ is as follows:

Theorem 4.1. *If $m \in \mathcal{M}_{\mathfrak{z}}$ and we define T_m by*

$$T_m f\widehat{\ }(\xi) = m(\xi) \cdot \hat{f}(\xi), \qquad f \in C_c^\infty(\mathbb{R}^3),$$

then we have the estimate

$$\|T_m(f)\|_{L^p(w)} \leq C\|f\|_{L^p(w)}$$

where C is independent of f whenever $1 < p < \infty$ and $w \in A^p(\mathfrak{z})$.

Proof. Since we know that $\mathrm{M}_{\mathfrak{z}}$ is bounded on $L^p(w)$ if and only if $w \in A^p(\mathfrak{z})$, we may apply Rubio's Extrapolation. This reduces our proof to the special case when $p = 2$. To prove this special case, we now estimate $\mathrm{S}_{\mathfrak{z}}(Tf)(x, y, z)$ where the area integral $\mathrm{S}_{\mathfrak{z}}$ is defined as

$$\mathrm{S}_{\mathfrak{z}}^2(Tf)(x, y, z) = \int\cdots\int_{\Gamma_{\mathfrak{z}}(x,y,z)} |T(f) * (\psi * \psi)_{s,t}(u, v, w)|^2 \frac{du\, dv\, dw\, ds\, dt}{s^3 t^3}$$

where $h_{s,t}(x, y, z)$ means $(st)^{-2} h(x/s, y/t, z/st)$ and $\psi(x, y, z) = \psi_1(|x|)\psi_2(|(y, z)|)$ with $\hat{\psi}_1$ and $\hat{\psi}_2$ vanishing identically near the origin.

Without loss of generality we may take $(x, y, z) = 0$ and consider $(u, v, w, s, t) \in \Gamma_{\mathfrak{z}}(0)$. Then

$$\left|T(f) * (\psi * \psi)_{s,t}(u, v, w)\right| = \left|(f * \psi_{s,t}) * (T(\psi_{s,t}))(u, v, w)\right|. \tag{‡}$$

It is easy to verify that since $\psi_1(|x|)\widehat{\ }$ and $\psi_2(|(y, z)|)\widehat{\ }$ have support away from the origin,

$$\begin{aligned} \left|T(\psi_{s,t})(x, y, z)\right| &\leq C_N \Big[\frac{1}{(1 + |x|)^N} \cdot \frac{1}{(1 + |(y, z)|)^N}\Big]_{s,t} \\ &= C_N \Phi_{s,t}(x, y, z), \end{aligned}$$

where N is fixed but can be taken arbitrarily large.

It follows that

$$(‡) \leq \left(|(f * \psi_{s,t})|^{u,v,w} * \Phi_{s,t}\right)(0, 0, 0),$$

where the upper u, v, w denotes translation, i.e. $Q^{u,v,w}(x, y, z)$ means $Q(x + u, y + v, z + w)$, and this implies that

$$\left|T(f) * (\psi * \psi)_{s,t}(u, v, w)\right| \leq C\mathrm{M}_{\mathfrak{z}}\left(\left|(f * \psi_{s,t})^{u,v,w}\right|\right)(0).$$

We have

$$\int_{\mathbb{R}^3} S^2_{\psi*\psi}(T(f))(x,y,z)w(x,y,z)\,dx\,dy\,dz$$
$$\leq C\int_{\mathbb{R}^3}\Big(\iint_{\Gamma_{\mathfrak{z}}(0)} \mathrm{M}^2_{\mathfrak{z}}\left[(f*\psi_{s,t})^{u,v,w}\right](x,y,z)\frac{du\,dv\,dw\,ds\,dt}{s^3t^3}\Big)w(x,y,z)\,dx\,dy\,dz$$
$$= C\iint_{\Gamma_{\mathfrak{z}}(0)}\Big(\int_{\mathbb{R}^3} \mathrm{M}^2_{\mathfrak{z}}\left[(f*\psi_{s,t})^{u,v,w}\right](x,y,z)w(x,y,z)\,dx\,dy\,dz\Big)\frac{du\,dv\,dw\,ds\,dt}{s^3t^3}.$$

Since $w \in A^2(\mathfrak{z})$, $\mathrm{M}_{\mathfrak{z}}$ is bounded on $L^2(w)$ and this last quantity is dominated by

$$C\iint_{\Gamma_{\mathfrak{z}}(0)}\int_{\mathbb{R}^3}\left|(f*\psi_{s,t})^{u,v,w}\right|^2(x,y,z)w(x,y,z)\,dx\,dy\,dz\frac{du\,dv\,dw\,ds\,dt}{s^3t^3}$$

and applying Fubini's Theorem this equals $C\|\mathrm{S}_{\mathfrak{z}}(f)\|^2_{L^2(w)}$. Therefore we have shown that whenever $w \in A^2(\mathfrak{z})$,

$$\|\mathrm{S}_{\mathfrak{z}}(Tf)\|_{L^2(w)} \leq C\|\mathrm{S}_{\mathfrak{z}}(f)\|_{L^2(w)}.$$

Applying the equivalence of the weighted L^2 norm of a function with that of its area integral (Section 3) we see that the proof of Theorem 4.1 is complete.

References

[1] A. Cordoba, Maximal functions, covering lemmas and Fourier multipliers, Proc. of Symposia in Pure Math. 35 (1979), 29–49.

[2] A. Cordoba and R. Fefferman, A geometric proof of the Strong Maximal Theorem, Ann. of Math. 102 (1975), 95–100.

[3] J. Duoandikoetxea, Multiple singular integrals and maximal functions along hypersurfaces, Ann. Inst. Fourier 36 (1986), 185–206.

[4] R. Fefferman, Strong differentiation with respect to measures, Amer. J. Math. 103 (1981), 33–40.

[5] R. Fefferman, Some weighted norm inequalities for Cordoba's maximal function, Amer. J. Math. 106 (1984), 1261–1264.

[6] R. Fefferman and E. M. Stein, Singular integrals on product spaces, Adv. in Math. 45 (1982), 117–143.

[7] R. Hunt, B. Muckenhoupt, and R. Wheeden, Weighted norm inequalities for the conjugate function and Hilbert transform, Trans. Amer. Math. Soc. 176 (1973), 227–251.

[8] B. Jawerth and A. Torchinsky, The strong maximal function with respect to measures, Stud. Math. 80 (1984), 261–285.

[9] A. Nagel and S. Wainger, L^2 boundedness of Hilbert transforms along surfaces and convolution operators homogeneous with respect to a multiple parameter group, Amer. J. Math. 99 (1977), 761–785.

[10] F. Ricci and E. M. Stein, Multiparameter singular integrals and maximal functions, Ann. Inst. Fourier 42 (1992), 637–670.

[11] J. L. Rubio de Francia, Factorization theory and A^p weights, Amer. J. Math. 106 (1984), 533–547.

Department of Mathematics
University of Chicago
Chicago, IL 60637, U.S.A.

Markov Matrices

S. R. Foguel[1]

Preface

We study, in these notes, the asymptotic behavior of P^n and $I + P + \cdots + P^n$ when P is an infinite Markov matrix.

In order to do that we prove the existence of an invariant measure for P.

Most of the results are classical but we use here a different approach: We do not use probability theory or measure theory but prove directly. In fact only some elementary facts about non negative series are used.

The discussion here is a modification of **[8]**: There we studied "Harris type Markov operators". By restriction to matrices one avoids all measure theoretic considerations. Thus the arguments are greatly simplified.

I. Definition and notation

Throughout these notes:

X is a *countable set*. The *Markov matrix* P, is a function $p(x, y)$, defined for $x \in X$ and $y \in X$, satisfying:

$$p(x, y) \geq 0, \quad \sum_{y \in X} p(x, y) \leq 1.$$

Define

$$(Pf)(x) = \sum_{y \in X} p(x, y) f(y) \quad (0 \cdot \infty = 0).$$

If $f \geq 0$ then $(Pf)(x) \leq \infty$ is well defined. If $f \in \ell_\infty$ then $\sum_{y \in X} p(x, y) f(y)$ converges absolutely and $\|Pf\|_\infty \leq \|f\|_\infty$.

Define

$$(uP)(x) = \sum_{y \in X} u(y) p(y, x) \quad (\infty \cdot 0 = 0).$$

1 Partially sponsored by the Edmund Landau Center for research in Mathematical Analysis, supported by the Minerva Foundation (Germany).

If $u \geq 0$ then $(uP)(x) \leq \infty$ is well defined. If $u \in \ell_1$ then $\sum_{y \in X} u(y)p(y, x)$ converges absolutely and $\|uP\|_1 \leq \|u\|_1$.

Put

$$\langle u, f \rangle = \sum_{x \in X} u(x) f(x).$$

If $u \geq 0$, $f \geq 0$ then $\langle u, f \rangle \leq \infty$. If $u \in \ell_1$, $f \in \ell_\infty$ then the sum converges absolutely and

$$|\langle u, f \rangle| \leq \|u\|_1 \, \|f\|_\infty.$$

By change of order of summation

$$\langle uP, f \rangle = \langle u, Pf \rangle$$

when $u \geq 0$, $f \geq 0$ or $u \in \ell_1$, $f \in \ell_\infty$.

Recall: If $0 \leq a_n(x) \uparrow a(x)$ then $\sum_{x \in X} a_n(x) \underset{n \to \infty}{\to} \sum_{x \in X} a(x)$.

Thus:

(a) If $0 \leq f_n(x) \uparrow f(x)$ then $(Pf_n)(x) \underset{n \to \infty}{\to} (Pf)(x)$.

(b) If $0 \leq u_n(x) \uparrow u(x)$ then $(u_n P)(x) \to (uP)(x)$.

If P_1 and P_2 are Markov matrices then so is $\alpha P_1 + \beta P_2$ provided $\alpha, \beta \geq 0$ and $\alpha + \beta \leq 1$.

Also if $q(x, y) = \sum_{z \in X} p_1(x, z) p_2(z, y)$ then $Q = (q(x, y))$ is a Markov matrix. Note:

$$Qf = P_1(P_2 f)$$

hence $Q1 = P_1(P_2 1) \leq P_1 1 \leq 1$. Also

$$uQ = (uP_1)P_2.$$

In particular P^n is well defined if we put $P^0 = I$: $If = f$, $i(x, y) = \delta(x, y)$. $P^{n+1} = PP^n$.

Denote

$$P^n = (p^{(n)}(x, y)).$$

II. Subinvariant sets

The characteristic function 1_A is defined by:

$$1_A(x) = \begin{cases} 1 & x \in A \\ 0 & x \notin A. \end{cases}$$

Thus

$$(P1_A)(x) = \sum_{y \in A} p(x, y).$$

In particular

$$\big(P1_{\{y_0\}}\big)(x) = p(x, y_0)$$
$$\big(1_{\{y_0\}}P\big)(x) = p(y_0, x).$$

Put $(I_A f)(x) = 1_A(x)f(x)$: Let P be a Markov matrix then $Q = I_A P$ is given by

$$q(x, y) = \begin{cases} p(x, y) & \text{if } x \in A \\ 0 & \text{if } x \notin A. \end{cases}$$

Also $R = PI_A$ is given by

$$r(x, y) = \begin{cases} p(x, y) & \text{if } y \in A \\ 0 & \text{if } y \notin A. \end{cases}$$

Thus if $S = I_A P I_A$ then $s(x, y) = p(x, y)$ for $x \in A,\ y \in A$. If either $x \notin A$ or $y \notin A$ then $s(x, y) = 0$. We call S the *"restriction of P to the set A"*.

Definition 2.1. P is called "irreducible" if given x_0, y_0 then $p^{(n)}(x_0, y_0) > 0$ for some $n \geq 0$.

Another way of putting this is the following. *P is irreducible if $\sum_{n=0}^{\infty} p^{(n)}(x, y) > 0$ for all $x, y \in X$.*

Now $p^{(n)}(x, y) = P^n 1_{\{y\}}(x)$. Hence $\sum_{n=0}^{\infty} P^n 1_{\{y\}} > 0$. Thus: *if $A \neq \emptyset$ and P is irreducible then $\sum_{n=0}^{\infty} P^n 1_A(x) > 0$ for every $x \in X$*: If $y \in A$ then $P^n 1_A \geq P^n 1_{\{y\}}$. Let $f \geq 0$ but not identically zero. Put $A = \{x : f(x) \geq \varepsilon\}$ where $\varepsilon > 0$ is so small that $A \neq \emptyset$. Then $f \geq \varepsilon 1_A$: If $x \in A$ then $f(x) \geq \varepsilon = \varepsilon 1_A(x)$, if $x \notin A$ then $f(x) \geq 0 = \varepsilon 1_A(x)$. Therefore *if $f \geq 0$, $f \neq 0$ and P is irreducible then $\sum_{n=0}^{\infty} P^n f(x) > 0$ for every $x \in X$*. If P is irreducible and $P1_A \leq 1_A$ then

$$P^n 1_A(x) \leq 1_A(x) = 0 \text{ whenever } x \in A'$$

or *a subinvariant set of an irreducible matrix is trivial.* (A' is the complement of A. The set A is called subinvariant if $P1_A \leq 1_A$.)

Actually this is equivalent to irreducibility but we shall not bother to prove it.

Let us consider now a "reducible" matrix:

Assume A is not trivial and

$$P1_{A'} \leq 1_{A'}.$$

This assumption is equivalent to $P1_{A'}(x) = 0$ whenever $x \in A$ (recall $P1_{A'}(x) \leq 1$ if $x \in A'$), or $(I_A P I_{A'})1 = 0$ but $I_A P I_{A'}$ is a Markov matrix. Thus

$$P1_{A'} \leq 1_{A'} \Leftrightarrow I_A P I_{A'} = 0.$$

Theorem 2.1. *Let $P1_{A'} \le 1_{A'}$ then*

$$I_A P^n = I_A P^n I_A = (I_A P I_A)^n.$$

Proof. $P^n 1_{A'} \le 1_{A'}$ for every $n \ge 1$. Hence $I_A P^n I_{A'} = 0$. Thus $I_A P^n = I_A P^n I_A + I_A P^n I_{A'} = I_A P^n I_A$. Let us prove the second equality by induction: it certainly holds for $n = 1$.

$$\begin{aligned} I_A P^{n+1} &= (I_A P^n) P = (I_A P^n I_A) P \\ &= (I_A P^n I_A) I_A P = (I_A P I_A)^n (I_A P I_A) \\ &= (I_A P I_A)^{n+1}. \end{aligned}$$ □

Thus if $P1_{A'} \le 1_{A'}$ then: *The restriction of P^n to A is the* nth *power of the restriction of P to A.*

III. Nonrecurrent states

Definition 3.1. $\Omega = \{f : 0 \le f \le 1,\ Pf \le f;\ P^n f \underset{n\to\infty}{\to} 0\}$.

$$Y = \bigcup_{f\in\Omega} \{x : f(x) > 0\}$$

$$Z = X - Y.$$

If $x \in Y$ we call x “a nonrecurrent state”.
If $x \in Z$ we call x “a recurrent state”.
If $X = Y$ we call P “a nonrecurrent matrix”.
If $X = Z$ we call P “a recurrent matrix”.

Theorem 3.1. *Let $y_0 \in Y$ then*

$$\sup_x \sum_{n=0}^{\infty} p^{(n)}(x, y_0) \le \text{const.} < \infty.$$

Proof. Find f with $0 \le f \le 1$, $f(y_0) > 0$, $Pf \le f$, $P^k f \to 0$. Then for some k $f(y_0) - (P^k f)(y_0) \ge \delta > 0$. Thus $1_{\{y_0\}} \le \frac{1}{\delta}(f - P^k f)$:

$$1 \le \frac{1}{\delta}\big(f(y_0) - (P^k f)(y_0)\big),$$

$$0 \le \frac{1}{\delta}\big(f(x) - (P^k f)(x)\big), \quad x \ne y_0.$$

Therefore

$$\sum_{n=0}^{N} p^{(n)}(x, y_0) = \sum_{n=0}^{N} P^n 1_{\{y_0\}}(x) \leq \frac{1}{\delta} \sum_{n=0}^{N} \big(P^n(I - P^k) f\big)(x)$$
$$= \frac{1}{\delta}\{f(x) + (Pf)(x) + \cdots + (P^{k-1} f)(x) - (P^{N+1} f)(x) - \cdots - (P^{N+k} f)(x)\}$$
$$\leq \frac{1}{\delta}\{f(x) + (Pf)(x) + \cdots + (P^{k-1} f)(x)\}$$
$$\leq \frac{k}{\delta}.$$

Let $N \to \infty$ to obtain the result. □

Theorem 3.2. *Let* $0 \leq u \in \ell_1$ *and* $y_0 \in Y$ *then* $\sum_{n=0}^{\infty} (uP^n)(y_0) < \infty$.

Proof. $\sum_{n=0}^{\infty} (uP^n)(y_0) = \langle \sum_{n=0}^{\infty} uP^n, 1_{\{y_0\}}\rangle = \langle u, \sum_{n=0}^{\infty} P^n 1_{\{y_0\}}\rangle < \infty$ since $u \in \ell_1$ and $\sum_{n=0}^{\infty} P^n 1_{\{y_0\}} \in \ell_\infty$. □

Corollary. *Let* $0 \leq u \in \ell_1$ *satisfy* $uP = u$ *then* $u(y_0) = 0$ *if* $y_0 \in Y$.

Proof. $\sum_{n=0}^{\infty} (uP^n)(y_0) = \sum u(y_0)$ but the sum converges by Theorem 3.2. □

For our last result we wish to show that we can restrict P to Z, or by Theorem 2.1 that $P1_Y \leq 1_Y$. Let us "count" Y: $Y = \{y_i\}$, $i \geq 1$. Find $f_i \in \Omega$ with $f_i(y_i) > 0$. Put

$$f_0 = \sum_{i=1}^{\infty} \frac{f_i}{2^i}.$$

If $Y = \emptyset$ put $f_0 = 0$. If Y is finite take a finite sum. Now $0 \leq f_0 \leq 1$, $Pf_0 = \sum_{i=1}^{\infty} \frac{Pf_i}{2^i} \leq \sum_{i=1}^{\infty} \frac{f_i}{2^i} = f_0$, and

$$(P^n f_0)(x) \leq \sum_{i=1}^{k} \frac{1}{2^i}(P^n f_i)(x) + \sum_{i=k+1}^{\infty} \frac{1}{2^i} \leq \sum_{i=1}^{k} \frac{1}{2^i}(P^n f_i)(x) + \frac{1}{2^k}.$$

Thus

$$\lim_{n\to\infty} (P^n f_0)(x) \leq \frac{1}{2^k} \quad \text{or} f_0 \in \Omega.$$

We have established:

Lemma 3.3. *There exists a function* f_0 *with* $f_0 \in \Omega$, $Y = \{x : f_0(x) > 0\}$.

Theorem 3.4. $P1_Y \leq 1_Y$.

Proof. Put $f_n = \min(nf_0, 1)$ then $0 \leq f_n \leq 1$, $f_n \uparrow 1_Y$. But

$$Pf_n \leq nPf_0 \leq nf_0.$$
$$Pf_n \leq P1 \leq 1.$$

Thus $Pf_n \leq f_n$ and as $n \to \infty$ $f_n \uparrow 1_Y$, $Pf_n \uparrow P1_Y$. □

Corollary. *If P is irreducible then either P is recurrent or P is nonrecurrent.*

IV. Recurrent states

In this chapter we study the set Z.

Lemma 4.1. *Let $0 \leq f \leq 1$ satisfy $Pf \leq f$ then $(P^k f)(x) = f(x)$ for all k, when $x \in Z$.*

Proof. $Pf \leq f \Rightarrow P^{n+1} f \leq P^n f$. Let $g = \lim P^n f$ then $Pg = g$ thus $f - g \in \Omega$. Therefore $f(x) - g(x) = 0$ if $x \in Z$ but

$$f(x) \geq (P^k f)(x) \geq g(x).$$ □

Corollary. $(P^k 1)(x) = 1$ *for all k if $x \in Z$.*

Theorem 4.2. *Let $0 \leq f(x) \leq \infty$ satisfy $Pf \leq f$. Then $(P^k f)(x) = f(x)$ for all k, when $x \in Z$.*

Proof. Let $f_n = \min(f, n)$ then $Pf_n \leq Pf \leq f$, $Pf_n \leq n$ so $Pf_n \leq f_n$. Apply Lemma 4.1 to $\frac{1}{n} f_n$ to conclude $(P^k f_n)(x) = f_n(x)$, for every k, when $x \in Z$.

Now

$$(P^k f)(x) = \lim_{n\to\infty} (P^k f_n)(x) = \lim_{n\to\infty} f_n(x) = f(x).$$ □

Theorem 4.3. *Let $f \geq 0$ then*

$$\sum_{n=0}^{\infty} (P^n f)(x) = \begin{cases} 0 \\ \infty \end{cases} \text{if } x \in Z.$$

Proof. Put $g(x) = \sum_{n=0}^{\infty} (P^n f)(x)$ then $(Pg)(x) = \sum_{n=1}^{\infty} (P^n f)(x) \leq g(x)$. Thus, if $x \in Z$ then

$$g(x) = (P^k g)(x) = \sum_{n=k}^{\infty} (P^n f)(x)$$

and if $\sum_{n=0}^{\infty}(P^n f)(x) < \infty$ then

$$\sum_{n=k}^{\infty}(P^n f)(x) \underset{k\to\infty}{\to} 0. \qquad \square$$

Corollary. *Let $x \in Z$. If $p^{(j)}(x, y) > 0$ for some integer $j \geq 0$ then $y \in Z$ and $\sum_{n=0}^{\infty} p^{(n)}(x, y) = \infty$. In particular if $x \in Z$ then $\sum_{n=0}^{\infty} p^{(n)}(x, x) = \infty$. The restriction of P to Z is recurrent.*

Proof. Put $f = 1_{\{y\}}$. Then $(P^n f)(x) = p^{(n)}(x, y)$. We can not have $y \in Y$ since that would imply:

$$p^{(j)}(x, y) = (P^j f)(x) \leq (P^j 1_Y)(x) \leq 1_Y(x) = 0.$$

Now $\sum_{n=0}^{\infty}(P^n f)(x) = \infty$ since $(P^j f)(x) \neq 0$. In particular

$$\sum_{n=0}^{\infty} p^{(n)}(x, x) = \sum (P^n 1_{\{x\}})(x) = \infty,$$

since $P^0 1_{\{x\}}(x) = 1$. Now, the restriction of P to Z satisfies, by Theorem 2.1, $(I_Z P I_Z)^n = I_Z P^n I_Z$. Hence

$$\sum_{n=0}^{\infty}(I_Z P I_Z)^n 1_{\{x\}}(x) = \infty \text{ if } x \in Z. \qquad \square$$

Corollary. *Let P be recurrent. Let $f \geq 0$ and put $A = \{x : \sum_{n=0}^{\infty}(P^n f)(x) > 0\}$. Then $P 1_A = 1_A$ and $\sum_{n=0}^{\infty}(P^n f)(x) = \infty$ if $x \in A$.*

Proof. Let $\varphi = \min(1, \sum_{n=0}^{\infty} P^n f)$ then $0 \leq \varphi \leq 1$ $P\varphi \leq \varphi$ so $P\varphi = \varphi$. But on A $\sum_{n=0}^{\infty}(P^n f)(x) = \infty$, thus $\varphi = 1_A$. $\square$

Theorem 4.4. *Let $u \geq 0$. Then $\sum_{n=0}^{\infty}(u P^n)(x)$ is either 0 or ∞ if $x \in Z$.*

Proof. Let $x \in Z$ and $\sum_{n=0}^{\infty} (uP^n)(x) < \infty$. Then

$$\begin{aligned} \infty > \langle \sum_{n=0}^{\infty} uP^n, 1_{\{x\}} \rangle &= \langle u, \sum_{n=0}^{\infty} P^n 1_{\{x\}} \rangle \\ &\geq \langle u, \sum_{n=k}^{\infty} P^n 1_{\{x\}} \rangle = \langle uP^k, \sum_{n=0}^{\infty} P^n 1_{\{x\}} \rangle \\ &\geq (uP^k)(x) \sum_{n=0}^{\infty} p^{(n)}(x, x). \end{aligned}$$

Now $\sum_{n=0}^{\infty} p^{(n)}(x, x) = \infty$. Hence $(uP^k)(x) = 0$, and this holds for every integer k. □

Corollary. *Let $x \in Z$. If $p^{(j)}(x_0, x) \neq 0$ for some integer $j \geq 0$, then*

$$\sum_{n=0}^{\infty} p^{(n)}(x_0, x) = \infty .$$

Proof. $(1_{\{x_0\}} P^n)(x) = p^{(n)}(x_0, x)$ and the sum is infinite since it is not zero. □

Corollary. *Let $X = Z$. If $u \geq 0$ then $A = \{x : \sum_{n=0}^{\infty} (uP^n)(x) > 0\}$ satisfies $P1_A = 1_A$. On A, $\sum_{n=0}^{\infty} (uP^n)(x) = \infty$.*

Proof. $0 = \langle uP^{k+1}, 1_{A'} \rangle = \langle uP^k, P1_{A'} \rangle$. Thus $P1_{A'}(x) = 0$ if $x \in A$ or $P1_{A'} \leq 1_{A'}$. Therefore, since P is recurrent, $P1_A = 1_A$. Now on A $\sum_{n=0}^{\infty} (uP)^n(x)$ is infinite since it is not zero. □

Theorem 4.5. *Let $0 \leq u < \infty$ satisfy $uP \leq u$. Then $(uP)(x) = u(x)$ whenever $x \in Z$.*

Proof. $\sum_{n=0}^{\infty} (u - uP)P^n \leq u < \infty$, thus, by Theorem 4.3, the sum vanishes for every $x \in Z$. □

The notation of recurrence and non recurrence are classical. We used here a method developed for operators on a topological space in **[5]**. It was suggested to us by S. Horowitz (oral communication) that one can use the same method for Markov operators.

V. Invariant functions of a recurrent matrix

Throughout this chapter we shall assume: P *is a recurrent matrix.*

Definition 5.1.

$$\Sigma_i = \{A : P1_A = 1_A\}$$

Thus if $A \in \Sigma_i$ then

$$\sum_{y \in A} p(x, y) = 1 \text{ if } x \in A$$

$$\sum_{y \in A} p(x, y) = 0 \text{ if } x \notin A.$$

Theorem 5.1. Σ_i *is a* σ *field.*

Proof. Clearly $\emptyset, X \in \Sigma_i$. If $A \in \Sigma_i$ then $A' \in \Sigma_i$ since $P1 = 1$. If $A, B \in \Sigma_i$ then, since $1_{A \cap B} = \min(1_A, 1_B)$ we have

$$P1_{A \cap B} = P(\min(1_A, 1_B)) \leq \min(P1_A, P1_B) = \min(1_A, 1_B) = 1_{A \cap B}.$$

Thus, since P is recurrent, $A \cap B \in \Sigma_i$. Finally let $A_n \in \Sigma_i$ and $A_n \uparrow A$ then:

$$P1_A = \lim_{n \to \infty} P1_{A_n} = \lim_{n \to \infty} 1_{A_n} = 1_A. \qquad \square$$

Definition 5.2. The matrix P is called ergodic if it is recurrent and $\Sigma_i = \{\emptyset, X\}$.

Now Σ_i is atomic (every σ field on X is). Restrict P to an atom of Σ_i: If P is recurrent then, with no loss of generality, we may assume that P is ergodic.

From the second corollary to Theorem 4.3 follows:

Theorem 5.2. *Let* P *be ergodic. If* $f \geq 0$, $f \neq 0$, *then* $\sum_{n=0}^{\infty} (P^n f)(x) = \infty$ *for every* x.

Take $f = 1_{\{y\}}$ to conclude:

$$\sum_{n=0}^{\infty} p^{(n)}(x, y) = \infty \text{ for any } x, y \in X.$$

From the second corollary to Theorem 4.4 follows:

Theorem 5.3. *Let* P *be ergodic. If* $u \geq 0$, $u \neq 0$, *then* $\sum_{n=0}^{\infty} (uP^n)(x) = \infty$ *for every* x.

Another important result about ergodic operators is:

Theorem 5.4. *Let* P *be ergodic. If* $0 \leq f(x) < \infty$ *and* $Pf = f$ *then* $f = \text{const}$.

Proof. It suffices to show that $\{x : f(x) > a\} \in \Sigma_i$.
Let us recall:

$$g^+(x) = \max(g(x), 0)$$
$$g^-(x) = \max(-g(x), 0).$$

Thus $f - a = (f - a)^+ - (f - a)^-$,

$$(f - a)^- = \max(a - f, 0) \le \max(a, 0) = a.$$

Therefore $P((f - a)^-) \le a$ too. Now $f + (f - a)^- = a + (f - a)^+ \Rightarrow f + P((f - a)^-) = a + P((f - a)^+)$ or

$$f - a = P((f - a)^+) - P((f - a)^-).$$

Therefore

$$(f - a)^- = \max(a - f, 0) \le P((f - a)^-).$$

Hence

$$P(a - (f - a)^-) \le a - (f - a)^-,$$

but P is recurrent and inequality implies equality:

$$P((f - a)^-) = (f - a)^- \text{ thus } (f - a)^+ = f - a + (f - a)^-$$

is invariant too. Now

$$P(\min(1, n(f - a)^+)) \le \min(1, n(f - a)^+)$$

and we must have equality. Finally

$$\min(1, n(f - a)^+) \underset{n\to\infty}{\uparrow} 1_{\{x : f(x) > a\}}. \qquad \square$$

Theorem 5.5. *Let P be ergodic. There exists, at most, a unique solution (up to multiplication by a constant) to:*

$$0 \le u < \infty, \; uP = u.$$

Moreover if $u \ne 0$ then $\{x : u(x) > 0\} = X$.

Proof.

$$\{x : u(x) > 0\} = \{x : \sum_{n=0}^{\infty} (uP^n)(x) > 0\} \in \Sigma_i$$

by Theorem 5.3, hence it is X. Let us prove the uniqueness: Let $0 \le u, v < \infty$ $uP = u$ $vP = v$. Choose $0 < h \in \ell_\infty$ with $\langle u, h\rangle < \infty$ and $\langle v, h\rangle < \infty$. Normalize so that

$$\langle u, h\rangle = \langle v, h\rangle < \infty.$$

Now

$$(u - v)^+ \le u, \qquad (u - v)^- \le v$$

thus

$$((u-v)^+)P^k \le uP^k = u < \infty$$
$$((u-v)^-)P^k \le vP^k = v < \infty.$$

Also

$$u + (u-v)^- = (u-v)^+ + v;$$

hence

$$u + ((u-v)^-)P = ((u-v)^+)P + v$$

or

$$u - v = ((u-v)^+)P - ((u-v)^-)P.$$

Thus

$$((u-v)^+)P \ge (u-v)^+.$$

Therefore

$$\begin{aligned}\langle ((u-v)^+)P &- (u-v)^+, (I+P+\cdots+P^n)h\rangle \\ &= \langle ((u-v)^+)(P-I)(I+P+\cdots+P^n), h\rangle \\ &= \langle ((u-v)^+)P^{n+1}, h\rangle - \langle ((u-v)^+), h\rangle \\ &\le \langle uP^{n+1}, h\rangle = \langle u, h\rangle < \infty.\end{aligned}$$

But $(I+P+\cdots+P^n)h(x) \underset{n\to\infty}{\to} \infty$ for every $x \in X$. Thus

$$((u-v)^+)P = (u-v)^+.$$

Hence the support of $(u-v)^+$ is trivial: either $u \le v$ or $v \le u$ and finally $u = v$ since $\langle u, h\rangle = \langle v, h\rangle$. □

VI. The main results

Lemma 6.1. *Let P be an ergodic matrix. Let $P = Q + R$ where Q, R are Markov matrices and $R \ne 0$. Then:*

$$Q^n 1 \underset{n\to\infty}{\downarrow} 0. \tag{1}$$

$$\sum_{n=0}^{\infty} Q^n R 1 = 1. \tag{2}$$

Proof. (1): $Q1 \le P1 = 1$, thus $Q^{n+1}1 \le Q^n 1$. Let $g = \lim_{n\to\infty} Q^n 1$ then $0 \le g \le 1$ and $Qg = g$. Thus $Pg \ge Qg = g$ and hence $P(1-g) \le 1-g$. Therefore $g = \text{const}$.

Now $Rg = Pg - Qg = g - g = 0$, hence $g = \text{ const } = 0$ (otherwise $R1 = 0$ thus $R = 0$).

(2): $\sum_{n=0}^{N} Q^n R1 = \sum_{n=0}^{N} Q^n(P-Q)1 = \sum_{n=0}^{N} Q^n(I-Q)1 = 1 - Q^{N+1}1 \underset{N\to\infty}{\uparrow} 1.$ □

Corollary. *If* $f \in \ell_\infty$ *then* $Q^n f \underset{n\to\infty}{\to} 0$.

Proof. $|Q^n f| \le Q^n |f| \le \|f\|_\infty \; Q^n 1 \underset{n\to\infty}{\to} 0.$ □

Throughout this chapter we shall use the following notation: $(h \otimes w)$ *is the Markov matrix defined by* $(h \otimes w)(x, y) = h(x)w(y)$, *where*

$$0 \le h(x) \le 1, \quad h \ne 0$$
$$0 \le w(x), \quad \sum_{x\in X} w(x) \le 1, \quad w \ne 0.$$

Note

$$\sum_{y\in X} (h \otimes w)(x, y) = \big(\sum_{y\in X} w(y)\big)h(x) \le h(x) \le 1.$$

Hence $h \otimes w$ is a Markov matrix. Also

$$((h \otimes w)f)(x) = \big(\sum_{y\in X} w(y)f(y)\big)h(x) = \langle w, f\rangle h(x)$$

or

$$(h \otimes w)f = \langle w, f\rangle h.$$

Similarly

$$u(h \otimes w) = \langle u, h\rangle w.$$

If P_1 is any Markov matrix then

$$P_1((h \otimes w)f) = \langle w, f\rangle P_1 h = (P_1 h \otimes w)f$$

or

$$P_1(h \otimes w) = (P_1 h) \otimes w.$$

Also

$$(h \otimes w)(P_1 f) = \langle w, P_1 f\rangle h = \langle wP_1, f\rangle h = (h \otimes wP_1)f$$

or

$$(h \otimes w)P_1 = h \otimes (wP_1).$$

Given P choose h, w with

$$P = h \otimes w + Q, \quad Q \ge 0. \tag{$*$}$$

Recall $h \otimes w \ne 0$. We may choose $w = 1_{\{y_0\}}$ and $h(x) \le p(x, y_0)$ or we may take $h = 1_{\{x_0\}}$ and $w(y) \le p(x_0, y)$. We can not have $p(x, y_0) = 0$ for all x:

$$P1_{\{y_0\}} = 0 \le 1_{\{y_0\}} \Rightarrow 1_{\{y_0\}} = 0$$

by recurrence, a contradiction.

We can not have $p(x_0, y) = 0$ for all y:

$$1_{\{x_0\}}P = 0 \le 1_{\{x_0\}} \Rightarrow 1_{\{x_0\}} = 0$$

by recurrence, a contradiction. Thus we may obtain $(*)$ with

$$h \otimes w \ne 0.$$

Lemma 6.2. *Let P be an ergodic matrix satisfying $(*)$. Then*

$$\sum_{n=0}^{\infty} Q^n h = \frac{1}{\langle w, 1\rangle}.$$

Proof. Use part (2) of Lemma 6.1. with $R = h \otimes w$:

$$1 = \sum_{n=0}^{\infty} Q^n (h \otimes w)1 = \langle w, 1\rangle \sum_{n=0}^{\infty} Q^n h.$$ □

Theorem 6.3. *Let P be an ergodic matrix. There exists a function v with $0 < v(x) < \infty$ for all x and $vP = v$. If P satisfies $(*)$ we may choose $v = \sum_{n=0}^{\infty} wQ^n$, in which case $\langle v, h\rangle = 1$.*

Proof. We saw that $Q^n 1 \downarrow 0$, hence Q is nonrecurrent. Thus $0 \le v(x) < \infty$. Now

$$\langle v, h\rangle = \Big\langle \sum_{n=0}^{\infty} wQ^n, h\Big\rangle = \Big\langle w, \sum_{n=0}^{\infty} Q^n h\Big\rangle = 1$$

by Lemma 6.2.

Hence $vP = vQ + v(h \otimes w) = \sum_{n=1}^{\infty} wQ^n + \langle v, h\rangle w = v$. Finally $v(x) > 0$ by Theorem 5.5. □

Lemma 6.4. *Let P be an ergodic matrix satisfying $(*)$. Then*

$$\sum_{n=0}^{\infty} Q^n \Big(\sum_{i=0}^{K} P^i h\Big) \in \ell_\infty.$$

Proof. $K = 0$ is proved in Lemma 6.2. Now if $f \ge 0$ and $\sum_{n=0} Q^n f \in \ell_\infty$ then

$$\sum_{n=0}^{\infty} Q^n Pf = \sum_{n=1}^{\infty} Q^n f + \langle w, f\rangle \sum_{n=0}^{\infty} Q^n h \in \ell_\infty.$$ □

Definition 6.1. Let P be an ergodic matrix satisfying $(*)$. Put

$$\Lambda = \{A : \sum_{n=0} Q^n 1_A \in \ell_\infty\}$$

and denote $C(A) = \left\| \sum_{n=0}^{\infty} Q^n 1_A \right\|_\infty$ when $A \in \Lambda$.

If

$$A = \{x : \sum_{i=0}^{K} (P^i h)(x) \geq \varepsilon\}$$

where $\varepsilon > 0$ then

$$1_A \leq \frac{1}{\varepsilon} \sum_{i=0}^{K} P^i h$$

or:

Lemma 6.5. *Let P be an ergodic matrix satisfying $(*)$. Then, for every $\varepsilon > 0$ and every integer K,*

$$\{x : \sum_{i=0}^{K} (P^i h)(x) > \varepsilon\} \in \Lambda.$$

In particular: Every finite set belongs to Λ.

Proof. It suffices to recall that

$$\sum_{n=0}^{\infty} (P^n h)(x) = \infty. \qquad \square$$

Theorem 6.6. *Let P be an ergodic matrix satisfying $(*)$. Let $A \in \Lambda$. If $f_1, f_2 \in \ell_\infty$ are supported on A and $\langle v, f_1 \rangle = \langle v, f_2 \rangle$ then*

$$\left\| \sum_{n=0}^{N} P^n (f_1 - f_2) \right\|_\infty \leq 2C(A) \left(\|f_1\|_\infty + \|f_2\|_\infty \right).$$

Proof. Recall $\sum_{n=0}^{\infty} Q^n 1_A \leq C(A)$. Thus

$$\sum_{n=0}^{\infty} Q^n f_i \leq \|f_i\|_\infty \sum_{n=0}^{\infty} Q^n 1_A \leq C(A) \|f_i\|_\infty, \quad i = 1, 2.$$

Now

$$\begin{aligned} \langle v, f_i \rangle = \langle \sum_{n=0}^{\infty} w Q^n, f_i \rangle &= \langle w, \Sigma Q^n f_i \rangle \\ &\leq C(A) \|f_i\|_\infty \langle w, 1 \rangle < \infty \quad i = 1, 2. \end{aligned}$$

Put

$$\tilde{f}_i = \sum_{n=0}^{\infty} Q^n f_i, \text{ then } \|\tilde{f}_i\|_\infty \leq C(A) \|f_i\|_\infty.$$

Thus $\tilde{f}_i \in \ell_\infty$ $i = 1, 2$. Now

$$(I - Q)\tilde{f}_i = \lim_{N\to\infty} (I - Q) \sum_{n=0}^{N} Q^n f_i$$
$$= f_i - \lim_{N\to\infty} Q^{N+1} f_i = f_i, \quad i = 1, 2,$$

by the corollary to Lemma 6.1.

Finally

$$(I - P)(\tilde{f}_1 - \tilde{f}_2) = (I - Q)(\tilde{f}_1 - \tilde{f}_2) - (h \otimes w)(\tilde{f}_1 - \tilde{f}_2)$$
$$= f_1 - f_2 - \langle w, \tilde{f}_1 - \tilde{f}_2 \rangle h.$$

But

$$\langle w, \tilde{f}_i \rangle = \langle w, \sum_{n=0}^{\infty} Q^n f_i \rangle = \langle \sum_{n=0}^{\infty} w Q^n, f_i \rangle = \langle v, f_i \rangle$$

and $\langle v, f_1 \rangle = \langle v, f_2 \rangle < \infty$. Hence

$$(I - P)(\tilde{f}_1 - \tilde{f}_2) = f_1 - f_2.$$

Therefore

$$\Big\| \sum_{n=0}^{N} P^n (f_1 - f_2) \Big\|_\infty = \Big\| \sum_{n=0}^{N} P^n (I - P)(\tilde{f}_1 - \tilde{f}_2) \Big\|_\infty$$
$$= \| \tilde{f}_1 - \tilde{f}_2 - P^{N+1}(\tilde{f}_1 - \tilde{f}_2) \|_\infty$$
$$\leq 2(\| \tilde{f}_1 \|_\infty + \| \tilde{f}_2 \|_\infty)$$
$$\leq 2C(A)(\| f_1 \|_\infty + \| f_2 \|_\infty). \qquad \square$$

Corollary. *Let P be an ergodic matrix then*

$$\sup_N \sup_x \Big| v(x_1) \sum_{n=0}^{N} p^{(n)}(x, x_2) - v(x_2) \sum_{n=0}^{N} p^{(n)}(x, x_1) \Big| \leq \text{const.} < \infty.$$

Proof. $\{x_1, x_2\} \in \Lambda$ by Lemma 6.4. Now

$$\langle v, v(x_2) 1_{\{x_1\}} - v(x_1) 1_{\{x_2\}} \rangle = 0.$$

$\square$

Divide by $\sum_{n=0}^{N} p^{(n)}(x, x_2)$ which tends to ∞ as $N \to \infty$ to obtain

$$\frac{\sum_{n=0}^{N} p^{(n)}(x, x_1)}{\sum_{n=0}^{N} p^{(n)}(x, x_2)} \underset{N\to\infty}{\to} \frac{v(x_1)}{v(x_2)}.$$

Theorem 6.7. *Let P be ergodic. If for some $y_0 \in X$, integer K and $\varepsilon > 0$ we have*

$$\frac{1}{K}\sum_{i=1}^{K} p^{(i)}(x, y_0) \geq \varepsilon$$

for all $x \in X$, then there exists a constant C such that

$$\sum_{y\in X}\Big|\frac{1}{N+1}\sum_{n=0}^{N} p^{(n)}(x, y) - \frac{v(y)}{\langle v, 1\rangle}\Big| \leq \frac{C}{N+1}$$

where $\langle v, 1\rangle < \infty$.

Proof. Choose $h(x) = p(x, y_0)$ and $w = \delta_{y_0}$. Then $P = h \otimes w + Q$. If $v = \sum_{n=0}^{\infty} wQ^n$ then $vP = v$ and $\langle v, h\rangle = 1$. Now

$$(P^n h)(x) = p^{(n+1)}(x, y_0)$$

hence the assumption of the theorem reads:

$$\sum_{i=0}^{K-1}(P^i h)(x) \geq K\varepsilon.$$

Thus

$$\langle v, 1\rangle \leq \frac{1}{K\varepsilon}\sum_{i=0}^{K-1}\langle v, P^i h\rangle = \frac{1}{K\varepsilon}\sum_{i=0}^{K-1}\langle vP^i, h\rangle = \frac{1}{\varepsilon} < \infty,$$

since $vP^i = v$.

Now $X \in \Lambda$ (Definition 6.1):

$$\sum_{n=0}^{\infty} Q^n 1 \leq \frac{1}{K\varepsilon}\sum_{n=0}^{\infty} Q^n\Big(\sum_{i=0}^{K-1} P^i h\Big) \in \ell_\infty$$

by Lemma 6.4.

Let $0 \leq f \in \ell_\infty$. Put $f_1 = f$, $f_2 = \frac{\langle v, f\rangle}{\langle v, 1\rangle}1$. Then $\|f_2\|_\infty \leq \|f\|_\infty = \|f_1\|_\infty$. Also $\langle v, f_1\rangle = \langle v, f_2\rangle$. Use Theorem 6.6 with $C = C(X)$:

$$4C\|f\|_\infty \geq \Big\|\sum_{n=0}^{N} P^n(f_1 - f_2)\Big\|_\infty = \Big\|\sum_{n=0}^{N} P^n f - (N+1)\frac{\langle v, f\rangle}{\langle v, 1\rangle}1\Big\|_\infty.$$

Divide by $N + 1$ and put in a matrix form:

$$\Big|\sum_{y\in X}\Big[\frac{1}{N+1}\sum_{n=0}^{N} p^{(n)}(x, y) - \frac{v(y)}{\langle v, 1\rangle}\Big]f(y)\Big| \leq \frac{4C\|f\|_\infty}{N+1}.$$

If $-1 \leq g \leq 1$ use the above for $f = g^+$ or $f = g^-$ to conclude:

If $-1 \le g \le 1$ then

$$\Big|\sum_{y\in X}\Big[\frac{1}{N+1}\sum_{n=0}^{N} p^{(n)}(x,y) - \frac{v(y)}{\langle v,1\rangle}\Big]g(y)\Big| \le \frac{8C}{N+1}$$

Fix x and choose $g(y) = \pm 1$ to obtain

$$\sum_{y\in X}\Big|\Big(\frac{1}{N+1}\sum_{n=0}^{N} p^{(n)}(x,y)\Big) - \frac{v(y)}{\langle v,1\rangle}\Big| \le \frac{8C}{N+1}. \qquad \square$$

If S is a linear operator on ℓ_∞ define

$$\|S\| = \sup\{\|Sf\|_\infty : \|f\|_\infty \le 1\}.$$

It is easy to see that if S is the matrix $\big(s(x,y)\big)$ then

$$\|S\| = \sup_x \sum_{y\in X} |s(x,y)|.$$

Thus the conclusion of the Theorem is:

$$\Big\|\frac{1}{N+1}\sum_{n=0}^{N} P^n - E\Big\| \le \frac{C}{N+1}$$

where $Ef = \frac{\langle v,f\rangle}{\langle v,1\rangle}$.

Let us outline a result about the convergence of P^n:

Let $p(x,y_0) \ge \varepsilon > 0$ for all $x \in X$, then $P = h \otimes w + Q$, $h(x) = p(x,y_0)$, $w = 1_{\{y_0\}}$. Hence $P = 1 \otimes w_1 + Q_1$, $w_1 = \varepsilon w$. Therefore:

If $v = \sum_{n=0}^{\infty} w_1 Q_1^n$, then $vP = v$, $\langle v,1\rangle = 1$. Also

$$Q_1 1 = P1 - \langle w_1, 1\rangle = 1 - \langle w_1, 1\rangle$$

or

$$\|Q_1\| = 1 - \langle w_1, 1\rangle.$$

It is easy to see, by an induction argument that $P^n = 1 \otimes w_1 + 1 \otimes (w_1 Q_1) + \cdots + 1 \otimes (w_1 Q_1^{n-1}) + Q_1^n$. Hence $\|P^n - E\| \le 2(1 - \langle w_1, 1\rangle)^n$ where $Ef = \langle v, f\rangle$.

Remark. The basic tool, in this chapter, was Equation $(*)$: $P = h \otimes w + Q$. In the case of "Harris type Markov operators" one uses instead Orey's Lemma (see **[11**, Theorem 2.1**]**) which is a weaker result and it requires a rather difficult measure theoretic consideration. Of course it deals with a much more general situation.

We used here the methods developed in **[8]**. The classical proof of Theorem 6.3 uses "Taboo probabilities" see **[2**, Chapter I.9**]**.

Theorem 6.6 was proved, for Harris type Markov operators, in **[12]**, **[10]** and **[1]**. Theorem 6.8 was proved in **[3]**. Other proofs are given in **[9]** and **[13]**.

VII. The adjoint matrix

Throughout this chapter we assume:

Assumption 7.1. $P1 = 1$. *There exists a function* v *such that:* $0 < v(x) < \infty$, $vP = v$.

If P is ergodic then Assumption 7.1. holds.

Definition 7.1. Let P satisfy Assumption 7.1. Define P^* (the adjoint matrix) by

$$p^*(x, y) = \frac{v(y)}{v(x)} p(y, x).$$

Theorem 7.1. *Let* P *satisfy Assumption* 7.1. *Then* P^* *is a Markov matrix and:*
(1) $P^*1 = 1$.
(2) $vP^* = v$.
(3) $P^{**} = P$.
(4) $(P^*f)(x) = \frac{1}{v(x)}\Big((vf)P\Big)(x)$.
(5) $(uP^*)(x) = v(x)(P\frac{u}{v})(x)$.
(6) $(P^*)^n = (P^n)^*$.
(7) $Y(P^*) = Y(P)$, $Z(P^*) = Z(P)$.
(8) *If* P *is ergodic then so is* P^*.
(9) $\langle P^*f, vg\rangle = \langle Pg, vf\rangle$ *if* $f, g \geq 0$.

Proof. (1): $\sum_{y \in X} p^*(x, y) = \frac{1}{v(x)} \sum_{y \in X} v(y)p(y, x) = 1$ since $vP = v$.

(2) :
$$(vP^*)(x) = \sum_{y \in X} v(y)p^*(y, x) = \sum_{y \in X} v(y)\frac{v(x)}{v(y)}p(x, y)$$
$$= v(x)\sum_{y \in X} p(x, y) = v(x)$$

since $P1 = 1$.

(3): $p^{**}(x, y) = \frac{v(y)}{v(x)} p^*(y, x) = \frac{v(y)}{v(x)}\frac{v(x)}{v(y)} p(x, y) = p(x, y)$.

(4) :
$$(P^*f)(x) = \sum_{y \in X} p^*(x, y)f(y) = \sum_{y \in X} \frac{1}{v(x)} v(y)f(y)p(y, x)$$
$$= \frac{1}{v(x)}\Big((vf)P\Big)(x).$$

(5) :
$$(uP^*)(x) = \sum_{y \in X} u(y)p^*(y, x) = \sum_{y \in X} u(y)\frac{v(x)}{v(y)}p(x, y)$$
$$= v(x)(P\frac{u}{v})(x).$$

(6): Induction on n:

$$\begin{aligned}\big((P^*)^{n+1} f\big)(x) &= \big(P^*(P^{*n} f)\big)(x)\\ &= P^*\big(\frac{1}{v}((vf)P^n)\big)(x)\\ &= \frac{1}{v(x)}\big((vf)P^{n+1}\big)(x) = \big((P^{n+1})^* f\big)(x).\end{aligned}$$

(7): Let $x_0 \in Y(P)$. Then $\sum_{n=0}^{\infty} p^{(n)}(x_0, x_0) < \infty$. Hence

$$\sum_{n=0}^{\infty} p^{*(n)}(x_0, x_0) = \frac{v(x_0)}{v(x_0)} \sum_{n=0}^{\infty} p^{(n)}(x_0, x_0) < \infty.$$

Therefore $x_0 \in Y(P^*)$ or $Y(P) \subset Y(P^*)$, and by (3) we have equality.

(8): If P is ergodic, then

$$\sum_{n=0}^{\infty} p^{(n)}(x, y) = \infty \quad \text{for every} \quad x, y \in X.$$

Now

$$p^{*(n)}(x, y) = \frac{v(y)}{v(x)} p^{(n)}(y, x).$$

So

$$\sum_{n=0}^{\infty} p^{*(n)}(x, y) = \infty \quad \text{for every} \quad x, y \in X.$$

(9): $\langle P^* f, vg\rangle = \langle f, (vg)P^*\rangle = \langle f, vPg\rangle = \langle Pg, vf\rangle$. □

Let us now consider *all* our results applied to P^*. From Theorem 3.1. follows:

Theorem 7.2. *If* $y_0 \in Y$ *then*

$$\sup_x \Big[\frac{1}{v(x)} \sum_{n=0}^{\infty} p^{(n)}(y_0, x)\Big] < \infty.$$

From Theorem 3.2. follows:

Theorem 7.3. *Let* $0 \leq u \in \ell_1$ *and* $y_0 \in Y$ *then*

$$\sum_{n=0}^{\infty} (P^n \frac{u}{v})(y_0) < \infty.$$

From Theorem 3.4. follows:

Theorem 7.4. $(v 1_Y) P \leq v 1_y$.

Let us now consider recurrent states. The corollary to Theorem 4.3. reads: *If* $x_0 \in Z(P)$ *and* $p^{(i)}(x_0, x) > 0$ *for some integer* $i \geq 0$ *then*

$$\sum_{n=0}^{\infty} p^{(n)}(x_0, x) = \infty,$$

which is nothing but the corollary to Theorem 4.4.

Consider now Chapter VI. The corollary to Theorem 6.6 becomes:

Theorem 7.6. *Let* P *be ergodic then*

$$\Big|\sum_{n=0}^{N}\Big(p^{(n)}(x_2, x) - p^{(n)}(x_1, x)\Big)\Big| \leq \text{const. } v(x).$$

Remark. If $v \in \ell_1$ then

$$\sum_{x\in X}\Big|\sum_{n=0}^{N}\Big(p^{(n)}(x_2, x) - p^{(n)}(x_1, x)\Big)\Big| \leq \text{const.}$$

Now Theorem 6.7. reads:

Theorem 7.7. *Let* P *be ergodic. If for some* $y_0 \in X$, *an integer* K *and an* $\varepsilon > 0$ *we have*

$$\frac{1}{K}\sum_{i=1}^{K}\frac{1}{v(x)}p^{(i)}(y_0, x) \geq \varepsilon$$

for all $x \in X$, *then there exists a constant* C *such that*

$$\sum_{y\in X} v(y)\Big|\frac{1}{N+1}\frac{1}{v(x)}\sum_{n=0}^{N} p^{(n)}(y, x) - \frac{1}{\langle v, 1\rangle}\Big| \leq \frac{C}{N+1}.$$

Bibliography

[1] A. Brunel, Chaînes abstraites de Markov vérifient une condition de Orey, Z. Wahrscheinlichkeitstheorie verw. Geb. 19 (1971), 323–329.

[2] K. L. Chung, Markov chains with stationary transition probabilities, Grundlehren Math. Wiss. 104, Springer-Verlag, Berlin 1960.

[3] W. Doeblin, Éléments d'une théorie générale de chaînes simples constantes de Markoff, Ann. École Norm. (3) 57 (1940), 61–111.

[4] S. R. Foguel, The ergodic theory of Markov processes, Van Nostrand, New York 1969.

[5] S. R. Foguel, Ergodic decomposition of a topological space, Israel J. Math. 7 (1969), 164–167.

[6] S. R. Foguel, Selected topics in the study of Markov operators, Carolina lecture series, Dept. of Math., University of North Carolina at Chapel Hill 1980.

[7] S. R. Foguel, Harris operators, Israel J. Math. 33 (1979), 281–309.

[8] S. R. Foguel, A new approach to the study of Harris type Markov operators, Rocky Mountain J. Math. 19 (1989), 491–512.

[9] S. Horowitz, Transition probabilities and contractions on L_∞, Z. Wahrscheinlichkeitstheorie verw. Geb. 24 (1972), 263–274.

[10] M. Métivier, Existence of an invariant measure and an Ornstein's ergodic theorem, Ann. Math. Statist. 40 (1969), 79–96.

[11] S. Orey, Limit theorems for Markov chain transition probabilities, Van Nostrand, New York 1971.

[12] D. S. Ornstein, Random walks I and II, Trans. Amer. Math. Soc. 138 (1969), 1–43, 45–60.

[13] K. Yosida and S. Kakutani, Operator theoretical treatment of Markoff's process and mean ergodic theorem, Ann. Math. 42 (1941), 188–228.

Institute of Mathematics
The Hebrew University
Givat Ram
91904 Jerusalem, Israel

Existence and Continuity of the Quadratic Variation of Strong Martingales

Nikos E. Frangos and Peter Imkeller

Abstract. We prove the existence (in the sense of L_0 convergence) of L_1-bounded strong martingales. The proof is through stochastic intergrals with respect to strong martingales. The continuity is an easy consequence of the fact that the Q.V. of a strong martingale is equal to the Q.V. of either of the one parameter 'marginal' martingales.

Let M be a strong martingale. We assume that M is regular (see Walsh **[12]**). By one parameter results we know that $[M]^1$ and $[M]^2$ exist. From this we conclude:

Proposition 1. *The quadratic variation of M exists if and only if for every 0-sequence of partitions $(J_m)_{m\in\mathbb{N}}$ of partitions $[0,1]^2$ and any $t\in[0,1]^2$, the sequence*

$$\Big(\sum_{j\in J_m}\Delta_{j\cap(0,t)}M\,\Delta_{j^1\cap(0,t)}M\Big)_{m\in\mathbb{N}}\quad (J=J^1\times J^2)$$

converges in $L^0(\Omega,\mathcal{F},P)$ or equivalently $(\Sigma_{j\in J_m}\Delta_{j\cap(0,t)}\Delta_{j^2\cap(0,t)}M)_{m\in\mathbb{N}}$ converges in $L^0(\Omega,\mathcal{F},P)$.

Proof. Let $(J_m)_{m\in\mathbb{N}}$ be an 0-sequence of partitions of $[0,1]^2$; we assume $t=1$. Then for any $m\in\mathbb{N}$

$$\sum_{j\in J_m}(\Delta_{j^1}M_{(.,1)})^2-\sum_{j\in J_m}(\Delta_j M)^2=2\sum_{j\in J_m}\Delta_j M\,\Delta_{j^1}M. \tag{1}$$

Since the first term on the left had side of (1) converges to $[M]^1$ by one-parameter results, the first equivalence assertion follows readily. The rest is symmetric. □

Following Proposition 1, we have to prove the convergence of the sequence $\sum_{j\in J_m}\Delta_j M\Delta_{j^1}M_{m\in\mathbb{N}}$ in $L^0(\Omega,\mathcal{F},P)$ for any 0-sequence $(J_m)_{m\in\mathbb{N}}$ of partitions of $[0,1]^2$, in order to establish the existence of the quadratic variation of M. In order to prove this we first have to study a stochastic L^0-integral which is defined on the following space of simple functions:

$\mathcal{E}_2=\{\sum_{j\in J}\alpha_j 1_j : \alpha_j$ is an $\mathcal{F}_{(1,s_2^j)}$-measurable step function$\}$, $(j\in J)$, J partition of $[0,1]^2$ i.e. the space of 2-previsible elementary functions. It is easy to show that $\mathcal{E}_2$ is a linear space.

Proposition 2. *For an* $Y_0 = \sum_{j\in J} \alpha_j 1_j \in \mathcal{E}_2$ *such that* $|Y_0| \leq 1$ *and any* $\lambda > 0$ *we have*

$$\lambda P(|\sum_{j\in J} \alpha_j \Delta_j M| > \lambda) \leq \|M_1\|_1.$$

Proof. Number $J_1 = \{K_i : 1 \leq i \leq |J_1|\}$, $J_2 = \{L_j : 1 \leq j \leq |J_2|\}$, in increasing order and set $J_{(i,j)} = K_i \times L_j$, $1 \leq (i, j) \leq (|J_1|, |J_2|)$. Then take the alphabetical order of these intervals following direction 1 and set

$$I_{|J^1|\cdot(l-1)+k} = J_{(k,l)}, \quad 1 \leq (k, l) \leq (|J_1|, |J_2|).$$

Moreover, take

$$g_m = V_{1\leq j\leq m} \mathcal{F}_t I_j, \quad N_m = \sum_{1\leq j\leq m} \Delta_{I_k} M, \quad \beta_m = \alpha_{I_m}, \quad I \leq m \leq |J|.$$

Then

$$(N_m, g_m)_{1\leq m\leq |J|}$$

is a martingale and

$$\sum_{1\leq m|J|} \beta_m \Delta_{(m-1,m)} N = \sum_{j\in J} \alpha_j \Delta_j M, M_1 = N_{|J|}.$$

Therefore the desired inequality follows from Burkholder **[4]**. □

Definition 1. Let X be a process in $[0, 1]^2$. Then the linear mapping

$$\int \cdot dX: \mathcal{E}_2 \to L^0(\Omega, \mathcal{F}, P), \quad Y_0 = \sum_{j\in J} \alpha_j 1_j \to \sum_{j\in J} \alpha_j \Delta_j X,$$

is called "*elementary stochastic integral of X*" on $\mathcal{E}_2$.

Corollary. *The set* $\{\int Y_0\, dM : Y_0 \in \mathcal{E}_2, |Y_0| \leq 1\}$ *is bounded in* $L^0(\Omega, \mathcal{F}, P)$.

We would like to apply Bichteler **[1]**, p. 55, Theorem (2.6), in order to show that M defines an L^0-stochastic integrator in the sense of

Definition 2. A process X is called an "*L^0-stochastic integrator*" if

$$\int \cdot dX: \mathcal{E}_2 \to L^0(\Omega, \mathcal{F}, P)$$

has an extension satisfying the dominated convergence theorem.

With $\mathcal{E}_2$ in place of $\mathcal{R}$ the proof of Theorem (2.1) of Bichteler **[1]**, p.51, goes through without essential changes. Also the proof of theorem (2.6) there can be taken over almost word by word. As a consequence of this and the right continuity in probability of M we have

Proposition 3. *The martingale M is an L^0-stochastic integrator.*

Now according to Bichteler **[1]**, p. 54, (2.4.1), and (2.4.2) all $\mathcal{P}^2$-measurable processes ($\mathcal{P}^2$ is the σ-algebra generated by $\mathcal{E}_2$) Y for which $\lim_{\lambda \to 0} G(\lambda Y) = 0$ are integrable, where G is defined in several steps (Bichteler **[1]**, p. 52):

$$s(f) = \inf\{\epsilon > 0 : P(|f| > \epsilon) \leq \epsilon\},$$

f is $\mathcal{F}$- measurable,

$$G(Y) = \sup\{s(\int XdM) : X \in \mathcal{E}_2,\ |X| \leq Y\},$$

for Y which are pointwise suprema of functions in $\mathcal{E}_2$,

$$G(Z) = \inf\{G(Y) : |Z| \leq Y,\ Y \text{ as above}\}.$$

Now we are in a position to deal with the limit of Proposition 1.

Proposition 4. *Let $(J_m)_{m \in \mathbb{N}}$ be an arbitrary 0-sequence of partitions of $[0,1]^2$. Then $(\sum_{j \in J} \Delta_j M \Delta_{j1} M)_{m \in \mathbb{N}}$ converges to $\int \Delta^1_{s_1} M_{(\cdot, s_2 -)}\, dM_s$ in $L^0(\Omega, \mathcal{F}, P)$.*

Proof. 1. Let $Y_t = 4 \sup_{s_2 < t_2, s_1 \in [0,1]} |M_s|, t \in [0,1]^2$.

Then $Y \in m(\mathcal{P}^2, \mathcal{L}(\mathcal{R}))$. To show that Y is integrable, we have to show that Y is G-finite. To this end, we consider a function $X \in \mathcal{E}_2$ such that $|X| \leq |Y|$. Then for $\epsilon > 0$, putting $X = \sum_{j \in J} \alpha_j 1_j$,

$$\begin{aligned} P(|\sum_{j \in J} \alpha_j \Delta_j M| > \epsilon) &\leq (|\sum_{j \in J} \langle -k, \alpha_j k \rangle \Delta_j M| > \epsilon) + P(\sup_{j \in J} |\alpha_j| < k) \\ &\leq 2k/\epsilon \|M_1\|_1 + P(\sup_{t \in [0,1]^2} |M_t| > k) \\ &\leq 2k/\epsilon \|M_1\|_1 + 2c/k \|M_1\|_1,\ k \in \mathcal{R}_+ \quad \text{(by Proposition 1).} \end{aligned}$$

This implies that $\lim_{\lambda \to 0} G(\lambda Y) = 0$, i.e. that Y is G-finite. Therefore Y is integrable.

2. First of all, 1 implies that the sequence $(Y_m)_{m \in \mathbb{N}}$ defined by

$$Y_m = \sum_{j \in J} \Delta_j M 1_j \quad m \in \mathbb{N},$$

is integrable (2-previsible and finite, due to $|Y_m| \leq Ym \in \mathbb{N}$) and converges pointwise to $\Delta^1_{s_1} M_{(\cdot, s_2)}$, bounded by Y. Hence the dominated convergence theorem applies and gives the desired conclusion, since $\int Y_m\, dM = \sum_{j \in J} \Delta_j M \Delta_{j1} M,\ m \in \mathbb{N}$. □

As a direct consequence of Propositions 1 and 4, our first result comes in.

Theorem 1. *The martingale M has a quadratic variation in $L^0(\Omega, \mathcal{F}, P)$.*

Proof. Combine Propositions 1 and 4. □

Also immediately from our procedure we obtain the continuity properties of the quadratic variation.

Theorem 2. 1. *If M does not have 0-jumps and 1-jumps, the $[M]$ does not have them either.*

2. *If M does not have 0-jumps and 2-jumps, then $[M]$ does not have them either.*

3. *If M is continuous, then so is $[M]$.*

Proof. If M has no 0- and 1-jumps $[M]^1_{(\cdot,1)}$ is continuous by one parameter result and $\int \Delta^1_{s_1} M_{(\cdot,s_2-)} dM = 0$, hence by Propositions 1 & 4 $[M]^1 = [M]$. Since $[M]$ is increasing, it has no 0- and 1-jumps. If M is continuous, we have in addition that $[M]^2_{(\cdot,1)}$ is continuous and by analogus result $[M]^2 = [M]$. Hence $[M]$ has no 2-jumps, thus it is continuous. □

References

[1] Bichteler, K., Stochastic integration and L^p-theory of of semimartingales, Ann. Probab. 9 (1981), 49–89.

[2] Brossard, J., Comparaison des "normes" L_p du processus croissant et de la variable maximale pour les martingales régulières à deux indices. Théorème local correspondant, Ann. Probab. 8 (1980), 1183–1188.

[3] Brossard, J., Régularité des martingales à deux indices et inégalités de normes, in: Processus Aléatoires à Deux Indices, Lecture Notes in Math. 863, pp. 91–121, Springer-Verlag, Berlin-Heidelberg-New York 1981.

[4] Burkhoder, D. L., Distribution function inequalities for martingales, Ann. Probab. 1 (1973), 19–42.

[5] Cairoli, R, Walsh, J. B., Stochastic integrals in the plane, Acta Math. 134 (1975), 111–183 .

[6] Davis, B. J., On the integrability of the martingale square function, Israel J. Math. 8 (1970), 187–190.

[7] Frangos, N. E., Imkeller, P., Some inequalities for strong martingales, Ann. Inst. H. Poincaré 24 (1988), 395–402.

[8] Gundy, R. F., Inégalités pour martingales à un et deux indices: l'éspace H^p, in: Ecole d'Eté de Probalités de Saint-Flour VIII-1978, Lecture Notes in Math. 774, pp. 251–331, Springer-Verlag, Berlin-Heidelberg-New York 1980.

[9] Imkeller, P., A stochastic calculus for continuous N-parameter strong martingales, Stochastic Process. Appl. 20 (1985), 1–40.

[10] Ledoux, M., Transformées de Burkholder et sommabilité de martingales à deux parametres, Math. Z. 181 (1982), 529–535.

[11] Neveu, J., Discrete-parameter martingales, North-Holland 1971.

[12] Walsh, J. B., Convergence and regularity of multiparameter strong martingales, Z. Wahrscheinlichkeitsheorie Verw. Geb. 46 (1974), 177–192.

[13] Wong, E., Zakai, M., Weak martingales and stochastic integrales in the plane, Ann. Probab. 4 (1976), 570–586.

Hofstra University
Department of Mathematics
Hempstead, N.Y. 11550, U.S.A.
and
Department of Statistics
Economics University of Athens
Patision 76 Athens, Greece

Mathematisches Institut
der Ludwig-Maximillians-Universität München
Theresienstraße 39
80333 München
Federal Republic of Germany

Convergence Rate in the Strong Law of Large Numbers for Markov Chains

Cheng Der Fuh and Tze Leung Lai

1. Introduction

Let $\{X_n,\ n \geq 0\}$ be a Markov chain on a measurable space $(D, \mathcal{A})$. If the chain is Harris recurrent and has an invariant probability measure π, then for every Borel function $f: D \to \mathbb{R}$ such that $\int |f(x)|\, d\pi(x) < \infty$ and for every $x \in D$,

$$P\{\lim_{n\to\infty} n^{-1} \sum_{t=1}^{n} f(X_t) = \mu \mid X_0 = x\} = 1, \tag{1}$$

where $\mu = \int f(x)\, d\pi(x)$, cf. **[AtNe]**. We study here the rate of almost sure convergence in (1). Let $S_n = \sum_{t=1}^{n} f(X_t)$. Specifically, we shall evaluate the order of magnitude, as $m \to \infty$, of

$$p_{x,\epsilon}(m) := P\{n^{-1} S_n - \mu > \epsilon \text{ for some } n \geq m \mid X_0 = x\}. \tag{2}$$

Replacing f by $-f$ in the argument also yields $P\{\mu - n^{-1} S_n > \epsilon \text{ for some } n \geq m \mid X_0 = x\}$.

For the special case of independent and identically distributed (i.i.d.) random variables X_n taking values in D, π is the distribution of X_1 and Siegmund **[S]** showed under certain assumptions that

$$p_{x,\epsilon}(m) \sim C_\epsilon m^{-1/2} \exp(-m\rho_\epsilon), \tag{3}$$

where C_ϵ and ρ_ϵ are positive constants depending on ϵ. In particular, $\rho_\epsilon = (\mu+\epsilon)\theta_0 - \Lambda(\theta_0)$, where $\Lambda(\theta) = \log E e^{\theta X_1}$ and θ_0 is defined as the solution (which is assumed to exist) of the equation $\Lambda'(\theta) = \mu + \epsilon$. Note that $p_{x,\epsilon}(m)$ does not depend on x in this i.i.d. case. We shall generalize (3) to the setting of Markov additive processes which are more general than the partial sum $\sum_{t=1}^{n} f(X_t)$ of a Markov chain.

2. Convergence rate in the ergodic theorem of Markov additive processes

Consider a Markov chain $\{X_n,\ n \geq 0\}$ on a general state space D with σ-algebra $\mathcal{A}$. Suppose that an additive component $S_n, n \geq 0$, taking values in $\mathbb{R}$, is adjoined to the chain such that $\{(X_n, S_n),\ n \geq 0\}$ is a Markov chain on $D \times \mathbb{R}$ and

$$P\{(X_1, S_1) \in A \times (B + s) \mid (X_0, S_0) = (x, s)\}$$
$$= P\{(X_1, S_1) \in A \times B \mid (X_0, S_0) = (x, 0)\} = P(x, A \times B)$$

for all $x \in D$, $s \in \mathbb{R}$, $A \in \mathcal{A}$ and $B \in \mathcal{B}$ ($:=$ Borel σ-algebra on $\mathbb{R}$). The chain $\{(X_n, S_n),\ n \geq 0\}$ is called a *Markov additive process*, with transition kernel P. Note that this includes $S_n = \sum_{t=1}^{n} f(X_t)$, considered in (1), as a special case.

To analyze large deviation probabilities of the form $P\{n^{-1}S_n \in B \mid X_0 = x\}$, Iscoe, Ney and Nummelin developed "twisting" transformations (cf. Section 4 of **[INN]**) for Markov additive processes satisfying the following condition: $\{X_n\}$ is aperiodic, irreducible with respect to some maximal irreducibility measure, and there exist positive constants $c_1 < c_2$, a probability measure ν on $\mathcal{A} \times \mathcal{B}$ and a positive integer m_0 such that

$$c_1 \nu(A \times B) \leq P^{m_0}(x, A \times B) \leq c_2 \nu(A \times B), \tag{4}$$

for all $x \in D$, $A \in \mathcal{A}$ and $B \in \mathcal{B}$, where P^m denotes the m-step transition function of $\{(X_n, S_n),\ n \geq 0\}$. Such Markov additive processes are called *uniformly recurrent*. For $\theta \in \mathbb{R}$, letting

$$\widehat{P}(x, A; \theta) = \int_{-\infty}^{\infty} e^{\theta s} P(x, A \times ds), \quad \widehat{\nu}(A; \theta) = \int_{-\infty}^{\infty} e^{\theta s} \nu(A \times ds), \tag{5}$$

it follows from (4) that $c_1 \widehat{\nu}(A; \theta) \leq \widehat{P}^{m_0}(x, A; \theta) \leq c_2 \widehat{\nu}(A; \theta)$. Let

$$\Theta = \{\theta \in \mathbb{R} : \widehat{\nu}(D; \theta) < \infty\}. \tag{6}$$

For $\theta \in \Theta$, define the linear operator $\widehat{P}(\theta)$ on the space of bounded measurable functions $f: D \to \mathbb{R}$ by $\widehat{P}(\theta) f(x) = \int f(y) \widehat{P}(x, dy; \theta)$. Then $\widehat{P}(\theta)$ has a maximal simple real eigenvalue $\lambda(\theta)$, with associated right eigenfunction $r(\cdot\,; \theta): D \to (0, \infty)$ and left eigenmeasure $\ell(\cdot\,; \theta): \mathcal{A} \to [0, \infty)$, normalized so that $\int r(x; \theta) \ell(dx; \theta) = 1$. Moreover, $r(\cdot\,; \theta)$ and $(d\ell/d\nu)(\cdot\,; \theta)$ are uniformly positive and bounded and

$$\widehat{P}^n(x, A; \theta) = \ell(A; \theta) r(x; \theta) \lambda^n(\theta)[1 + O(\delta^n(\theta))], \tag{7}$$

with $0 < \delta(\theta) < 1$ (cf. Lemma 3.1 of **[INN]**). Define the "twisting" transformation

$$P_\theta(x, dy \times ds) = \frac{r(y; \theta)}{r(x; \theta)} e^{-\Lambda(\theta) + \theta s} P(x, dy \times ds), \quad \text{where } \Lambda = \log \lambda. \tag{8}$$

Then P_θ is the transition kernel of a Markov additive process $\{(X_n^{(\theta)}, S_n^{(\theta)}),\ n \geq 0\}$, cf. Section 4 of **[INN]**. In Section 3 we shall use these twisting transformations to prove the following result on the convergence rate in the ergodic theorem for uniformly recurrent Markov additive processes.

Theorem 1. *Let $\{(X_n, S_n),\ n \geq 0\}$ be a uniformly recurrent Markov additive process such that $\Theta \neq \{0\}$, where Θ is defined in (6) and ν is the probability measure in (4). Let $\epsilon > 0$ and $\mu = \int_{\mathbb{R}\times D\times D} sP(x, dy \times ds)\, d\pi(x)$, where π is the stationary distribution of the Markov chain $\{X_n, n \geq 0\}$. Assume furthermore that*

(i) $\exists\, \theta_0 < \theta_1$ *in the interior of Θ such that $\Lambda'(\theta_0) = \mu + \epsilon$ and $\Lambda(\theta_1) = \theta_1(\mu + \epsilon)$,*

and that either

(ii) *ν_S is a lattice distribution supported by $L := \{0, \pm h, \pm 2h, \ldots\}$ and $\mu + \epsilon \in L$, where ν_S is the probability measure defined by $\nu_S(B) = \nu(D \times B)$ for all $B \in \mathcal{B}$,*

or

(ii$'$) *ν_S is nonlattice.*

Let $\rho_\epsilon = (\mu+\epsilon)\theta_0 - \Lambda(\theta_0)$. Then for every $A \in \mathcal{A}$ such that $\pi(A) > 0$, there exists a positive constant $c(\epsilon, A)$ such that as $m \to \infty$,

$$P\{X_m \in A \text{ and } \sup_{n\geq m} n^{-1}S_n > \mu + \epsilon \mid X_0 = x\} \sim c(\epsilon, A)r(x; \theta_0)m^{-1/2}\exp(-m\rho_\epsilon),$$

uniformly in $x \in D$.

The proof of Theorem 1 will be given in Section 3. Note that $\rho_\epsilon = \Delta^*(\mu+\epsilon)$, where $\Delta^*(s) = \sup_\theta(\theta s - \Lambda(\theta))$ is the convex conjugate of Λ, cf. **[INN]**. For the special case of an i.i.d. sequence $\{X_n, n \geq 1\}$ and $S_n = \sum_{t=1}^{n} X_t$, $\Lambda(\theta) = \log Ee^{\theta S_1}$ and we can define ν in (4) by $\nu(A \times B) = P\{S_1 \in A \cap B\}$ so that $\nu_S(B) = P\{S_1 \in B\}$. In this case, (i) and (ii) or (ii$'$) reduce to Siegmund's conditions (cf. Theorem 1 of **[S]**, in which $\mu = 0$).

3. Proof of Theorem 1

The proof of Theorem 1 uses two twisting transformations and Lemma 1 below. Following Siegmund's proof for the i.i.d. case in **[S]**, the first twisting transformation takes $\theta = \theta_0$ in (8), for which $\bar{S}_n := S_n - n(\mu + \epsilon)$ has zero drift under P_{θ_0}, and the second twisting transformation takes $\theta = \theta_1$ in (8), for which $\bar{\Lambda}(0) = 0 = \bar{\Lambda}(\theta_1)$, where $\bar{\Lambda}(\theta) = \Lambda(\theta) - \theta(\mu + \epsilon)$. Here and in the sequel, we use P_θ to denote the probability measure under which the Markov additive process (X_n, S_n) has transition kernel (8) and use E_θ to denote expectation under P_θ. The following lemma generalizes Lemma 1 of **[S]** from sums of i.i.d. random variables to Markov additive processes.

Lemma 1. *With the same notation and assumptions as in Theorem 1, let $\bar{S}_n = S_n - n(\mu+\epsilon)$ and $\bar{\Lambda}(\theta) = \Lambda(\theta) - \theta(\mu+\epsilon)$. Denote the stationary distribution of $\{X_n\}$ under P_{θ_0} by π_0 and define $\sigma^2 = \bar{\Lambda}''(\theta_0)$, $\delta = \bar{\Lambda}(\theta_1) - \bar{\Lambda}(\theta_0)$ and $i_A = \int_A (r(y; \theta_1)/r(y; \theta_0))\, d\pi_0(y)$.*

(a) *Under condition* (ii′) *of Theorem* 1, *for every* $s \le 0$ *and* $A \in \mathcal{A}$,

$$P_{\theta_1}\{X_m \in A,\ \bar{S}_m \le s \mid X_0 = x\} \sim \frac{i_A r(x;\theta_0)\exp\{-\delta m + (\theta_1-\theta_0)s\}}{\sigma\sqrt{2\pi m}\, r(x;\theta_1)(\theta_1-\theta_0)}$$

as $m \to \infty$.

(b) *Under condition* (ii) *of* Theorem 1, *for every* $A \in \mathcal{A}$ *and* $k = 0, -1, -2, \ldots$,

$$P_{\theta_1}\{X_m \in A,\ \bar{S}_m \le hk \mid X_0 = x\} \sim \frac{h i_A r(x;\theta_0)\exp\{-\delta m + (\theta_1-\theta_0)hk\}}{\sigma\sqrt{2\pi m}\, r(x;\theta_1)(1-e^{-(\theta_1-\theta_0)h})}$$

as $m \to \infty$.

Proof. First note from (8) that

$$\begin{aligned} &P_{\theta_1}\{X_m \in A,\ \bar{S}_m \le s \mid X_0 = x\} \\ &\quad = \frac{r(x;\theta_0)e^{-\delta m}}{r(x;\theta_1)} E_{\theta_0}[g(X_m)e^{(\theta_1-\theta_0)\bar{S}_m} I_{\{X_m \in A,\ \bar{S}_m \le s\}} \mid X_0 = x], \end{aligned} \tag{9}$$

where $g(y) = r(y;\theta_1)/r(y;\theta_0)$. Since g is a bounded positive function (cf. Lemma 3.1 of **[INN]**), there exists a positive constant K such that $0 < g(x) \le K$ for all $x \in D$. Take any $\eta > 0$ and define $a_0 = 0, a_j = a_{j-1} + \eta$ for $1 \le j \le K_0 := [\eta^{-1}K] + 1$. Letting $A_j = \{x \in A : a_{j-1} < g(x) \le a_j\}$, note that

$$\begin{aligned} &a_{j-1}E_{\theta_0}[e^{(\theta_1-\theta_0)\bar{S}_m} I_{\{X_m \in A_j,\ \bar{S}_m \le s\}} \mid X_0 = x] \\ &\qquad \le E_{\theta_0}[g(X_m)e^{(\theta_1-\theta_0)\bar{S}_m} I_{\{X_m \in A_j,\ \bar{S}_m \le s\}} \mid X_0 = x] \\ &\qquad \le a_j E_{\theta_0}[e^{(\theta_1-\theta_0)\bar{S}_m} I_{\{X_m \in A_j, \bar{S}_m \le s\}} \mid X_0 = x]. \end{aligned} \tag{10}$$

For every $j \in \{1, \ldots, K_0\}$ such that $\pi_0(A_j) > 0$, we can apply Lemma 2(a) below and integration by parts to show that for $s \le 0$,

$$\begin{aligned} &E_{\theta_0}\{I_{\{X_m \in A_j\}} E_{\theta_0}[e^{(\theta_1-\theta_0)\bar{S}_m} I_{\{\bar{S}_m \le s\}} \mid X_m \in A_j, X_0 = x] \mid X_0 = x\} \\ &\qquad \sim \pi_0(A_j)e^{(\theta_1-\theta_0)s}/\{\sigma\sqrt{2\pi m}(\theta_1-\theta_0)\} \end{aligned} \tag{11}$$

when condition (ii′) holds, and apply Lemma 2(b) to show that for $k = 0, -1, -2, \ldots$,

$$\begin{aligned} &E_{\theta_0}\{I_{\{X_m \in A_j\}} E_{\theta_0}[e^{(\theta_1-\theta_0)\bar{S}_m} I_{\{\bar{S}_m \le hk\}} \mid X_m \in A_j, X_0 = x] \mid X_0 = x\} \\ &\qquad \sim \pi_0(A_j)e^{(\theta_1-\theta_0)hk} h/\{\sigma\sqrt{2\pi m}(1-e^{-(\theta_1-\theta_0)h})\} \end{aligned} \tag{12}$$

when condition (ii) holds. These arguments, therefore, are similar to those of Siegmund in the proof of Lemma 1 of **[S]**. Moreover, note that

$$\sum_{j=1}^{K_0} a_{j-1}\pi_0(A_j) \le i_A \le \sum_{j=1}^{K_0} a_j \pi_0(A_j), \quad \sum_{j=1}^{K_0}(a_j - a_{j-1})\pi_0(A_j) \le \eta. \tag{13}$$

Furthermore, since

$$\sup_{x \in D, A \in \mathcal{A}} |P_{\theta_0}(X_m \in A \mid X_0 = s) - \pi_0(A)| = O(\gamma^m) \text{ for some } 0 < \gamma < 1, \tag{14}$$

by uniform recurrence, and since $0 \leq e^{(\theta_1-\theta_0)\bar{S}_m} I_{\{\bar{S}_m \leq 0\}} \leq 1$, it follows that

$$E_{\theta_0}[e^{(\theta_1-\theta_0)\bar{S}_m} I_{\{X_m \in A_j,\, \bar{S}_m \leq 0\}} \mid X_0 = x] = O(\gamma^m) \text{ if } \pi_0(A_j) = 0. \tag{15}$$

From (9)–(15), the desired conclusion follows.

Lemma 2. *Using the same notation as that in Theorem* 1 *and Lemma* 1, *let* $A \in \mathcal{A}$ *be such that* $\pi_0(A) > 0$.

(a) *Assume that* ν_S *is nonlattice. Then as* $n \to \infty$,

$$P_{\theta_0}\{\bar{S}_m \leq s \mid X_0 = x,\, X_m \in A\} = \Phi(\frac{s}{\sigma\sqrt{m}}) - \frac{d(x, A)}{\sqrt{m}} \cdot (1 - \frac{s^2}{\sigma^2 m})\phi(\frac{s}{\sqrt{m}}) + o(\frac{1}{\sqrt{m}}),$$

where $d(x, A)$ *is a constant depending only on* x *and* A, *and* Φ *and* ϕ *denote the distribution function and density function of the standard normal distribution.*

(b) *Assume that* ν_S *is a lattice distribution supported by* L. *Then as* $m \to \infty$, *for* $s \in L$,

$$P_{\theta_0}\{\bar{S}_m = s \mid X_0 = x,\ X_m \in A\} = \frac{h}{\sigma\sqrt{m}}\phi(\frac{s}{\sigma\sqrt{m}}) + o(\frac{1}{\sqrt{m}}).$$

Proof. Let $\psi_{m,x,A}(t) = E\{\exp(it\bar{S}_m) \mid X_0 = x, X_m \in A\}$. It can be shown by arguments similar to those of Nagaev **[N1]**, **[N2]** and Jensen **[J1]** that as $t \to 0$ and $m \to \infty$,

$$\psi_{m,x,A}(\frac{t}{\sigma\sqrt{m}}) = e^{-t^2/2}\{1 + \sum_{j=1}^{k} \alpha_j (\frac{it}{\sigma})^j m^{-j/2+1}\} + o(m|t/\sqrt{m}|^k) + O(|t|\delta^m), \tag{16}$$

for every $k \geq 1$, in which $0 < \delta < 1$ and the α_j depend on x and A. Suppose that ν_S is nonlattice. Then by an argument similar to that of Jensen **[J2]**, it can be shown that

$$|\psi_{m,x,a}(t)| \leq \{(\frac{c_1}{c_2})^2 \Big| \int_{\mathbb{R}\times D} e^{its}\nu(dy \times ds)\Big| + \big(1 - (\frac{c_1}{c_2})^2\big)\}^{[n/2m_0]}, \tag{17}$$

where c_1, c_2 and m_0 are given in (4). In view of (16) and (17), the proof of the desired conclusion in (a) is similar to that in **[F]**, p. 539, for sums of i.i.d. random variables.

Suppose that ν_S is lattice. It has been shown by an argument similar to Lemma 3.1 of **[N1]** that for every $\eta > 0$ there exists $\lambda > 0$ such that for $\eta \leq |t| \leq 2\pi/h - \eta$ and all large m, $|\psi_{m,x,A}(t)| \leq e^{-\lambda m}$. In view of this and (16), the proof of the desired conclusion in (b) is similar to that in **[F]**, pp. 517–518, for sums of i.i.d. random variables.

Proof of Theorem 1. Define $T_m = \inf\{n \geq m : \bar{S}_n > 0\}$ $(\inf \emptyset = \infty)$, where $\bar{S}_n = S_n - n(\mu + \epsilon)$. Note that

$$\begin{aligned} &P\{X_m \in A, \sup_{n\geq m} n^{-1} S_n > \mu + \epsilon \mid X_0 = x\} \\ &\quad = P\{X_m \in A, \bar{S}_m > 0 \mid X_0 = x\} + P\{X_m \in A, \ m < T_m < \infty \mid X_0 = x\} \\ &\quad = E_{\theta_0}\Big\{\frac{r(x;\theta_0)}{r(X_m;\theta_0)} e^{m\bar{\Lambda}(\theta_0) - \theta_0 \bar{S}_m} I_{\{X_m \in A, \bar{S}_m > 0\}} \mid X_0 = x\Big\} \\ &\quad\quad + E_{\theta_1}\Big\{\frac{r(x;\theta_1)}{r(X_{T_m};\theta_1)} e^{-\theta_1 \bar{S}_{T_m}} I_{\{X_m \in A, \, m < T_m < \infty\}} \mid X_0 = x\Big\} := e_1(\theta_0) + e_2(\theta_0), \text{ say,} \end{aligned}$$

since $\bar{\Lambda}(\theta_1) = \Lambda(\theta_1) - \theta_1(\mu + \epsilon) = 0$.

To analyze the term $e_1(\theta_0)$, we use an idea similar to that used in the proof of Lemma 1 with $g(y)$ defined here by $g(y) = 1/r(y;\theta_0)$, again using the events $A_j = \{x \in A : a_{j-1} < g(x) \leq a_j\}$. An argument involving Lemma 2, similar to that used the proof of Lemma 1, can be used to show that

$$e_1(\theta_0) \sim \frac{C_1}{\sqrt{m}} r(x;\theta_0) e^{m\bar{\Lambda}(\theta_0)} \tag{18}$$

under either (ii) or (ii′), with the constant C_1 represented differently under (ii) and (ii′).

To analyze the term $e_2(\theta_0)$, we note that Lemma 1 implies that for every $B \subset A$ and $s \leq 0$,

$$P_{\theta_1}\{X_m \in B, \ \bar{S}_m \leq s \mid X_0 = x, \ X_m \in A, \ \bar{S}_m \leq 0\} \to (i_B/i_A) e^{(\theta_1 - \theta_0)s} \text{ as } m \to \infty.$$

Therefore, analogous to Lemma 2 of **[S]**, the limiting conditional distribution under P_{θ_1} of $(X_m, \bar{S}_m)$ given that $X_0 = x$, $X_m \in A$ and $\bar{S}_m \leq 0$ can be expressed as

$$(\theta_1 - \theta_0) e^{(\theta_1 - \theta_0)s} r(y;\theta_1)(i_A r(y;\theta_0))^{-1} d\pi_0(y) ds \quad (s \leq 0, \ y \in D)$$

when the nonlattice condition (ii′) holds, and can be expressed as

$$(1 - e^{-(\theta_1 - \theta_0)h}) e^{(\theta_1 - \theta_0)kh} r(y;\theta_1)(i_A r(y;\theta_0))^{-1} d\pi_0(y) \quad (k = 0, -1, \ldots, y \in D)$$

when the lattice condition (ii) holds. Moreover, since $T_m = \inf\{n \geq m : \bar{S}_n > 0\}$, it follows from the strong Markov property that

$$E_{\theta_1}\Big(\frac{e^{-\theta_1 \bar{S}_{T_m}}}{r(X_{T_m};\theta_1)} \mid \bar{S}_m = s, \ X_m = y\Big) = E_{\theta_1}\Big(\frac{e^{-\theta_1(\bar{S}_{\tau(|s|)} - |s|)}}{r(X_{\tau(|s|)};\theta_1)} \mid X_0 = y\Big), \quad s \leq 0,$$

where $\tau(a) = \inf\{n \geq 1 : \bar{S}_n > a\}$. Using these results and Lemma 1, it can be shown that when (ii$'$) holds,

$$e_2(\theta_0) \sim \frac{r(x;\theta_0)e^{-\delta m}}{\sigma\sqrt{2\pi m}} \int\limits_{y\in A, s\leq 0} e^{(\theta_1-\theta_0)s}\frac{r(y;\theta_1)}{r(y;\theta_0)} \times E_{\theta_1}\Big(\frac{e^{-\theta_1(\bar{S}_{\tau(|s|)}-|s|)}}{r(X_{\tau(|s|)};\theta_1)} \mid X_0 = y\Big)\, d\pi_0(y)ds.$$

Likewise, when (ii) holds,

$$e_2(\theta_0) \sim \frac{hr(x;\theta_0)e^{-\delta m}}{\sigma\sqrt{2\pi m}} \int\limits_{y\in A} \sum_{k\leq 0} e^{(\theta_1-\theta_0)hk}\frac{r(y;\theta_1)}{r(y;\theta_0)} \times E_{\theta_1}\Big(\frac{e^{-\theta_1(\bar{S}_{\tau(h|k|)}-h|k|)}}{r(X_{\tau(h|k|)};\theta_1)} \mid X_0 = y\Big)\, d\pi_0(y).$$

Since $\bar{\Lambda}(\theta_1) = 0$, $\delta = \bar{\Lambda}(\theta_1) - \bar{\Lambda}(\theta_0) = \bar{\Lambda}(\theta_0)$ and therefore $-\delta m = m\bar{\Lambda}(\theta_0)$. Combining (18) with the above results for $e_2(\theta_0)$ yields

$$\begin{aligned} P\{X_m \in A,\ \sup_{n\geq m} n^{-1}S_n > \mu+\epsilon \mid X_0 = x\} &= e_1(\theta_0) + e_2(\theta_0) \\ &\sim c(\epsilon, A)r(x;\theta_0)m^{-1/2}e^{m\bar{\Lambda}(\theta_0)}, \end{aligned}$$

noting that $\bar{\Lambda}(\theta_0) = -\rho_\epsilon$.

Remark. The preceding proof shows that

$$P\{X_m \in A,\ S_m > m(\mu+\epsilon) \mid X_0 = x\} \sim C(\epsilon, A)r(x;\theta_0)m^{-1/2}e^{-m\rho_\epsilon}, \tag{19}$$

cf. (18). Iscoe, Ney and Nummelin **[INN]** have shown that there exist constants $c_1(\epsilon, A, x)$ and $c_2(\epsilon, A, x)$ such that

$$c_1(\epsilon, A, x)m^{-1/2}e^{-m\rho_\epsilon} \leq P\{X_m \in A,\ S_m > m(\mu+\epsilon) \mid X_0 = x\} \leq c_2(\epsilon, A, x)e^{-m\rho_\epsilon}, \tag{20}$$

for all n. Comparison of (20) with (19) shows that the lower bound of (20) is asymptotically sharp.

References

[AtNe] K. B. Athreya and P. Ney, Some aspects of ergodic theory and laws of large numbers for Harris-recurrent Markov chains, Colloquia Mathematica Societatis János Bolyai, pp. 41–56, Budapest 1980.

[F] W. Feller, An Introduction to Probability Theory and Its Applications, Vol. 2, Second Edition, Wiley, New York 1971.

[INN] I. Iscoe, P. Ney and E. Nummelin, Large deviations of uniformly recurrent Markov additive processes, Adv. Appl. Math. 6 (1985), 373–412.

[J1] J. L. Jensen, A note on asymptotic expansions for Markov chains using operator theory, Adv. Appl. Math. 8 (1987), 377–392.

[J2] J. L. Jensen, Saddlepoint expansions for sums of Markov dependent variables on a continuous state space, Probab. Theory and Related Fields 89 (1991), 181–199.

[N1] S. V. Nagaev, Some limit theorems for stationary Markov chains, Theory Probab. Appl. 2 (1957), 378-406.

[N2] S. V. Nagaev, More exact statements of limit theorem for homogeneous Markov chains, Theory Probab. Appl. 6 (1961), 62–81.

[S] D. Siegmund, Large deviation probabilities in the strong law of large numbers, Z. Wahrsch. Verw. Gebiete 31 (1975), 107–113.

Institute of Statistical Science
Academia Sinica
Taipai, Taiwan
Republic of China

Department of Statistics
Stanford University
Stanford, CA 94305, U.S.A.

A Mean Ergodic Theorem for $\frac{1}{N}\Sigma_{n=1}^{N} f(T^n x)g(T^{n^2}x)$

Hillel Furstenberg and Benjamin Weiss[1]

0. Introduction

Let T be a measure preserving automorphism of a measure space $(X, \mathcal{B}, \mu)$ with $\mu(X) < \infty$. Expressions of the form

$$\frac{1}{N}\sum_{n=1}^{N} f_1(T^{p_1(n)}x) f_2(T^{p_2(n)}x) \dots f_k(T^{p_k(n)}x) \tag{*}$$

where the $p_j(n)$ are integer sequences converging to ∞, and $f_1, \dots, f_k \in L^\infty(X, \mathcal{B}, \nu)$, may be called "non-conventional" ergodic averages, and one can inquire as to their convergence in any of the usual modes of convergence. Limits of expressions of the form (*) play a role in the ergodic theoretic proof of Szeméredi's theorem on arithmetic progressions and its generalizations (**[FKO]**, **[BL]**), but they are of interest in their own right. Limit theorems for such averages have been obtained for $k = 2, 3$ and $p_1(n) = an$, $p_2(n) = bn$, $p_3(n) = cn$, and for norm convergence in $L^2(X, \mathcal{B}, \mu)$. For $k = 2$, and $p_1(n)$, $p_2(n)$ linear, J. Bourgain has proved a pointwise convergence theorem (**[Bo]**), while for $k = 1$ and $p_1(n)$ any polynomial, Bourgain has also obtained a pointwise theorem. In the present paper we shall obtain an L^2-convergence result for $k = 2$, $p_1(n)$, $p_2(n)$ polynomials of degree 1 and 2 respectively. The case $p_1(n) = n$, $p_2(n) = n^2$ is entirely typical and our discussion will focus on this case. Our discussion however will include as a corollary the existence of limits also in the case $p_1(n) = n^2+an$, $p_2(n) = n^2 + bn$.

When the transformation T is assumed to be weakly mixing, Bergelson has obtained a very general L^2-convergence theorem for (*) for arbitrary polynomials sequences $p_j(n)$ (**[Be]**). His result gives that, assuming (as we may) that the polynomials $p_j(t)$ differ from one another by polynomials of positive degree, and T is weakly mixing, the limit in (*) exists in L^2 and is constant.

Simple examples show that without the hypothesis of weak mixing, the limit in (*) need not be constant. Now the absence of weak mixing implies the existence of "group rotation" factors for the system $(X, \mathcal{B}, \mu, T)$. When a group rotation factor is present there may also be more complex factors which can be described as skew products of group rotations, and systems obtained by a succession of such skew products. These systems are called ***distal systems***. Distal factors of a given system are indicated by

1 Research supported by BSF contract 92-00065/1

the presence of eigenfunctions as well as generalized eigenfunctions for the operator $T: L^2(X, \mathcal{B}, \mu) \to L^2(X, \mathcal{B}, \mu)$ induced by the transformation $T: X \to X$. It can be shown that the behavior of averages (*) for any system of polynomial exponents $p_j(n)$ and for arbitrary measure preserving systems can be reduced to that for an appropriate distal factor of the system. Thus Bergelson's result in **[Be]** follows from this general result, since in the weak mixing case, the distal factors are all trivial (**[F2]**).

When the limiting behavior of a non-conventional ergodic average for a system $(X, \mathcal{B}, \mu, T)$ can be "reduced" (in a manner to be made clear) to that of a factor system $(Y, \mathcal{D}, \nu, T)$, we shall say that the latter is a ***characteristic factor*** of the former for the ***scheme*** $\{p_1(n), \dots, p_k(n)\}$. One approach to studying non-conventional ergodic averages consists in trying to find efficient characteristic factors. The arguments presented here will show that for any ergodic T, a characteristic factor for $\{n, n^2\}$ is a group rotation in a procyclic group. Since in this case the averages $\frac{1}{N}\sum T^n f\, T^{n^2} g$ clearly converge in L^2, we will have convergence in the general case for this particular non-conventional ergodic average.

We will not be able to obtain the "efficient" characteristic factor for $\{n, n^2\}$ directly. We first consider schemes $\{an, bn\}$ and show that for these, a characteristic factor (in fact, the smallest one) consists of the ***Kronecker*** factor, namely, the largest group rotation factor of $(X, \mathcal{B}, \mu, T)$. We will also need to consider schemes $\{an, bn, cn\}$ and for these we show that a characteristic factor is a two-stage distal factor with the skewing function satisfying a certain functional equation. It will not be needed for our discussion but it may be shown that this functional equation implies that the factor in question has as its space a homogeneous space of a two-step nilpotent group, and the measure preserving transformation is a group rotation on this homogeneous space.

This analysis will be relevant for us since it will turn out that a characteristic factor for all $\{an, bn, cn\}$ is also a characteristic factor for $\{n, n^2\}$. Thus the behavior of $\frac{1}{N}\sum T^n f T^{n^2} g$ will be effectively reduced to special skew product systems. From here a further reduction will enable us to replace the "nilpotent" factor by a simple procyclic group factor.

We should remark that this sketch is something of an oversimplification, but it does clarify the strategy that is used here.

We should take note of the fact that for any non-conventional average (*), the existence of limits either pointwise or in $L^2(X)$ for a general measure preserving case will follow by ergodic decomposition once it is known in the ergodic case. For this reason we will often implicitly assume that our systems are ergodic. We will also assume that our measure spaces are separable so that we may take them to be Lebesgue spaces and when convenient, we take them to be compact metric spaces.

A tool that will be repeatedly used will be a Hilbert space version of van der Corput's lemma in equidistribution theory. This will be the main tool in enabling us to effect a reduction from one measure preserving system to a factor system.

1. A Hilbert space van der Corput lemma

The following appears in **[Be]** and is due to Bergelson. We reproduce the proof here for completeness.

Lemma 1.1. *Let $\{u_n\}$ be a bounded sequence of vectors in a Hilbert space $\mathcal{H}$ with inner product $\langle\ ,\ \rangle$. Let*

$$\gamma_m = \lim_{N\to\infty} \frac{1}{N} \sum_{n=1}^{N} \langle u_n, u_{n+m}\rangle$$

assuming this exists, and in any case let

$$\gamma'_m = \limsup_{N\to\infty} \frac{1}{N} \Big| \sum_{n=1}^{N} \langle u_n, u_{n+m}\rangle \Big|.$$

If either $\frac{1}{M}\sum_{m=1}^{M} \gamma_m \to 0$ or $\frac{1}{M}\sum_{m=1}^{M} \gamma'_m \to 0$ then $\|\frac{1}{N}\sum_{n=1}^{N} u_n\| \to 0$.

Proof. Since $\{u_n\}$ is bounded, if N is large with respect to M, $\frac{1}{N}\sum_1^N u_n$ will be close to $\frac{1}{N}\frac{1}{M}\sum_{n=1}^{N}\sum_{m=1}^{M} u_{n+m}$. Using the fact that in $\mathcal{H}$, $\|\frac{1}{N}\sum_1^N y_n\|^2 \le \frac{1}{N}\sum_1^N \|y_n\|^2$, we can write

$$\Big\|\frac{1}{N}\frac{1}{M}\sum_{n=1}^{N}\sum_{m=1}^{M} u_{n+m}\Big\|^2 \le \frac{1}{N}\sum_{n=1}^{N}\frac{1}{M^2}\sum_{m_1,m_2=1}^{M} \langle u_{m+m_1}, u_{n+m_2}\rangle.$$

Letting $N \to \infty$ we find

$$\limsup \Big\|\frac{1}{N}\sum_{1}^{n} u_n\Big\|^2 \le \frac{1}{M^2}\sum_{m_1,m_2=1}^{M} \gamma_{m_1-m_2}$$

as well as

$$\limsup \Big\|\frac{1}{N}\sum_{1}^{N} u_n\Big\|^2 \le \frac{1}{M^2}\sum_{m_1,m_2=1}^{M} \gamma'_{m_1-m_2}.$$

Letting $M \to \infty$ we obtain the desired result. □

We illustrate the use of this lemma in the following result which will be needed in our subsequent analysis.

Lemma 1.2. *Let $\Phi(x)$ be a measurable function on $(X, \mathcal{B}, \mu)$ with $|\Phi(x)| = 1$ and such that for each $\zeta \in \mathbb{C}$, $\mu\{\Phi^{-1}(\zeta)\} = 0$. Then for any bounded sequence $\xi(n)$,*

$$\frac{1}{N}\sum_{1}^{N} \xi(n)\Phi(x)^n \to 0 \tag{1.1}$$

in norm in $L^2(X, \mathcal{B}, \mu)$.

We set $u_n = \xi(n)\Phi(x)^n$. Then

$$\langle u_n, u_{n+m}\rangle = \xi(n)\overline{\xi(n+m)} \int \overline{\Phi(x)^m}\, d\mu(x).$$

Now $\int \overline{\Phi(x)^m}\, d\mu(x) = \hat{\pi}(m)$ where π is the distribution of $\Phi(x)$ on the unit circle and $\hat{\pi}$ is its Fourier transform. By Wiener's theorem $\frac{1}{M}\sum |\hat{\pi}(m)| \to 0$ since π is non-atomic. Since $\gamma'_m = O(|\hat{\pi}(m)|)$ the result follows from Lemma 1.1. □

Another simple application of Lemma 1.1 is the following, which is also a special case of our main theorem.

Proposition 1.3. *In any measure preserving system* $(X, \mathcal{B}, \mu, T)$, *if* $f \in L^2(X, \mathcal{B}, \mu)$ *and* a, b *are integers, the limit*

$$\frac{1}{N}\sum_{n=1}^{N} T^{an^2+bn} f \tag{1.2}$$

exists in $L^2(X, \mathcal{B}, \mu)$ *as* $N \to \infty$.

Proof. Let $\mathcal{P} \subset L^2(X, \mathcal{B}, \mu)$ be the closure of the subspace of functions $\{g : \exists m$ with $T^m g = g\}$. We will show that the limit (1.2) exists both for $f \in \mathcal{P}$ and $f \perp \mathcal{P}$. In the former case this is obvious. In the latter case we can see that the limit is 0. For setting $u_n = T^{an^2+bn} f$,

$$\langle u_n, u_{n+m}\rangle = \int f T^{2anm}(T^{am^2+bm}\bar{f})\, d\mu$$

so that

$$\lim \frac{1}{N}\sum_{n=1}^{N}\langle u_n, u_{n+m}\rangle = \int f \lim \frac{1}{N}\sum_{n=1}^{N}(T^{2am})^n (T^{am^2+bm}\bar{f})\, d\mu$$
$$= \int fg\, d\mu$$

with $T^{2am} g = g$ so that $g \in \mathcal{P}$. By Lemma 1.1, the limit in question vanishes. □

2. The asymptotic behavior of $\frac{1}{N}\Sigma T^{an} f\, T^{bn} g$

Let $(X, \mathcal{B}, \mu, T)$ be an ergodic system. If T is not weakly mixing there will be non-trivial eigenfunctions $\varphi: X \to \mathbb{C}$, $T\varphi = \lambda\varphi$. It is well known that there is a group rotation system (Z, α), i.e., Z is a compact abelian (additive) group, $\alpha \in Z$ generates a dense cyclic subgroup of Z, and there is a measurable map $\pi: X \to Z$ satisfying: $\pi(Tx) = \pi(x) + \alpha$, a.e. π is measure preserving from $(X, \mathcal{B}, \mu)$ to (Z, Borel sets, Haar measure), and such that φ is an eigenfunction for $(X, \mathcal{B}, \mu, T)$ if and only if it has the form $\varphi = c\chi \circ \pi$ for some constant c and some character χ on Z. An ergodic system of the form (compact abelian group, Z, Borel sets, Haar measure, translation by fixed element α of the group) will be referred to as a ***Kronecker system***, and will be denoted (Z, α). The particular Kronecker system occurring in the preceding remarks will be called *the Kronecker factor* of $(X, \mathcal{B}, \mu, T)$. T is weakly mixing iff its Kronecker factor reduces to a single element group.

We note that Z is a ***monothetic*** group, that is, a group having a dense cyclic subgroup. In the sequel we will use the fact that for a monothetic group Z, the subgroup $mZ, m \in \mathbb{Z} \setminus \{0\}$, is of finite index $\leq |m|$.

$L^2(Z)$ is spanned by characters of Z, and lifting these to X via π, we obtain the subspace spanned by all eigenfunctions of T. Thus we may identify the discrete spectrum component of $L^2(X, \mathcal{B}, \mu)$ with $L^2(Z)$. Let us denote by $f \to \tilde{f}$ the orthogonal projection in $L^2(X, \mathcal{B}, \mu)$ of f to the subspace $L^2(Z) \circ \pi$. We regard $\tilde{f}$ as a function on Z. We now have the following:

Theorem 2.1. *Let $(X, \mathcal{B}, \mu, T)$ be any ergodic system and (Z, α) its Kronecker factor. Let $a \neq b$ be two integers, then if $f, g \in L^\infty(X, \mathcal{B}, \mu)$*

$$\lim_{N\to\infty} \frac{1}{N} \sum_{n=1}^{N} f(T^{an}x) g(T^{bn}x) \tag{2.1}$$

exists in $L^2(X, \mathcal{B}, \mu)$ and equals

$$\lim_{N\to\infty} \frac{1}{N} \sum_{n=1}^{N} \tilde{f}(z + na\alpha)\tilde{g}(z + nb\alpha) = \int_Z \tilde{f}(z + a\theta)\tilde{g}(z + b\theta)\, d\theta \tag{2.2}$$

with $z = \pi x, \pi$ the map $X \to Z$ for the Kronecker factor (Z, α).

Proof. The existence of the limit as well as the equality in (2.2) is obvious for $\tilde{f}, \tilde{g}$ linear combinations of characters on Z, and it is established in general for $L^2(Z)$ using the fact that the latter are dense in $L^4(Z)$. It follows that in order to prove the theorem it suffices to prove the equality in $L^2(X, \mathcal{B}, \mu)$ of the limits in (2.1) and (2.2) and for this it suffices to show that

$$\frac{1}{N} \sum_{1}^{N} T^{an} f T^{bn} g \to 0 \tag{2.3}$$

in $L^2(X)$ if either $\tilde{f}$ or $\tilde{g}$ vanishes.

Let us assume $\tilde{f} = 0$. We will apply Lemma 1.1. Let $u_n = T^{an} f\, T^{bn} g$ and suppose $b > a$. We have

$$\langle u_n, u_{n+m} \rangle = \int f T^{am} \bar{f}\, T^{(b-a)n} (g T^{bm} \bar{g})\, d\mu.$$

The average $\gamma_m = \lim_{N\to\infty} \frac{1}{N}\Sigma_1^N \langle u_n, u_{n+m}\rangle$ exists by the ergodic theorem and can be written

$$\gamma_m = \int f\, T^{am}\bar{f}\, P_{b-a}(gT^{bm}\bar{g})\, d\mu,$$

where P_{b-a} is the projection operator onto the $T^{(b-a)}$-invariant functions of $L^2(X)$. If $T^{(b-a)}$ is also ergodic then P_{b-a} is the projection to constants. Otherwise it can be expressed as an integral operator with kernel

$$\ell \sum_{i=1}^{\ell} 1_{A_i}(x)1_{A_i}(y)$$

where $X = \bigcup_{i=1}^{\ell} A_i$ is the partition of X to $T^{(b-a)}$-invariant sets. We obtain

$$\gamma_m = \ell \iint f(x)f(y)\, T^{am}\bar{f}(x)\, T^{bm}\bar{g}(y)\Sigma 1_{A_i}(x)1_{A_i}(y)\, d\mu(x)d\mu(y)$$

and averaging over m,

$$\lim \frac{1}{M}\sum_{1}^{M} \gamma_m = \iint f(x)f(y)\Sigma 1_{A_i}(x)1_{A_i}(y)H(x,y)\, d\mu(x)d\mu(y) \qquad (2.4)$$

where $H(T^a x, T^b y) = H(x,y)$. Since the point spectrum subspace of $T^a \times T^b$ on $L^2(X\times X, \mathcal{B}\times\mathcal{B}, \mu\times\mu)$ is the tensor product of the corresponding subspaces of $L^2(X,\mathcal{B},\mu)$ for T^a and T^b, $H(x,y)$ can be expanded as a sum of expressions $\varphi_i(x)$ $\psi_i(y)$ where φ_i and ψ_i are eigenfunctions for T. Since $T^{(b-a)}1_{A_i} = 1_{A_i}$ it also follows that 1_{A_i} is a combination of finitely many eigenfunctions for T. We thus find that

$$1_{A_i}(x)\int H(x,y)1_{A_i}(y)f(y)\, d\mu(y)$$

is a combination of eigenfunctions of T, that is, it belongs to $L^2(Z)\circ\pi$. Since $\tilde{f} = 0$, $f \perp L^2(Z)\circ\pi$ so that (2.4) vanishes. Applying Lemma 1.1 we have $\frac{1}{N}\sum T^{an} f\, T^{bn} g \to 0$. □

The foregoing theorem provides an example of what we referred to in the Introduction as a ***characteristic factor***. Namely, the asymptotic behavior of $\frac{1}{N}\sum_{n=1}^{N} T^{an} f T^{bn} g$ for arbitrary (ergodic) systems is reduced to its behavior for Kronecker systems, and in any given case, the limit is determined by projecting the functions onto the Kronecker factor and evaluating the limit there. We now formalize these notions.

3. Characteristic factors

Let $(X, \mathcal{B}, \mu, T)$ and $(Y, \mathcal{D}, \nu, T)$ be two measure preserving systems. A measurable, measure-preserving map $\pi: X \to Y$ defines a ***homomorphism*** of $(X, \mathcal{B}, \mu, T)$ to $(Y, \mathcal{D}, \nu, T)$ if for almost every $x \in X$, $\pi(Tx) = T\pi(x)$. In this case we say $(Y, \mathcal{D}, \nu, T)$ is a *factor* of $(X, \mathcal{B}, \mu, T)$, the mapping π being implicit in this relationship. We also say in this case that $(X, \mathcal{B}, \mu, T)$ is an ***extension*** of $(Y, \mathcal{D}, \nu, T)$.

Composition with π imbeds $L^2(Y)$ isometrically into $L^2(X)$. This enables us to define the operator $f \to E(f \mid Y)$ taking $L^2(X)$ to $L^2(Y)$ by setting $E(f \mid Y) \circ \pi =$ image of $f \in L^2(X)$ under orthogonal projection to $L^2(Y) \circ \pi$. The operator $f \to E(f \mid Y)$ can also be described as follows. The measure μ on $(X, \mathcal{B})$ can be ***disintegrated*** to measures μ_y, parametrized by $y \in Y$, so that

i) $y \to \mu_y$ is a measurable map of Y to the space of probability measures on $(X, \mathcal{B})$,

ii) for almost all y, $\mu_y\{\pi^{-1}(y)\} = 1$,

iii) $\int_X f\,d\mu = \int_Y \left[\int f\,d\mu_y\right] d\nu(y)$,

iv) for almost all y, $T(\mu_y) = \mu_{Ty}$.

We then have $E(f \mid Y)(y) = \int f\,d\mu_y$. Note that (iv) implies

$$TE(f \mid Y) = E(Tf \mid Y). \tag{3.1}$$

We shall often use the symbol of the underlying space of a system to denote the system, so that $X, Y, \ldots$ respectively represent the systems $(X, \mathcal{B}, \mu, T)$, $(Y, \mathcal{D}, \nu, T), \ldots$. Thus a map $\pi: X \to Y$ as above will be spoken of as a homomorphism of X to Y.

It will be convenient to identify $L^2(Y)$ with its lift $L^2(Y) \circ \pi \subset L^2(X)$. Thus factors of a system X correspond to certain invariant subspaces of $L^2(X)$.

We now define the notion of a ***characteristic factor***.

Definition. If $\{p_1(n), p_2(n), \ldots, p_k(n)\}$ are k integer-valued sequences, and Y is a factor of a system X, we say that Y is a ***characteristic factor for the scheme*** $\{p_1(n), \ldots, p_k(n)\}$, if whenever $f_1, \ldots, f_k \in L^\infty(X)$ we have

$$\frac{1}{N}\sum_{n=1}^{N} T^{p_1(n)} f_1 \ldots T^{p_k(n)} f_k - \frac{1}{N}\sum_{n=1}^{N} T^{p_1(n)} E(f_1 \mid Y) \ldots T^{p_k(n)} E(f_k \mid Y) \to 0 \tag{3.2}$$

in $L^2(X)$.

Y is a ***partial characteristic factor*** (with respect to $J \subset \{1, 2, \ldots, k\}$) if

$$\frac{1}{N}\sum_{n=1}^{N} T^{p_1(n)} f_1 \ldots T^{p_k(n)} f_k - \frac{1}{N}\sum_{n=1}^{N} T^{p_1(n)} \tilde{f}_1 \ldots T^{p_k(n)} \tilde{f}_k \to 0 \tag{3.3}$$

in $L^2(X)$, where $\tilde{f}_j = f_j$ for $j \notin J$ and $\tilde{f}_j = E(f_j \mid Y)$ for $j \in J$.

Theorem 2.1 can now be formulated as the assertion that for any ergodic system, its Kronecker factor is characteristic for the schemes $\{an, bn\}$, a, b distinct integers.

If (Z, α) is the Kronecker factor of a system X, then $L^2(Z) \hookrightarrow L^2(X)$ is the subspace spanned by eigenfunctions in $L^2(X)$ of the operator T. If X is a factor of a system $\tilde{X}$, clearly the Kronecker factor Z of X is a factor of the Kronecker factor $\tilde{Z}$ of $\tilde{X}$.

4. A partial characteristic factor for $\{n, n^2\}$

Our object is to prove

Theorem A. *For any measure preserving system $(X, \mathcal{B}, \mu, T)$ and $f, g \in L^\infty(X)$, the averages $\frac{1}{N}\sum_{n=1}^N T^n f\, T^{n^2} g$ converge in $L^2(X)$.*

As remarked in the Introduction it will suffice to prove this for ergodic systems. We also note that the result is clearly true for a system X if it is true for an extension $\tilde{X}$ of X. We will later introduce the notion of ***normality*** of an ergodic system and prove that every ergodic system has a normal extension. The bulk of our discussion will be the proof of Theorem A for normal systems.

We begin by describing a characteristic factor of any system for the scheme $\{n^2 + an, n^2 + bn\}$.

Lemma 4.1. *Let $(X, \mathcal{B}, \mu, T)$ be a measure preserving system, and let $(Y, \mathcal{D}, \nu, T)$ be a characteristic factor of $(X, \mathcal{B}, \mu, T)$ for all schemes $\{rn, sn, tn\}$ with $t = r + s$ and $r, s > 0$. Then $(Y, \mathcal{D}, \nu, T)$ is a characteristic factor of $(X, \mathcal{B}, \mu, T)$ for all schemes $\{n^2 + an, n^2 + bn\}$ for integers $a \neq b$.*

Proof. We want to show that for $f, g \in L^\infty(X)$, we have in $L^2(X)$

$$\frac{1}{N}\sum_1^N T^{n^2+an} f T^{n^2+bn} g - \frac{1}{N}\sum_1^N T^{n^2+an} E(f \mid Y) T^{n^2+bn} E(g \mid Y) \to 0. \quad (4.1)$$

We can rewrite (4.1) in terms of $f - E(f \mid Y)$ and $g - E(g \mid Y)$ so that it suffices to prove

$$\frac{1}{N}\sum_1^N T^{n^2+an} f\, T^{n^2+bn} g \to 0 \quad (4.2)$$

if either $E(f \mid Y) = 0$ or $E(g \mid Y) = 0$. We are given that for all $f_1, f_2, f_3 \in L^\infty(X)$,

$$\frac{1}{N}\sum_1^N T^{rn} f_1 T^{sn} f_2 T^{tn} f_3 - \frac{1}{N}\sum_1^N T^{rn} E(f_1 \mid Y) T^{sn} E(f_2 \mid Y) T^{tn} E(f_3 \mid Y) \to 0. \quad (4.3)$$

Hence if some $E(f_i \mid Y) = 0$, $\frac{1}{N}\sum_1^N T^{rn} f_1 T^{sn} f_2 T^{tn} f_3 \to 0$.

Assume now $E(f \mid Y) = 0$ and suppose that $b > a$.

We apply Lemma 1.1 with $u_n = T^{n^2+an} f\ T^{n^2+bn} g$. Then

$$\langle u_n, u_{n+m}\rangle = \int T^{n^2+an} f T^{n^2+bn} g T^{(n+m)^2+a(n+m)}\,\bar{f} T^{(n+m)^2+b(n+m)}\bar{g}\,d\mu$$
$$= \int f T^{(b-a)n} g T^{2mn}(T^{m^2+am}\,\bar{f}) T^{2mn+(b-a)n}(T^{m^2+bm}\,\bar{g})\,d\mu.$$

Take $r = (b-a),\ s = 2m,\ t = 2m + (b-a)$ and apply (4.3). Then since

$$E(T^{m^2+am}\,\bar{f} \mid Y) = T^{m^2+am}\overline{E(f \mid Y)} = 0,$$

we have

$$\frac{1}{N}\sum_{n=1}^{N} T^{(b-a)n} g T^{2mn}(T^{m^2+am}\,\bar{f}) T^{2mn+(b-a)n}(T^{m^2+am}\,\bar{g}) \to 0.$$

provided $m \neq 0,\ (a-b)/2$. This gives

$$\gamma_m = \lim \frac{1}{N}\sum_{n=1}^{N}\langle u_n, u_{n+m}\rangle = 0$$

for all but two values of m. By Lemma 1.1, $\frac{1}{N}\sum_1^N u_n \to 0$. A similar argument applies to the case $a > b$, and by symmetry also to the case $E(g \mid Y) = 0$. This completes the proof. □

We can apply Lemma 4.1 to obtain a partial reduction for $\frac{1}{N}\sum T^n f T^{n^2} g$.

Lemma 4.2. *Let $(X, \mathcal{B}, \mu, T)$ and $(Y, \mathcal{D}, \nu, T)$ be as in the foregoing lemma. Then for any bounded measurable $f,\ g$ on X,*

$$\frac{1}{N}\sum_{n=1}^{N} T^n f T^{n^2} g - \frac{1}{N}\sum_{n=1}^{N} T^n f T^{n^2} E(g \mid Y) \to 0. \tag{4.4}$$

Thus Y is a partial characteristic factor of $\{n, n^2\}$.

Proof. We may assume $E(g \mid Y) = 0$. Again apply Lemma 1.1 with $u_n = T^n f T^{n^2} g$. Then

$$\langle u_n, u_{n+m}\rangle = \int T^n f T^{n^2} g T^{n+m}\,\bar{f} T^{(n+m)^2}\,\bar{g}\,d\mu$$
$$= \int f T^m\,\bar{f} T^{n^2-n} g T^{n^2+(2m-1)n}(T^{m^2}\,\bar{g})\,d\mu.$$

By Lemma 4.1,

$$\frac{1}{N}\sum T^{n^2-n} g T^{n^2+(2m-1)n}(T^{m^2}\,\bar{g}) \to 0$$

in $L^2(X, \mathcal{B}, \mu)$ for every $m \neq 0$. Thus once again

$$\lim \frac{1}{N} \sum_{n=1}^{N} \langle u_n, u_{n+m} \rangle = 0$$

for almost all m, and by Lemma 1.1,

$$\frac{1}{N} \sum_{n=1}^{N} T^n f T^{n^2} g \to 0.$$

□

We can reformulate Lemma 4.2 as follows.

Lemma 4.3. *If Y is a factor of X characteristic for all schemes $\{rn, sn, tn\}$ with $r, s > 0$, $t = r + s$, then Y is a partial characteristic factor (with respect to n^2) of X for the scheme $\{n, n^2\}$.*

The succeeding sections will be devoted to identifying characteristic factors for the schemes $\{rn, sn, tn\}$ as required. When this has been made as explicit as possible we shall be able to use Lemma 4.3 to prove Theorem A.

5. Isometric extensions

Kronecker systems may be characterized as those for which the transformation T acts as an isometry on the space X which can be taken to be a compact metric space. The relativization of this leads to the notion of an ***isometric extension***. The details of this and its relationship to the structure of distal systems can be found in **[F1]** and **[Z]**.

Let $(X, \mathcal{B}, \mu, T)$ be an ergodic measure preserving system and let $(Y, \mathcal{D}, \nu, T)$ be a factor. We consider the ring $L^\infty(Y) = L^\infty(Y, \mathcal{D}, \nu)$ as a subring of functions on X, and we can speak of a subspace $V \subset L^2(X, \mathcal{B}, \mu)$ as being a ***finite rank module*** over $L^\infty(Y)$ if for some $\varphi_1, \ldots, \varphi_k \in V$ each function of V can be expressed as $\sum u_i(y)\varphi_i(x)$ for some $u_i \in L^\infty(Y)$. We say $(X, \mathcal{B}, \mu, T)$ is an ***isometric extension*** of $(Y, \mathcal{D}, \nu, T)$ if $L^2(X, \mathcal{B}, \mu)$ is spanned by finite rank modules over $L^\infty(Y)$. It can be shown that in this case there is a homogeneous space $M = G/H$ of a compact group G, and a measurable map $\rho: Y \to G$ so that $(X, \mathcal{B}, \mu, T)$ is isomorphic to the skew product $(Y \times M, \tilde{\mathcal{B}}, \nu \times m_M, \tilde{T})$, where m_M is the G-invariant measure on M and $\tilde{T}(y, u) = (Ty, \rho(y)u)$, and $\tilde{\mathcal{B}}$ is an appropriate σ-algebra. Conversely any such skew product will lead to an isometric extension. Since G is compact it preserves a metric on G/H so one can suppose that G acts by isometries. Hence the term ***isometric***.

Thus, if we take $M = S^1$, the unit circle in $\mathbb{C}$, $\rho: Y \to S^1$ a measurable map, and if we set $X = Y \times S^1$ with $T(y, \zeta) = (Ty, \rho(y)\zeta)$, then $L^2(X)$ is spanned by the rank 1 modules $V_k = \{u(y)\zeta^k\}$ and these are T-invariant since $T(u(y)\zeta^k) = u(Ty)\rho(y)^k\zeta^k$. It follows that the skew product system on X is an isometric extension of the system on Y.

We return to the general case of a isometric extension. If we iterate the transformation T on X we obtain

$$T^n(y,u) = (T^n y, \rho_n(y)u)$$

where

$$\rho_n(y) = \rho(T^{n-1}y)\rho(T^{n-2}y)\dots\rho(Ty)\rho(y). \tag{5.1}$$

ρ_n satisfies the ***cocyle equation*** $\rho_{n+m}(y) = \rho_n(T^m y)\rho_m(y)$. $\rho_n: Y \to G$ is, in the strict sense, a ***one-cocycle*** for the action of $\mathbb{Z}$ on functions from Y to G. We shall refer interchangeably to ρ_n and (by abuse of language) to the single function $\rho = \rho_1$ as a ***cocycle*** determining the isometric extension $X = Y \times G/H \to Y$. We shall sometimes denote the system on X determined by the cocycle ρ by $Y \times_\rho G/H$.

In general if $(X, \mathcal{B}, \mu, T)$ is an ergodic system and $(Y, \mathcal{D}, \nu, T)$ is a factor $\pi: X \to Y$, we may consider the subspace of $L^2(X, \mathcal{B}, \mu)$ spanned by T-invariant finite rank modules over $L^\infty(Y)$. This subspace, if it is not all of $L^2(X, \mathcal{B}, \mu)$ will in any case be the subspace of functions defined by some factor $(\hat{Y}, \hat{\mathcal{B}}, \hat{\nu}, T)$ between X and Y:

$$X \overset{\hat{\pi}}{\to} \hat{Y} \overset{\omega}{\to} Y$$

with $\omega \circ \hat{\pi} = \pi$. The system on $\hat{Y}$ will be the ***maximal isometric extension*** of Y in X.

It will be noted for later reference that the maximal isometric extension of $(Y, \mathcal{D}, \nu, T)$ in $(X, \mathcal{B}, \mu, T)$ is the same for T as it is for any power of T.

Among isometric extensions by homogeneous spaces $M = G/H$ we have those for which $H = \{1\}$ so that M is itself a group. These will be called ***group extensions***. If $(X, \mathcal{B}, \mu, T)$ is an isometric extension of $(Y, \mathcal{D}, \nu, T)$, say $X = Y \times_\rho G/H$ with $T(y,u) = (Ty, \rho(u)u)$ where $\rho: Y \to G$, then we can use the same cocyle ρ to give a group extension: $\tilde{X} = Y \times_\rho G$, where now $T(y,g) = (Ty, \rho(y)g)$.

We now present a situation in which isometric extensions play an important role. Let $(X_i, \mathcal{B}_i, \mu_i, T_i)$, $i = 1, \dots, k$, be measure preserving systems and let $(Y_i, \mathcal{D}_i, \nu_i, T_i)$, $i = 1, \dots, k$ be corresponding factors. A measure μ on $\prod X_i$ defines a ***joining*** of the systems on X_i if it is invariant under $T_1 \times \dots \times T_k$ and maps onto μ_j under the natural map $\prod X_i \to X_j$. We will say that a joining μ on $\prod X_i$ is a ***conditional product joining relative to*** $\prod Y_i$ if

$$\mu = \int \mu_{1,y_1} \times \mu_{2,y_2} \times \dots \times \mu_{k,y_k} d\nu(y_1, \dots, y_k) \tag{5.3}$$

where the μ_{i,y_i} represent the disintegration of μ_i with respect to ν_i as described in §3, and ν is the projection of μ on $\prod Y_i$.

The following result is proved in **[F1]**.

Theorem 5.1. *For* $i = 1, 2, \dots, k$, *let* $(Y_i, \mathcal{D}_i, \nu_i, T_i)$ *be a factor of* $(X_i, \mathcal{B}_i, \mu_i, T_i)$ *and assume each* $(X_i, \mathcal{B}_i, \mu_i, T_i)$ *has finitely many ergodic components. Let* $(\hat{Y}_i, \hat{\mathcal{D}}_i, \hat{\nu}_i, T_i)$ *be the maximal isometric extension of* $(Y_i, \mathcal{D}_i, \nu_i, T_i)$ *in* $(X_i, \mathcal{B}_i, \mu_i, T_i)$. *If* μ *is a measure on* $\prod X_i$ *defining a joining which is a conditional product relative* *to* $\prod Y_i$, *then almost every ergodic component of* μ *is a conditional product relative to* $\prod \hat{Y}_i$.

Another way of stating this conclusion is that if $F \in L^2(\prod X_i, \prod \mathcal{B}_i, \mu)$ is invariant under $T_1 \times T_2 \times \cdots \times T_k$ then $\exists \Phi \in L^2(\prod \hat{Y}_i, \prod \hat{\mathcal{D}}_i, \hat{\nu})$ for $\hat{\nu}$ the image of μ on $\prod \hat{Y}_i$ so that $F(x_1, x_2, \ldots, x_k) = \Phi(\hat{\pi}_1(x_1), \hat{\pi}_2(x_2), \ldots, \hat{\pi}_k(x_k))$.

A special case of the foregoing is to take an ergodic system, $(X', \mathcal{B}', \mu', T)$, and to let $(X_i, \mathcal{B}_i, \mu_i) = (X', \mathcal{B}', \mu')$ for $i = 1, 2$ and $T_1 = T^a$, $T_2 = T^b$ and let Y_i be 1-point spaces. Then $\hat{Y}_i = Z$ where (Z, α) is the Kronecker factor of $(X', \mathcal{B}', \mu', T)$. We take $\mu = \mu' \times \mu'$ to obtain a (conditional) product joining. Theorem 5.1 in this case gives the familiar result that invariant functions for $T^a \times T^b$ on $X' \times X'$ come from invariant functions on $Z \times Z$ under the translation $(z_1, z_2) \to (z_1 + a\alpha, z_2 + b\alpha)$. These functions are those constant on cosets modulo the subgroup $\{(a\theta, b\theta) : \theta \in Z\}$.

We note here for later use that the condition (5.3) can also be written:

$$\int_{\prod X_i} f_1(x_1) f_2(x_2) \ldots f_k(x_k)\, d\mu(x_1, x_2, \ldots, x_k)$$

$$= \int_{\prod Y_i} \{E(f_1 \mid Y_1) E(f_2 \mid Y_2) \ldots E(f_k \mid Y_k)\}\, d\nu(y_1, y_2, \ldots, y_k). \qquad (5.4)$$

6. A characteristic factor for $\{rn, sn, tn\}$

Let $(X, \mathcal{B}, \mu, T)$ be an ergodic system and let us fix non-zero integers r, s, t. Let (Z, α) be the Kronecker factor of $(X, \mathcal{B}, \mu, T)$ and let $(\hat{Z}, \hat{\mathcal{D}}, \hat{\nu}, T)$ denote the maximal isometric extension of (Z, α) in $(X, \mathcal{B}, \mu, T)$.

Theorem 6.1. *$\hat{Z}$ is a characteristic factor of X for $\{rn, sn, tn\}$; that is,*

$$\frac{1}{N}\sum_{n=1}^{N} T^{rn} f T^{sn} g T^{tn} h - \frac{1}{N}\sum_{n=1}^{N} T^{rn} E(f|\hat{Z}) T^{sn} E(g \mid \hat{Z}) T^{tn} E(h \mid \hat{Z}) \to 0$$

in $L^2(X, \mathcal{B}, \mu)$ for $f, g, h \in L^\infty(X, \mathcal{B}, \mu)$.

Proof. It suffices to show that if $E(f \mid \hat{Z}) = 0$ then

$$\frac{1}{N}\sum_{n=1}^{N} T^{rn} f T^{sn} g T^{tn} h \to 0.$$

We apply Lemma 1.1 with $u_n = T^{rn} f T^{sn} g T^{tn} h$.

$$\langle u_n, u_{n+m} \rangle = \int (f T^{rm} \bar{f}) T^{(s-r)n} (g T^{sm} \bar{g}) T^{(t-r)n} (h T^{tm} \bar{h}) d\mu.$$

Since by Theorem 2.1

$$\lim_{N\to\infty} \frac{1}{N} \sum_{n=1}^{N} T^{(s-r)n} G T^{(t-r)n} H$$

exists in $L^2(X, \mathcal{B}, \mu)$ for bounded G, H, we have that

$$\gamma_m = \lim_{N\to\infty} \frac{1}{N} \sum_{n=1}^{N} \langle u_n, u_{n+m} \rangle$$

exists. Moreover, by the same theorem,

$$\begin{aligned} \gamma_m &= \int_Z E(fT^{rm}\bar{f} \mid Z)(z)\{\int_Z E(gT^{sm}\bar{g} \mid Z)(z+(s-r)\theta) \\ &\qquad\qquad E(hT^{tm}\bar{h} \mid Z)(z+(t-r)\theta)d\theta\}dz \\ &= \iint E(fT^{rm}\bar{f} \mid Z)(z+r\theta)E(gT^{sm}\bar{g} \mid Z)(z+s\theta)E(hT^{tm}\bar{h} \mid Z)(z+t\theta)\, dz d\theta. \end{aligned} \tag{6.1}$$

Define a measure $\tilde{\mu}$ on $X \times X \times X$ by setting

$$\begin{aligned} \int f_1(x_1) f_2(x_2) f_3(x_3)\, d\tilde{\mu} = \int E(f_1 \mid Z)(z_1) E(f_2 \mid Z)(z_2) \\ E(f_3 \mid Z)(z_3)\, d\nu(z_1, z_2, z_3) \end{aligned} \tag{6.2}$$

with $\nu =$ Haar measure on the subgroup $W_{r,s,t} = \{(z+r\theta, z+s\theta, z+t\theta), z, \theta \in Z\}$ of Z^3. Then $\tilde{\mu}$ will define a joining of $(X, \mathcal{B}, \mu, T^r)$, $(X, \mathcal{B}, \mu, T^s)$ and $(X, \mathcal{B}, \mu, T^t)$ which is a conditional product joining relative to (the measure ν on) Z^3. Moreover we will have

$$\gamma_m = \int_{X^3} f(x_1)g(x_2)h(x_3)T^{rm} f(x_1) T^{sm} g(x_2) T^{tm} h(x_3)\, d\tilde{\mu}. \tag{6.3}$$

Rewriting (6.2) in terms of the disintegration of μ with respect to Z, and using (3.1) we can easily check that $\tilde{\mu}$ is invariant under $T^r \times T^s \times T^t$. We can apply the ergodic theorem to (6.3) to obtain

$$\lim \frac{1}{M} \sum_{1}^{M} \gamma_m = \int f(x_1)g(x_2)h(x_3)D(x_1, x_2, x_3)\, d\tilde{\mu}(x_1, x_2, x_3) \tag{6.4}$$

for D some $L^2(\tilde{\mu})$ function invariant under $T^r \times T^s \times T^t$. But by Theorem 5.1, $D(x_1, x_2, x_3) = \hat{D}(\hat{x}_1, \hat{x}_2, \hat{x}_3)$ where $x \to \hat{x}$ denotes the map of $X \to \hat{Z}$, the maximal isometric extension of Z in X.

Now by (6.2), if $\hat{K}(\hat{x}_1, \hat{x}_2, \hat{x}_3) = K_1(\hat{x}_1)K_2(\hat{x}_2)K_3(\hat{x}_3)$ then

$$\int f(x_1)g(x_2)h(x_3)\hat{K}(\hat{x}_1, \hat{x}_2, \hat{x}_3)\, d\tilde{\mu}$$
$$= \int E(fK_1 \mid Z)(z_1)E(gK_2 \mid Z)(z_2)E(hK_3 \mid Z)(z_3)\, d\nu$$
$$= \int E\{E(fK_1 \mid \hat{Z}) \mid Z)\}(z_1)E(gK_2 \mid Z)(z_2)E(hK_3 \mid Z)(z_3)\, d\nu = 0$$

since $E\{E(fK_1 \mid \hat{Z}) \mid Z\} = E\{K_1 E(f \mid \hat{Z}) \mid Z\} = 0$. But then it follows that for an arbitrary function $\hat{K}(\hat{x}_1, \hat{x}_2, \hat{x}_3)$

$$\int f(x_1)g(x_2)h(x_3)\hat{K}(\hat{x}_1, \hat{x}_2, \hat{x}_3)\, d\tilde{\mu} = 0$$

and hence each $\gamma_m = 0$. By Lemma 1.1 we obtain the desired result. □

We end this section with the following definition:

Definition. An ergodic system is said to be ***normal*** if the maximal isometric extension of its Kronecker factor is a group extension.

Thus in the case of a normal system, the behavior of $\frac{1}{N}\sum T^{rn}fT^{sn}gT^{rn}h$ (and ultimately the behavior of $\frac{1}{N}\sum T^n fT^{n^2}g$) can be reduced to the situation where X is a group extension of its Kronecker factor.

7. Ergodic decomposition of group extensions and the Mackey group

Consider an ergodic system $(Y, \mathcal{D}, \nu, T)$ and a group extension defined by a cocycle ρ. ρ is a measurable map of Y to G, and $T(y, g) = (Ty, \rho(y)g)$ gives a transformation of $Y \times G$ to itself. The measure $\nu \times m_G$ (m_G = Haar measure on G) is invariant on $Y \times G$, and may or may not be ergodic. We are interested in studying the ergodic invariant measures that map to ν under $Y \times G \to Y$.

We can reparameterize $Y \times G$ replacing (y, g) by $\Phi(y, g) = (y, \varphi(y)g)$ for some $\varphi\colon Y \to G$. The transformation $T' = \Phi T \Phi^{-1}$ has the form $T'(y, g) = (Ty, \rho'(y)g)$ with $\rho'(y) = \varphi(Ty)\rho(y)\varphi(y)^{-1}$. The two cocycles ρ, ρ' will be called ***equivalent***, (or ***cohomologous***) and it is clear that equivalent cocycles lead to isomorphic dynamical systems.

If a cocycle ρ takes values in a proper subgroup $H \subset G$ then it is clear that the measure $\nu \times m_G$ will not be ergodic, since any function $\eta(y, g) = \psi(Hg)$ for ψ on $H \setminus G$ is an invariant function for T. It follows that if ρ is equivalent to a cocycle with values in a proper subgroup, then the product measure $\nu \times m_G$ will again not be ergodic for T. The following proposition describes the ergodic measures for T on $Y \times G$ (extending ν) in terms of the range of equivalent cocycles.

Proposition 7.1. *For any cocycle $\rho: Y \to G$ there is a closed subgroup $H \subset G$ uniquely determined up to conjugacy, so that*

(i) *there is a cocycle $\rho'(y) = \varphi(Ty)\rho(y)\varphi(y)^{-1}$ equivalent to ρ taking values in H,*

(ii) *the transformation $T': Y \times G \to Y \times G, T'(y, g) = (Ty, \rho'(y)g)$ has ergodic invariant measures $\nu \times m_{H\gamma}$, where $m_{H\gamma}$ is Haar measure m_H translated by γ to the right,*

(iii) *any ergodic T'-invariant measure on $Y \times G$ has the form $\mu = \nu \times m_{H\gamma}$ for some coset $H\gamma \in H \setminus G$ and the ergodic T-invariant measures on $Y \times G$ are obtained by applying Φ^{-1} to the ergodic T'-invariant measures. The group H is called the* ***Mackey group*** *of the extension $Y \times G$.*

Note that since T' acts ergodically on $\nu \times m_H$, the cocycle ρ' (and therefore ρ) cannot be equivalent to one taking values in a smaller subgroup. Thus H can be thought of as the minimal range of ρ.

The proof follows from a general theory developed by G. Mackey, but we give a brief sketch. Take a version of Y on a compact metric space and consider some ergodic measure μ extending ν. A point is called generic for a system on a compact space if the ergodic theorem holds for that point and all continuous functions. An ergodic measure has almost all points generic. For each $\gamma \in G$ the transformation $S_\gamma(y, g) = (y, g\gamma)$ commutes with T and takes T-invariant measures to T-invariant measures. Let $H = \{\gamma : S_\gamma\mu = \mu\}$. For $\gamma \in H$, S_γ permutes the generic points of μ, and moreover if x and $S_\gamma x$ are both generic points of μ then $\gamma \in H$. It follows that for any y the (y, g) generic for μ have the form $(y, \varphi(y)^{-1}h)$, $h \in H$, for some (measurable) $\varphi: Y \to G$. Since this accounts for almost all points with respect to μ, we see that $\rho(y)\varphi(y)^{-1}H \subset \varphi(Ty)^{-1}H$ so that (i) is satisfied by ρ'. If $\Phi(y, g) = (y, \varphi(y)g)$ then $\Phi\mu = \nu \times m_H$ is T'-invariant. We get other T-ergodic measures by applying S_γ to μ, and since we then exhaust all possible generic points this gives all the ergodic measures extending ν. Applying Φ we obtain all T'-ergodic measures extending ν as described in (iii).

We shall give several lemmas which are consequences of Proposition 7.1. In the sequel the Mackey group will continue to play a major role.

Suppose $(X, \mathcal{B}, \mu, T)$ is an isometric extension of a system $(Y, \mathcal{D}, \nu, T)$ so that $X = Y \times_\rho M$ where M is a homogeneous metric space. If we express M as a homogeneous space of G, $M = G/H$, then T is given on X by $T(y, u) = (Ty, \rho(y)u)$ for a measurable map $\rho: Y \to G$. We can then form a group extension: $\tilde{X} = Y \times_\rho G$, with $T: \tilde{X} \to \tilde{X}$ given by $T(y, g) = (Ty, \rho(y)g)$. Generally there is more than one way of choosing G. We now have

Lemma 7.2. *If $(X, \mathcal{B}, \mu, T)$ is an ergodic isometric extension of $(Y, \mathcal{D}, \nu, T)$ and $X = Y \times M$, then it is possible to express $M = G/H$ and $X = Y \times_\rho G/H$ so that the corresponding group extension $\tilde{X} = Y \times_\rho G$ is also ergodic.*

Proof. It will be convenient to take a version of the dynamical system $(Y, \mathcal{D}, \nu, T)$ where Y is a compact space and $\mathcal{D}$ is the Borel σ-algebra.

Suppose $M = L/K$ and that T on $Y \times M$ is determined by the cocycle $\rho: Y \to L$. Form the group extension $T: Y \times_\rho L \to Y \times_\rho L$, with the same cocycle. In general if $\pi: X_1 \to X_2$ is a continuous map of compact spaces commuting with transformations $T_i: X_i \to X_i$, then if μ_2 is T_2-invariant on X_2, there will exist T_1-invariant μ_1 on X_1 with $\pi\mu_1 = \mu_2$. Moreover, the set of such μ_1 is compact convex, and if μ_2 is ergodic for T_2, then the extreme measures in this set are ergodic for T_1. Applying this to the map $Y \times L \to Y \times L/K = X$ we find that there is an ergodic measure $\tilde{\mu}$ on $Y \times L$ so that X is a factor of $Y \times L$, and $\tilde{\mu}$ maps to ν under $Y \times L \to Y$. So by Proposition 7.1 the system $(Y \times L, \tilde{\mu}, T)$ is isomorphic to $(Y \times G, \nu \times m_G, T)$ where $G \subset L$ is the corresponding Mackey group. Since $Y \times G \to Y \times M$ takes $\nu \times m_G$ to $\nu \times m_M$ where m_M is the L-invariant measure on M, it follows that G is transitive on M so that M is a homogeneous space of G. □

Lemma 7.3. *Let $X = Y \times G$ be an ergodic group extension of a system Y, and let $X \xrightarrow{\omega} W \xrightarrow{\pi} Y$ be homomorphisms for a factor W of X intermediate between X and Y. Then X is a group extension of W.*

Proof. Let $\rho: Y \to G$ be the cocycle defining the group extension $X = Y \times G$. Let $\hat{\rho}: W \to G$ be defined by $\hat{\rho} = \rho \circ \pi$ and use $\hat{\rho}$ to define a group extension $W \times_{\hat{\rho}} G$ of W. Let H be the Mackey group of this extension, so that there is an ergodic subsystem of $W \times G$ isomorphic to $W \times H$. Denote by μ_X, μ_W, μ_Y the measures for the systems on X, W and Y respectively. By Proposition 7.1, an ergodic T-invariant measure on $W \times G$ that maps to μ_W under $W \times G \to W$ provides a system isomorphic to $W \times H$. Now consider the maps $\sigma: X \to W \times G$ and $\tau: W \times G \to X$ defined by $\sigma(y, g) = (\omega(y, g), g)$ and $\tau(w, g) = (\pi(w), g)$. Using the fact that ω is a homomorphism so that $T\omega(y, g) = \omega(Ty, \rho(y)g)$, we check that σ and τ are homomorphisms. Hence $\sigma(\mu_X)$ is an ergodic T-invariant measure on $W \times G$ projecting onto $\omega(\mu_X) = \mu_W$ on W. Thus $\sigma(\mu_X) = \tilde{\mu}$ defines a system isomorphic to $W \times H$. On the other hand σ is 1-1 since $\tau\sigma(y, g) = \tau(\omega(y, g), g) = (\pi\omega(y, g), g) = (y, g)$. This shows that $X \to W$ is a group extension. □

8. Normal systems

At the end of §6 we defined a ***normal ergodic system*** as one for which the maximal isometric extension $\hat{Z}$ of its Kronecker factor Z is a group extension. In this section we will show that every ergodic system X is a factor of some normal system. It will follow that in order to prove Theorem A we may confine our attention to normal systems.

Our construction of a normal extension of a given system will involve a passage to the inverse limit of a sequence of systems. We need several lemmas relating to inverse limits.

Lemma 8.1. *Let* $X_1 \overset{\theta_1}{\leftarrow} X_2 \overset{\theta_2}{\leftarrow} X_3 \overset{\theta_2}{\leftarrow} \cdots$ *be an inverse sequence of measure preserving systems and let* $Z_1 \overset{\varphi_1}{\leftarrow} Z_2 \overset{\varphi_2}{\leftarrow} Z_3 \overset{\varphi_3}{\leftarrow} \cdots$ *be the corresponding sequence of Kronecker factors. Then* $Z_\infty = \underset{\leftarrow}{\lim} Z_n$ *is the Kronecker factor of* $X_\infty = \underset{\leftarrow}{\lim} X_n$.

Proof. We have to show that every eigenfunction on X_∞ is measurable over Z_∞. But if $Tf = \lambda f$, then $TE(f|X_n) = E(Tf \mid X_n) = \lambda E(f \mid X_n)$, and $E(f \mid X_n)$ is measurable over Z_n. Now $f = \lim E(f \mid X_n)$, so that f is measurable over Z_∞. □

Let $(X, \mathcal{B}, \mu, T)$ be a system and $(Y, \mathcal{D}, \nu, T)$ a factor with the homomorphism $\pi: X \to Y$. We denote by $X \times_Y X$ the fiber product $\{(x, x') : \pi(x) = \pi(x')\}$. We define the relative product measure

$$\tilde{\mu} = \int \mu_y \times \mu_y d\nu(y)$$

for the disintegration $\mu = \int \mu_y d\nu(y)$ of μ relative to Y. We now obtain a measure preserving system $(X \times_Y X, \mathcal{B} \times \mathcal{B}, \tilde{\mu}, T)$ with $T(x, x') = (Tx, Tx')$ and we denote this system by $X \times_Y X$ as well.

Consider 4 systems, $(X, \mathcal{B}, \mu, T)$, $(Y, \mathcal{D}, \nu, T)$, $(X', \mathcal{B}', \mu', T)$, $(Y', \mathcal{D}', \nu', T)$, and a commutative diagram

$$\begin{array}{ccc} X & \overset{\theta}{\rightarrow} & X' \\ \pi \downarrow & & \downarrow \pi' \\ Y & \overset{\varphi}{\rightarrow} & Y' \end{array}$$

where θ, φ, π and π' are homomorphisms. Note that if φ exists, it is uniquely determined by θ, π, π'. We have decompositions $\mu = \int \mu_y \, d\nu(y)$, $\mu' = \int \mu'_{y'} \, d\nu'(y')$. We now say that θ defines a ***homomorphism of the pair*** (X, Y) ***to*** (X', Y') if for a.e. $y \in Y$

$$\theta(\mu_y) = \mu_{\varphi(y)},$$

i.e., if θ is measure preserving on each fibre $\pi^{-1}\{y\}$. An equivalent condition is that regarding $L^2(X'), L^2(Y), L^2(Y')$ as subspaces of $L^2(X)$, we have

$$E(L^2(X')|Y) \subset L^2(Y'). \tag{8.1}$$

Note that the composition of pair homomorphisms is a pair homomorphism. It is easy to verify that if $\theta(X, Y) \to (X', Y')$ is a homomorphism, then we also have an induced homomorphism $\tilde{\theta}: X \times_Y X \to X' \times_{Y'} X'$, where $\tilde{\theta}(x_1, x_2) = (\theta(x_1), \theta(x_2))$.

Lemma 8.2. *Let* $\{(X_n, Y_n), \theta_n: (X_{n+1}, Y_{n+1}) \to (X_n, Y_n)\}$ *be an inverse sequence of homomorphisms of pairs of systems, and let* $X_\infty = \underset{\leftarrow}{\lim} X_n$, $Y_\infty = \underset{\leftarrow}{\lim} Y_n$. *Then*

$$\underset{\leftarrow}{\lim} X_n \times_{Y_n} X_n = X_\infty \times_{Y_\infty} X_\infty.$$

Proof. We must first check that the natural map $X_\infty \to X_n$ defines a homomorphism of pairs $(X_\infty, Y_\infty) \to (X_n, Y_n)$. For this we verify (8.1) noting that $E(f \mid Y_\infty) =$

$\lim E(f \mid Y_n)$ by martingale convergence. Thus we have a homomorphism $X_\infty \times_{Y_\infty} X_\infty \to X_n \times_{Y_n} X_n$ composing properly with $\tilde{\theta}_n$. We can identify $X_\infty \times_{Y_\infty} X_\infty$ with $\underset{\leftarrow}{\lim} X_n \times_{Y_n} X_n$ provided this is true set-theoretically. But this is manifest. □

Lemma 8.3. *If X is ergodic and $\theta: X \to X'$ is a homomorphism and Z, Z' are the Kronecker factors of X, X' respectively, then θ defines a homomorphism of pairs $\theta: (X, Z) \to (X', Z')$.*

Proof. We take $f \in L^2(X')$ and show that $E(f \mid Z) \in L^2(Z')$. Now $E(f \mid Z) = \Sigma\langle f, \chi\rangle\chi$ as χ ranges over the characters of Z. If $\langle f, \chi\rangle \neq 0$ and $T\chi = \lambda\chi$, then $\lim \frac{1}{N}\sum_1^N T^n f T^n \bar{\chi} \neq 0$ so that $\lim \frac{1}{N}\Sigma\lambda^{-n}T^n f \neq 0$. But the latter is an eigenfunction for λ in $L^2(X')$, and if λ is in the discrete spectrum of T on $L^2(X')$, X must belong to $L^2(Z')$. It follows that $E(f \mid Z) \in L^2(Z')$. □

An isometric extension $X \to Y$ determines an invariant function on $X \times_Y X$, namely the metric on fibres of X over Y. Conversely invariant functions on $X \times_Y X$ can be seen to come from the maximal isometric extension $\hat{Y}$ of Y inside X. This gives

Lemma 8.4. *Let Z be the Kronecker of X, and let $\hat{Z}$ be its maximal isometric extension in X. Let Y be any intermediate factor, $X \to Y \to Z$. Then Y is an extension of $\hat{Z}$ iff the T-invariant functions on $X \times_Z X$ are measurable over $Y \times_Z Y$.*

Let $(X_1, \mathcal{B}_1, \mu_1)$, $(X_2, \mathcal{B}_2, \mu_2)$ be two measure spaces. A measure $\tilde{\mu}$ on $(X_1 \times X_2, \mathcal{B}_1 \times \mathcal{B}_2)$ is a ***coupling*** of μ_1 and μ_2 if it maps to μ_1 and μ_2 under the coordinate maps. A coupling is a ***graph coupling*** if we have a measure preserving map $\theta: X_1 \to X_2$ so that $\tilde{\mu} = (\mathrm{id} \times \theta)(\mu_1)$, where $\mathrm{id} \times \theta: X_1 \to X_1 \times X_2$. Assume now that X_1 and X_2 are extensions of a system Y. We consider couplings of μ_1 and μ_2 on the relative product $X_1 \times_Y X_2$, and there we say that a coupling is a ***graph coupling*** if there is a map $\theta: X_1 \to X_2$ taking the fibre in X_1 over $y \in Y$ to the fibre in X_2 over y, and such that $\tilde{\mu} = (\mathrm{id} \times \theta)(\mu_1)$.

Lemma 8.5. *Let $Y \to Z$ be an ergodic isometric extension. It is a group extension iff almost all of the ergodic components of the measure $\tilde{\mu}$ on $Y \times_Z Y$ are graphs.*

Proof. Suppose first that $Y = Z \times_\rho G$ is a group extension. $Y \times_Z Y$ can be expressed as $Z \times G \times G$ and it is clear that for each $\gamma \in G$, $\{(z, g, g\gamma) \mid z \in Z,\ g \in G\}$ is an invariant set carrying an invariant measure which is the image of the product measure on Y. These are ergodic and they integrate to the relative product measure on $Y \times_Z Y$.

Suppose now that $Y = Z \times_\rho G/H$ where we can assume, by Lemma 7.2 that $Z \times_\rho G$ is also ergodic. The ergodic components of $Z \times G \times G$ map to the ergodic components of $Z \times G/H \times G/H$ under the natural map. Assume the latter are graphs. Then the measure on the graph of $g \to g\gamma$ in $G \times G$ is carried to a graph on $G/H \times G/H$ for almost every γ. It is not hard to see that this implies that if m_H denotes Haar measure on H, $m_H\gamma$ maps to a point measure on G/H for every γ. But this implies that H is normal and so G/H is itself a group. □

Lemma 8.6. *Let $\{(Y_n, Z_n), \theta_n\colon (Y_{n+1}, Z_{n+1}) \to (Y_n, Z_n)\}$ be an inverse sequence of pair homomorphisms of ergodic systems. Assume that for each n, Y_n is a group extension of Z_n. Then $Y_\infty = \varprojlim Y_n$ is a group extension of $Z_\infty = \varprojlim Z_n$.*

Proof. Clearly Y_∞ is an isometric extension of Z_∞ so we can apply Lemma 8.5. The ergodic components of $Y_\infty \times_{Z_\infty} Y_\infty$ map to ergodic components of $Y_n \times_{Z_n} Y_n$ and since these are graphs for all n, the same is true for the measure on $Y_\infty \times_{Z_\infty} Y_\infty$. □

Lemma 8.7. *Let $\{(X_n, Z_n), \theta_n\colon X_{n+1} \to X_n\}$ be an inverse sequence of pair homomorphisms of ergodic systems with Z_n the Kronecker factor of X_n. Let $\hat{Z}_n$ denote the maximal isometric extension of Z_n in X_n. Then θ_n induces a homomorphism of $\hat{Z}_{n+1} \to \hat{Z}_n$ and $\varprojlim \hat{Z}_n$ is the maximal isometric extension of $\varprojlim Z_n$ in $\varprojlim X_n$.*

Proof. Consideration of finite rank modules makes it evident that we have a natural map $\hat{\theta}_n\colon \hat{Z}_{n+1} \to \hat{Z}_n$ induced by $\theta_n\colon X_{n+1} \to X_n$. Since Z_n is also the Kronecker factor of $\hat{Z}_n$ we have $\hat{\theta}_n$ defining a pair homomorphism $(\hat{Z}_{n+1}, Z_{n+1}) \to (\hat{Z}_n, Z_n)$. We can therefore identify $\varprojlim \hat{Z}_n \times_{Z_n} \hat{Z}_n$ with $\hat{Z}_\infty \times_{Z_\infty} \hat{Z}_\infty$ where $Z_\infty = \varprojlim Z_n$, $\hat{Z}_\infty = \varprojlim \hat{Z}_n$. Let $X_\infty = \varprojlim X_n$. It is clear that $\hat{Z}_\infty$ is an isometric extension of Z_∞ since $L^2(\hat{Z}_\infty)$ is spanned by $\cup L^2(\hat{Z}_n)$. We need to show it is the maximal isometric extension inside X_∞. We apply Lemma 8.4; so we show that an invariant function in $L^2(X_\infty \times_{Z_\infty} X_\infty)$ is in $L^2(\hat{Z}_\infty \times_{Z_n} X_n)$. But such a function maps to an invariant function in $L^2(X_n \times_{Z_n} X_n)$, and so this projection is in $L^2(\hat{Z}_n \times_{Z_n} \hat{Z}_n)$. By Lemma 8.2, the original function is in $L^2(\hat{Z}_\infty \times_{Z_\infty} \hat{Z}_\infty)$. □

We can now prove:

Theorem 8.8. *Every ergodic system is a factor of a normal ergodic system.*

Proof. We shall construct inductively an inverse sequence of systems $X_0 \leftarrow X_2 \leftarrow X_4 \leftarrow X_6 \leftarrow \cdots$ with X_0 the given system and where $\varprojlim X_{2n}$ will be the normal system. More precisely we construct the following chain of systems:

$$\begin{array}{ccccccc}
X_0 \longleftarrow & & & X_2 \longleftarrow & & & X_4 \longleftarrow \cdots \\
\downarrow & & \swarrow & \downarrow & & \swarrow & \downarrow \\
\hat{Z}_0 \leftarrow & Z_0 \times & G_0 \leftarrow & \hat{Z}_2 \leftarrow & Z_2 \times & G_2 \leftarrow & \hat{Z}_4 \leftarrow \cdots \\
\downarrow & \swarrow & \downarrow & \downarrow & \swarrow & \downarrow & \downarrow \\
Z_0 \leftarrow & & Z_1 \leftarrow & Z_2 & \leftarrow & Z_3 \leftarrow & Z_4 \leftarrow \cdots
\end{array}$$

Here Z_{2n} is the Kronecker factor of X_{2n}, $\hat{Z}_{2n}$ is its maximal isometric extension in X_{2n}. We write $\hat{Z}_{2n} = Z_{2n} \times G_{2n}/H_{2n}$ so that the group extension $Z_{2n} \times G_{2n}$ is ergodic (Lemma 7.2). Z_{2n+1} is the Kronecker factor of $Z_{2n} \times G_{2n}$ and by Lemma 7.3, $Z_{2n} \times G_{2n}$ is a group extension of Z_{2n+1}. At each stage the construction proceeds by letting X_{2n+2} be an ergodic joining of $Z_{2n} \times G_{2n}$ and X_{2n}. This gives us the entire chain as described.

Now we set $X_\infty = \varprojlim X_{2n}$, $Z_\infty = \varprojlim Z_n$, and $\hat{Z}_\infty = \varprojlim Z_{2n} \times G_{2n} = \varprojlim \hat{Z}_{2n}$. From the second description of $\hat{Z}_\infty$ it follows by Lemma 8.7 that $\hat{Z}_\infty$ is the maximal isometric extension of Z_∞ in X_∞. By Lemma 8.1, Z_∞ is the Kronecker factor of X_∞. Finally, the sequence $\{(Z_{2n} \times G_{2n}, Z_{2n+1})\}$ is an inverse sequence of pairs because Z_{2n+1} is the Kronecker factor of $Z_{2n} \times G_{2n}$, and since the latter is a group extension of Z_{2n+1}, we have by Lemma 8.6, that $\hat{Z}_\infty$ is a group extensions of Z_∞. This completes the proof. □

9. Abelian extensions of Kronecker systems

In view of Theorem 8.8 we can henceforth restrict our attention to normal systems. We assume then that $(X, \mathcal{B}, \mu, T)$ is a normal system so that $X \to Z \times_\rho G \to Z$ with Z the Kronecker factor of X and $Z \times_\rho G$ its maximal isometric extension in X. $Z \times_\rho G$ is a characteristic factor for all $\{rn, sn, tn\}$. In this section we shall show that we can reduce the characteristic factor further, replacing $Z \times_\rho G$ by $Z \times_{\bar{\rho}} H$ where H is an abelian group-theoretic factor of G.

This will be based on the following group-theoretic lemma:

Lemma 9.1. *Let G_1, G_2, G_3 be three groups, and denote by π_{ij}, $1 \le i < j \le 3$, the natural homomorphism: $G_1 \times G_2 \times G_3 \to G_i \times G_j$. Let L be a subgroup of $G_1 \times G_2 \times G_3$ satisfying $\pi_{ij}(L) = G_i \times G_j$ for all $i < j$. Then:*

(i) *There exists an abelian group H and homomorphisms $\psi_i\colon G_i \to H$, $i = 1, 2, 3$ so that $L = \{(g_1, g_2, g_3) \mid \psi_1(g_1)\psi_2(g_2)\psi_3(g_3) = 1\}$.*

(ii) *If G_i' denotes the commutator subgroup of G_i, $i = 1, 2, 3$, then $L \supset G_1' \times G_2' \times G_3'$.*

Proof. We will show that L is a normal subgroup of $G_1 \times G_2 \times G_3$. Let $L_{12} = G_1 \times G_2 \times \{1\} \cap L$ and let L_{13}, L_{23} be defined similarly. We claim each L_{ij} is normal in $G_1 \times G_2 \times G_3$. Consider $\tilde{g} = (g_1, g_2, 1)$ in L_{12}, and take any $\tilde{\gamma} = (\gamma_1, \gamma_2, \gamma_3) = G_1 \times G_2 \times G_3$. For some $\delta \in G_3$, $\lambda = (\gamma_1, \gamma_2, \delta) \in L$ and so $\tilde{\gamma}\tilde{g}\tilde{\gamma}^{-1} = (\gamma_1 g_1 \gamma_1^{-1}, \gamma_2 g_2 \gamma_2^{-1}, 1) = \lambda\tilde{g}\lambda^{-1} \in L_{12}$. Next we claim $L = L_{12} \cdot L_{13}$. For if $(g_1, g_2, g_3) \in L$ and we can find $(h, 1, g_3) \in L_{13}$ for some $h \in G_1$, we can then express $(g_1, g_2, g_3) = (g_1 h^{-1}, g_2, 1)(h, 1, g_3)$ where $(g_1 h^{-1}, g_2, 1)$ necessarily belongs to L and hence to L_{12}. This shows that L is normal in $G_1 \times G_2 \times G_3$. Let $H = G_1 \times G_2 \times G_3 / L$. Define $\psi_1\colon G_1 \to H$ by setting $\psi_1(g_1) = (g_1, 1, 1,)L \in H$, and let ψ_2, ψ_3 be defined similarly. Clearly $(g_1, g_2, g_3) \in L \Leftrightarrow \psi_1(g_1)\psi_2(g_2)\psi_3(g_3) = 1$. Now each map $\psi_i\colon G_i \to H$ is onto and moreover for $i \neq j$, $\psi_i(g_i)\psi_j(g_j) = \psi_j(g_j)\psi_i(g_i)$. It follows that H is abelian. This proves (i) and (ii) is an immediate consequence. □

We now want to produce an abelian group H, a homomorphic image of G, so that $Z \times H$ will be a factor of $Z \times_\rho G$ and will still be a characteristic factor for $\{rn, sn, tn\}$. We shall denote the group extension $Z \times_\rho G$ by V.

We need to evaluate

$$\lim_{N\to\infty}\sum_{1}^{N} T^{rn} f\, T^{sn} g T^{tn} h \tag{9.1}$$

for $f, g, h \in L^\infty(V)$. We repeat the calculation in the proof of Theorem 6.1, this time for $f, g, h \in L^\infty(V)$. We find that the limit in (9.1) will be 0 if

$$\int f(x_1)g(x_2)h(x_3)\, D(x_1, x_2, x_3)\, d\tilde{\mu}(x_1, x_2, x_3) = 0 \tag{9.2}$$

where $\tilde{\mu}$ is a measure on V^3 invariant under $T^r \times T^s \times T^t$ as in (6.2), and D is a $T^r \times T^s \times T^t$-invariant function in $L^2(\tilde{\mu})$. We wish to analyze such invariant functions more closely. We do this in the following context.

Let (Z, α) be a Kronecker system, let G be a compact group, ρ a cocycle $\rho: Z \to G$, and let T be defined on $V = Z \times G$ by $T(z, g) = (z + \alpha, \rho(z)g)$. Let s, t be integers $\neq 0$. Let $W = W_{r,s,t} \subset Z^3$ be defined as the subgroup

$$W_{r,s,t} = \{(z + rz', z + sz', z + tz') : z, z' \in Z\}.$$

Let $\tilde{\mu}$ be Haar measure on $W \times G^3$ which we identify with a subset of V^3 by identifying $(z_1, z_2, z_3; g_1, g_2, g_3) \in W \times G^3$ with $(z_1, g_1; z_2, g_2; z_3, g_3) \in V^3$. $\tilde{\mu}$ is then invariant under $T^r \times T^s \times T^t$.

Lemma 9.2. *Let $F(z_1, g_1; z_2, g_2; z_3; g_3)$ be a function in $L^2(\tilde{\mu})$ invariant under $T^r \times T^s \times T^t$. Let G' be the commutator of G and let $\varphi: (Z \times G)^3 \to (Z \times G/G')^3$ be the natural map. Then there exists a function F' on $(Z \times G/G')^3$ so that $F = F' \circ \varphi$.*

Proof. The statement of the lemma is equivalent to the assertion that almost all ergodic components $\tilde{\mu}$ with respect to the transformation $T^r \times T^s \times T^t$ are invariant under the action of $G' \times G' \times G'$ on V^3. Let $Z_{r,s,t} = \{rz', sz', tz') : z' \in Z\}$. The ergodic components of $T^r \times T^s \times T^t$ on $W_{r,s,t} \subset Z^3$ are the cosets of $Z_{r,s,t}$. For each $z \in Z$ we need the ergodic decomposition of $T^r \times T^s \times T^t$ on the $G \times G \times G$-extension of the Kronecker system on the coset $(z, z, z) + Z_{r,s,t}$. We obtain a Mackey group $L_z \subset G^3$ defined up to conjugacy. Because we are dealing with closed subgroups of a compact group, the space of these conjugacy classes is a metric space and the map $z \to [L_z]$ is measurable, where $[L_z]$ denotes the conjugacy class of the subgroup L_z. Since $T \times T \times T$ commutes with $T^r \times T^s \times T^t$ it follows that $[L_z] = [L_{z+\alpha}]$. By ergodicity, there is a single subgroup L and $[L_z] = [L]$ for almost all z.

We now claim that if $\pi_{ij}: G^3 \to G^2$, $1 \le i < j \le 3$ are the natural projections, then $\pi_{ij}(L) = G^2$. Write $Z_{r,s} = \{(rz', sz') : z' \in Z\}$. Then the statement that $\pi_{12}(L_z) = G^2$ is the assertion that the Mackey group of $T^r \times T^s$ on $\big((z, z) + Z_{r,s}\big) \times G^2$ relative to $(z, z)+Z_{r,s}$ is G^2. Equivalently, this says that $T^r \times T^s$ is ergodic on $\big((z, z)+Z_{r,s}\big)\times G^2$. That this is true for a.e. z amounts to the statement that $T^r \times T^s$-invariant function with respect to $\pi_{12}(\tilde{\mu})$ are lifted from invariant function on the base space, $Z_{1,1} + Z_{r,s}$. But this statement is true by the remarks at the end of §5 for $T^r \times T^s$-invariant functions with respect to the full product measure $\mu_V \times \mu_V$ on $V \times V$. One can see, however, that $Z_{1,1} + Z_{r,s} = \{(z_1, z_2) \in Z^2 : z_2 - z_1 \in (r-s)Z\}$ and since $(r-s)Z$ is of finite index in Z (see §2), $Z_{1,1} + Z_{r,s}$ is of finite index in Z^2. This shows that $\pi_{12}(\tilde{\mu})$ is absolutely continuous with respect to $\mu_V \times \mu_V$, and hence $L^2(\pi_{12}(\tilde{\mu})) \subset L^2(\mu_V \times \mu_V)$. It follows that for a typical L_z, and hence for L, $\pi_{12}(L_z) = \pi_{12}(L) = G^2$. The same is true for π_{23} and π_{12} and so by Lemma 8.1 (ii), $L \supset G' \times G' \times G'$. This in turn implies that almost all ergodic components of $\tilde{\mu}$ for $T^r \times T^s \times T^t$ are invariant under $G' \times G' \times G'$ and this proves the lemma. □

We return now to evaluating the limit in (9.1) for $f, g, h \in L^\infty(V)$, where V is a G-extension of its Kronecker factor Z. Our conclusion is that this limit is 0 whenever (9.2) holds and where we now know that the function

$$D(x_1, x_2, x_3) = D(z_1, g_1 G'; z_2, g_2 G'; z_3, g_3 G').$$

Assume now that $\int_{G'} f(z, \gamma\gamma')d\gamma' = 0$ for all z, γ. We will then obtain the vanishing of

$$\int f(z_1, \gamma_1) g(z_2, \gamma_2) h(z_3, \gamma_3) D(y_1, g_1 G'; y_2, g_2 G'; y_3, g_3 G')\, d\tilde{\mu}$$

from which it follows that $\frac{1}{N}\sum T^{rn} f\, T^{sn} g\, T^{tn} h \to 0$. Similarly if $\int_{G'} g(z, \gamma\gamma')d\gamma' = 0$ or if $\int_{G'} h(z, \gamma\gamma')d\gamma' = 0$. Now let $Y = Z \times G/G'$. We conclude that in (9.1) we can replace f by $E(f \mid Y) = \int_{G'} f(z, \gamma\gamma')d\gamma'$ and similarly for g and h.

We have thereby proved

Theorem 9.3. *Any normal ergodic system X has a factor Y which is an* ***abelian*** *group extension of a Kronecker factor, $Y = Z \times_{\bar{\rho}} H$, and which is a characteristic factor for all schemes $\{rn, sn, tn\}$*

10. $\mathcal{C}-\mathcal{L}$ cocycles and $\mathcal{C}-\mathcal{L}$ functions

In Theorem 9.3 we have reduced the asymptotic behavior of $\frac{1}{N}\sum T^{rn} f\, T^{sn} g\, T^{tn} h$ for $f, g, h \in L^\infty(X)$ to that of $f, g, h \in L^\infty(Y)$ where Y is an abelian group extension of a Kronecker system. We will now reduce this still further. In this section we will assume $t = r + s$.

Write $Y = Z \times H$ where H is abelian, and let $\rho: Z \to H$ determine $T: Y \to Y$ by $T(z, k) = (z + \alpha, \rho(z)h)$. We have seen in §§6, 9 that to determine a characteristic factor of Y for $\{rn, sn, tn\}$, we need to analyze the $T^r \times T^s \times T^t$ invariant function $D(y_1, y_2, y_3)$ in $L^2(\tilde{\mu})$ where $\tilde{\mu}(= \tilde{\mu}_{r,s,t})$ is Haar measure of $W \times H^3 \subset Y^3$ where $W = \{(z + rz', z + sz', z + tz') : z, z' \in Z\}$. We have also seen that there is a subgroup $L \subset H^3$ such that the $T^r \times T^s \times T^t$-invariant functions are invariant under rotation by element of L. L can be described as follows: there are 3 homomorphisms ψ_1, ψ_2, ψ_3 of H to an abelian group J so that $L = \{(h_1, h_2, h_3) : \psi_1(h_1)\psi_2(h_2)\psi_3(h_3) = 1\}$. Now let $H_0 = \bigcap_{i=1}^{3} \ker \psi_i$. Then $H_0 \times H_0 \times H_0 \subset L$, and if we write $D(y_1, y_2, y_3) = D(z_1, h_1 ; z_2, h_2 ; z_3, h_3)$ then the latter function can be written $D(z_1, h_1 H_0 ; z_2, h_2 H_0 ; z_3, h_3 H_0)$. From this it follows that if $f(y) = f(z, h)$ satisfies $\int_{H_0} f(z, hh')\, dh' = 0$ then $\int f(y_1)g(y_2)h(y_3)D(y_1, y_2, y_3)\, d\tilde{\mu}(y_1, y_2, y_3) = 0$ which implies that $\frac{1}{N}\sum T^{rn} f\, T^{sn} g\, T^{tn} h \to 0$. It follows that in this average, $f(y) = f(z, h)$ may be replaced by $\int_{H_0} f(z, hh_0)\, dh_0$. The same is true for $g(y)$ and $h(y)$ and we conclude that H may be replaced by H/H_0. So we may assume that $H_0 = \{1\}$.

By Pontryagin duality, the homomorphism $H \to (S^1)^{\hat{H}}$ defined by

$$h \to (\ldots, \omega(h), \ldots)$$

where ω ranges over the character group $\hat{H}$ is 1-1. The factor $Z \times H$ of $\tilde{X}$ is therefore the "join" of factors of the form $Z \times S^1$, where for each $\omega \in \hat{H}$ a system is defined in $Z \times S^1$ by the cocycle $\omega \circ \rho \colon Z \to S^1$. Moreover since we may assume $\cap \ker \psi_i = \{1\}$, the characters of the form $\chi \circ \psi_i$, $\chi \in \hat{J}$, $i = 1, 2, 3$, span $\hat{H}$. It follows that a characteristic factor of X for $\{rn, sn, tn\}$ may be described by taking the join of factors $Z \times S^1$ defined by cocycles $\chi \circ \psi_i \circ \rho$. We now study those cocycles.

We return to $Y = Z \times H$ and the $T^r \times T^s \times T^t$-invariant measure $\tilde{\mu} = \tilde{\mu}_{r,s,t}$ on Y^3 and its ergodic decomposition. These ergodic components lie over cosets $W_z = (z, z, z) + Z_{r,s,t} \subset Z^3$, and for each of these cosets we may apply Proposition 7.1 to determine the nature of the ergodic components lying above it in Y^3. Namely above W_z we have an H^3-extensions defined by the cocycle $\tilde{\rho}(z_1, z_2, z_3) = (\rho_r(z_1), \rho_s(z_2), \rho_t(z_3))$ where $\rho_i(g)$ is given by (5.1) and the Mackey subgroup describing the ergodic components is L (for almost every z). There exists therefore a function $\varphi \colon W_z \to H \times H \times H$ so that

$$\varphi(z_1 + r\alpha, z_2 + s\alpha, z_3 + t\alpha)\tilde{\rho}(z_1, z_2, z_3)\varphi(z_1, z_2, z_3)^{-1} \in L.$$

Let us write $\tilde{z}$ for (z_1, z_2, z_3) and $\tilde{\alpha}$ for $(r\alpha, s\alpha, t\alpha)$ and set $\varphi = (\varphi_1, \varphi_2, \varphi_3)$. The foregoing characterization of L now gives us

$$[\psi_1 \circ \rho_r(z_1)][\psi_2 \circ \rho_s(z_2)][\psi_3 \circ \rho_t(z_3)] = \Phi(\tilde{z} + \tilde{\alpha})\Phi(\tilde{z})^{-1} \tag{10.1}$$

where $\Phi(\tilde{z}) = \psi_1 \circ \varphi_1(\tilde{z})\psi_2 \circ \varphi_2(\tilde{z})\psi_3 \circ \varphi_3(\tilde{z}) \in J$. Apply a character $\chi \in \hat{J}$ to (10.1) and we obtain

$$[\chi \circ \psi_1 \circ \rho_r(z_1)][\chi \circ \psi_2 \circ \rho_s(z_2)][\chi \circ \psi_3 \circ \rho_t(z_3)] = \frac{F_\chi(\tilde{z} + \tilde{\alpha})}{F_\chi(\tilde{z})} \tag{10.2}$$

for $F_\chi \colon (z, z, z) + Z_{r,s,t} \to S^1$. By the ergodicity of $\tilde{z} \to \tilde{z} + \tilde{\alpha}$ on the coset in question we see that F_χ is unique up to a constant multiple. The left side of (10.2) is now defined for all of $W_{r,s,t} = \{(z + rz', z + sz', z + tz') : z, z' \in Z\}$ and it is not hard to show that if it has the form (10.2) (it is a coboundary) on ergodic components of $W_{r,s,t}$, it has that form for a measurable F_χ on all of $W_{r,s,t}$.

We now wish to study factors $Z \times S^1$ determined by cocycles $\sigma = \chi \circ \psi_i(\rho)$, $\chi \in \hat{J}$, $i = 1, 2, 3$. We will focus on $i = 1$ since the other cases are perfectly analogous. T is defined on $Z \times S^1$ by $T(z, \zeta) = (z + \alpha, \sigma(z)\zeta)$ and iterating this r times we have $T^r(z, \zeta) = (z + r\alpha, \sigma(z + (r-1)\alpha) \dots \sigma(z)\zeta) = (z + r\alpha, \chi \circ \psi_1 \circ \rho_r(z)\zeta)$ so that $\sigma_r(z) = \chi \circ \psi_1 \circ \rho_r(z)$. The cocycle $\sigma_r \colon Z \to S^1$ then satisfies

$$\sigma_r(z_1)p(z_2)q(z_3) = \frac{F(z_1 + r\alpha, z_2 + s\alpha, z_3 + t\alpha)}{F(z_1, z_2, z_3)} \tag{10.3}$$

for almost all $(z_1, z_2, z_3) \in W_{r,s,t}$, where p, q are measurable functions on Z to X^1 and $F = F_\chi$.

We wish to convert (10.3) to an equation involving only (z_1, z_2). To do this we let $\Delta = Z_{1,1} \cap Z_{r,s} \subset Z^2$. Letting $W_{r,s} = Z_{1,1} + Z_{r,s}$ there is a natural isomorphism $W_{r,s}/Z_{r,s} \cong Z_{1,1}/\Delta$. This maps cosets modulo $Z_{r,s}$, i.e., ergodic components of $T^r \times T^s$ in $W_{r,s}$, to $Z_{1,1}/\Delta$. Take a measurable cross-section $Z_{1,1}/\Delta \to Z_{1,1}$. Putting this together we obtain a measurable map $\theta \colon W_{r,s} \to Z$ so that for $(z_1, z_2) \in W_{r,s}$, $(z_1 - \theta(z_1, z_2), z_2 - \theta(z_1, z_2)) \in Z_{r,s}$, and $\theta(z_1 + rz', z_2 + sz') = \theta(z_1, z_2)$. Now set

$$u(z_1, z_2) = z_1 + z_2 - \theta(z_1, z_2).$$

Writing $z_1 - \theta(z_1, z_2) = rw$, $z_2 - \theta(z_1, z_2) = sw$ we have $u(z_1, z_2) = \theta(z_1, z_2) + tw$, so that

$$z_1 = \theta(z_1, z_2) + rw, \; z_2 = \theta(z_1 + z_2) + sw, \; u(z_1, z_2) = \theta(z_1, z_2) + tw.$$

It follows that $(z_1, z_2, u(z_1, z_2)) \in W_{r,s,t}$. Now let $\Delta' = \{\delta \in Z : (\delta, \delta) \in \Delta\}$. If for $\delta \in \Delta'$ we write $\theta_\delta(z_1, z_2) = \theta(z_1, z_2) + \delta$ and let $u_\delta(z_1, z_2) = z_1 + z_2 - \theta_\delta(z_1, z_2)$, then u_δ will behave just as u, and $(z_1, z_2, u_\delta(z_1, z_2)) \in W_{r,s,t}$.

Lemma 10.1. *The map of $W_{r,s} \times \Delta' \to W_{r,s,t}$ given by*

$$(z_1, z_2, \delta) \to (z_1, z_2, u_\delta(z_1, z_2))$$

is onto and measure preserving with respect to Haar measure on the groups.

Proof. Let $(z_1, z_2, z_3) \in W_{r,s,t}$ so that $(0, 0, z_3 - u(z_1, z_2)) \in W_{r,s,t}$. We can write $z + rz' = 0$, $z + sz' = 0$, $z + tz' = z_3 - u(z_1, z_2)$. Now $z = -rz' = -sz'$ so that $z \in \Delta'$. Also $2z + tz' = 0$ so that $z + tz' = -z = z_3 - u(z_1, z_2)$. Hence $z_3 = u(z_1, z_2) - z = u_z(z_1, z_2)$. This proves that the map is onto. Since $(z_1, z_2) \to (z_2, z_2, u(z_1, z_2))$ defines a cross-section of the homomorphism $W_{r,s,t} \to W_{r,s}$, it is clear that our map is measure preserving. □

Lemma 10.2. $u_\delta(z_1 + r\alpha, z_2 + s\alpha) = u_\delta(z_1, z_2) + t\alpha$

Proof. This follows from the fact that θ is constant on cosets modulo $Z_{r,s}$. □

Now return to (10.3) and for $v \in Z$, replace $(z_1, z_2, z_3) \in W_{r,s,t}$ by $(z_1 + (r - t)v,\ z_2 + (s - t)v,\ z_3) \in W_{r,s,t}$. Dividing the two expressions we obtain

$$\frac{\sigma_r(z_1 + (r-t)v)}{\sigma_r(z_1)} \frac{p(z_2 + (s-t)v)}{p(z_2)} = \frac{F_v(z_1 + r\alpha,\ z_2 + s\alpha, z_3 + t\alpha)}{F_v(z_1, z_2, z_3)} \tag{10.4}$$

with

$$F_v(z_1, z_2, z_3) = \frac{F(z_1 + (r-t)v,\ z_2 + (s-t)v, z_3)}{F(z_1, z_2, z_3)}.$$

For every v, (10.4) is valid almost everywhere in $W_{r,s,t}$. We will now substitute $u_\delta(z_1, z_2)$ for z_3 in (10.4). Using Lemma 10.1, we conclude that the resulting equality will be valid for some δ and a.e. (z_1, z_2) in $W_{r,s}$. If we now set, for that δ, $G_v(z_1, z_2) = F_v(z_1, z_2, u_\delta(z_1, z_2))$, then using Lemma 10.2 we obtain

$$\frac{\sigma_r(z_1 + (r-t)v)}{\sigma_r(z_1)} \frac{p(z_2 + (s-t)v)}{p(z_2)} = \frac{G_v(z_1 + r\alpha, z_2 + s\alpha)}{G_v(z_1, z_2)}. \tag{10.5}$$

Next we wish to isolate z_1 from z_2.

Lemma 10.3. *Let $(X, \mathcal{B}, \mu, T)$ and $(Y, \mathcal{D}, \nu, S)$ be two ergodic systems, and suppose that f, g are measurable functions with values in S^1 satisfying a.e.*

$$f(x)g(y)H(x, y) = H(Tx, Sy), \tag{10.6}$$

and $H(x, y)$ is not 0 a.e. Then there exist functions $K: X \to S^1$, $L: Y \to S^1$ and constants c', c'' so that

$$f(x) = c' \frac{K(Tx)}{K(x)}, \quad g(y) = c'' \frac{L(Sy)}{L(y)}. \tag{10.7}$$

Proof. Form S^1-extensions of X and Y as follows:

$$T(x, \zeta) = (Tx, f(x)^{-1}\zeta), \quad S(y, \eta) = (Sy, g(y)^{-1}\eta).$$

Let $\tilde{X} = X \times S^1$, $\tilde{Y} = Y \times S^1$. On $\tilde{X} \times \tilde{Y}$ we have an invariant function:

$$\zeta\eta H(x, y) \overset{T\times S}{\to} f(x)^{-1}\zeta\ g(y)^{-1}\eta H(Tx, Sy) = \zeta\eta H(x, y).$$

This function must be a combination of products of eigenfunctions on each space:

$$\zeta\eta H(x, y) = \Sigma \Psi_n(x, \zeta)\Phi_n(y, \eta) \tag{10.8}$$

where $T(\Phi_n(x, \zeta)) = c'_n \Phi_n(x, \zeta)$, $S(\Phi_n(y, \eta)) = c'^{-1}_n \Phi_n(y, \eta)$. We can write $\Psi_n(x, \zeta) = \sum \psi_{nm}(x)\zeta^m$, $\Phi_n(y, \eta) = \sum \varphi_{nm}(y)\eta^m$. Substituting in (10.8) we find that

$$\zeta\eta H(x, y) = \sum \psi_{n1}(x)\varphi_{n1}(y)\zeta\eta$$

and so some $\psi_{n1}(x) \not\equiv 0$, $\varphi_{n1}(y) \not\equiv 0$. But

$$T(\sum \psi_{nm}(x)\zeta^m) = c'_n \sum \psi_{nm}(x)\zeta^m \Rightarrow \psi_{n1}(Tx)f(x) = c'_n \psi_{n1}(x).$$

By ergodicity of T, since $|f(x)| = 1$, $|\psi_{n1}(x)|$ is a constant $\neq 0$. Setting $K(x) = \psi_{n1}(x)^{-1}$ we obtain the first half of (10.7) and the second follows in the same way. □

We apply Lemma 10.3 to the systems $(Z, r\alpha)$ and $(Z, s\alpha)$ and to the equality (8.5). The lemma doesn't apply directly because $(Z, r\alpha)$ and $(Z, s\alpha)$ may not be ergodic and because (10.5) is valid in $W_{r,s}$ which is a subset of $Z \times Z$. However $j\alpha + rZ$ is ergodic under $z \to z + r\alpha$, and $j\alpha + sZ$ is ergodic under $z \to z + s\alpha$. Now we have the following:

Lemma 10.4. *If $q = |r - s|$ then $qZ \times qZ \subset W_{r,s}$.*

Proof. We wish to write (qz_1, qz_2) as $(z + rz', z + sz')$. This can be done if $q(z_1 - z_2)$ can be written as $(r - s)z'$. For this we choose $z' = \mp(z_1 - z_2)$. □

It follows that $(j\alpha + qrZ) \times (j\alpha + qsZ) \subset W_{r,s} \cap (j\alpha + rZ) \times (j\alpha + sZ)$. But qrZ and qsZ are of finite index in Z and so (10.5) is valid for (z_1, z_2) in a subset of positive measure of $(j\alpha + rZ) \times (j\alpha + sZ)$. If we let

$$G'_v(z_1, z_2) = \begin{cases} G_v(z_1, z_2) & \text{for } (z_1, z_2) \in W_{r,s} \\ 0 & \text{for } (z_1, z_2) \notin W_{r,s} \end{cases}$$

then

$$G'_v(z_1 + r\alpha, z_2 + s\alpha) = f(z_1)g(z_2)\, G'_v(z_1, z_2)$$

in $(j\alpha + rZ) \times (j\alpha + sZ)$ where f and g are the functions appearing in the left-hand side of (10.5). We can now apply Lemma 10.3 to obtain for each $j = 0, 1, \ldots, r - 1$,

$$\frac{\sigma_r(z + (r - t)v)}{\sigma_r(z)} = c_j \frac{K_j(v, z + r\alpha)}{K_j(v, z)}, \quad z \in j\alpha + rZ.$$

Here c_j and K_j depend on the coset of $j\alpha + rZ$ modulo rZ, and we can write $c_j = c_j(z) = \Lambda(z + rZ)$. The functions K_j can be combined to a single function on Z so that including the dependence on v we have

$$\frac{\sigma_r(z + (r - t)v)}{\sigma_r(z)} = \Lambda_v(z + rZ) \frac{K_v(z + r\alpha)}{K_v(z)}.$$

This type of functional equation was studied by Conze and Lesigne **[CL]**.

We consider next the question of measurability of Λ_v and K_v as functions of v. For this we have

Proposition 10.5. *Let $T: X \to X$ be ergodic, and $B(X, S^1)$ the measurable S^1-valued functions. If $\theta \to f_\theta(x)$ is a Borel measurable mapping from a compact space Θ into $B(X, S^1)$ and if for each θ, $\exists$ constant Λ_θ and $K_\theta \in B(X, S^1)$ such that*

$$f_\theta(x) = \Lambda_\theta \frac{K_\theta(Tx)}{K_\theta(x)}$$

then there is a Borel measurable choice for K_θ.

Proof. Suppose Λ_θ, K_θ and Λ'_θ, K'_θ are two solutions. Then $\frac{K'_\theta(Tx)}{K'_\theta(x)} = \frac{\Lambda_\theta}{\Lambda'_\theta}\frac{K_\theta(Tx)}{K_\theta(x)}$ or

$$\left(\frac{K'_\theta(Tx)}{K_\theta(Tx)}\right) = \left(\frac{\Lambda_\theta}{\Lambda'_\theta}\right)\left(\frac{K'_\theta(x)}{K_\theta(x)}\right).$$

Thus $K'_\theta(x)/K_\theta(x)$ is an eigenfunction for T. Let $\bar{B} = B/\text{constants}$, and $f \to \bar{f}$ the canonical projection of B to $\bar{B}$. We can write

$$\bar{f_\theta} = \bar{K}_\theta \circ T/\bar{K}_\theta.$$

Let $\phi\colon \bar{B} \to \bar{B}$ denote the mapping

$$\phi(\bar{K}) = \bar{K} \circ T/\bar{K}.$$

By what we have just said ϕ is a ***countable to one*** mapping that is clearly Borel measurable with respect to the standard Borel structure on $\bar{B}$. By Lusin's theorem, $\phi(\bar{B})$ is a Borel subset of $\bar{B}$ and there is a section, i.e., a map $\psi\colon \phi(\bar{B}) \to \bar{B}$ such that $\phi \circ \psi = \mathrm{id}\,|_{\phi(\bar{B})}$ (see **[Lu]**, Chapter IV).

Now apply ψ to $\bar{f_\theta}$ to get $\bar{K}_\theta = \psi(\bar{f_\theta})$. This is Borel measurable as a mapping from Θ to $\bar{B}$, and this $\bar{K}_\theta$ will give back f_θ. The measurability of Λ_θ follows as it is the quotient of measurable mappings. □

The foregoing now motivates the following definition.

Definition. Let (Z, α) be a Kronecker system. A cocycle $S_n(z)$ or $S(z)$ (where S_n is generated by $S_1 = S$) with values in S^1 is a ***CL-cocycle*** if for some m, ℓ we have for $z, v \in Z$,

$$\frac{S_\ell(z + mv)}{S_\ell(z)} = \Lambda_v(z + \ell Z)\frac{K_v(z + \ell\alpha)}{K_v(z)} \tag{10.9}$$

with Λ and K measurable functions with values in S^1.

Note that if $S(z)$ is a CL-cocycle for particular ℓ and m, it is so also for multiples ℓ' of ℓ and m' of m. This is clear for m'; for ℓ' note that (10.9) implies for $p > 1$

$$\begin{aligned}\frac{S_{p\ell}(z + mv)}{S_{p\ell}(z)} &= \frac{S_\ell(z + mv)S_\ell(z + mv + \ell\alpha)\ldots S_\ell(z + mv + (p-1)\ell\alpha)}{S_\ell(z)S_\ell(z + \ell\alpha)\ldots S_\ell(z + (p-1)\ell\alpha)} \\ &= \Lambda_v(z + \ell Z)^p\frac{K_v(z + p\ell\alpha)}{K_v(z)}.\end{aligned}$$

From this observation we obtain

Proposition 10.6. *The CL-cocycles on (Z, α) form a group.*

Finally we formulate

Definition. Let $(X, \mathcal{B}, \mu, T)$ be an ergodic system and (Z, α) its Kronecker factor. A function $\zeta \in L^\infty(X, \mathcal{B}, \mu)$ with values in S^1 is called a ***CL-function*** if $T\zeta = S\zeta$ where S is a function on Z defining a CL-cocycle.

The previous proposition implies that the product of CL-functions is a CL-function and so the linear combinations of CL-functions form an algebra (closed under conjugation). The subspace of $L^2(X)$ spanned by these functions therefore determines a factor X_{CL} of X, for which $L^2(X_{\mathrm{CL}})$ lifts to this subspace. We can now summarize the foregoing discussion in the following theorem.

Theorem 10.7. *Let $(X, \mathcal{B}, \mu, T)$ be a normal ergodic system. X has a factor X_{CL} so that $L^2(X_{\mathrm{CL}}) \hookrightarrow L^2(X)$ is spanned by CL-functions, and X_{CL} is a characteristic factor of X for all schemes, $\{rn, sn, (r+s)n\}$.*

11. Degenerate CL-functions

We have already noted that in the definition of CL-cocycles,

$$\frac{S_\ell(z+mv)}{S_\ell(z)} = \Lambda_v(z+\ell Z)\frac{K_v(z+\ell\alpha)}{K_v(z)}, \qquad (10.9\,bis)$$

Λ_v and K_v are not uniquely determined. Clearly we can replace K_v by $K_v\phi$ where $\phi(z+\ell\alpha) = \lambda\varphi(z)$, at the same time replacing Λ_v by $\lambda^{-1}\Lambda_v$. Modulo these changes the expression in (10.9) is unique. This gives us a relationship between the Λ and K for v_1, v_2 and $v_1 + v_2$. For we can write

$$\frac{S_\ell(z+m(v_1+v_2))}{S_\ell(z)} = \frac{S_\ell(z+m(v_1+v_2))}{S_\ell(z+mv_2)}\frac{S_\ell(z+mv_2)}{S_\ell(z)}$$

$$= \Lambda_{v_1}(z+mv_2+\ell Z)\Lambda_{v_2}(z+\ell Z)\frac{K_{v_1}(z+mv_2+\ell\alpha)}{K_{v_1}(z+mv_2)}\cdot\frac{K_{v_2}(z+\ell Z)}{K_{v_2}(z)}.$$

Hence up to the modifications mentioned we will have

$$\Lambda_{v_1+v_2}(z+\ell Z) = \Lambda_{v_1}(z+mv_2+\ell Z)\Lambda_{v_2}(z+\ell Z)$$
$$K_{v_1+v_2}(z) = K_{v_1}(z+mv_2)K_{v_2}(z) \qquad (11.1)$$

Definition. A CL-function ζ is ***degenerate*** if it is defined over the Kronecker factor of the system.

Lemma 11.1. *If ζ is a CL-function for the CL-cocycle satisfying* (10.9) *and if $\Lambda_v \equiv 1$, then ζ is degenerate.*

Proof. First note that if $mv_1 = mv_2$ so that

$$\frac{K_{v_1}(z+\ell\alpha)}{K_{v_1}(z)} = \frac{K_{v_2}(z+\ell\alpha)}{K_{v_2}(z)},$$

then K_{v_1} and K_{v_2} are proportional in each coset $z+\ell Z$. Treating each coset individually we can assume that K_v depends only on mv.

Write

$$\frac{S_\ell(z+mv)}{S_\ell(z)} = \frac{K(mv, z+\ell\alpha)}{K(mv, z)} \qquad (11.2)$$

We apply Lemma 10.3 to (11.2) regarding $K(mv, z)$ as a function of z and $z+mv$, and we conclude that $S_\ell(z) = \lambda \frac{M(z+\ell\alpha)}{M(z)}$ for some $M(z)$ with values in S^1. Write

$$T^\ell \zeta = S_\ell(z)\zeta = \lambda M(z+\ell\alpha)\frac{\zeta}{M(z)}.$$

It follows that $\zeta/M(z)$ is an eigenfunction of T^ℓ and therefore is defined over the Kronecker factor Z. This implies that ζ is defined over Z and is therefore degenerate. □

We use this to conclude that $\Lambda_v(z+\ell Z)$ as a function of v has a continuous distribution of values on S^1.

Lemma 11.2. *Let ζ be a CL-function for a CL-cocycle S_n satisfying* (10.9), *and assume that $\Lambda_v(z+\ell Z)$ as a function on $Z \times Z/\ell Z$ takes on some value in S^1 on a set of positive measure. Then ζ is degenerate.*

Proof. We must have $\Lambda_v(z+\ell Z)$ constant for $v \in A$ for some $A \subset Z$ of positive measure for all z in some coset $j\alpha + \ell Z$, $0 \le j < \ell$. We may assume that A is in a fixed coset modulo ℓZ so that if $z \in j\alpha + \ell Z$ then $z + m(v_2 - v_2) \in j\alpha + \ell Z$ whenever $v_1, v_2 \in A$. Now apply (11.1) to $v_2 - v_1$ and v_1 with $v_1, v_2 \in A$. Then a choice of $\Lambda_{v_2 - v_1}$ is given by $\Lambda_{v_1}(j\alpha + \ell Z)/\Lambda_{v_1}(j\alpha + \ell Z) = 1$. But $A - A$ contains an open neighborhood of θ in Z. There is therefore an integer ℓ' so that for $v \in \ell' Z$, $\Lambda_v(j\alpha + \ell Z) = 1$. We can assume $\ell | \ell'$. We compute $S_{\ell'}$ from S_ℓ and we conclude that

$$\frac{S_{\ell'}(z+mv)}{S_{\ell'}(z)} = \frac{K_v(z+\ell'\alpha)}{K_v(z)}$$

for $z \in j\alpha + \ell Z$. We then conclude as in Lemma 11.1 that

$$T^{\ell'}\left(\frac{\zeta}{M(z)}\right) = \lambda\left(\frac{\zeta}{M(z)}\right)$$

holds in some coset of $\ell' Z$. But this implies that in that coset ζ is defined as a function on Z. Use $T\zeta = S(z)\zeta$ to conclude that ζ is a function on Z throughout. □

12. The existence of $\lim_{N\to\infty} \frac{1}{N}\Sigma T^n f\, T^{n^2} g$

We will need the following lemma.

Lemma 12.1. *Let $(X, \mathcal{B}, \mu, T)$ be an ergodic measure preserving system, let $\varphi(x)$ be defined on the Kronecker factor of X with values in S^1, and assume that the distribution of φ on S^1 has no atoms (i.e., is a continuous measure). Let $f \in L^2(X, \mathcal{B}, \mu)$ and let*

$\xi(n)$ be an arbitrary bounded sequence. Then for any integers a, b, c, with $a \neq 0$,

$$\frac{1}{N}\sum_{n=1}^{N}\xi(n)\varphi(x)^{an^2+bn}T^{cn}f \to 0$$

in $L^2(X, \mathcal{B}, \mu)$.

Proof. We use Lemma 1.1. Let $u_n = \xi(n)\phi(x)^{an^2+bn}T^{cn}f$ in $L^2(X)$. then

$$\begin{aligned}\langle u_n, u_{n+m}\rangle &= \xi(n)\overline{\xi(n+m)}\int \varphi(x)^{an^2+bn}\varphi(x)^{-a(n+m)^2-b(n+m)}T^{cn}f\; T^{c(n+m)}\bar{f}d\mu \\ &= \xi(n)\overline{\xi(n+m)}\int \varphi(x)^{-2anm-am^2-bm}T^{cn}(fT^{cm}\bar{f})d\mu \\ &= \xi(n)\overline{\xi(n+m)}\int \varphi(x)^{-2anm-am^2-bm}T^{cn}E(fT^{cm}\bar{f}|Z)d\mu. \end{aligned} \tag{12.1}$$

We show that $\frac{1}{N}\Sigma\langle u_n, u_{n+m}\rangle \to 0$ for all $m \neq 0$ by approximating $E(fT^{cm}\bar{f}|Z)$ by a linear combination of eigenfunctions of T. Now if we write $\psi(z)$ in the place of $E(fT^{cm}\bar{f}|Z)$, where ψ is an eigenfunction, then $T^{cn}\psi(z) = \lambda^{cn}\psi(z)$ for some λ. Our assertion will follow if we show that

$$\frac{1}{N}\sum_{n=1}^{N}\xi(n)\overline{\xi(n+m)}[\varphi(x)^{-2am}\lambda^c]^n \to 0.$$

But this follows from Lemma 1.2 by our hypothesis on $\varphi(x)$. □

We turn now to the average $\frac{1}{N}\sum_1^N T^n fT^{n^2}g$. We're going to use the results of §3 according to which we can replace g in this expression by $E(g|Y')$ for an appropriate system Y'. This system should be a characteristic factor for $\{rn, sn, tn\}$ where $t = r+s$, and $r, s > 0$. In the previous sections we have identified this as X_{CL}. What is important now is simply that $L^2(X_{\mathrm{CL}})$ is spanned by CL-functions with respect to its Kronecker factor Z. More precisely we obtain a dense subset of $L^2(X_{\mathrm{CL}})$ by forming linear combinations of functions of the form $g_1(z)$ and $g_2(z)\zeta$, where ζ is a non-degenerate CL-function. These in turn can be reduced to the form $\psi(z)$ and $\psi(z)\zeta$ where $\psi(z)$ is an eigenfunction and ζ is a CL-function for a non-degenerate CL-cocycle.

To achieve the result indicated in the heading of this section we need to show the convergence in $L^2(X)$ of

$$\text{(a)}\quad \frac{1}{N}\sum_{1}^{N}\lambda^{n^2}T^n f$$

$$\text{(b)}\quad \frac{1}{N}\sum_{1}^{N}\lambda^{n^2}T^n f\; T^{n^2}\zeta \tag{12.2}$$

where λ is the eigenvalue for ψ. Now the existence of a limit in (a) is known, and can also easily be deduced from Lemma 1.1; so we focus on (b).

ζ is a non-degenerate CL-function, and we have $T\zeta = S(z)\xi$ for S a non-degenerate CL-cocycle. Hence we have for some ℓ, m,

$$\frac{S_\ell(z+mv)}{S_\ell(z)} = \Lambda_v(z+\ell Z)\frac{K_v(z+\ell\alpha)}{K_v(z)}. \tag{10.9 bis}$$

We have already noted that if S_n is a CL-cocycle for ℓ, m, it is also one for ℓ', m' where $\ell|\ell'$, $m|m'$. So we can assume $m = \ell$ in (10.9):

$$\frac{S_\ell(z+\ell v)}{S_\ell(z)} = \Lambda_v(z+\ell Z)\frac{K_v(z+\ell\alpha)}{K_v(z)}. \tag{12.3}$$

We shall replace the average in (12.2)(b) by

$$\frac{1}{N\ell}\sum_{n=1}^{N}\sum_{j=0}^{\ell-1}\lambda^{(j+n\ell)^2}T^{n\ell+j}f\,T^{(j+n\ell)^2}\zeta$$

and we shall show that for each j the average over n converges to 0 in $L^2(X)$. We evaluate $T^{(j+n\ell)^2}\zeta = T^{\ell^2n^2+2j\ell n}T^{j^2}\zeta$, using the relationship between ζ and S: $T^m\zeta = S_m(z)\zeta$. This gives

$$T^{(j+n\ell)^2}\zeta = S_{\ell^2n^2+2j\ell n}(z+j^2\alpha)\,T^{j^2}\zeta.$$

Our result will thus follow if we have proved:

Lemma 12.2. *If S_n is a CL-cocycle on a Kronecker system Z satisfying* (12.3) *where $\Lambda_v(z+\ell Z)$ takes on each value in S^1 on a set $(v, g+\ell Z)$ of measure 0 in $Z\times Z/\ell Z$, then for any $f\in L^2(X)$ for an extension X of Z, and any bounded sequence $\eta(n)$ and any $\beta\in Z$,*

$$\frac{1}{N}\sum_{n=1}^{N}\eta(n)S_{a\ell n^2+b\ell n}(z+\beta)T^{n\ell}f(x)\to 0 \tag{12.4}$$

in $L^2(X)$. Here $z=\pi(x)$ where $\pi\colon X\to Z$ is the map of X to its Kronecker factor.

Proof. We will obtain convergence to 0 in $L^2(X)$ by showing it separately in each ergodic component of T^ℓ, so that $z\in j\alpha+\ell Z$ for a fixed j. We shall need a modification of (12.3). For each $u\in\ell Z$ choose $v\in Z$ so that $u=\ell v$. Suppose this to be done in a measurable way. Now set $\tilde\Lambda_u=\Lambda_v$ and $\tilde K_u = K_v$. We then have for $u\in\ell Z$:

$$\frac{S_\ell(z+u)}{S_\ell(z)} = \tilde\Lambda_u(z+\ell Z)\frac{\tilde K_u(z+\ell\alpha)}{\tilde K_u(z)}. \tag{12.3a}$$

We shall now use (12.3a) in which we will fix z and regard u as the variable. (12.3a) is valid for almost all pairs of (z,v) so for almost all z it is valid for almost all u. We now replace $z+\beta$ in (12.4) by $\theta+u$ where for each coset of ℓZ an appropriate θ has been chosen and is regarded as fixed.

Consider the function $\tilde{K}_u(z)$. We can write

$$\tilde{K}_u(z) = \sum_{\chi \in \hat{Z}} k_\chi(u)\chi(z). \tag{12.5}$$

We shall write $L^2(u, z)$ for $L^2(Z \times Z)$ and $L^2(u)$ or $L^2(z)$ for $L^2(Z)$ where in the latter, a function of both u and z is regarded as a function of u or of z respectively. For a given $\varepsilon > 0$, we replace the infinite sum in (12.5) by a finite sum

$$\tilde{K}_u^\varepsilon(z) = \sum_{\cdot\chi \in \hat{Z}}{}' k_\chi(u)\chi(z) \tag{12.6}$$

so that $\|\hat{K}_u^\varepsilon(z) - \tilde{K}_u(z)\|^2_{L^2(u,z)} < \varepsilon$.

Let $m \in \mathbb{N}$. $S_{m\ell}(z) = S_\ell(z)S_\ell(z + \ell\alpha)\ldots S_\ell(z + (m-1)\ell\alpha)$ and by (12.3a) we obtain

$$\frac{S_{m\ell}(z+u)}{S_{m\ell}(z)} = \tilde{\Lambda}_u(z + \ell Z)^m \frac{\tilde{K}_u(z + m\ell\alpha)}{\tilde{K}_u(z)} \tag{12.7}$$

which gives

$$S_{a\ell n^2+b\ell n}(z+u) = S_{a\ell n^2+b\ell n}(z)\tilde{\Lambda}_u(z+\ell Z)^{an^2+bn}\overline{\tilde{K}_u(z)}\tilde{K}_u(z + a\ell n^2\alpha + b\ell n\alpha).$$

If we replace $\tilde{K}_u$ by $\tilde{K}_u^\varepsilon + (\tilde{K}_u - \tilde{K}_u^\varepsilon)$ we obtain

$$\begin{aligned} S_{a\ell n^2+bn}(z+\ell u) = {} & S_{a\ell n^2+b\ell n}(z)\tilde{\Lambda}_u(z+\ell Z)^{an^2+bn} \\ & \overline{\tilde{K}_u(z)}\tilde{K}_u^\varepsilon(z + a\ell n^2\alpha + b\ell n\alpha) + R_n^\varepsilon(z, u), \end{aligned} \tag{12.8}$$

where (since $|S|$, $|\Lambda|$, $|\tilde{K}_u| = 1$)

$$|R_n^\varepsilon(z, u)| = L_u^\varepsilon(z + a\ell n^2\alpha + b\ell n\alpha)$$

and $L_u^\varepsilon(z) = |\tilde{K}_u(z) - \tilde{K}_u^\varepsilon(z)|$. Form

$$Q^\varepsilon(u, z) = \lim_{N\to\infty} \frac{1}{N}\sum_{n=1}^N L_u^\varepsilon(z + a\ell n^2\alpha + b\ell n\alpha)^2.$$

This limit exists in $L^1(z)$ for any integrable function on Z, approximating by linear combinations of characters. We then have

$$\begin{aligned} \limsup \Big\|\frac{1}{N}\sum_1^N R_n^\varepsilon(z, u)\Big\|^2_{L^2(u)} &\le \limsup \frac{1}{N}\sum_1^N \|R_n^\varepsilon(z, u)\|^2_{L^2(u)} \\ &= \limsup \frac{1}{N}\sum_1^N \int L_u^\varepsilon(z + a\ell n^2\alpha + b\ell n\alpha)^2\, du \\ &= \int Q^\varepsilon(u, z)\, du. \end{aligned}$$

Now $\int\int Q^\varepsilon(u,z)\,dudz = \int\int L_u^\varepsilon(z)\,dudz < \varepsilon$. If now $\varepsilon < \delta/\ell$, then in any specified coset in Z modulo ℓZ we can find $z = \theta$ so that $\int Q^\varepsilon(u,\theta)du < \delta$. We now consider the average in (12.4),

$$\frac{1}{N}\sum \eta(n) S_{a\ell n^2+b\ell n}(z+\beta)T^{\ell n} f(x), \tag{12.9}$$

with $z+\beta$ replaced by $\theta+u$ for θ satisfying the foregoing inequality. (12.9) can now be written

$$\frac{1}{N}\sum_{n=1}^{N} \eta(n) S_{a\ell n^2+b\ell n}(\theta+u)T^{\ell n} f(x)$$

$$= \frac{1}{N}\sum_{n=1}^{N} \eta(n) S_{a\ell n^2+b\ell n}(\theta)\tilde{\Lambda}_u(\theta+\ell Z)^{an^2+bn}\overline{\tilde{K}_u(\theta)}\tilde{K}_u^\varepsilon(\theta+a\ell n^2\alpha+b\ell n\alpha)T^{\ell n} f(x)$$

$$+ \frac{1}{N}\sum_{n=1}^{N} \eta(n) R_n^\varepsilon(\theta,u)T^{\ell n} f(x). \tag{12.10}$$

In proving the lemma, it suffices to do so for a dense set of $f \in L^2(X)$ and so we may assume $f \in L^\infty(X)$. The $L^2(X)$ norm squared of the second term on the right hand side of (12.10) is therefore comparable to $\|\frac{1}{N}\sum_1^N |R_n^\varepsilon(\theta,u)|\,\|_{L^2(u)}^2$ and by our assumption on θ, the limsup of this is less than a preassigned $\delta > 0$. We next show that the first term converges to 0 in $L^2(X)$. Using (12.6) we can replace $\tilde{K}_u^\varepsilon(z)$ by $k(u)\chi(z)$ for some character $\chi \in \hat{Z}$. As a Fourier coefficient of the bounded function $\tilde{K}_u(z)$, the function $k(u)$ is also bounded. Letting $z = \theta + a\ell n^2\alpha + b\ell n\alpha$ we have replaced the K_u^ε-term in (12.10) by $\chi(\theta)\chi(\alpha)^{a\ell n^2\alpha+b\ell n\alpha}k(u)$. Since $k(u)$ and $\overline{K_u(\theta)}$ are bounded and independent of n, we are left with showing that

$$\frac{1}{N}\sum_{n=1}^{N} \eta(n) S_{a\ell n^2+b\ell n}(\theta)\{\tilde{\Lambda}_u(\theta+\ell Z)\chi(\ell\alpha)\}^{an^2+bn}T^{\ell n} f(x) \to 0. \tag{12.11}$$

Note that, since S_n is a non-degenerate CL-cocycle, the function $\tilde{\Lambda}_u(\theta+\ell Z)$ as a function of u has a non-atomic distribution and the same is true for $\tilde{\Lambda}_u(\theta+\ell Z)\chi(\ell\alpha)$.

(12.11) now has the form

$$\frac{1}{N}\sum_{n=1}^{N} \xi(n)\varphi(u)^{an^2+bn}T^{\ell n} f(x). \tag{12.12}$$

This average is being considered for x in a particular ergodic component of T^ℓ so that the original variable z is in the corresponding coset of ℓZ. We have substituted $\theta+u$ for $z+\beta$ with θ chosen so that $u \in \ell Z$. Since $z = \pi(x)$ we will have $u = \pi(x)+\gamma$ for some fixed γ. If we write $\varphi(u) = \varphi(\pi(x)+\gamma)$ then (12.12) has precisely the form of the average treated in Lemma 12.1, except that we are considering this in $L^2(X')$ where X' is an ergodic component of X for T^ℓ. Since each component has positive measure we can apply Lemma 12.1 and find that the first part of the right hand side of (12.10) converges to 0 in $L^2(X)$. Since for large N the second part has norm $< \delta$, and $\delta > 0$ was arbitrary, this completes the proof of Lemma 12.2. □

We have thereby proved our main result:

Theorem A. *For any measure preserving system* $(X, \mathcal{B}, \mu, T)$, *if* $f, g \in L^\infty(X, \mathcal{B}, \mu)$ *then the average*

$$\frac{1}{N}\sum_{n=1}^{N} f(T^n x)g(T^{n^2}x) \tag{12.13}$$

converges in norm in $L^2(X, \mathcal{B}, \mu)$ *as* $N \to \infty$.

Remark. 1.) It is not hard to extend the foregoing discussion to evaluate the limit in (12.13). The result is that we can identify (in the ergodic case) the characteristic factor of $\{n, n^2\}$ as the "procyclic factor", namely the factor of X generated by functions $\varphi \in L^2(X, \mathcal{B}, \mu)$ which are fixed by some power of T. (This is the factor of the Kronecker factor Z of X corresponding to eigenvalues which are roots of unity.) If we denote this factor by X_{cyc} then we can write

$$\lim_{N\to\infty}\frac{1}{N}\sum_{1}^{N} T^n f\, T^{n^2} g = \lim_{N\to\infty}\frac{1}{N}\sum_{1}^{N} T^n E(f \mid X_{\text{cyc}})T^{n^2}E(g \mid X_{\text{cyc}}). \tag{12.14}$$

2.) Our treatment of the non-conventional average $\frac{1}{N}\sum T^n f\, T^{n^2} g$ has involved consideration also of averages $\frac{1}{N}\sum T^{n^2+an} f\, T^{n^2+bn} g$. If we follow this through we can prove the analogous result:

$$\lim \frac{1}{N}\sum T^{n^2+an} f\, T^{n^2+bn} g$$
$$= \lim \frac{1}{N}\sum T^{n^2+an} \cdot E(f \mid X_{\text{cyc}})T^{n^2+bn}E(g \mid X_{\text{cyc}}). \tag{12.15}$$

Bibliography

[Be] Bergelson, V., Weakly mixing PET, Ergodic Theory Dynam. Systems 7 (1987), 337–349.

[BL] Bergelson, V. and Leibman, A., Polynomical van der Waerden and Szemëredi Theorems, J. Amer. Math. Soc., to appear.

[Bo] Bourgain, J., Pointwise ergodic theorems for arithmetic sets, Inst. Hautes Études Sci. Publ. Math. 69 (1989), 5–45.

[CL] Conze, J. P. and Lesigne, E., Sur un théorème ergodique pour des mesures diagonales, C. R. Acad. Sci. Paris 306 Série 1 (1988), 491–493.

[F1] Furstenberg, H., Ergodic behavior of diagonal measures and a theorem of Szeméredi on arithmetic progressions, J. d'Analyse Math. 31 (1977), 204–256.

[F2] Furstenberg, H., Recurrent ergodic structures and Ramsey theory, ICM-90, Kyoto, Japan, 1990, pp. 1057–1069.

[FKO] Furstenberg, H., Katznelson, Y., and Ornstein, D., The ergodic theoretical proof of Szemėredi's theorem, Proc. Sympos. Pure Math. 39 (1983), Part 2, 217–242.

[Lu] Lusin, N., Leçons sur les Ensembles Analytiques, Paris 1930.

[Z] Zimmer, R., Extensions of ergodic group actions, Illinois J. Math. 20 (1976), 373–409.

Landau Center for Analysis
Institute of Mathematics
The Hebrew University of Jerusalem
91904 Jerusalem, Israel

Convergence of Ergodic Averages

Roger L. Jones[1] *and Máté Wierdl*[2]

Abstract. In this paper we consider almost everywhere convergence properties of various ergodic averages. For example certain moving averages which are known to have the "strong sweeping out property" have Cesàro averages which converge a.e. Also, we give a method of construction new universally good sequences (that is, along which the pointwise ergodic theorem holds,) from previously known universally good sequences.

1. Introduction and notation

Throughout the paper (X, Σ, m) denotes a σ-finite measure space, and τ denotes a measurable, measure preserving mapping from X to itself. In some cases we will also assume the transformation τ is invertible. More general operators such as a positive contraction on $L^p(X)$, p fixed, and $L^1 - L^\infty$ contractions will be denoted by T.

There have been many attempts to generalize the classical ergodic theorem, some successful, and others. unsuccessful. For example, in 1975 Akcoglu and del Junco **[2]** showed that the moving averages

$$A_n f(x) = \frac{1}{[\![\sqrt{n}]\!]} \sum_{k=0}^{[\![\sqrt{n}]\!]} f(\tau^{n+k} x) \tag{1.1}$$

can diverge a.e. even for $f \in L^\infty(X)$. (Here and below, $[\![x]\!]$ will denote the greatest integer n satisfying $n \leq x$.) Subsequent to the work of Akcoglu and del Junco, Schwartz **[21]** considered averages such as

$$A_n f(x) = \frac{1}{n} \sum_{k=0}^{n-1} f(\tau^{n^2+k} x), \tag{1.2}$$

and showed that these averages can also diverge.

In this paper we will be concerned with convergence properties of a variety of averaging operators including those mentioned above.

1 Partially supported by NSF Grant 9302012

2 Partially supported by NSF Grant

Definitions. For a probability measure μ supported on the integers we define the associated operators μf by

$$\mu f(x) = \sum_{k \in \mathbb{Z}} \mu(k) f(\tau^k x).$$

The sequence of measures (μ_n) supported on $\mathbb{Z}$ is *dissipative* if for every $k \in \mathbb{Z}$ we have $\lim_{n\to\infty} \mu_n(k) = 0$. The sequence of measures (μ_n) supported on $\mathbb{Z}$ is said to be *universally good* in L^p if and only if in every measure preserving system (X, Σ, m, τ) the associated sequence $(\mu_n f)$ converges a.e. for every $f \in L^p$. If (μ_n) is the sequence of Cesàro averages along a sequence (a_k) of integers, and if (μ_n) is universally good, then we say that the sequence (a_n) is universally good.

In **[18]** Rosenblatt considered the binomial averages

$$b_n f(x) = \frac{1}{2^n} \sum_{k=0}^{n} \binom{n}{k} f(\tau^k x). \tag{1.3}$$

He showed that these averages can diverge a.e. and in fact for τ ergodic, that these averages have the "strong sweeping out property". (The statement that (b_n) has the strong sweeping out property means that for any dynamical system (X, Σ, m, τ), where τ is an ergodic aperiodic measure preserving point transformation and (X, Σ, m) is a probability space, for $\epsilon > 0$ there is a set E such that $m(E) < \epsilon$, $\limsup_{n\to\infty} b_n \chi_E(x) = 1$ a.e. and $\liminf_{n\to\infty} b_n \chi_E(x) = 0$ a.e. Thus these averages are very badly behaved.)

More recently averages such as (1.1), (1.2) and (1.3) were studied by Bellow, Jones and Rosenblatt **[4]**. (See also the paper by Rosenblatt and Wierdl, **[20]**.) In **[4]** necessary and sufficient conditions on sequences $\{(n_k, \ell_k)\}_{k=1}^{\infty}$ are given so that averages of the form

$$B_k f(x) = \frac{1}{\ell_k} \sum_{j=0}^{\ell_k - 1} f(\tau^{n_k + j} x) \tag{1.4}$$

converge a.e. for $f \in L^p(X)$, $p \geq 1$. These conditions imply that there are subsequences (n_k) so that the averages (1.1), (1.2) and (1.3), taken along these subsequences, converge a.e. (For example, we can take $n_k = 2^{2^k}$ for such a subsequence.)

Remark. In **[11]** it is shown that for $p > 1$, the conditions sufficient for the a.e. convergence of the averages (1.4) using transformations induced by measure preserving point transformations also imply convergence when T is an $L^1 - L^\infty$ contraction or a positive contraction on L^p for fixed p.

In the case where the averages (1.4) are bad, it is shown in **[4]** that for τ ergodic, they have the "strong sweeping out property". Additional examples of operators with the "strong sweeping out property" include the averages (1.1), (1.2) and (1.3) above. See also **[1]** and **[20]** for further examples.

It is natural to ask what "good properties" (if any), in addition to the convergence along certain subsequences, hold for the "bad" operators such as (1.1), (1.2) and (1.3).

It turns out that in many cases these operators still do have certain good properties, and, as in the case of the averages (1.4), these "good" properties remain even in the situation when the transformation τ is replaced by an arbitrary positive contraction on $L^p(X)$. The "good" properties of these operators is the subject of Section 2 below. In the last part of the paper we consider subsequence ergodic theorems. That is, we consider averages of the form

$$\mu_n f(x) = \frac{1}{n}\sum_{k=1}^{n} f(\tau^{a_k}x)$$

where (a_k) is a sequence of integers. The following sequences are known to be universally good for L^p, $p > 1$: the sequence of s-th powers, (k^s), for a positive integer s; (**[8]**, **[10]**) the sequence of prime numbers; (**[9]**, **[23]**) sequences of the form $([\![k^\delta]\!])$ with fixed $\delta > 0$; $([\![\frac{k^2}{\ln k}]\!])$ or $([\![k \ln k]\!])$; (**[24]**) and the sequence of s-th powers of primes (**[17]**, **[24]**).

In this paper we will show a simple way to construct new universally good sequences from sequences previously known to be universally good. For example, we will show that sequences like the set $\{k^3 + p_u \mid u = 1, 2, \ldots, k;\ k = 1, 2, \ldots\}$, where p_u is the u-th prime number, arranged in increasing order, and the set $\{[\![k^\sigma]\!] + [\![u^\delta]\!] \mid u = 1, 2, \ldots, k;\ k = 1, 2, \ldots\}$ for $\delta > 0$ and $\sigma - 1 \geq \delta$, arranged in increasing order, are universally good for L^p, $p > 1$.

2. Convergence results

Let $b_k f(x)$ denote the binomial averages (1.3) considered by Rosenblatt **[18]**. He showed that these averages have the "strong sweeping out property". However, he also showed that the Cesàro averages of these binomial averages converge a.e. To see this let $C_N(f)$ denote these Cesàro averages,

$$C_N f(x) = \frac{1}{N}\sum_{k=1}^{N} b_k f(x) = \frac{1}{N}\sum_{n=1}^{N}\frac{1}{2^n}\sum_{k=0}^{n}\binom{n}{k} f(\tau^k x).$$

Define the $L^1 - L^\infty$ contraction S by $S(f)(x) = \frac{1}{2}(f(x) + f(\tau x))$. Note that $b_k f(x) = S^k(f)(x)$. Consequently, by the Dunford–Schwartz ergodic theorem, we know the averages $C_N f(x)$ converge a.e. for all $f \in L^1(X)$. Note that this remains true if τ is replaced by a general $L^1 - L^\infty$ contraction. It turns out that this type of situation occurs in many other cases where the original sequence of averages diverges.

Let $B_k f(x)$ denote the moving averages (1.4). Consider the Cesàro averages of these moving averages. Define

$$C_N(n_k, \ell_k) f(x) = C_N f(x) = \frac{1}{N}\sum_{k=1}^{N} B_k f(x) = \frac{1}{N}\sum_{k=1}^{N}\frac{1}{\ell_k}\sum_{j=0}^{\ell_k - 1} f(\tau^{n_k + j}x).$$

We know that for many choices of $\{(n_k, \ell_k)\}_{k=1}^{\infty}$, the averages $B_k f$ are badly behaved. However, as in the binomial case considered above, their Cesàro averages can be much better. For general moving averages we have the following result.

Theorem 2.1. *For each choice of* $\{(n_k, \ell_k)\}_{k=1}^{\infty}$ *with the property that* $\lim_{k\to\infty} \ell_k = \infty$, *there is a subsequence* (m_j) *(which depends only on the sequence* $\{(n_k, \ell_k)\}_{k=1}^{\infty}$*) such that the subsequence of Cesàro averages,* $(C_{m_j}(n_k, \ell_k) f)$*, converges* a.e. *for all* $f \in L^p(X)$, $1 < p \le \infty$. *In addition, the result remains true if for fixed* p, *the operator induced by* τ *is replaced by an operator* T *which is a positive contraction on* $L^p(X)$, *or for* T *a Dunford–Schwartz operator.*

Remark. In **[13]** examples of sequences $\{(n_k, \ell_k)\}_{k=1}^{\infty}$ are constructed to have the property that the full sequence of Cesàro averages C_N can fail to converge a.e., and in fact can have the strong sweeping out property. Thus Theorem 2.1 gives the best result of this type that can be true in general.

Proof of Theorem 2.1. A sequence of probability measures (μ_n) supported on the integers is said to have asymptotically trivial transforms if for each $\delta > 0$, the sequence $(\hat{\mu}_n)$ converges uniformly to zero on $\{\gamma : |\gamma| = 1,\ |\gamma - 1| \ge \delta\}$. (We define $\hat{\mu}(\gamma)$ by $\hat{\mu}(\gamma) = \sum \mu(k)\gamma^k$ for $|\gamma| = 1$.) Bellow, Jones and Rosenblatt **[5]** show that if (μ_n) has asymptotically trivial transforms, then there is a subsequence (m_j) such that $\lim_{j\to\infty} \mu_{m_j} f(x)$ exists a.e. for all $f \in L^p(X)$, $p > 1$. Thus in the measure preserving case it will be enough to show that the measures defining the averages C_N have asymptotically trivial transforms. To see this fix $\delta > 0$ and assume $|\gamma - 1| > \delta$, $|\gamma| = 1$. Note that

$$\begin{aligned}
\widehat{C}_N(\gamma) &= \frac{1}{N}\sum_{k=1}^{N}\frac{1}{\ell_k}\sum_{j=0}^{\ell_k-1}\gamma^{n_k+j} \\
&= \frac{1}{N}\sum_{k=1}^{N}\gamma^{n_k}\frac{1}{\ell_k}\sum_{j=0}^{\ell_k-1}\gamma^{j} \\
&= \frac{1}{N}\sum_{k=1}^{N}\gamma^{n_k}\frac{1}{\ell_k}\frac{1-\gamma^{\ell_k}}{1-\gamma} \\
&\le \frac{1}{N}\sum_{k=1}^{N}\frac{1}{\ell_k}\frac{2}{|1-\gamma|} \\
&\le \frac{1}{N}\sum_{k=1}^{N}\frac{2}{\ell_k\delta}.
\end{aligned}$$

The last inequality holds because $|1 - \gamma| \ge \delta$. From the assumption that $\lim_{k\to\infty} \ell_k = \infty$, clearly $\frac{1}{N}\sum_{k=1}^{N}\frac{1}{\ell_k}$ converges to zero. Thus we have asymptotically trivial transforms, and the result follows for the measure preserving case.

The results in **[5]** used above are for measure preserving point transformations. In the course of the proof given there, it is shown that with the transformation being translation on $\mathbb{Z}$, if a sequence of measures (μ_n) has asymptotically trivial transforms, then there is a subsequence (m_j) such that the maximal function $\sup_j |\mu_{m_j}(\varphi)|$ is a bounded operator from $\ell^p(Z)$ to $\ell^p(\mathbb{Z})$, and

$$S(\varphi)(n) = \Big(\sum_{j=1}^{\infty} | \mu_{m_j}(\varphi)(n) - A_{m_j}(\varphi)(n) |^2\Big)^{1/2}$$

is a bounded operator from $\ell^2(\mathbb{Z})$ to $\ell^2(\mathbb{Z})$, where $A_k(\varphi)(n)$ denotes the usual Cesàro average: $A_k(\varphi)(n) = \frac{1}{k}\sum_{j=1}^{k}\varphi(n+j)$. If we consider the associated operators on $L^p(X)$, this together with techniques from **[12]** imply that for T an L^1-L^∞ contraction, or a positive contraction on L^p,

$$\sum_{j=1}^{\infty} \| \mu_{m_j}(f) - A_{m_j}(f) \|_p^p < \infty.$$

Thus for a.e. x we have

$$\sum_{j=1}^{\infty} | \mu_{m_j}(f)(x) - A_{m_j}(f)(x) |^p < \infty.$$

Hence

$$\lim_{j\to\infty} | \mu_{m_j}(f)(x) - A_{m_j}(f)(x) | = 0,$$

and, since the usual Cesàro averages converge, it follows that $\lim_{j\to\infty}\mu_{m_j}(f)(x)$ exists a.e. Thus, as in the measure preserving case, convergence of $C_{m_j}f$ follows from the fact that the measures C_{m_j} have asymptotically trivial transforms. □

Because of the methods used to prove Theorem 2.1, we do not obtain any information regarding what happens in L^1. However in certain cases, and in particular for the cases mentioned in the introduction, there are results for L^1.

Theorem 2.2. *Let*

$$B_k(f)(x) = \frac{1}{[\![\sqrt{k}]\!]} \sum_{j=k}^{k+[\![\sqrt{k}]\!]-1} T^j(f)(x),$$

where T is a positive contraction on L^p, $p>1$ fixed. Write

$$C_n f(x) = \frac{1}{n}\sum_{k=1}^{n} B_k(f)(x).$$

Then $C_n f(x)$ converges a.e. *for all $f\in L^p(X)$. If T is an L^1-L^∞ contraction then the averages $C_n f(x)$ converge* a.e. *for all $f\in L^p(X)$, $p\geq 1$.*

Remark. The averages (B_k) considered above are the averages (1.1), which were shown to diverge a.e. by Akcoglu and del Junco. ([**2**])

Proof. We first consider the case when T is a positive contraction of L^p, for a fixed $p > 1$. Using the definition of B_k and changing the order of summation, we can write

$$C_n f(x) = \frac{1}{n}\sum_{k=1}^{n} \eta(k) T^k f(x) + \frac{1}{n}\sum_{k=n}^{n+[\![\sqrt{n}]\!]-1} d(n,k) T^k f(x),$$

where $0 \le d(n,k) \le 1$ and $|1-\eta(k)| \to 0$. If we denote by $M_n f(x)$ the usual ergodic averages, we have, since T is a positive operator, for fixed $\epsilon > 0$, $m = m(\epsilon)$, and $f \ge 0$,

$$\begin{aligned} |\, M_n f(x) - C_n f(x)\,| &\le \frac{1}{n}\sum_{k=1}^{n} |\, 1-\eta(k)\,| T^k f(x) + \frac{1}{n}\sum_{k=n}^{n+[\![\sqrt{n}]\!]-1} d(n,k) T^k f(x) \\ &\le \frac{c}{n}\sum_{k=1}^{m} T^k f(x) + \epsilon \frac{1}{n}\sum_{k=1}^{n} T^k f(x) + \frac{1}{n}\sum_{k=n}^{n+[\![\sqrt{n}]\!]-1} T^k f(x) \\ &= P_n f + \epsilon M_n f + R_n f. \end{aligned}$$

By Akcoglu's ergodic theorem, we know that $\lim_{n\to\infty} M_n f(x)$ exists, denote this limit by $\bar{f}(x)$. It is clear that $P_n f \to 0$, $\epsilon M_n f \to \epsilon \bar{f}$ and finally

$$R_n f = \frac{n + [\![\sqrt{n}]\!]}{n} M_{n+[\![\sqrt{n}]\!]} f - M_n f \to 0\,.$$

Hence $C_n f \to \bar{f}$. For $L^1 - L^\infty$ contractions, we use the same argument with the fact that an $L^1 - L^\infty$ contraction can be dominated by a positive $L^1 - L^\infty$ contraction. □

Remark. Note that this argument did not use Banach's Principle. While in this case we could have taken such an approach, (it is not hard to see that the maximal function in this case is dominated by a multiple of the usual maximal function, and convergence on a dense class is obvious) we will see more complicated examples later where convergence for the usual dense classes is not clear.

Before giving further examples we introduce the following lemma on summation.

Lemma 2.3. *Suppose that $(d(u))$ and $(g(u))$, are sequences of positive numbers. Let (N_k) be a strictly increasing sequence of positive integers going to infinity. Assume that for any fixed positive k,*

$$d(u) = d(t) \text{ and } g(u) = g(t) \text{ whenever } N_{k-1} < u, t \le N_k. \tag{2.1}$$

For $n \ge 1$ set $D(n) = \sum_{u=1}^{n} d(u)$, $G(n) = \sum_{u=1}^{n} g(u)$, and assume that

$$\lim_{k\to\infty} G(N_k) = \infty \text{ and } \lim_{k\to\infty} D(N_k) = \infty. \tag{2.2}$$

For a complex sequence $F(u),\ u = 1, 2, \ldots,$ *define*

$$M_k F = \frac{1}{D(N_k)} \sum_{u=1}^{N_k} d(u) F(u);$$

$$C_k F = \frac{1}{G(N_k)} \sum_{u=1}^{N_k} g(u) F(u).$$

Consider the following two conditions.

i) $g(n)/d(n),\ n = 1, 2, \ldots,$ *is decreasing.*

ii) $g(n)/d(n),\ n = 1, 2, \ldots,$ *is increasing and* $\sup_k \frac{g(N_k)D(N_k)}{G(N_k)d(N_k)} = c < \infty$.

a) *If conditions* i) *or* ii) *is satisfied, and* $\lim_{k\to\infty} M_k F = \alpha < \infty$ *then* $\lim_{k\to\infty} C_k F = \alpha$.

b) *If condition* i) *holds then* $\sup_k |C_k F| \leq \sup_k |M_k F|$.

c) *If condition* ii) *holds then* $\sup_k |C_k F| \leq (2c - 1) \sup_k |M_k F|$.

Proof. (We only prove Part a) with assumption ii), the other assertions are proved similarly.) Assume, with no loss of generality, that the sequence $F(u),\ u = 1, 2, \ldots$ is a sequence of non-negative real numbers, and $0 \leq \alpha < \infty$. Set $B_n F = \sum_{u=1}^{n} d(u) F(u)$, $B_0 F = G(0) = D(0) = 0$. Using summation by parts we see that

$$\begin{aligned}
C_k F &= \frac{1}{G(N_k)} \sum_{u=1}^{N_k} \frac{g(u)}{d(u)} d(u) F(u) \\
&= \frac{1}{G(N_k)} \sum_{u=1}^{N_k} \frac{g(u)}{d(u)} (B_u F - B_{u-1} F) \\
&= \frac{1}{G(N_k)} \sum_{u=1}^{N_k - 1} B_u F \Big(\frac{g(u)}{d(u)} - \frac{g(u+1)}{d(u+1)} \Big) + \frac{g(N_k)}{G(N_k) d(N_k)} B_{N_k} F \\
&= \frac{1}{G(N_k)} \sum_{i=1}^{k-1} B_{N_i} F \Big(\frac{g(N_i)}{d(N_i)} - \frac{g(N_{i+1})}{d(N_{i+1})} \Big) + \frac{g(N_k)}{G(N_k) d(N_k)} B_{N_k} F \\
&= \frac{1}{G(N_k)} \sum_{i=1}^{k-1} M_i F\ D(N_i) \Big(\frac{g(N_i)}{d(N_i)} - \frac{g(N_{i+1})}{d(N_{i+1})} \Big) + \frac{g(N_k) D(N_k)}{G(N_k) d(N_k)} M_k F.
\end{aligned}$$

Taking $F(u) = 1,\ u = 1, 2, \ldots,$ we have $B_{N_i} F = D(N_i)$ and $C_k F = 1$. Thus we have

$$1 = \frac{1}{G(N_k)} \sum_{i=1}^{k-1} D(N_i) \Big(\frac{g(N_i)}{d(N_i)} - \frac{g(N_{i+1})}{d(N_{i+1})} \Big) + \frac{g(N_k) D(N_k)}{G(N_k) d(N_k)}.$$

For $1 < m < k$, write

$$\begin{aligned}
&\frac{1}{G(N_k)}\sum_{i=1}^{k-1} M_i F\, D(N_i)\Big(\frac{g(N_i)}{d(N_i)} - \frac{g(N_{i+1})}{d(N_{i+1})}\Big) + \frac{g(N_k)D(N_k)}{G(N_k)d(N_k)} M_k F \\
&\quad = \frac{1}{G(N_k)}\sum_{i=1}^{m-1} M_i F\, D(N_i)\Big(\frac{g(N_i)}{d(N_i)} - \frac{g(N_{i+1})}{d(N_{i+1})}\Big) \\
&\qquad + \frac{1}{G(N_k)}\sum_{i=m}^{k-1} M_i F\, D(N_i)\Big(\frac{g(N_i)}{d(N_i)} - \frac{g(N_{i+1})}{d(N_{i+1})}\Big) + \frac{g(N_k)D(N_k)}{G(N_k)d(N_k)} M_k F \\
&\quad = P + Q + R.
\end{aligned}$$

Choose m so large that $\alpha - \epsilon < M_i F < \alpha + \epsilon$ whenever $i \geq m$, next let k_1 be so large that if $k > k_1$ then $-\epsilon < P < \epsilon$. Since $g(u)/d(u)$ is increasing, $D(N_i)$ $\Big(\frac{g(N_i)}{d(N_i)} - \frac{g(N_{i+1})}{d(N_{i+1})}\Big)$ is non-positive. Using the fact that $\frac{g(N_k)D(N_k)}{G(N_k)d(N_k)} \leq c < \infty$, we can estimate $P + Q + R$ by

$$\begin{aligned}
&(\alpha+\epsilon)\Big(\frac{1}{G(N_k)}\sum_{i=m}^{k-1} D(N_i)\Big(\frac{g(N_i)}{d(N_i)} - \frac{g(N_{i+1})}{d(N_{i+1})}\Big) + \frac{g(N_k)D(N_k)}{G(N_k)d(N_k)}\Big) - (2c+1)\epsilon \\
&< P + Q + R \\
&< (\alpha-\epsilon)\Big(\frac{1}{G(N_k)}\sum_{i=m}^{k-1} D(N_i)\Big(\frac{g(N_i)}{d(N_i)} - \frac{g(N_{i+1})}{d(N_{i+1})}\Big) + \frac{g(N_k)D(N_k)}{G(N_k)d(N_k)}\Big) + (2c+1)\epsilon.
\end{aligned}$$

There is an integer k_2 such that if $k > k_2$ we can include the initial terms of the sum, and make an additional error of less than ϵ. Thus we can write

$$\begin{aligned}
&(\alpha+\epsilon)\Big(\frac{1}{G(N_k)}\sum_{i=1}^{k-1} D(N_i)\Big(\frac{g(N_i)}{d(N_i)} - \frac{g(N_{i+1})}{d(N_{i+1})}\Big) + \frac{g(N_k)D(N_k)}{G(N_k)d(N_k)}\Big) - 2c\epsilon \\
&\quad < P + Q + R \\
&\quad < (\alpha-\epsilon)\Big(\frac{1}{G(N_k)}\sum_{i=1}^{k-1} D(N_i)\Big(\frac{g(N_i)}{d(N_i)} - \frac{g(N_{i+1})}{d(N_{i+1})}\Big) + \frac{g(N_k)D(N_k)}{G(N_k)d(N_k)}\Big) + 2c\epsilon.
\end{aligned}$$

Using the fact that the inner sum is 1, we see that

$$(\alpha+\epsilon) - 2c\epsilon < P + Q + R < (\alpha-\epsilon) + 2c\epsilon.$$

Since ϵ was arbitrary, we are done.

The other statements in the lemma are proved in a similar way. □

We now consider applications of this lemma.

Theorem 2.4. *For T a positive contraction on $L^p(X)$, p fixed, $1 < p < \infty$, the averages*

$$\frac{1}{N}\sum_{k=0}^{N-1} T^{[\![\sqrt{k}]\!]} f(x) \tag{2.3}$$

converge a.e. *for all $f \in L^p(X)$. If T is an $L^1 - L^\infty$ contraction (in particular, if T is induced by a measure preserving point transformation), then we have convergence for $f \in L^p(X)$, $p \geq 1$.*

Remark. Note that for U_t an aperiodic measure preserving flow, the averages $A_n f(x) = \frac{1}{N}\sum_{k=1}^{N} f(U_{\sqrt{k}}x)$ can diverge a.e. even for $f \in L^\infty(X)$. This was first proved in **[7]**. In **[1]** and **[13]** it is shown that in fact this sequence is $\frac{6}{\pi^2}$ sweeping out, that is given $\epsilon > 0$ there is a set E such that $m(E) < \epsilon$ while $\limsup_{n\to\infty} A_n\chi_E(x) > \frac{6}{\pi^2}$ a.e.

Proof of Theorem 2.4. Let $d(u) = 1$, $N_k = k$ and $g(u) = 2u + 1$. An application of Lemma 2.3 shows that convergence of the usual Cesàro averages implies convergence of

$$\frac{1}{n^2}\sum_{k=0}^{n-1}(2k+1)T^k f(x). \tag{2.4}$$

Note that if $N = n^2$ for some positive integer n then we can write the averages in (2.3) as

$$\begin{aligned}\frac{1}{N}\sum_{k=0}^{N-1} T^{[\![\sqrt{k}]\!]} f(x) &= \frac{1}{n^2}\sum_{k=0}^{n-1}\sum_{j=k^2}^{(k+1)^2-1} T^{[\![\sqrt{j}]\!]} f(x) \\ &= \frac{1}{n^2}\sum_{k=0}^{n-1}(2k+1)T^k f(x).\end{aligned}$$

Now, for an arbitrary N write $N = n^2 + j$, $0 \leq j < 2n + 1$. Then

$$\begin{aligned}&\left|\frac{1}{n^2+j}\sum_{k=0}^{n^2+j-1} T^{[\![\sqrt{k}]\!]} f(x) - \frac{1}{n^2}\sum_{k=0}^{n^2-1} T^{[\![\sqrt{k}]\!]} f(x)\right| \\ &\quad\leq \left|\left(\frac{1}{n^2+j} - \frac{1}{n^2}\right)n^2\frac{1}{n^2}\sum_{k=0}^{n^2-1} T^{[\![\sqrt{k}]\!]} f(x)\right| + \left|\frac{1}{n^2+2}\sum_{k=n^2}^{n^2+j-1} T^{[\![\sqrt{k}]\!]} f(x)\right| \\ &\quad\leq \frac{j}{n^2+j}\left|\frac{1}{n^2}\sum_{k=0}^{n^2-1} T^{[\![\sqrt{k}]\!]} f(x)\right| + \left|\frac{j}{n^2+j}T^n f(x)\right|.\end{aligned}$$

Now clearly both terms go to 0 since $j < 2n + 1$. □

Theorem 2.5. *Let* $\{(n_k, \ell_k)\}_{k=1}^{\infty} = \{(k^2, k)\}_{k=1}^{\infty}$. *(This gives the averages in* (1.2) *considered by Schwartz* **[21]**.*) Let*

$$B_k f(x) = \frac{1}{k} \sum_{j=0}^{k-1} T^{(k^2+j)} f(x),$$

and

$$C_n f(x) = \frac{1}{n} \sum_{k=1}^{n} B_k(f)(x),$$

with T *a positive contraction on* $L^p(X)$, p *fixed,* $1 < p < \infty$; *then the averages* $C_n f(x)$ *converge* a.e. *for all* $f \in L^p(X)$. *If* T *is an* $L^1 - L^\infty$ *contraction, (in particular, if* T *is induced by a measure preserving point transformation), then we have convergence for all* $f \in L^p(X)$, $p \geq 1$.

Proof. We will apply Lemma 2.3. Consider the sequence $1^2,\ 2^2,\ 2^2+1,\ 3^2,\ 3^2+1,\ 3^2+2,\ \dots,\ k^2,\ k^2+1, \dots,\ k^2+k-1,\ (k+1)^2,\ \dots$. Write a_u for the u th element of this sequence. Using the notation of Lemma 2.3, let $F(u) = T^{a_u} f(x)$, $g(u) = \frac{1}{k}$ if $k^2 \leq a_u < k^2 + k$ for some $k > 0$, that is $g(u) = 1/[\![\sqrt{a_u}]\!]$, and $d(u) = 1$. Define $n_j = j$, $N_0 = 0$, and $N_k = N_{k-1} + n_k$. Then $G(N_k) = k$ and $D(N_k) = k(k+1)/2$. Note that $g(u)/d(u) = 1/[\![\sqrt{a_u}]\!]$ is decreasing. Thus condition i) of Lemma 2.3 is satisfied. With the notation of Lemma 2.3 we have that

$$M_k F = M_k f(x) = \frac{2}{k(k+1)} \sum_{j=1}^{k} \sum_{u=j^2}^{j^2+j-1} T^u f(x).$$

A multiple of the usual ergodic averages dominate these averages. To see that we have convergence on a dense class, consider the class $\{h + g - Tg : h \text{ is invariant}, g \in L^p(X)\}$. Then

$$\begin{aligned} \left| M_k f(x) - h(x) \right| &= \left| \frac{2}{k(k+1)} \sum_{j=1}^{k} \sum_{u=j^2}^{j^2+j-1} T^u (g(x) - Tg(x)) \right| \\ &= \left| \frac{2}{k(k+1)} \sum_{j=1}^{k} (T^{j^2} g(x) - T^{j^2+j} g(x)) \right| \\ &\leq \frac{2}{k} \sum_{j=1}^{k} \left| \frac{T^{j^2} g(x)}{j} \right| + \left| \frac{T^{j^2+j} g(x)}{j} \right|. \end{aligned}$$

We now show that for any sequence (n_j) we have $|\frac{T^{n_j} g(x)}{j}|$ converging to zero a.e. The result will then follow because we have a Cesàro average of terms that converge to zero a.e. To show that $|\frac{T^{n_j}g(x)}{j}|$ converges to zero a.e. we consider

$$\int_X \sum_{j=1}^{\infty} \left|\frac{T^{n_j}g(x)}{j}\right|^p dx = \sum_{j=1}^{\infty} \int_X \left|\frac{T^{n_j}g(x)}{j}\right|^p dx \le \sum_{j=1}^{\infty} \frac{1}{j^p} \|g\|_p^p < \infty.$$

Because for a.e. x, $\sum_{j=1}^{\infty} |\frac{T^{n_j}g(x)}{j}|^p$ is finite, the terms must converge to zero.

The maximal inequality together with the convergence on the dense class implies $M_k f(x)$ converges a.e. for all $f \in L^p(X)$. From this and Lemma 2.3, we see that

$$C_k f(x) = \frac{1}{k} \sum_{j=1}^{k} \frac{1}{j} \sum_{u=j^2}^{j^2+j-1} T^u f(x)$$

converges a.e. for all $f \in L^p(X)$. For T an $L^1 - L^\infty$ contraction, the same argument works, using the fact that we can dominate an $L^1 - L^\infty$ contraction by a positive $L^1 - L^\infty$ contraction. (See **[13]** page 159.) □

Theorem 2.6. *Fix p, $1 < p < \infty$. Let T be an $L^1 - L^\infty$ contraction. Let $\{(n_k, \ell_k)\}_{k=1}^{\infty}$ be such that $\ell_k \to \infty$, and such that from each block $(n_k, n_k + \ell_k)$ we can select a point s_k with the property that the averages $\frac{1}{N}\sum_{k=1}^{N} T^{s_k}(f)(x)$ satisfy the strong maximal inequality*

$$\| \sup_N \frac{1}{N} \sum_{k=1}^{N} T^{s_k}(f)(x) \|_p \le c_p \|f\|_p$$

for all $f \in L^p(X)$. Then the averages

$$C_N f(x) = \frac{1}{N} \sum_{k=1}^{N} \frac{1}{\ell_k} \sum_{j=n_k}^{n_k+\ell_k-1} T^j f(x)$$

converge a.e. *for all $f \in L^p(X)$. If $\ell_k \ge k$, the same holds if T is positive contraction on L^p, p fixed, $1 < p < \infty$.*

Proof. We have $C_N f(x) \le \frac{1}{N}\sum_{k=1}^{N} T^{s_k}(f^\star)(x)$, where $f^\star(x)$ is the two sided ergodic maximal function. Let $C^\star f(x) = \sup_N |C_N f(x)|$, then

$$C^\star f(x) = \sup_N \frac{1}{N} \sum_{k=1}^{N} T^{s_k}(f^\star)(x).$$

However we have assumed that there exists a constant c_p such that

$$\| \sup_N \frac{1}{N} \sum_{k=1}^{N} T^{s_k}(g)(x) \|_p \le c_p \|g\|_p$$

for all $g \in L^p(X)$, $p > 1$. Thus we have

$$\|C^\star f\|_p \leq \| \sup_N \frac{1}{N} \sum_{k=1}^{N} T^{s_k}(f^\star)(x)\|_p \leq c_p \|f^\star\|_p \leq 2c_p \|f\|_p,$$

since we know that $\|f^\star\|_p \leq 2\|f\|_p$. Consequently, the maximal function is a bounded operator from $L^p(X)$ to itself, and, as before, we have convergence on the dense class $\{f : f(x) = h(x) + g(x) - Tg(x),\ h \text{ invariant},\ g \in L^\infty(X)\}$. In the case of a positive contraction on L^p, p fixed, we use the same argument to control the maximal function, and use the dense class $\{h + g - Tg : h \text{ invariant},\ g \in L^p(X)\}$, together with the fact that for $\ell_k \geq k$, $T^{n_k} g/\ell_k$ converges to zero a.e. □

Example 2.7. Let $\{(n_k, \ell_k)\}_{k=1}^\infty = \{(k^4, k)\}_{k=1}^\infty$. By **[4]** the moving averages associated with this sequence can diverge, and in fact have the strong sweeping out property. However, for the Cesàro averages of these moving averages, $C_N f(x)$, we have $\lim_{N\to\infty} C_N f(x)$ exists a.e. for all $f \in L^p$, $p > 1$. To see this we let $s_k = k^4$ in Theorem 2.6, and use Bourgain's result **[10]** (together with the extensions in **[11]**) to establish the necessary maximal inequality for the averages $\frac{1}{N}\sum_{k=1}^N T^{k^4}(f)(x)$. The result then follows from Theorem 2.6.

Example 2.8. If $\{(n_k, \ell_k)\}_{k=1}^\infty = \{([\![k^\delta]\!], k)\}_{k=1}^\infty$, with $\delta > 1$ and not an integer, then using results of Wierdl **[23]**, and Theorem 2.6, we have *a.e* convergence of the Cesàro averages, $C_N f(x)$, if $f \in L^p$, $p > 1$, and T an $L^1 - L^\infty$ contraction.

Note that it is possible to obtain the conclusion of Theorem 2.6, even if we perturb a sequence such as n^4 by a small amount, so that we do not have convergence, or even a maximal inequality along the perturbed sequence. In particular, let $(n_k, \ell_k) = (k^4 + \delta_k, k)$ where (δ_k) is a sequence satisfying $0 \leq \delta_k \leq \log k$. It is shown in **[13]** (see also **[1]**) that we can find a sequence (δ_k), $0 \leq \delta_k \leq \log k$, such that the averages $A_N f(x) = \frac{1}{N}\sum_{k=1}^N f(\tau^{k^4+\delta_k} x)$ satisfy the strong sweeping out property. However the maximal function associated with the moving averages indexed by $(k^4 + \delta_k,\ k)$ is controlled by the maximal function associated with the moving averages indexed by (k^4, k). This maximal function is an $L^p(X)$ bounded operator. Thus the Cesàro averages of these moving averages satisfy a maximal inequality and convergence on a dense class follows as in the proof of Theorem 2.6.

Remark. Using the result of Example 2.7, and Lemma 2.3, we can also obtain a subsequence theorem. We know that if T and p are as in Example 2.7, then the sequence of averages

$$\frac{1}{N}\sum_{k=1}^{N} \frac{1}{k} \sum_{j=0}^{k-1} T^{k^4+j}(f)(x)$$

converges a.e. for all $f \in L^p(X)$. Letting a_u be the u th element in the sequence

$$1^4,\ 2^4,\ 2^4+1,\ 3^4,\ 3^4+1,\ 3^4+2,\ 4^4,\ \ldots, n^4,\ \ldots, n^4+n-1,\ \ldots,$$

we claim that (a_u) is universally good in $L^p(X)$, $p > 1$. Let $g(u) = 1$, $d(u) = [\![a_u^{1/4}]\!]^{-1}$, $F(u) = T^{a_u}(f)(x)$, $G(n) = n(n+1)/2$, and $D(n) = n$. Using the notation of Lemma 2.3, we see that $M_k F$ is the average considered above. From Example 2.7 we know these averages converge a.e. Thus condition ii) of Lemma 2.3 is satisfied. From this we conclude that the averages

$$\frac{2}{n(n+1)} \sum_{k=1}^{n} \sum_{j=0}^{k-1} T^{k^4+j}(f)(x)$$

converge a.e. for all $f \in L^p(X)$. The convergence of these averages implies that we have a maximal inequality along the sequence (a_u). Using this maximal inequality together with the fact that, as in the proof of Theorem 2.6, we have convergence on the dense class $\{g + f - Tf : g \text{ invariant and } f \in L^p\}$, we conclude that (a_u) is good universal in L^p. Similar "block averages" were considered by Bellow and Losert **[6]** but they averaged over much longer blocks, their blocks were from n_k to $n_k + \ell_k$ where (n_k) and (ℓ_k) satisfy: $\ell_k < n_{k+1} - n_k$ and $\ell_k > Cn_{k-1}$. With their conditions they were able to show that for T induced by a measure preserving transformation, a.e. convergence holds for all $f \in L^1(X)$. If we knew that the sequence (k^4) was good universal in $L^1(X)$, we would be able to prove an $L \log L$ result for such T and the above "block average".

We will now develop the machinery which will allow us to study more complicated examples.

Definition 2.9. Let D denote the set of a.e. finite measurable functions on the measure space (X, Σ, m). Let $G \subset D$ and let $(T_{n,k})$ be a double sequence of $G \to D$ operators.

a) The double sequence $(T_{n,k})$ is called dissipative if and only if for each $k \geq 1$ and for each $g \in G$, .

$$\lim_{n\to\infty} T_{n,k} g = 0$$

almost everywhere.

b) We call the sequence of operators $\{\tau_n\}$ defined by

$$\tau_n g = \sum_{k=1}^{\infty} T_{n,k} g$$

the associated averaging sequence (provided the operators are well defined).

With this notation we can state and prove a quite general lemma which will be useful for proving results on subsequences. This lemma is similar to a result of Sucheston **[22]**.

Lemma 2.10. *Let (X, Σ, m) be a measure space. Denote by D the set of* a.e. *finite measurable functions. Let φ be a Young function, (For $x \geq 0$, $\varphi(x)$ is non-negative, convex, and satisfies $\varphi(0) = 0$.) and let $(L_\varphi, \|\ \|_\varphi)$ be the corresponding Orlicz space. Let $(T_{n,k})$ be a dissipative double sequence of linear operators which are continuous in measure, positive, and map $L_\varphi \to D$. Let (τ_n) be the associated averaging sequence.*

Assume for every $f \in L_\varphi$ *we have* $\tau_n f \to \bar{f}$ a.e. *for some* $\bar{f} \in D$. *Suppose that* $(g_k) \subset L_\varphi$ *is a sequence of functions with the property that* $g_k \to g$ *a.e. for some* $g \in L_\varphi$, *and that*

$$\| \sup_k |g_k| \, \|_\varphi = C < \infty, \tag{2.5}$$

then

$$\sum_{k=1}^{\infty} T_{n,k} g_k \to \bar{g} \text{ a.e.}$$

Before giving the proof of this lemma, we make some remarks and give some examples of how it can be used.

Remark. Example 2.14 below will show that condition (2.5) or at least some similar condition is required for the lemma to be true.

Remark. The sequence (g_k) can be taken from a subspace of L_φ. For example take $(M_k f)$ to be the usual ergodic averages and $f \in L \log L$. Fix a sequence (ℓ_k) going to infinity. Let $g_k = M_{\ell_k} f$. Then we know that g_k converges a.e. and $\| \sup_k |g_k| \, \|_1 \leq C < \infty$. For T an $L^1 - L^\infty$ contraction take $T_{n,k} = \frac{1}{n} T^k$ for $1 \leq k \leq n$. Consider the averages $C_n f(x) = \frac{1}{n} \sum_{k=1}^n T^k M_{\ell_k} f(x)$. These are Cesàro averages of the block averages associated with the sequence $\{(k, \ell_k)\}_{k=1}^\infty$. Because of our assumption, condition (2.5) of Lemma 2.10 is satisfied, and hence by Lemma 2.10, these averages converge a.e. We can also view these averages as weighted averages, but because the sequence (ℓ_k) is not necessarily increasing, the weights may behave quite irregularly.

For $(\ell_k) = ([\![\sqrt{k}]\!])$ this gives a different approach to the averages considered in Theorem 2.2, however in this case the conclusion obtained using Lemma 2.10 is slightly weaker than the conclusion obtained earlier by other methods.

We will now give results which allow us to combine two good sequences to form a new good sequence.

Theorem 2.11. *Let* $p > 1$ *and let* T, S *be positive contractions of* L^p. *Fix* $\delta > 0$ *and* $\sigma > 0$. *Let* (k_n) *be any sequence of positive integers tending to infinity. Denote*

$$B_k f = \frac{1}{k} \sum_{u=1}^{k} T^{[\![u^\delta]\!]} f \text{ and } C_n f = \frac{1}{n} \sum_{u=1}^{k} S^{[\![u^\sigma]\!]} B_{k_u} f.$$

Then the averages $C_n f$ *converge* a.e. *for* $f \in L^p$. *(Note that if we take* $\delta = 1$, $\sigma = 4$ *and* $k_u = u$, *then we have a different proof of the result in Example 2.7.)*

Proof. We will use Lemma 2.10 with $g_k = B_k f$. Thus what we need to know is that $\lim_{k \to \infty} B_k f$ exists and is finite a.e. In the case when T is induced by a measure preserving point transformation this follows from Bourgain **[10]** when δ is an integer, and from Wierdl **[24]** if δ is not an integer. For T a positive contraction use these results and the results in **[12]**.

Remark. If we assume that δ and σ are not integers, then we can take $f \in L\log^2 L$. It is likely that an examination of Bourgain's proof **[10]** would give the same if δ and σ are integers.

Corollary 2.12. *Let $p > 1$. Let $\delta > 0$, and $\sigma - 1 \geq \delta$, then the set $\{[\![k^\sigma]\!] + [\![u^\delta]\!] : u = 1, 2, \ldots, k;\ k = 1, 2, \ldots\}$ arranged in increasing order, gives a universally good sequence for L^p. This remains true even if we assume only that T is a positive contraction on L^p.*

Remark. The reason for the assumption that $\sigma - 1 \geq \delta$ is so that the numbers

$$[\![k^\sigma]\!] + [\![u^\delta]\!] : u = 1, 2, \ldots, k$$

all fall between $[\![k^\sigma]\!]$ and $[\![(k+1)^\sigma]\!]$, at least for k large enough.)

Proof. Apply Theorem 2.11, and Lemma 2.3. □

In the same way we can also prove:

Theorem 2.13. *Let $p > 1$ and let $\delta > 2$. Denote p_n the n-th prime. Then the set $\{[\![k^\delta]\!] + p_u : u = 1, 2, \ldots, k;\ k = 1, 2, \ldots\}$ arranged in increasing order is universally good for L^p.*

Proof of Lemma 2.10. We can assume that X has finite measure. First we write

$$\sum_{k=1}^{\infty} T_{n,k} g_k = \sum_{k=1}^{\infty} T_{n,k} g + \sum_{k=1}^{\infty} T_{n,k}(g_k - g).$$

Since $\sum_{k=1}^{\infty} T_{n,k} g = \tau_n g \to \bar{g}$ a.e., we see that we can assume $g = \bar{g} = 0$. We will be done if we can prove that for any $\lambda > 0$, we have

$$m\Big(\lim_{N\to\infty} \sup_{n \geq N} |\sum_{k=1}^{\infty} T_{n,k} g_k| > \lambda\Big) = 0.$$

First, by our assumption (2.5), the dominated convergence theorem, and the fact that $g_k \to 0$ a.e., we have

$$\lim_{k_0\to\infty} \| \sup_{k\geq k_0} |g_k| \|_\varphi = 0.$$

We also have that $\tau^\star f$ defined by $\tau^\star f = \sup_n |\tau_n f|$ is continuous in measure. From these facts it follows that for any $\epsilon > 0$,

$$\lim_{k_0\to\infty} m(\tau^\star(\sup_{k\geq k_0} |g_k|) > \epsilon) = 0.$$

For fixed k_0 let us estimate

$$\begin{aligned}
m\Big(\lim_{N\to\infty}\sup_{n\geq N}|\sum_{k=1}^{\infty}T_{n,k}g_k|>\lambda\Big)
&\leq m\Big(\lim_{N\to\infty}\sup_{n\geq N}|\sum_{k=1}^{k_0-1}T_{n,k}g_k|>\frac{\lambda}{2}\Big)+m\Big(\lim_{N\to\infty}\sup_{n\geq N}|\sum_{k=k_0}^{\infty}T_{n,k}g_k|>\frac{\lambda}{2}\Big)\\
&\leq \sum_{k=1}^{k_0-1}m\Big(\lim_{N\to\infty}\sup_{n\geq N}|T_{n,k}g_k|>\frac{\lambda}{2k_0}\Big)+m\Big(\sup_n|\sum_{k=k_0}^{\infty}T_{n,k}g_k|>\frac{\lambda}{2}\Big)\\
&\leq m\Big(\sup_n|\sum_{k=k_0}^{\infty}T_{n,k}g_k|>\frac{\lambda}{2}\Big)\\
&\leq m\Big(\tau^\star(\sup_{k\geq k_0}|g_k|)>\frac{\lambda}{2}\Big),
\end{aligned}$$

and we know $m(\tau^\star(\sup_{k\geq k_0}|g_k|)>\frac{\lambda}{2})\to 0$. □

We now construct an example on a probability space (X,Σ,m) with the property that there exists a dissipative double sequence $(T_{n,k})$ of linear bounded and positive $L^1\to L^1$ operators such that for each $f\in L^1$, $(\tau_n f)$ converges a.e. to 0, but for some a.e. convergent sequence $\{g_k\}\subset L^1$, we have

$$\sup_n\sum_{k=1}^{\infty}T_{n,k}g_k=+\infty \text{ a.e.}$$

This example shows that condition (2.5), which is not satisfied in this case, or a condition similar to (2.5), is required for the conclusion of Lemma 2.10 to hold.

Example 2.14. Let $X=[0,1)$, and denote addition mod 1 by $\oplus$. For each $n\geq 1$, define $m_n=[\![\log_2 n]\!]$ and define j_n by $j_n=n-2^{m_n}$. Thus $0\leq j_n<2^{m_n}$. Define the family of operators $T_{n,k}$ by

$$T_{n,k}f(x)=\begin{cases}\frac{1}{m_n2^{m_n}}f(x\oplus\frac{k}{2^{m_n}}), & \text{if } 4^{m_n}+j_n2^{m_n}\leq k<4^{m_n}+(j_n+1)2^{m_n}\\ 0, & \text{otherwise.}\end{cases}$$

Then

$$\tau_n f(x)=\frac{1}{m_n2^{m_n}}\sum_{k=0}^{2^{m_n}-1}f(x\oplus\frac{k}{2^{m_n}}).$$

This average is just a Riemann sum multiplied by $1/m_n$. We know that as $n\to\infty$ the associated Riemann sums converge a.e. for all $f\in L^1(X)$. The additional factor of $1/m_n$ forces these averages to converge to zero a.e.

For each choice of n, consider k that satisfy

$$4^{m_n}+j_n2^{m_n}\leq k<4^{m_n}+(j_n+1)2^{m_n}$$

and for such k define g_k by

$$g_k(x) = 3^{m_n} \chi_{(0,\frac{1}{2^{m_n}})}(x \oplus \frac{j_n}{4^{m_n}}).$$

For values of k for which g_k is not defined by this procedure, let $g_k = 0$. With these choices we see that for each m we have

$$\begin{aligned}
&\sup_{2^m \le n < 2^{m+1}} \sum_{k=1}^{\infty} T_{n,k} g_k(x) \\
&\ge \sup_{2^m \le n < 2^{m+1}} \frac{1}{m2^m} \sum_{k=4^m+j_n 2^m}^{4^m+j_n 2^m+2^m-1} 3^m \chi_{(0,\frac{1}{4^m})}(x \oplus \frac{j_n}{4^m} \oplus \frac{k}{2^m}) \\
&\ge \frac{3^m}{m2^m} \text{ a.e.}
\end{aligned}$$

Thus

$$\limsup_{n\to\infty} \sum_{k=1}^{\infty} T_{n,k} g_k(x) = \infty \text{ a.e.}$$

We now show that condition (2.5) is not satisfied. For each n and k such that $4^{m_n} + j_n 2^{m_n} \le k < 4^{m_n} + (j_n+1)2^{m_n}$ we see that $g_k(x) \le 3^{m_n} \chi_{(0,\frac{1}{2^{m_n}})}(x)$, and hence $\{g_k\}$ converges to zero a.e. We also see that $\|g_k\|_1 \le \frac{3^{m_n}}{4^{m_n}}$, and thus converges to zero. However if we consider $\| \sup_k g_k \|_1$ we see that

$$\begin{aligned}
&\| \sup\{g_k(x) : 4^m + j_n 2^m \le k < 4^m + (j_n+1)2^m, \ 2^m \le n < 2^m\} \|_1 \\
&\ge 3^m (\frac{1}{2^m} - \frac{1}{2^{m+1}}) \\
&= \frac{3^m}{2^{m+1}},
\end{aligned}$$

and hence $\| \sup_k \ g_k \|_1 = \infty$, and condition (2.5) was not satisfied. □

References

[1] Akcoglu, M, Bellow, A., Jones, R., Losert, V., Reinhold-Larsson, K. and Wierdl, M., The strong sweeping out property for lacunary sequences, Riemann sums, convolution powers and related matters, preprint, 61 pages, to appear in Ergodic Theory and Dynamical Systems.

[2] Akcoglu, M, and del Junco, A., Convergence of averages of point transformations, Proc. Amer. Math. Soc. 49 (1975), 265–266.

[3] Bellow, A., On bad universal sequences in ergodic theory (II), in: Measure theory and its Applications, pp. 74–78, Lecture Notes in Math. 1033, Springer-Verlag, Berlin-New York 1983.

[4] Bellow, A., Jones, R., and Rosenblatt, J., Convergence for moving averages, Ergodic Theory Dynamical Systems 10 (1990), 43–62.

[5] Bellow, A., Jones, R., and Rosenblatt, J., Almost everywhere convergence of weighted averages, Math. Ann. 293 (1992), 399–426.

[6] Bellow, A. and Losert, V., On sequences of density zero in ergodic theory, Conference in modern analysis and probability, Contemp. Math. 26, pp. 49–60, Amer. Math. Soc., Providence 1984.

[7] Bergelson, V., Boshernitzan, M., and Bourgain, J., Some results on non-linear recurrence, J. d'Analyse Math. 62 (1994), 29–46

[8] Bourgain, J., On the maximal ergodic theorem for certain subsets of the integers, Israel J. Math. 61 (1988) 39–72.

[9] Bourgain, J., An approach to pointwise ergodic theorems, GAFA-Seminar 1986/87, Lecture Notes in Math. 1317, pp. 204–223, Springer-Verlag, Berlin-New York 1987.

[10] Bourgain, J., Pointwise ergodic theorems for arithemtic sets, Inst. Hautes Études Sci. Publ. Math. 69 (1989), 5–45.

[11] Jones, R., Olsen, J., Subsequence pointwise ergodic theorems for operators in L^p, Israel J. Math. 77 (1992), 33–54.

[12] Jones, R., Olsen, J., and Wierdl, M., Subsequence ergodic theorems for L^p contractions, Trans. Amer. Math. Soc. 331 (1992), 837–850.

[13] Jones, R., and Wierdl, M., Convergence and divergence of ergodic averages, Ergodic Theory Dynamical Systems 14 (1994), 515–535.

[14] Krengel, U., Ergodic Theorems, de Gruyter Stud. Math. 6, Berlin-New York 1985.

[15] Krengel, U., On the individual ergodic theorem for subsequences, Ann. Math. Statist. 42 (1971), 1091–1095.

[16] Lind, D., Locally compact measure preserving flows, Adv. Math. 15 (1975), 175–193.

[17] Nair, R., On polynomials in primes and J. Bourgain's circle method approach to ergodic theorems, Ergodic Theory Dynamical Systems 11 (1991), 485–499.

[18] Rosenblatt, J., Ergodic group actions, Arch. Math. 47 (1986), 263–269.

[19] Rosenblatt, J., Universally bad sequences in ergodic theory, in: Almost Everywhere Convergence II, A. Bellow and R. Jones, ed., Academic Press, New York 1991.

[20] Rosenblatt, J., and Wierdl, M., A new maximal inequality and its applications, Ergodic Theory Dynamical Systems 12 (1992), 509–558.

[21] Schwartz, M., Polynomially moving ergodic averages, Proc. Amer. Math. Soc. 103 (1988), 252–254.

[22] Sucheston, L., On one-parameter proofs of almost sure convergence of multi-parameter processes, Z. Wahrscheinlichkeitstheorie verw. Geb. 63 (1983), 43–49.

[23] Wierdl, M., Pointwise ergodic theorems along the prime numbers, Israel J. Math. 64 (1988), 315–336.

[24] Wierdl, M., Almost everywhere convergence and recurrence along subsequences in ergodic theory, Ph.D Thesis, Ohio State University, 1989.

Department of Mathematics
DePaul University
2219 North Kenmore
Chicago, IL 60614-3504, U.S.A.

Department of Mathematical Sciences
University of Memphis
Memphis, TN 38152-0001, U.S.A.

Bourgain's Entropy Criteria

Michael Lacey[1]

1. The entropy criteria

Jean Bourgain **[4]** has introduced a simple necessary condition for the pointwise convergence of operators, which has come to be called Bourgain's Entropy Criteria. It has proved to be an interesting condition, for as a general principle, it unifies divergence in a broad class of examples. The original paper demonstrates the usefulness of the criteria in several examples. Since then, it has been used to disprove convergence in many more instances although some of the the proofs in e.g. **[2]**, **[10]**, and **[12]** may not mention the criteria, because the authors went on to use more constructive techniques tailored to the problem at hand.

But again, the remarkable feature of criteria is its broad applicability, and so it deserves a careful elucidation. I offer this note as such. Another exposition is **[17]**, although they do not explicitly treat the proof of Theorem 1.5 below.

The framework for the results is a probability space (X, μ), on which act a sequence of positive, L^p isometries T_j, with $T_j 1 = 1$ for all j. Under these assumptions, disjoint indicator sets map to disjoint indicator sets, hence for all j and f,

$$(T_j f)^2(x) = T_j f^2(x) \qquad \text{a.e. } (x). \tag{1.1}$$

Assume that the T_j satisfy an L^p Ergodic Theorem.

$$\frac{1}{N}\sum_{j=1}^{J} T_j f \xrightarrow{L^p} \int f\, d\mu \qquad f \in L^p(\mu), \quad \text{for all } 1 \le p < \infty. \tag{1.2}$$

This kind of condition is reminiscent of the "mixing" or "ergodic" assumptions of the theorems of Stein **[16]** and Sawyer **[15]**.

Let S_n be a sequence of bounded linear operators on $L^2(\mu)$. We assume that they commute with the T_j:

$$S_n T_j = T_j S_n \qquad \text{for all } n \text{ and } j. \tag{1.3}$$

As is explained in Section 3 of **[4]**, these assumptions can be verified in a variety of contexts.

Given $f \in L^2(\mu)$, for all $\delta > 0$, define the entropy numbers $N_f(\delta)$ to be the minimal number of L^2 balls of radius δ needed to cover the set $\{S_n f : n \ge 0\}$.

The criteria come in two parts, the first valid for L^p, $1 \le p < \infty$.

---

1 Research partially supported by an NSF grant, and an NSF postdoctoral Fellowship.

Theorem 1.4. *If for some $1 \le p < \infty$ and all $f \in L^p(\mu)$, we have*

$$\sup_n |S_n f(x)| < +\infty \quad a.e.$$

then there is the uniform entropy estimate

$$\sup_{\delta>0} \delta\sqrt{\log N_f(\delta)} \le C\|f\|_2 .$$

The conclusion above arises from the theory of Gaussian processes, which I think is intrinsic in the theorem. Note that the assumption weakens as p tends to infinity. At infinity, however, we have to strengthen the hypotheses a little, and the conclusion is weaker. Nevertheless, the next result is more useful than the first.

Theorem 1.5. *If the L^2 norms $\|S_n\|_2$ are uniformly bounded in n, and for all $f \in L^\infty(\mu)$, $\lim_{n\to\infty} S_n(x)f$ exists a.e. then there is the uniform entropy estimate*

$$N_f(\delta) \le C(\delta)\|f\|_2 \qquad \text{for all } \delta > 0 .$$

Remarks. (a) The criteria is used this way: for the given operators S_n, find a $\delta > 0$ so that the entropy numbers $N_f(\delta)$ are unbounded as f ranges over the unit ball of $L^2(X)$. Note, that is just a requirement on L^2 norms! The conclusion then, is that for some bounded f, the sequence $S_n f$ will not converge pointwise. Again, see **[4]** for some examples.

(b) Because of the weaker conclusion, it's clear that the second Theorem is not the only limiting case of the first Theorem. The proof will show that in Theorem 1.4, we need only assume that

$$\sup_n |S_n f(x)| < +\infty \quad \text{a.e. } (x) \quad \text{for all } f \in L^{\psi_2}(\mu) ,$$

where L^{ψ_2} denotes the Orlicz space given by the function $\psi_2(x) = e^{x^2} - 1$. Its appearance is again explained by general results in the theory of Gaussian processes. We discuss this at the end of Section 2.

(c) Actually in the first theorem, the criteria admits a distinction between the pointwise boundedness of $\sup_n |S_n f(x)|$ and the convergence of $S_n f$. See Section 3.

(d) In proving divergence, one naturally wants to identify the nicest function with the worst divergence properties. One way, indeed the typical way, that these Theorems have been used is to employ the L^∞ criteria above to verify that convergence fails, and then to use other methods to sharpen the divergence result. In Section 4, we try to identify the "worst" divergence behavior that the failure of the entropy criteria will guarantee.

(e) In most instances, if divergence occurs, it occurs for a set of $f \in L^p$ that contains a dense G_δ set. Thus one could weaken the hypotheses above further to only require that $\sup_n |S_n f|$ be finite a.e. for quasi-every f. But there probably is not much to be gained from pursuing this.

(f) An important idea in the proof of these results is that one can identify, in a canonical way, subsets of a Hilbert space with Gaussian processes. There is a similar identification between subsets of $L^p(\mu)$ and p-stable processes, for $1 \leq p < 2$. This suggests that there is an entropy criteria for L^p, for $1 \leq p < 2$. Michel Weber **[19]** has investigated these results.

(g) Other authors have made interesting observations about necessary conditions for convergence, or divergence. The papers of **[16]** and **[15]**, which we have already mentioned, are elegantly presented in the first section of Garsia's book **[8]**. This presentation has influenced, among many, the interesting paper of del Junco and Rosenblatt **[5]**.

In Section 2, we will review some aspects of Gaussian processes which we need for the proofs. The proof of Theorem 1.4 will be taken up in Section 3, and therein we give slightly stronger versions of the theorem. The proof of Theorem 1.5 is in Section 4. A note about our proofs: one could rework the proofs to use less of the theory in Section 2, but we have not done this because it would encumber the presentation, and more importantly, at least for Theorem 1.4, Gaussian processes are intrinsic in the theorem. It would be foolish not to exploit them.

2. Gaussian processes

We review some well-known features of Gaussian processes, so that we can reference them without ambiguity in the proofs. The reader should consult other sources for more background material and details. Three references are **[9]**, **[11]**, and **[1]**.

Let (Ω, P) be a probability space, and let E denote expectation with respect to P. A random variable g on Ω is Gaussian with mean Eg and variance $\sigma^2 = E(g - Eg)^2$, if for all x

$$P(g - Eg > x) = \frac{1}{\sqrt{2\pi\sigma^2}} \int_x^\infty e^{-y^2/2\sigma^2} dy \simeq \frac{\sigma}{x\sqrt{2\pi}} e^{-x^2/2\sigma^2}, \quad \text{as } x \to \infty. \tag{2.1}$$

We will henceforth assume that g has mean zero. From this, one can directly calculate the moment generating function of g,

$$Ee^{\lambda g} = e^{\lambda^2\sigma^2/2}, \tag{2.2}$$

as well as see that

$$(E|g|^p)^{1/p} \leq C\sqrt{pE|g|^2} \qquad 1 < p < \infty. \tag{2.3}$$

If $g_1, g_2, \ldots, g_n$ are independent *Gaussian* random variables with mean zero and variance 1, then $\sum_i a_i g_i$ is again Gaussian with mean zero and variance $\sum_i a_i^2$.

A stochastic process $(G_t : t \in T)$ on a countable index set T is said to be Gaussian if all finite linear combinations $\sum_t a_t G_t$ are Gaussian random variables. We will always assume that G_t has mean zero. A Gaussian random variable is characterized by its mean and variance, thus G_t is characterized by its covariance functional $\operatorname{cov}(G_s, G_t)$.

Equivalently, G_t is characterized by the variance functional $\sigma^2(t) = EG_t^2$, and the intrinsic pseudometric[1]

$$d_G(s,t) = (E|G_s - G_t|^2)^{1/2}.$$

Building upon this, one can see that associated with any subset A of a separable Hilbert space $\mathcal{H}$, there is a Gaussian process $\{G_t : t \in A\}$ with $\sigma(t) = \|t\|_{\mathcal{H}}$ and $d_G(s,t) = \|s-t\|_{\mathcal{H}}$.

These processes have a lot of structure. One instance of this we formulate in the result below, known as the Belyev Dichotomy.

Lemma 2.4. *Fix a Gaussian process G_t on T, and let B denote either the Banach space $\ell^\infty(T)$ (all bounded functions on T) or $C(T, d_G)$ (all bounded functions on T continuous with respect to the d_G metric). Then $P(G \in B)$ equals either 0 or 1.*

Thus returning to subsets A of a Hilbert space, A is said to be a GB set if the canonical process G_t associated with A has a.s. bounded paths; and A is a GC set if the sample paths of G_t are a.s. continuous with respect to the intrinsic metric described above.

Characterizations of the boundedness and continuity of Gaussian processes has been the object of a great deal of effort. We will use just a part of this now complete theory. First of all, when G_t is a.s. bounded, more is true:

Lemma 2.5. *If G_t is a.s. bounded on an index set T, then $E\sup_t |G_t| < \infty$, and moreover, there is the following concentration of measure phenomena: for all $x > 0$,*

$$P\big(\big|\sup_t |G_t| - E\sup_t |G_t|\big| > Kx\big) < e^{-(x/\delta)^2} \quad \text{where } \delta^2 = \sup_t EG_t^2.$$

This conclusion of the lemma is a special case of Borel's Inequality **[3]**. The proof there demonstrates the connection between this inequality and the isoperimetric inequality for the sphere. Also see the proof in **[13]**.

Thus a.s. continuity of the process G_t with respect to the metric d_G is characterized by the same metric. It was the study of continuity that lead Dudley **[6]** to introduce the metric entropy numbers into the study of Gaussian processes. Let

$$N(T,d,\delta) = \text{the minimal number of } d \text{ balls of radius } \delta \text{ needed to cover } T.$$

We will always assume that the metric space is compact, so that the entropy numbers will be finite. the conclusion of Theorem 1.4 is simply Sudakov's estimate, a necessary condition for the boundedness of Gaussian processes.

Lemma 2.6. *For an absolute constant K,*

$$\sup_{\delta>0} \delta\sqrt{\log N(T,d_G,\delta)} \le K\, E \sup_{s,t\in T} |G_s - G_t|. \tag{2.7}$$

---

1 By a pseudo-metric, we mean that the distance function satisfies the usual metric properties, but it is allowed to identify two distinct points.

This condition was long recognized to be in a certain sense unimprovable. Dudley gave a sufficient condition for continuity,

$$E \sup_{s,t} |G_s - G_t| \leq K \int_0^\infty \sqrt{\log N(T, d_G, \delta)}\, d\delta\,.$$

That is, the finiteness of the integral above implies a.s. continuity for G_t. In the case of G_t being stationary, finiteness of the integral is also necessary for a.s. continuity, as was established by Fernique **[7]**. Finally, Talagrand **[18]** has completely characterized boundedness and continuity with an analytical condition which is more stringent than Sudakov's estimate, and more general than Dudley's.

We need one more property, namely a comparison result between processes, known as the Sudakov–Fernique inequality.

Lemma 2.8. *Let G_t and H_t be two centered Gaussian processes on the same index set T. If $d_G(s,t) \leq d_H(s,t)$ for all $s, t \in T$, then*

$$E \sup_{s,t\in T} |G_t - G_s| \leq E \sup_{s,t\in T} |H_t - H_s|\,.$$

We have been making oblique references to the Orlicz space L^{ψ_2}, which we'll now say a little about. Recall that $\psi_2(x) = e^{x^2} - 1$ is a symmetric convex function on the real line. Set

$$\|f\|_{\psi_2} = \inf\{C > 0 : E\psi_2(f/C) \leq 1\}\,.$$

We take the inf of the empty set to be ∞. Then $\|\cdot\|_{\psi_2}$ is a norm, and we let L^{ψ_2} be the set of functions for which $\|f\|_{\psi_2}$ is finite.

A Gaussian random variable is a canonical member of L^{ψ_2}: Just look at the definition of a Gaussian random variable, (2.1). Moreover, if G_t is a Gaussian process with a.s. bounded paths on T, then $\sup_t |G_t|$ is in L^{ψ_2}, by Lemma 2.5.

A fact we perhaps should have explained earlier is where the $\sqrt{\log N}$ terms come from. The best explanation is the

Lemma 2.9. *Suppose that $\|g_i\|_{\psi_2} \leq 1$. Then*

$$E \sup_{1\leq i\leq N} |g_i| \leq K\sqrt{\log N + 3}\,.$$

Proof. By the convexity of ψ_2,

$$\psi_2\big(E \sup_{1\leq i\leq N} |g_i|\big) \leq \sum_{i\leq N} E\psi_2(g_i) \leq N\,.$$

But $\psi_2^{-1}(N) \leq K\sqrt{\log N + 3}$ for all integers N. □

The lemma above has a converse: If the g_i are i.i.d. mean zero, variance one Gaussians, then note that for any $\varepsilon > 0$,

$$\begin{aligned} E \sup_{1\le i\le N} |g_i| &\ge \sqrt{(2-\varepsilon)\log N}\, P(|g_i| \ge \sqrt{(2-\varepsilon)\log N} \text{ for some } 1 \le i \le N) \\ &\ge \sqrt{(2-\varepsilon)\log N}(1 - P(|g_1| < \sqrt{(2-\varepsilon)\log N})^N) \\ &\ge \sqrt{(2-2\varepsilon)\log N} \quad \text{for large } N . \end{aligned}$$

From this calculation, it follows that

$$P\Big(\sup_{1\le i,j\le N} |g_i - g_j| > 2\sqrt{(2-\varepsilon)\log N}\Big) \simeq 1 \quad \text{for large } N . \tag{2.10}$$

This will be used in Section 4.

Also, let us see how to prove Sudakov's estimate, Lemma 2.6, from the Sudakov–Fernique inequality. (This is relevant to the proof in Section 4.) Let G_t be a Gaussian process on an index set T. Given $\delta > 0$, set $S \subset T$ to be a maximal subset of T so that any two distinct points in S are at least a distance δ apart with respect to the d_G metric. Then, for any point $t \in T$, there is a point $s \in S$ for which $d_G(s,t)$ is no more than δ, indeed if this is not the case, we could add the point t to S, contradicting the maximality of S. This means that

$$\text{Cardinality of } S \ge N(T, d_G, \delta) .$$

Let H_s for $s \in S$ be i.i.d. Gaussian random variables with mean zero and variance $(\delta/2)^2$. Then, on the set S, the metric d_H is less than d_G. Hence by Lemma 2.8,

$$\begin{aligned} E \sup_{s,t\in T} |G_s - G_t| &\ge E \sup_{s,t\in S} |G_s - G_t| \\ &\ge C E \sup_{s\in S} |H_s| \\ &\ge C\delta\sqrt{\log(\text{cardinality}(S))} \\ &\ge C\delta\sqrt{\log N(T, d_G, \delta)} . \end{aligned}$$

This proves the Lemma.

The ψ_2 norm has an equivalent formulation in terms of p norms.

$$\sup_{p>1} p^{-1/2} \|f\|_p \simeq \|f\|_{\psi_2} . \tag{2.11}$$

This can be seen on the one hand by using the Taylor expansion of the exponential, and on the other, for $f \in L^{\psi_2}$, writing

$$E|f|^p = p \int_0^\infty x^{p-1} P(|f| > x)\, dx \le p \int_0^\infty x^{p-1} \exp\Big(\frac{x^2}{K\|f\|^2_{L^{\psi_2}}}\Big)\, dx .$$

3. The case of a finite p

For $f \in L^2(\mu)$, let $G_f(n)$ be the Gaussian process associated with $\{S_n f : n \geq 0\}$, which is a subset of $L^2(\mu)$ by assumption. Also, denote the metric by $d_f(m,n) = \|S_m f - S_n f\|_2$. We prove

Theorem 3.1. *If for some finite p, we have* $\sup_n |S_n f(x)| < \infty$ *a.e.* (x) *for all* $f \in L^p(\mu)$, *then the set* $\{S_n f : n \geq 0\}$ *is a GB set uniformly in* f:

$$E \sup_n |G_f(n)| \leq K\|f\|_2 \qquad \text{for all } f \in L^2(\mu)\,. \tag{3.2}$$

More specifically, if the last inequality fails, then there is an $f \in L^{\psi_2}(\mu)$ *for which*

$$\sup_n |S_n f| = \infty \quad a.e. \tag{3.3}$$

Notice that Theorem 1.4 follows from (3.2) and Sudakov's estimate, Lemma 2.6. Also, "$f \in L^p(\mu)$" above, and in the next Theorem, can be replaced by "$f \in L^{\psi_2}(\mu)$" throughout, thereby weakening the hypotheses.

In the case where the S_n's are converging, we can prove a little more.

Theorem 3.4. *If for some finite p, we have* $\lim_{n\to\infty} S_n f(x)$ *exists a.e.* (x) *for all* $f \in L^p$, *then* $\{S_n f : n \geq 0\}$ *is a GC set uniformly in* f: *for all* $\delta > 0$,

$$E \sup_{d_f(m,n)<\delta} |G_f(m) - G_f(n)| \leq K(\delta)\|f\|_2\,, \tag{3.5}$$

where $K(\delta) \to 0$ *as* $\delta \to 0$. *In particular,*

$$\sup_{\|f\|_2 \leq 1} \delta\sqrt{\log N_f(\delta)} \leq K(\delta) \quad \textit{for all } \delta > 0\,.$$

The proof of the first theorem is the simplest, and most direct. Read it before any of the other proofs.

Proof of Theorem 3.1. Let us first show how the Gaussian processes enter the picture. Let g_j be independent Gaussian random variables with mean zero, and variance 1. Let $f \in L^p(\mu)$, where $p > 2$. (The hypothesis of the theorem weakens as p increases, hence we are free to assume that $p > 2$.) Define

$$F(x,\omega) = \frac{1}{\sqrt{J}} \sum_{j=1}^{J} g_j(\omega) T_j f(x)\,. \tag{3.6}$$

We can estimate p norms of F, using (2.3).

$$\int_X E|F(x)|^p d\mu \leq (C\sqrt{p})^p \int_X \Big(\frac{1}{J}\sum_{j=1}^{J} |T_j f(x)|^2\Big)^{p/2} d\mu$$

$$\leq (C\sqrt{p})^p \int_X \Big(\frac{1}{J}\sum_{j=1}^{J} T_j|f|^2(x)\Big)^{p/2} d\mu \qquad (3.7)$$

$$\leq 2(C\sqrt{p})^p \Big(\int_X |f|^2 d\mu\Big)^{p/2}.$$

The last inequality holds for large J by the $L^{p/2}$ Ergodic Theorem. Of course, here we are assuming that $f \in L^p$, but throughout the proof, we will be able to assume that f is bounded, as L^∞ is dense in any L^p. Also note the use of elementary properties of Gaussian random variables, and (1.1) above.

We next observe that for appropriate J, the Gaussian process $S_n F(x, \cdot)$ approximates G_f:

$$\begin{aligned} E|S_n F(x,\omega) - S_m F(x,\omega)|^2 &= E\Big|\frac{1}{\sqrt{J}}\sum_{j=1}^{J} g_j(\omega)(S_n - S_m)T_j f(x)\Big|^2 \\ &= \frac{1}{J}\sum_{j=1}^{J} |(S_n - S_m)T_j f(x)|^2 \\ &= \frac{1}{J}\sum_{j=1}^{J} T_j|(S_n - S_m) f(x)|^2 \\ &\xrightarrow{L^2(\mu)} \|S_n f - S_m f\|_2^2 \\ &= E|G_f(n) - G_f(m)|^2 . \end{aligned}$$

Here, we have used the commuting property, (1.3), and the Ergodic Theorem again. So, we see that for all $f \in L^2(\mu)$, integers N and $\varepsilon > 0$, we can choose J so large that

$$\mu\big\{x : \big|E \sup_{n\leq N} |G_f(n)| - E \sup_{n\leq N} |S_n F(x)|\,\big| > \varepsilon\big\} < \varepsilon .$$

This is the approximation that we shall use.

Proceed with a proof by contradiction, using a straight forward construction. Assume that

$$\sup_{\|f\|_2=1} E \sup_n |G_f(n)| = \infty .$$

Then there are bounded functions f_k with $\|f_k\|_2 = 2^{-k}$, and $E \sup_n |G_{F_k}(n)| > 2^k$. Let g_{jk} be Gaussian random variables with mean zero and variance one, independent in both j and k. Choose J_k so large that for $F_k = \frac{1}{\sqrt{J_k}} \sum_{j=1}^{J_k} g_{jk} T_j f_k$, we have

$$\Big(\int_X E|F_k|^p \, d\mu\Big)^{1/p} < C\sqrt{p}2^{-k} \text{ and } \mu\{x : E \sup_n |S_n F_k| > 2^{k-1}\} > 1 - 2^{-k} .$$

Then, by the Borel–Cantelli Lemma applied on (X, μ),

$$\mu\{x : E \sup_n |S_n F_k| > 2^{k-1} \text{ for all sufficiently large } k\} = 1 .$$

And if we set $F = \sum_k F_k$, we have that

$$\Big(\int_X E|F(x,\omega)|^p \, d\mu\Big)^{1/p} \leq \sum_k \Big(\int_X E|F_k(x,\omega)|^p \, d\mu\Big)^{1/p} \leq C\sqrt{p} ,$$

so that $f \in L^p(\mu)$ a.s. Also, for almost all x, by the Sudakov–Fernique inequality (Lemma 2.8),

$$E \sup_n |S_n F| \geq E \sup_n |S_n F_k| \geq 2^{k-1} \qquad \text{for all sufficiently large } k .$$

That is, $E \sup_n |S_n F| = \infty$ a.e. (x). But by the 0-1 Law, Lemma 2.4, we see that $\sup_n |S_n F| = \infty$ a.e. with respect to $\mu \otimes P$. Therefore, by Fubini's Theorem there is an ω for which $F(\cdot, \omega) \in L^p(\mu)$, and $\sup_n |S_n F(\cdot, \omega)| = \infty$ a.e. (μ). This finishes the proof. □

We have only argued that $F \in L^p(X)$ above, but by using the moment generating function identity (2.2), we can see that $f \in L^{\psi_2}(X)$. This is done in the proof of Theorem 1.5 in the next section. In fact, the reader should move to the next section before reading the proof of Theorem 3.4, which we present next. The two proofs are quite similar in their details, so we have only given the more important L^∞ case a careful treatment.

For the proof of Theorem 3.4, we need the following continuity result.

Lemma 3.8. *Assume that the operators S_n are bounded operators on $L^2(X)$. If $\lim_n S_n f(x)$ exists a.e. (x) for all $f \in L^p(X)$ and some finite p, then for all $\varepsilon > 0$ there is a $\delta > 0$ so that*

$$\mu\Big\{ \sup_{d_f(m,n) \leq \delta} |S_m f - S_n f| > \varepsilon \Big\} < \varepsilon \quad \textit{whenever } \|f\|_p \leq 1 .$$

The proof is a simple modification of the proof of Lemma 4.4.

Proof of Theorem 3.4. Assume that the conclusion of the theorem is false. Then there is an $\varepsilon > 0$ so that for all $\delta > 0$ we can choose a bounded f of $L^2(X)$ norm 1, for which

$$P\big(\sup_{d_f(m,n) < \delta} |G_f(m) - G_f(n)| > \varepsilon \big) \geq 3/4 .$$

We can do this because of the concentration of measure phenomena, Lemma 2.5.

But then, defining F as in (3.6), and choosing J sufficiently large, we can have

$$\mu \otimes P\Big(\sup_{d_f(m,n) < \delta} |S_m F(x,\omega) - S_n F(x,\omega)| > \varepsilon \Big) \geq 1/2 .$$

So by Fubini's Theorem, we can select on ω for which

$$\mu\big(\sup_{d_f(m,n)<\delta} |S_m F(x,\omega) - S_n F(x,\omega)| > \varepsilon\big) \geq 1/2\,,$$

which contradicts the conclusion of Lemma 3.8. □

4. The case of p infinite

We promised a form of the entropy criteria which shows how bad divergence can be for bounded functions.

Theorem 4.1. *Let $S_n: L^2(X) \to L^2(X)$ have uniformly bounded operator norms. Suppose that for some finite δ*

$$\sup_{\|f\|_2 \leq 1} \delta\sqrt{\log N_f(\delta)} = \infty\,. \tag{4.2}$$

Then for all $\varepsilon > 0$ there is a function f so that $\|f\|_\infty \leq 1$,

$$\int |f|\,d\mu \leq \varepsilon\,, \quad \limsup_{n\to\infty} \sup_{m>n} |S_m f(x) - S_n f(x)| \geq \delta/\sqrt{2} \;\; a.e.(x)\,. \tag{4.3}$$

In particular, $\lim_n S_n f(x)$ does not exists a.e. (x).

We will need the next Lemma, a variant of which is left as an exercise (with no hints) in **[4]**.

Lemma 4.4. *Let $S_n: L^p(\mu) \to L^0(\mu)$ be bounded linear operators,*[2] *where $1 \leq p < \infty$. Suppose that there is a $\eta > 0$ so that for all M and all $\varepsilon > 0$ we can find a function g so that $\|g\|_\infty \leq 1$;*

$$\int |g|\,d\mu \leq \varepsilon \text{ and } \mu\{\sup_{m,n>M} |S_m g - S_n g| \geq \eta - \varepsilon\} \geq 1 - \varepsilon\,. \tag{4.5}$$

Then for any $\varepsilon > 0$ there is a function f bounded by 1, for which (4.3) holds, but with $\delta/\sqrt{2}$ replaced by $\eta/2$.

Proof. The proof will be a category argument. We work in the unit ball U of $L^\infty(\mu)$, but the metric will be the L^1 metric, under which U is a complete separable metric space. For any $0 < a < 1 \wedge \eta/4$, consider the set

$$F_a = \big\{f \in U : \mu\{\limsup_n \sup_{m>n} |S_m f - S_n f| \leq \eta/2 - a\} > a/2\big\}\,.$$

The lemma will be established by showing that the set is of the first category in U. Indeed, then the set $\widetilde{F} = \bigcup_{k=1}^{\infty} F_{1/k}$ is still of the first category, so that $U \setminus \widetilde{F}$ is dense in U. Taking a function in this last set which is also in a small neighborhood around 0 proves the lemma.

---

2 Only continuity into L^0, the space of measurable functions, is required for this lemma.

The set F_a is contained in $\bigcup_{1\le M<\infty} \bigcap_{M<N<\infty} F_{M,N}$, where

$$F_{M,N} = \{f \in U : \mu\{ \sup_{M\le m,n\le N} |S_m f - S_n f| \le \eta/2 - a\} > a/2\}.$$

These sets are closed by the continuity of the S's, thus $F_M = \bigcap_{M\le N<\infty} F_{M,N}$ is closed. We need to check that the F_M have empty interior. But this is clear, because for all $f \in F_M$, we can choose g as in (4.5), with $\varepsilon < a/4$. Then

$$\begin{aligned}\sup_{M\le m,n} &|S_m(f+g)(x) - S_n(f+g)(x)| \\ &\ge \sup_{M\le m,n} |S_m g(x) - S_n g(x)| - \sup_{M\le m,n} |S_m f(x) - S_n f(x)| \\ &\ge \eta/2 - \varepsilon + a > \eta/2\end{aligned}$$

on a set of x's of measure at least

$$\mu\{ \sup_{M<m,n} |S_m f(x) - S_n f(x)| \le \eta/2 - a\} - \varepsilon.$$

Now, $f \in F_M$, so $f \in F_{M,N}$ for each $N > M$. This means that the first term above is strictly bigger than $a/2$. Hence for $\varepsilon > 0$ sufficiently small, the bound above is strictly bigger than $a/2$. As a consequence, $f + g \notin F_M$, hence the set has empty interior, and so F is a set of the first category. □

We will now turn to the proof proper, but the reader is urged to understand the proof of Theorem 3.1 before beginning the next proof.

Proof of Theorem 4.1. Fix a $\delta > 0$ satisfying (4.2). That means that there are sets I of integers of arbitrarily large cardinality, and a L^2-norm 1, bounded function f for which

$$\|S_n f - S_m f\|2 \ge \delta \qquad \text{for all } m, n \in I.$$

From this, we want to verify (4.5). Concerning the parameter M occurring in (4.5), note that we can assume that each member of I is at least as big a M, hence we need not refer to M again.

What does the inequality above say about the associated Gaussian process G_f? Fix some small $\varepsilon > 0$. Let g_j be independent mean zero variance one Gaussians, and notice that for m and n in I,

$$\|G_f(m) - G_g(n)\|_2 = \|S_m f - S_n f\|_2 \ge (\delta/\sqrt{2})|g_m - g_n|_2.$$

Thus by the Sudakov–Fernique inequality, and (2.10),

$$E \sup_{m,n\in I} |G_f(m) - G_f(n)| \le (\delta/\sqrt{2}) E \sup_{m,n\in I} |g_m - g_n| \simeq 2\delta\sqrt{\log \sharp I}.$$

Consequently, with F as in (3.6), we can choose J so large that $E\int_X |F|^2 d\mu \le 1+\varepsilon$, and

$$\mu\otimes P\Big\{\sup_{m,n\in I} |S_m F - S_n F| > 2\delta\sqrt{(1-\varepsilon)\log\sharp I}\Big\} \simeq 1\,. \tag{4.6}$$

The set I can be taken arbitrarily large, so that F, or rather $F/\sqrt{(2-\varepsilon)\log\sharp I}$ satisfies the two condition in (4.5).

Our task is to truncate F. This is a relatively simple procedure, for $\|F\|_{L^{\psi_2}(\mu\otimes P)} \simeq 1$. This is almost contained in (3.7), but an alternate derivation is available from the moment generating function identity (2.2):

$$\int_X E e^{\lambda F}\, d\mu = \int_X \exp\Big(\frac{\lambda^2}{2}\frac{1}{J}\sum_{j=1}^{J} T_j|f|^2\Big)\, d\mu \le 2e^{\lambda^2/(2-\varepsilon)}\,.$$

The last line again follows from the mean Ergodic Theorem and that f is bounded. Thus, by an exponential Chebysheff inequality,

$$\mu\otimes P\{F>u\} \le 2e^{-\lambda u+\lambda^2/(2-\varepsilon)} \le e^{-(2-\varepsilon)u^2/4} \qquad \text{for } \lambda = (2-\varepsilon)u/2\,.$$

Hence, if we let $F_1 = F\mathbf{1}_{\{|F|>\sqrt{(2+2\varepsilon)\log\sharp I}\}}$, then $\|F_1\|_{L^2(\mu\otimes P)} < (\sharp I)^{(1-\varepsilon')}$, so that

$$E\int_X \sup_{n\in I}|S_n F_1|\,d\mu \le E\int_X \sum_{n\in I}|S_n F_1|\,d\mu \le C(\sharp I)^{-\varepsilon'}\,, \tag{4.7}$$

because we are assuming that the S_n are uniformly bounded operators on $L^2(\mu)$.

The function

$$F_0(x,\omega) = \frac{F(x,\omega)-F_1(x,\omega)}{\sqrt{(2+2\varepsilon)\log\sharp I}}$$

is the one we want. By construction, $\|F_0\|_\infty \le 1$. We also have from (4.6) and (4.7) that

$$\mu\otimes P(\sup_{m,n}|S_m F_0 - S_n F_0| > \sqrt{2}\delta) \simeq 1\,,$$

and since $E\int_X |F|^2 d\mu \le 1+\varepsilon$,

$$E\int_X |F_0|^2 d\mu \le (\log\sharp I)^{-1}\,.$$

Therefore, for $\sharp I$ large, there is some ω for which $F_0(\cdot,\omega)$ satisfies (4.5) with η replaced by $\sqrt{2}\delta$. Therefore, Lemma 4.4 shows that (4.3) holds. □

Note added in proof. The proof of Lemma 4.4 is wrong. It must be replaced by a far more subtle argument in an article by A. Bellow and R. Jones, A Banach principle for L^∞, to appear in Adv. Math.

References

[1] R. J. Adler, An introduction to continuity, extrema and related topics for general Gaussian processes. Lecture Notes-Monograph Series, Vol. 12, Inst. Math. Statist., Hayward 1990.

[2] A. Bellow, R. L. Jones, and J. Rosenblatt, Almost everywhere convergence of convolution powers, in: Almost everywhere convergence (Columbus, Ohio, 1988), pp. 99–120, Academic Press, Boston 1989.

[3] C. Borell, The Brunn–Minkowski inequality in Gauss space, Invent. Math. 30 (1975), 205–216.

[4] J. Bourgain, Almost sure convergence and bounded entropy, Israel J. Math. 63 (1988), 79–97.

[5] A. del Junco and J. Rosenblatt, Counterexamples in ergodic theory and number theory, Math. Ann. 245 (1979), 185–197.

[6] R. M. Dudley, Sizes of compact subsets of Hilbert space and continuity of Gaussian processes, J. Funct. Anal. 1 (1967), 66–103.

[7] X. Fernique, Régularité des trajectories des fonctions aléatoires gaussiennes, in: Ecole d'été de Probabilités de Saint-Flour IV, 1974, Lecture Notes in Math. 480, pp. 1–96, Springer Verlag, Berlin-Heidelberg-New York 1975.

[8] A. Garsia, Topics in Almost Everywhere Convergence, Markham Publishing Company, Chicago 1970.

[9] N. C. Jain and M. B. Marcus, Continuity of sub-Gaussian processes, Advances in Probability 4, M. Dekker, New York 1974.

[10] R. L. Jones and M. Wierdl, Convergence and divergence of ergodic averages, Ergodic Theory Dynam. Systems 14 (1994), 515–535.

[11] J. P. Kahane, Some random series of functions, 2nd ed., Cambridge Univ. Press, Cambridge 1985.

[12] M. T. Lacey, K. Petersen, D. Rudolph and M. Wierdl, Random ergodic theorems with universally representative sequences, to appear in Ann. Inst. H. Poincaré. Probab. Statist.

[13] G. Pisier, Probabilistic methods in the geometry of Banach spaces, in: Probability and Analysis, pp. 167–241, Lecture Notes in Math. 1206, Springer-Verlag, Berlin-Heidelberg-NewYork 1986

[14] J. Rosenblatt, Universally bad sequences in ergodic theory, in: Almost Everywhere Convergence II, (Evanston, Illinois, 1989), pp. 227–245, Academic Press, Boston MA 1991.

[15] S. Sawyer, Maximal inequalities of weak type, Ann. Math. 84 (1966), 157–173.

[16] E. M. Stein, On limits of sequences of operators, Ann. Math. 77 (1961), 140–170.

[17] D. Schneider and M. Weber, Une remarque sur un théorème de Bourgain, in: Séminare de Probabilitiés XXVII, pp. 202–206, Lecture Notes in Math. 1557, Springer-Verlag, Berlin-Heidelberg-New York 1993.

[18] M. Talagrand, Regularity of Gaussian processes, Acta Math. 159 (1987), 99–149.

[19] M. Weber, Opérateurs réguliers dans les espaces L^p, in: Séminare de Probabilitiés XXVII, pp. 207–215, Lecture Notes in Math. 1557, Springer-Verlag, Berlin-New York 1993.

Department of Mathematics
Indiana University
Bloomington, IN 47405-4301, U.S.A.

Self-Intersections of Closed Geodesics on a Negatively Curved Surface: Statistical Regularities

Steven P. Lalley

Abstract. For compact surfaces of negative curvature it is shown that most closed geodesics of length $\approx \ell$ have about $C\ell^2$ self-intersections, for some constant $C > 0$, and these self-intersections are approximately equidistributed on the surface. For surfaces of constant negative curvative $-\kappa$ the value of the constant C is $\kappa/2\pi^2(g-1)$, where g is the genus.

1. Introduction and statement of main results

The geodesic flow on a compact, negatively curved surface is perhaps the simplest example of a smooth flow for which typical orbits exhibit "random" behavior. Periodic orbits (i.e., closed geodesics) are *not* typical—there are only countably many, and they account for only a set of measure zero in the phase space. Nevertheless, the overall randomness of the flow is reflected in the sequence of periodic orbits (see, e.g., **[L 1]**, **[L 2]**) in certain ways. The purpose of this note is to record some statistical regularities in the self-intersections of closed geodesics.

Let S be a compact, C^∞ Riemannian manifold of dimension 2 with strictly negative curvature at every point. There are countably many closed geodesics $\gamma_1, \gamma_2, \dots$ (one in each free homotopy class) with lengths $\ell_1 \leq \ell_2 \leq \cdots$. A celebrated result of Margulis **[M]** states that if $\pi(t) = \max\{n : \ell_n \leq t\}$ then as $t \to \infty$

$$\pi(t) \sim \frac{e^{ht}}{ht}$$

where $h > 0$ is the topological entropy of the geodesic flow (for surfaces with constant curvature -1, $h = 1$).

Define s_n to be the number of (transversal) self-intersections of the closed geodesic γ_n.

Theorem 1. *For each compact, negatively curved surface S there is a constant $C = C_S > 0$ such that for every $\varepsilon > 0$,*

$$\lim_{t\to\infty} \pi(t)^{-1} |\{n \leq \pi(t) : |s_n - C\ell_n^2| < \varepsilon \ell_n^2\}| = 1 \tag{1.1}$$

If S has constant curvature $-\kappa$ and genus $g \geq 2$ then

$$C_S = \frac{\kappa}{2\pi^2(g-1)}. \tag{1.2}$$

Thus, "most" closed geodesics of length $\approx \ell$ have "about" $C_S\ell^2$ self-intersections. This result is consistent with the statement that "typical" closed geodesics are similar statistically to "generic" geodesics. Observe that if one randomly threw down ℓ geodesic segments of length 1 on S then the number of intersections would be about $C\ell^2$.

Theorem 1 seems to contradict Theorem 5 of **[P]**, which states that most closed geodesics of length $\approx \ell$ have about $\tilde{C}\ell$ self-intersections. Apparently, the formula in remark (iii), p. 212 of **[P]** is incorrect: the singularity in the zeta function is not a simple pole, as stated, but rather a logarithmic singularity, so the Ikehara Tauberian theorem does *not* apply.

If $s_n > 0$ define α_n to be the probability distribution on S that assigns mass $\frac{1}{s_n}$ to each point of self-intersection of γ_n; if $s_n = 0$ define α_n to be the zero measure.

Theorem 2. *For each compact, negatively curved surface S there is a Borel probability measure α on S such that for each continuous $f: S \to \mathbb{R}$ and $\varepsilon > 0$*

$$\lim_{t\to\infty} \pi(t)^{-1} \sum_{n=1}^{\pi(t)} 1\left\{ \left| \int f\, d\alpha_n - \int f\, d\alpha \right| < \varepsilon \right\} = 1. \tag{1.3}$$

If S has constant curvature then $\alpha = \nu$ = normalized area measure on S.

Thus most closed geodesics on S of length about ℓ experience about $C_S\ell^2$ self-intersections *and* these self-intersections are approximately distributed according to α. In fact the proof will show that α is the projection to S of the maximum entropy invariant probability measure for the geodesic flow. Theorem 1 is proved in Section 3, Theorem 2 in Section 4. Both theorems are in fact corollaries of the strong equidistribution result, Theorem 7, of **[L$_1$]**. This result is explained in Section 2.

Simple closed geodesics (closed geodesics with no self-intersections) are of interest for both topological and number-theoretic reasons (see **[S]**). Theorem 2 shows that simple closed geodesics are atypical of closed geodesics. Results of Birman and Series **[BS]** suggest that at least for a noncompact surface with finite area and free fundamental group, the number of simple closed geodesics with period $\leq t$ grows no faster than polynomially in t; thus simple closed geodesics are *very* atypical. It would be interesting to have an asymptotic formula for simple closed geodesics analogous to Margulis' formula.

2. Equidistribution of closed geodesics

We begin by discussing Theorem 7 of **[L₁]**, from which Theorems 1–2 will be deduced.

Let $A = A(i, j)$ be an irreducible, aperiodic $\ell \times \ell$ matrix of zeros and ones, where $\ell \geq 2$, and define

$$\Sigma_A = \{x \in \bigtimes_{n=-\infty}^{\infty} \{1, 2, \ldots, \ell\} : A(x_n, x_{n+1}) = 1 \ \forall n\}.$$

The shift $\sigma: \Sigma_A \to \Sigma_A$ is defined by $(\sigma x)_n = x_{n+1}$. Let $r: \Sigma_A \to \mathbb{R}$ be a strictly positive function which is Lipschitz relative to the metric d_ρ on Σ_A defined by

$$d_\rho(x, y) = \sum_{n=-\infty}^{\infty} 1\{x_n \neq y_n\}\rho^{|n|}$$

for some $\rho \in (0, 1)$. Define the suspension space

$$\Sigma_A^r = \{(x, s) : x \in \Sigma_A \text{ and } 0 \leq s \leq r(x)\}$$

with the points $(x, r(x))$ and $(\sigma x, 0)$ identified. The *suspension flow* (Σ_A^r, σ_t^r), $-\infty < t < \infty$, is defined as follows. Starting at any $(x, s) \in \Sigma_A^r$, move at unit speed up the fiber (x, s'), $s \leq s' \leq r(x)$, until reaching $(x, r(x))$, then jump instantaneously to $(\sigma x, 0)$ and proceed up the vertical fiber $(\sigma x, s')$, etc. Equivalently,

$$\sigma_t^r(x, s) = (x, s + t) \quad \forall\, 0 \leq s \leq s + t \leq r(x),$$
$$\sigma_{t+t'}^r = \sigma_t^r \circ \sigma_{t'}^r.$$

Observe that the periodic orbits of the suspension flow (Σ_A^r, σ^r) are precisely those orbits which pass through some $(x, 0)$ with x a periodic sequence. Since there are countably many periodic sequences, there are countably many periodic orbits of (Σ_A^r, σ^r). These may be labelled $\gamma_1^*, \gamma_2^*, \ldots$ and there periods $\ell_1^* \leq \ell_2^* \leq \cdots$. The distribution ν_n^* of γ_n^* may be defined by

$$\nu_n^*(B) = \int_0^{\ell_n} 1_B(\gamma_n^*(t))\, dt$$

where $B \subset \Sigma_A^r$ is a Borel set. Set $\pi^*(t) = \max\{n : \ell_n \leq t\}$.

Theorem 7. ([L₁]) *As* $t \to \infty$,

$$\pi^*(t) \sim e^{ht}/ht$$

where h *is the topological entropy of* (Σ_A^r, σ^r). *For each continuous* $f: \Sigma_A^r \to \mathbb{R}$ *and each* $\varepsilon > 0$, *as* $t \to \infty$

$$\pi^*(t)^{-1} \big|\{n \leq \pi^*(t) : \big| \int f\, d\nu_n^* - \int f\, d\nu^* \big| < \varepsilon\}\big| \longrightarrow 1,$$

where ν^* *is the maximum entropy invariant measure for* (Σ_A^r, σ^r).

Now consider the geodesic flow $\Phi = \Phi_t$ on the unit tangent bundle $T_1 S$ of a compact, negatively curved surface. This flow is of Anosov type **[A]**, hence also Axiom A, so the results of **[B 2]** (also **[R]**) apply. Thus, there exists a suspension flow (Σ_A^r, σ^r) and a continuous map $\pi\colon \Sigma_A^r \to T_1 S$ such that

(a) *π is surjective;*

(b) *π is at most N to 1 for some $N < \infty$;*

(c) *$\pi \circ \sigma_t^r = \Phi_t \circ \pi$ for all t;*

(d) *all but finitely many of the periodic orbits $\{\bar{\gamma}_n\}$ of Φ have the property that $\pi^{-1}(\bar{\gamma}_n)$ consists of a single periodic orbit of σ^r with the same least period.*

See **[L 3]**, Section 1 for an explanation of (d). Note that the periodic orbits $\bar{\gamma}_n$ of Φ are just the lifts to $T_1 S$ of the closed geodesics γ_n. For each $\bar{\gamma}_n$ the probability distribution $\bar{\nu}_n$ may be defined by

$$\int_{T_1 S} f \, d\bar{\nu}_n = \frac{1}{\ell_n} \int_0^{\ell_n} f(\bar{\gamma}_n(t)) \, dt$$

for $f\colon T_1 S \to \mathbb{R}$ continuous. By (a)–(d) above, $\bar{\nu}_n$ pulls back (via π^{-1}) to $\nu^*_{m_n}$, with only finitely many exceptions. Furthermore, the maximum entropy invariant probability measure $\bar{\nu}$ for the geodesic flow Φ pulls back to the maximum entropy measure ν^* for the suspension flow σ^r. Therefore, Theorem 1 implies

Corollary 1. *For each continuous $f\colon T_1 S \to \mathbb{R}$ and each $\varepsilon > 0$,*

$$\lim_{t\to\infty} \frac{1}{\pi(t)} \left|\left\{ n \le \pi(t) : \left| \int f \, d\bar{\nu}_n - \int f \, d\bar{\nu} \right| < \varepsilon \right\}\right| = 1.$$

3. The number of self-intersections

Recall that $\bar{\nu}$ is the maximum entropy invariant probability measure for the geodesic flow on $T_1 S$, and that ν is the induced measure on S.

Lemma 1. *For any nonempty, open set $U \subset T_1 S$, $\bar{\nu}(U) > 0$. For any orbit $\bar{\gamma}(t)$ of the flow, if $\Gamma = \{\bar{\gamma}(t) : -\infty < t < \infty\}$ then $\bar{\nu}(\Gamma) = 0$.*

Note. This is part of the folklore, but the proof is not easy to find.

Proof. It suffices to prove corresponding statements for an arbitrary suspension flow, by (a)–(c) of Section 2. Let ν^* be the maximum entropy measure for a suspension flow (Σ_A^r, σ^r); define a probability measure μ on Σ_A by

$$\mu(F) = \nu^*\{(x, s) : x \in F \text{ and } 0 \le s \le r(x)\}.$$

Then μ is the equilibrium state (sometimes called Gibbs state) for the function $-hr(x)$, by **[BR]**, Prop. 3 (see **[B 3]**, Ch. 1 for the definition).

We will prove that any Gibbs state μ is nonatomic. From this it follows immediately that ν^* assigns measure zero to each individual orbit of the flow. Suppose μ has an atom x, i.e., $\mu(\{x\}) = \rho > 0$. Since μ is σ-invariant ([**B**$_1$], Theorem (1.2)), $\mu(\{\sigma^n x\}) = \rho \ \forall\, n$; as μ is a finite measure, it must be that $x = \sigma^n x$ for some $n \geq 1$. But μ is mixing ([**B**$_1$], Prop. (1.14)) so it must be that $\mu(\{x\}) = 1$ and $x = \sigma x$. This is impossible, however, because by [**B**$_1$] Theorem (1.2) a Gibbs state μ gives positive mass to every cylinder set.

It remains to show that $\nu^*(W) > 0$ for every open $W \subset \Sigma_A^r$. If W is open then it contains a rectangle $R = F \times [k/m, (k+1)/m)$, where k, m are positive integers and F is a cylinder set $F = \{y \in \Sigma_A : y_n = x_n \ \forall\, |n| \leq n_*\}$. By [**B**$_1$] Theorem (1.2), $\mu(F) > 0$. Since ν^* is σ_t^r-invariant,

$$\nu^*(F \times [0/m, 1/m)) = \nu^*(F \times [1/m, 2/m)) = \cdots$$

(with the obvious convention about what happens when you get to the "ceiling" $\{(x, r(x)) : x \in \Sigma_A\}$). Let k_* be the smallest integer larger than $\min_{x\in F} r(x)$; then

$$\mu(F) \leq \sum_{i=0}^{k_*-1} \nu^*(F \times [i/m, (i+1)/m)) = k_* \nu^*(R)$$

so $\nu^*(R) > 0$. □

Lemma 2. *For any geodesic $\gamma(t)$, if $G = \{\gamma(t) : -\infty < t < \infty\}$ then $\nu(G) = 0$.*

Proof. The maximum entropy measure $\bar{\nu}$ is ergodic for the geodesic flow on $T_1 S$ (this follows from (a)–(c) of Section 2, because the maximum entropy measure for a suspension flow is unique and consequently ergodic). Hence, if $\nu(G) > 0$ then for $\bar{\nu}$-a.e. orbit $\bar{\varphi}(t) = (\varphi(t), \varphi'(t))$ of the geodesic flow

$$\lim_{T\to\infty} T^{-1} \int_0^T 1_G(\varphi(t))\, dt = \nu(G) > 0.$$

Now $\varphi(t)$ has at most countably many transversal intersections with G, and each transversal intersection contributes zero to the integral $\int_0^T 1_G(\varphi(t))\, dt$. Therefore $\varphi(t)$ must intersect G tangentially. But $\varphi(t)$ and $\gamma(t)$ are both geodesics, so it follows that for some $s \in \mathbb{R}$, $\varphi(t) = \gamma(t+s) \ \forall\, t$. This is impossible, because by Lemma 1 $\bar{\nu}$ assigns zero mass to the orbit $\{\bar{\gamma}(t) = (\gamma(t), \gamma'(t))\}$. □

Each $(x, v) \in T_1 S$ determines a geodesic $\gamma(t)$ in S emanating from $\gamma(0) = x$ and with $\gamma'(0) = v$. For any $\delta > 0$ let $G_\delta(x, v)$ be the segment $\{\gamma(t) : 0 \leq t < \delta\}$ of this geodesic, considered as a subset of S. Let $\bar{\nu} \times \bar{\nu}$ denote the product measure on $T_1 S \times T_1 S$ determined by $\bar{\nu} \times \bar{\nu}(B_1 \times B_2) = \bar{\nu}(B_1)\bar{\nu}(B_2)$. Define

$$U_\delta = \{((x, v), (x', v')) : G_\delta(x, v) \cap G_\delta(x', v') \neq \phi\};$$

then $\bar{\nu} \times \bar{\nu}(U_\delta)$ is the probability that two independent, randomly chosen geodesic segments of length δ cross. (Note that the $\bar{\nu} \times \bar{\nu}$-probability of a nontransversal intersection is zero, by Lemma 1.)

Lemma 3. *For each* $\delta > 0$, $\bar{\nu} \times \bar{\nu}(U_\delta) > 0$.

Proof. Each U_δ, $\delta > 0$, contains a nonempty open subset of $T_1S \times T_1S$, consequently also a rectangle $A \times B$ where A, B are nonempty open subsets of T_1S. Therefore, by Lemma 1, U_δ has positive $\bar{\nu} \times \bar{\nu}$-measure. □

When $\bar{\mu}$ is close to $\bar{\nu}$ in the weak-$*$ topology then $\bar{\mu} \times \bar{\mu}$ is close to $\bar{\nu} \times \bar{\nu}$ in the weak-$*$ topology, and consequently $|\bar{\mu} \times \bar{\mu}(U_\delta) - \bar{\nu} \times \bar{\nu}(U_\delta)|$ should be small. The following lemma justifies this assertion (cf. **[Bi]**, Theorem 2.1, statement (v)).

Lemma 4. *Let* ∂U_δ *denote the (topological) boundary of* U_δ *in* $T_1S \times T_1S$. *Then*

$$\bar{\nu} \times \bar{\nu}(\partial U_\delta) = 0.$$

Proof. If $((x, v), (x', v')) \in \partial U_\delta$ then an endpoint of $G_\delta(x, v)$ lies on $G_\delta(x', v')$, or an endpoint of $G_\delta(x', v')$ lies on $G_\delta(x, v)$. Consequently, to prove the lemma it suffices to show that for each $(x', v') \in T_1S$,

$$\bar{\nu}\{(x, v) : G_\delta(x, v) \text{ has an endpoint on } G_\delta(x', v')\} = 0.$$

For $(x, v) \in T_1S$, one endpoint of $G_\delta(x, v)$ is x; call the other x_*. By Lemma 2, $\bar{\nu}\{(x, v) : x \in G_\delta(x', v')\} = 0$. But $\bar{\nu}$ is invariant under time reversal, and the segment $G_\delta(x, v)$ reversed has initial endpoint x_*, so $\bar{\nu}\{(x, v) : x_* \in G_\delta(x, v)\} = 0$. □

Geodesics, being smooth curves, look like straight lines in the small. Therefore, if $\delta > 0$ is sufficiently small then two geodesic segments $G_\delta(x, v)$ and $G_\delta(x', v')$ of length δ will have at most one transverse intersection (this also uses the compactness of T_1S). Choose δ_0 sufficiently small that this is true for all $0 < \delta < \delta_0$. For the remainder of this section, fix δ with $0 < \delta < \delta_0$.

Let γ_n be a closed geodesic with length ℓ_n. Recall that s_n is the number of (transversal) self-intersections of γ_n, and that $\bar{\nu}_n$ is the distribution of $\bar{\gamma}_n = (\gamma_n, \gamma_n')$ in T_1S. For $k = 1, 2, \ldots$ let $m_k (= m_k^{(n)})$ be the least integer $\geq k\ell_n/\delta$. Define

$$\begin{aligned}
x_{j,k} &= x_{j,k}^{(n)} = \gamma_n(j\delta/k), \quad j = 0, 1, 2, \ldots, m_k - 1;\\
v_{j,k} &= v_{j,k}^{(n)} = \gamma_n'(j\delta/k), \quad j = 0, 1, 2, \ldots, m_k - 1;\\
\bar{\nu}_{n,k} &= \text{ uniform probability distribution on } \{(x_{j,k}, v_{j,k}) : j = 0, 1, \ldots, m_k - 1\}
\end{aligned}$$

(i.e., $\bar{\nu}_{n,k}$ is the probability measure that puts mass $1/m_k$ on each $x_{j,k}, v_{j,k}$, $j = 0, 1, \ldots, m_k - 1$).

Lemma 5. *weak-*$*$ $\lim_{k\to\infty} \bar{\nu}_{n,k} = \bar{\nu}_n$.

This is immediate from the definition of the weak-$*$ topology and elementary properties of the Riemann integral.

Lemma 6. *For each* $k = 1, 2, \ldots$,

$$(m_k^2/2k^2)(\bar{\nu}_{n,k} \times \bar{\nu}_{n,k})(U_\delta) - 2m_k/k \leq s_n \leq (m_k^2/2k^2)(\bar{\nu}_{n,k} \times \bar{\nu}_{n,k})(U_\delta).$$

Proof. Define G_j to be the segment of γ_n from $x_{j,k}$ to $x_{j+1,k}$ for $j = 0, 1, \ldots, m_k - 2$, and G_{m_k-1} to be the segment of γ_n from $x_{m_k-1,k}$ to $x_{0,k}$. Note that if $0 \leq j \leq m_k - 2$ then $G_j = G_{\delta/k}(x_{j,k}, v_{j,k})$, whereas $G_{m_k-1} = G_r(x_{m_k-1,k}, v_{m_k-1,k})$ for some $0 < r \leq \delta/k$. Since $\delta < \delta_0$, each intersection $G_i \cap G_j$, $i \neq j$, consists of at most a single point. Consequently,

$$s_n = \frac{1}{2} \sum_{(i,j):i \neq j} \sum 1\{G_i \cap G_j \neq \phi\}.$$

Next, define $G_i^* = G_\delta(x_{i,k}, v_{i,k})$. If $0 \leq i < m_k - k$ then $G_i^* = G_i \cup G_{i+1} \cup \cdots \cup G_{i+k-1}$, whereas if $m_k - k \leq i \leq m_k - 1$ then $G_i^* \supset G_i \cup G_{i+1} \cup \cdots \cup G_{i+k-1}$ but $G_i^* \subset G_i \cup G_{i+1} \cup \cdots \cup G_{i+k}$ (with the convention that if $i + j \geq m_k$ then $i + j$ should be reduced mod m_k). As before, each intersection $G_i^* \cap G_j^*$, $|i - j| \geq k$, consists of at most one point. Each G_i is contained in k distinct G_j^*, so by the result of the previous paragraph,

$$s_n \leq \frac{1}{2k^2} \sum_{(i,j):|i-j| \geq k} \sum 1\{G_i^* \cap G_j^* \neq \phi\} \leq \frac{m_k^2}{2k^2} (\bar{\nu}_{n,k} \times \bar{\nu}_{n,k})(U_\delta).$$

Moreover, each G_i intersects only k distinct G_j^* unless $0 \leq i < k$, in which case G_i may intersect $k + 1$ distinct G_j^*. Consequently,

$$\left(\frac{1}{2k^2} \sum_{(i,j):|i-j| \geq k} \sum 1\{G_i^* \cap G_j^* \neq \phi\} \right) - s_n \leq \frac{2km_k}{2k^2} = \frac{m_k}{k}.$$

Finally,

$$\frac{m_k^2}{2k^2} (\bar{\nu}_{n,k} \times \bar{\nu}_{n,k})(U_\delta) - \frac{1}{2k^2} \sum_{(i,j):|i-j| \geq k} \sum 1\{G_i^* \cap G_j^* \neq \phi\}$$

$$= \frac{1}{2k^2} \sum_{i=0}^{m_k - 1} (2k - 1) = \frac{(2k-1)m_k}{2k^2} \leq \frac{m_k}{k}. \qquad \square$$

Proof of (1.1). Corollary 1 implies that for most closed geodesics γ_n with length $\ell_n \leq t$ the distribution $\bar{\nu}_n$ is close to $\bar{\nu}$ in the weak-$*$ topology. Consequently, by Lemma 5, for k sufficiently large $\bar{\nu}_{n,k}$ is close to $\bar{\nu}$ and therefore $\bar{\nu}_{n,k} \times \bar{\nu}_{n,k}$ is close to $\bar{\nu} \times \bar{\nu}$ in the weak-$*$ topology. It follows, by Lemma 4 and **[Bi]**, Theorem 2.1, that $|\bar{\nu}_{n,k} \times \bar{\nu}_{n,k}(U_\delta) - \bar{\nu} \times \bar{\nu}(U_\delta)|$ is small. Thus, by Lemma 6, s_n/ℓ_n^2 is close to

$$C = \bar{\nu} \times \bar{\nu}(U_\delta)/\delta^2 > 0. \qquad \square$$

Since $\delta < \delta_0$ was arbitrary in this argument, the quantity $\bar{\nu} \times \bar{\nu}(U_\delta)/\delta^2$ is independent of δ, and so C can, in principle, be evaluated by letting $\delta \to 0$.

Lemma 7. *If S has constant negative curvature then*

$$\lim_{\delta \to 0} \bar{\nu} \times \bar{\nu}(U_\delta)/\delta^2 = \frac{4}{2\pi \operatorname{area}(S)} \tag{3.1}$$

Observe that $\text{area}(S) = 4\pi(g-1)/\kappa$ where $g = \text{genus}(S)$ and $-\kappa = \text{curvature}$, by the Gauss–Bonnet theorem, so Lemma 7 implies (1.2).

Proof of Lemma 7. The measure $\bar{\nu}$ has a simple structure for a surface S of constant curvature. Elements of T_1S may be represented as (x, θ), where $x \in S$ and $-\pi \le \theta < \pi$; in these coordinates

$$d\bar{\nu}(x, \theta) = d\nu(x)d\theta/2\pi$$

where $d\nu(x)$ is the normalized surface area measure on S.

Recall that $\bar{\nu} \times \bar{\nu}(U_\delta)$ is the probability that two independent, randomly chosen geodesic segments of length δ will intersect. To prove (3.1) it suffices to show that for any fixed geodesic segment G_δ of length δ

$$\bar{\nu}\{(x, \theta) : G_\delta(x, \theta) \cap G_\delta \neq \phi\} \sim 4\delta^2/2\pi \text{ area}(S)$$

as $\delta \to 0$ (uniformly for all choices of G_δ). To compute this probability, condition on the value of θ (the angle with G_δ); then the set of x for which $G_\delta(x, \theta) \cap G_\delta \neq \phi$ is approximately a rhombus of side δ and interior angle θ (at least when δ is small). Consequently,

$$\bar{\nu}\{(x, \theta) : G_\delta(x, \theta) \cap G_\delta \neq \phi\} \sim \int_{-\pi}^{\pi} \delta^2 |\sin\theta| \, d\theta/2\pi \text{ area}(S). \qquad \square$$

4. Distribution of self-intersection points

In Section 3 we showed that $\bar{\nu} \times \bar{\nu}(U_\delta) = C\delta^2$ for all sufficiently small $\delta > 0$. Thus we may define Borel probability measures μ_δ on $T_1S \times T_1S$ by

$$\mu_\delta(B) = \bar{\nu} \times \bar{\nu}(U_\delta \cap B)/C\delta^2$$

for Borel sets $B \subset T_1S$. Since $T_1S \times T_1S$ is compact, the Helly selection theorem (cf. **[RS]**, Theorem IV.21, which they call the "Banach–Alaoglu" theorem) implies that there is a weak-$*$ convergent subsequence μ_{δ_n}, $\delta_n \to 0$. Define

$$\bar{\alpha} = \text{weak-}* \lim_{n\to\infty} \mu_{\delta_n}$$

(it will not matter which subsequence δ_n is used).

The measure $\bar{\alpha}$ induces on $S \times S$ a measure α via the natural projection $p \times p \colon T_1S \to S \times S$, in particular,

$$\alpha(B) = \bar{\alpha}((p \times p)^{-1}(B)), \quad B \subset S \text{ Borel}.$$

The measure α is supported by the "diagonal" $D = \{(x, x) : x \in S\}$, because μ_δ is supported by the set U_δ, and if $\big((x, v), (x', v')\big) \in U_\delta$ then distance $(x, x') \le 2\delta$. Thus α may be regarded as a probability measure on S. We will not bother to distinguish between S and D in the subsequent arguments.

Let V be an open subset of $S \times S$; define

$$V_\delta^+ = \{(x_1, x_2) \in S \times S : \text{distance}(x_i, V) < \delta\},$$
$$V_\delta^- = \{(x_1, x_2) \in V : \text{distance}(x_i, V^c) > \delta\}.$$

Lemma 8. *If* $\alpha(\partial V) = 0$ *then*

$$\lim_{n\to\infty} \mu_{\delta_n}((p \times p)^{-1}(V_{2\delta_n}^+)) = \alpha(V)$$

and

$$\lim_{n\to\infty} \mu_{\delta_n}((p \times p)^{-1}(V_{\delta_n}^-)) = \alpha(V).$$

Proof. This is a standard argument. For any $\varepsilon > 0$ there exist open $V_1 \subset V \subset V_2$ such that $\alpha(V_2 \setminus V_1) < \varepsilon$ and $V_1 \subset V_\delta^-$, $V_2 \supset V_\delta^+$ for some $\delta > 0$ (this follows from the dominated convergence theorem), and also such that $\alpha(\partial V_2) = \alpha(\partial V_1) = 0$. Since $\mu_{\delta_n} \to \bar{\alpha}$ it then follows (**[Bi]**, Th. 2.1 (v)) that

$$\mu_{\delta_n}((p \times p)^{-1}(V_1)) \longrightarrow \alpha(V_1),$$
$$\mu_{\delta_n}((p \times p)^{-1}(V_2)) \longrightarrow \alpha(V_2).$$

The result then follows by monotonicity, because for large n, $V_1 \subset V_{\delta_n}^- \subset V \subset V_{2\delta_n}^+ \subset V_2$. □

Now let γ_n be a closed geodesic with length ℓ_n, and let $\bar{\nu}_{n,k}, m_k$, etc. be as in Lemma 6. If V is any closed subset of $S \times S$ let $s_n(V)$ be the number of self-intersection points x of γ_n such that $(x, x) \in V$.

Lemma 9. *For each* $k = 1, 2, \ldots,$

$$(m_k^2/2k^2)(\bar{\nu}_{n,k} \times \bar{\nu}_{n,k})(U_\delta \cap (p \times p)^{-1}(V_\delta^-)) - 2m_k/k$$
$$\leq s_n(V) \leq (m_k^2/2k^2)(\bar{\nu}_{n,k} \times \bar{\nu}_{n,k})(U_\delta \cap (p \times p)^{-1}(V_\delta^+)).$$

Proof. This is virtually the same as the proof of Lemma 6. The only novelty is in keeping track of where the geodesic segments cross. Observe that if geodesic segments G_1, G_2 of length δ intersect transversally at x such that $(x, x) \in V$ then the (initial) endpoints x_1, x_2 of G_1, G_2 are such that $(x_1, x_2) \in V_\delta^+$. Similarly, if G_1, G_2 intersection at x and the initial endpoints x_1, x_2 are such that $(x_1, x_2) \in V_\delta^-$, then $(x, x) \in V$. □

Proof of Theorem 2. Let V be any open subset of $S \times S$ such that $\alpha(\partial V) = 0$. By Theorem 1, Lemma 5, and Lemma 9, for any $\varepsilon > 0$ there is a $t_\varepsilon < \infty$ large enough that for all $t \geq t_\varepsilon$ and all but at most $\varepsilon\pi(t)$ closed geodesics γ_n with length $\leq t$,

$$C\mu_\delta((p \times p)^{-1}(V_\delta^-)) - \varepsilon \leq s_n(V)/\ell_n^2 \leq C\mu_\delta((p \times p)^{-1}(V_{2\delta}^+)) + \varepsilon.$$

The quantities ε and δ are both arbitrary, but affect how large t must be so that the preceding statement is true. By letting $\delta \to 0$ through the subsequence δ_n and appealing to Lemma 8, we see that for any $\varepsilon > 0$ there exists $t_\varepsilon < \infty$ such that for all $t \geq t_\varepsilon$ and all but at most $\varepsilon\pi(t)$ closed geodesics γ_n of length $\leq t$,

$$C\alpha(V) - 2\varepsilon \leq s_n(V)/\ell_n^2 \leq C\alpha(V) + 2\varepsilon.$$

Together with Theorem 1, this implies that for most closed geodesics γ_n with $\ell_n \leq t$ the distribution α_n of self-intersection points is close to α in the weak-$*$ topology. □

References

[A] Anosov, D.V. (1967), Geodesic flows on closed Riemannian manifolds with negative curvature, Proc. Steklov Inst. 90, Amer. Math. Soc., Providence 1969.

[Bi] Billingsley, P. (1968), Convergence of Probability Measures, Wiley, New York.

[BS] Birman, J. and Series, C. (1984), An algorithm for simple curves on surfaces, J. London Math. Soc. (2) 29, 331–342.

[B 1] Bowen, R. (1975), Equilibrium States and the Ergodic Theory of Anosov Diffeomorphisms, Lecture Notes in Math. 470, Springer-Verlag, Berlin-Heidelberg-New York 1975.

[B 2] Bowen, R. (1973), Symbolic dynamics for hyperbolic flows, Amer. J. Math. 95, 429–450.

[BR] Bowen, R. and Ruelle, D. (1975), Ergodic theory of Axiom A flows, Invent. Math. 29, 181–202.

[L 1] Lalley, S.P. (1989), Renewal theorems in symbolic dynamics, with applications to geodesic flows, noneuclidean tessellations and their fractal limits, Acta Math. 163, 1–55.

[L 2] Lalley, S.P. (1987), Distribution of periodic orbits of symbolic and Axiom A flows, Adv. Appl. Math. 8, 154–193.

[L 3] Lalley, S.P. (1989), Closed geodesics in homology classes on surfaces of variable negative curvature, Duke Math. J. 58, 795–821.

[M] Margulis, G. (1969), Applications of ergodic theory to the investigation of manifolds of negative curvature, Funct. Analysis Appl. 3, 335–336.

[P] Pollicott, M. (1985), Asymptotic distribution of closed geodesics, Israel J. Math. 52, 209–224.

[R] Ratner, M. (1973), Markov partitions for Anosov flows on n-dimensional manifolds, Israel J. Math. 15, 92–114.

[RS] Reed, M., and Simon, B. (1980), Methods of Mathematical Physics I. Functional Analysis, Academic Press, New York.

[S] Series, C. (1985), The geometry of Markov numbers, Math. Intelligencer 7, 20–29.

Department of Statistics
Purdue University
West Lafayette, IN 47907, U.S.A.

A Note on the Law of the Iterated Logarithm for Weighted Sums of Independent Identically Distributed Random Variables

Deli Li and M. Bhaskara Rao

Abstract. A triangular array of real numbers and a sequence of independent identically distributed random variables are provided for which the law of the iterated logarithm fails.

1. Introduction

Let $X, X_n, n \geq 1$ be a sequence of independent identically distributed random variables such that $EX = 0$ and $EX^2 = 1$. The Hartman–Wintner Law of the Iterated Logarithm for the sequence $S_n = \sum_{i=1}^n X_i, n \geq 1$ of partial sums is well known. More precisely,

$$\limsup_{n\to\infty} S_n/(2n \log\log n)^{1/2} = 1 \text{ a.s.},$$

and

$$\liminf_{n\to\infty} S_n/(2n \log\log n)^{1/2} = -1 \text{ a.s.}$$

Let $a_{n,k}, n \geq 1, 1 \leq k \leq n$ be a triangular array of real numbers. There is an extensive literature available to study the nature of

$$\limsup_{n\to\infty} \Big(\sum_{k=1}^n a_{nk} X_k\Big)/(2n \log\log n)^{1/2},$$

and

$$\liminf_{n\to\infty} \Big(\sum_{k=1}^n a_{nk} X_k\Big)/(2n \log\log n)^{1/2},$$

for a variety of triangular arrays of real numbers. For a sample of papers, see Chow and Lai (1973), Chow and Teicher (1974), Lacey (1989), Lai and Wei (1982), Stadtmuller (1984), Taylor (1978), Teicher (1985), Thrum (1987) and Tomkins (1974). Tomkins (1974) established the following elegant result. Let f be a Lipschitz function of order one defined on the unit interval $[0, 1]$, i.e., $|f(x) - f(y)| \leq K|x - y|$ for all x, y in $[0, 1]$ for some non-negative constant K. Then

$$\limsup_{n\to\infty} \Big(\sum_{k=1}^n f(k/n) X_k\Big)/(2n \log\log n)^{1/2} = \Big(\int_0^1 f^2(x)\,dx\Big)^{1/2} \text{ a.s.},$$

and

$$\liminf_{n\to\infty}\Big(\sum_{k=1}^{n} f(k/n)X_k\Big)/(2n\log\log n)^{1/2} = -\Big(\int_0^1 f^2(x)\,dx\Big)^{1/2} \text{ a.s.}$$

Lai and Wei (1989) established the veracity of the above result for any Lipschitz function f of order $q \geq 1/2$.

On the other hand, Li, Rao and Wang (1990) established the following result which seems to say that for almost all choices of weights the Law of the Iterated Logarithm fails, using what they call the Law of the Logarithm. Let $A_{nk},\ n \geq 1,\ 1 \leq k \leq n$ be an array of independent identically distributed real random variables taking values in a bounded Borel subset of the real line with a non-degenerate common distribution. Then for almost all realizations $a_{nk},\ n \geq 1,\ 1 \leq k \leq n$ of $A_{nk},\ n \geq 1,\ 1 \leq k \leq n$, the following hold.

$$\limsup_{n\to\infty}\Big(\sum_{k=1}^{n} a_{nk}X_k\Big)/(2n\log\log n)^{1/2} = \infty \text{ a.s.},$$

and

$$\liminf_{n\to\infty}\Big(\sum_{k=1}^{n} a_{nk}X_k\Big)/(2n\log\log n)^{1/2} = -\infty \text{ a.s.}$$

In a presentation of these results at the conference, Professor J. Rosenblatt raised the problem of constructing specific arrays of real numbers for which the Law of the Iterated Logarithm fails. Even though there is an abundance of such arrays available as typified by the result of Li, Rao and Wang (1990), examples seem hard to come by. The purpose of this note is to construct an array of real numbers for which the Law of the Iterated Logarithm fails.

2. An example

We fix a sequence $X_n,\ n \geq 1$ of independent identically distributed random variables each with the same normal distribution with mean zero and variance unity. We define a sequence $A_n,\ n \geq 1$ of Hadamard matrices defined inductively as follows:

$$A_{n+1} = \begin{pmatrix} A_n & -A_n \\ A_n^T & A_n^T \end{pmatrix}$$

for $n = 1, 2, \ldots,$ where $A_1 = (1)$, a matrix of order 1×1. These matrices have the following properties.

(a) The order of the matrix A_n is $2^{n-1} \times 2^{n-1},\ n = 1, 2, \ldots.$

(b) Every entry in A_n is either $+1$ or $-1,\ n = 1, 2, \ldots.$

(c) For each $n \geq 1,\ A_n A_n^T = A_n^T A_n = 2^{n-1} I_{(n)}$, where $I_{(n)}$ is the identity matrix of appropriate order, i.e., of order $2^{n-1} \times 2^{n-1}$.

Let $\mathcal{D} = (d_{nk})$ be the matrix of order $\infty \times \infty$ with the following structure.

$$\mathcal{D} = \begin{pmatrix} A_1 & 0 & 0 & \dots \\ 0 & A_2 & 0 & \dots \\ 0 & 0 & A_3 & \dots \\ \vdots & \vdots & \vdots & \ddots \end{pmatrix}$$

The entries of $\mathcal{D}$ have the following properties.

(d) $d_{nk} = 0$ for every $k \geq 2n,\ n \geq 1$.

(e) $n/2 \leq \sum_{k=1}^{2n} d_{nk}^2 \leq 2n$ for every $n \geq 1$.

For the sequence $X_n,\ n \geq 1$ of random variables under consideration, it is well known that

$$\limsup_{n\to\infty} X_n/(2\log n)^{1/2} = 1 \text{ a.s.}$$

Let $\sigma_n^2 = \sum_{k=1}^{2n} d_{nk}^2$ and $Y_n = \sum_{k\geq 1} d_{nk}X_k,\ n \geq 1$. Since the rows of $\mathcal{D}$ are orthogonal, it is clear that the sequence $Y_n/\sigma_n,\ n \geq 1$ is a sequence of independent identically distributed random variables each normally distributed with mean zero and variance unity. By what we have observed above, we have

$$\limsup_{n\to\infty} Y_n/(2n\log n)^{1/2} \geq \limsup_{n\to\infty}[Y_n/\sigma_n]/[2(\log n)^{1/2}] = 2^{-1/2} \text{ a.s.}$$

From this, it follows that

$$\limsup_{n\to\infty}\Big[\sum_{k\geq 1} d_{nk}X_k\Big]/[2n\log\log n]^{1/2} = \infty \text{ a.s.}$$

But the weights are not exactly of a triangular array type. Further, the weights are not all non-negative. If one requires an example of a triangular array of positive numbers, one needs to make some adjustments. Let for each $n \geq 1$ and $k \geq 1$,

$$a_{2n,k} = 2 + d_{nk}$$
$$a_{2n-1,k} = 2.$$

The array $a_{nk},\ n \geq 1,\ k \geq 1$ of real numbers has the following properties.

(a) $1 \leq a_{nk} \leq 3$ for every n and k.

(b) $n \leq \sum_{k=1}^{n} a_{nk}^2 \leq 9n$ for every n.

Further,

$$\begin{aligned}\limsup_{n\to\infty}\Big|\sum_{k=1}^{n} a_{nk}X_k\Big|/(2n\log n)^{1/2} &\geq \limsup_{n\to\infty}\Big|\sum_{k=1}^{2n} d_{nk}X_k\Big|/\big(4n\log(2n)\big)^{1/2} \\ &\quad - \limsup_{n\to\infty}\Big|\sum_{k=1}^{n} 2X_k\Big|/(2n\log n)^{1/2} \\ &\geq 2^{-1/2} + 0 \text{ a.s.}\end{aligned}$$

It follows that

$$\limsup_{n\to\infty} \Big| \sum_{k=1}^{n} a_{nk} X_k \Big| / (2n \log\log n)^{1/2} = \infty \text{ a.s.}$$

Acknowledgement. The second author's research was partly supported by US Army Research Office under Grant # DAAH04-93-G-0030 and NSF-EPSCOR grant.

References

[1] Chow, Y. S. and Lai, T. L. (1973), Limiting behavior of weighted sums of independent random variables, Ann. Probab. 1, 810–814.

[2] Chow, Y. S. and Teicher, H. (1973), Iterated Logarithm laws for weighted averages, Z. Wahrscheinlichkeitstheorie verw. Geb. 26, 87–94.

[3] Lacey, M. (1989). Laws of the iterated Logarithm for partial sum processes indexed by functions, J. Theoret. Probab. 2, 377–398.

[4] Lai, T. L. and Wei, C. Z. (1982), A law of the Iterated Logarithm for double arrays of independent random variables with application to regression and time series models, Ann. Probab. 10, 320–335.

[5] Li, D., Rao, M. B. and Wang, X. C. (1995), On the strong law of large numbers and the law of the logarithm for weighted sums of independent random variables with multidimensional indices, J. Multivariate Analysis 52, 181–198.

[6] Stadtmüller, U. (1984), A note on the law of iterated logarithm for weighted sums of random variables, Ann. Probab. 12, 35–44.

[7] Taylor, R. L. (1978), Stochastic convergence of weighted sums of random elements in linear spaces, Lecture Notes in Math. 672, Springer-Verlag, New York.

[8] Teicher, H. (1985), Almost certain convergence in double arrays, Z. Wahrscheinlichkeitstheorie verw. Geb. 69, 331–345.

[9] Thrum, R. (1987), A remark on almost sure convergence of weighted sums, Prob. Theory and Related Fields 75, 425–430.

[10] Tomkins, R. J. (1974), On the law of the iterated logarithm for double sequences of random variables, Z. Wahrscheinlichkeitstheorie verw. Geb. 30, 303–314.

Department of Mathematics
Brock University
St. Catharines, Ontario L2S 3A1, Canada

Department of Statistics
North Dakota State University
Fargo, ND 58105, U.S.A.

Besicovitch Functions and Weighted Ergodic Theorems for LCA Group Actions

Michael Lin and James Olsen

Abstract. Let G be an amenable group, $\{\mu_n\}$ an ergodic sequence of probabilities. We show that bounded 1-Besicovitch functions (with respect to $\{\mu_n\}$) are r-Besicovitch. For G abelian, we obtain Fourier expansions for 2-Besicovitch functions.

Let $\{\mu_n\}$ be an ergodic sequence on a σ-compact LCA group, and let $T(t)$ be a weakly almost periodic representation of G in a Banach space X. If w is 1-Besicovitch, then $\int w(t)T(t)x\,d\mu_n(t)$ converges strongly for every $x \in X$. The limit is identified when $w(t)$ is 2-Besicovitch. If $w(t)$ is weakly almost periodic, the limit is the same for all ergodic sequences.

For G abelian and $\{A_n\}$ a Følner sequence satisfying Tempelman's conditions, $\lim_{n\to\infty} \frac{1}{\lambda(A_n)} \int_{A_n} w(t) f(\theta_t x)\,d\lambda(t)$ exists a.e. for any measure preserving group action $\{\theta_t\}_{t\in G}$ in (Ω, Σ, m), $f \in L_1(\Omega, m)$ and $w(t)$ bounded 1-Besicovitch with respect to $\{\frac{1}{\lambda(A_n)} \mathbf{1}_{A_n} d\lambda\}$. If $w(t)$ is weakly almost periodic, the limit is the same for all such $\{A_n\}$.

1. Introduction

Let G be a locally compact σ-compact group with right Haar measure λ. A bounded continuous function f is *(weakly) almost periodic* if $\{\delta_t * f : t \in G\}$ is (weakly) conditionally compact in $C(G)$. The set of (weakly) almost periodic functions is denoted by (W)AP(G). AP(G) and WAP(G) are closed subspaces of the space UCB$_\ell(G)$ of left uniformly continuous bounded functions **[E]**.

For a regular probability μ on G, the convolution operator defined by $\mu * f(t) = \int f(ts)\,d\mu(s)$ is a Markov operator, having λ as a σ-finite invariant measure. The right translation operators $(\delta_s * f)(t) = f(ts)$ yield representations of G in $L_p(G, \lambda)$, $1 \le p \le \infty$, and in $C(G)$, UCB$_\ell(G)$, AP(G) or WAP(G). Except for $C(G)$ and $L_\infty(G)$, the representation is strongly continuous in each of the other function spaces, and the convolution operator is an average of the representation.

More generally, Let X be a Banach space, and let $T\colon G \to B(X)$ be a strongly continuous bounded operator representation of G. For a probability μ, we define the μ- average of the representation by $U_\mu x = \int T(t)x\,d\mu(t)$, which is well defined in the strong topology of X. Moreover, $U_\mu x = \int T(t)x\,d\mu(t)$ is defined for any finite signed measure on G, and we have $U_{\mu*\nu} = U_\mu U_\nu$, where the convolution of two measures on G is defined by

$$\mu * \nu(A) = \iint \mathbf{1}_A(ts)\,d\mu(t)\,d\nu(s).$$

Let $\{\mu_n\}$ be a sequence of probabilities of G.

Definitions. (1) The sequence $\{\mu_n\}$ *converges weakly to invariance* if for every $g \in C(G)$ and $t \in G$ we have $\lim_n \int (g - \delta_t * g)\, d\mu_n = 0$ (**[D]**).

(2) The sequence $\{\mu_n\}$ is *ergodic* if for every $f \in L_1(G, \lambda)$ and every $t \in G$ we have $\lim_n \|\mu_n * (f - \delta_t * f)\|_1 = 0$.

It was observed in **[Li-W]** that if G is not compact, weak convergence to invariance is equivalent to $\|t\mu_n - \mu_n * \delta_t\| \to 0$ for every $t \in G$, an ergodic sequence need not converge weakly to invariance even for G abelian, convergence to invariance always implies ergodicity, and existence of an ergodic sequence implies amenability of G. The ergodic sequences (of densities) as defined in **[T-3]** are weakly convergent to invariance. Every amenable σ-compact group has a sequence converging weakly to invariance - take $d\mu_n = \lambda(A_n)^{-1} \mathbf{1}_{A_n} d\lambda$, where $\{A_n\}$ is a Følner sequence.

Definition. A *(weakly) almost periodic representation* of G is a bounded continuous representation of G such that $\mathrm{Or}(x) = \{T(t)x : t \in G\}$ is (weakly) conditionally compact for all $x \in X$.

Thus, any bounded continuous representation in a reflexive space is weakly almost periodic. Clearly, $\{y : \mathrm{Or}(y)$ is (weakly) conditionally compact $\}$ is a closed invariant subspace, and the restriction of the representation to this subspace is (weakly) almost periodic.

Ryll-Nardzewski **[RN-1]** proved that if $T(t)$ is a weakly almost periodic representation of G, then

$$X = \{y : T(t)y = y \ \forall t \in G\} \oplus \mathrm{clm} \bigcup_{t \in G} \big(I - T(t)\big)X. \tag{1.1}$$

It was shown in **[Li-W]** that in that case, for every ergodic sequence $\{\mu_n\}$ we have $\|U_{\mu_n}x - Px\| \to 0$, where P is the projection on the set of common fixed points corresponding to (1.1).

Following Tempelman **[T-3]**, in this paper we study *weighted ergodic theorems*, that is, we look for functions $w(t)$ on G abelian such that for an ergodic sequence $\{\mu_n\}$ and a weakly almost periodic representation $T(t)$ we have strong convergence of

$$\int w(t)T(t)x\, d\mu_n(t).$$

When we have a group action $\{\theta_t\}$ by measure preserving transformations, we are also interested in pointwise convergence.

2. Besicovitch functions on amenable groups

For f measurable on G and $\{\mu_n\}$ a sequence of probabilities on G we define

$$\|f\|_{r,\{\mu_n\}}^r = \limsup_{n\to\infty} \int |f|^r \, d\mu_n.$$

Then for $1 \le r < \infty$, $\|\cdot\|_{r,\{\mu_n\}}$ is a seminorm on the set of functions f such that $\|f\|_{r,\{\mu_n\}} < \infty$. We will call this seminorm "the r-seminorm relative to $\{\mu_n\}$". We will say that f is r-Besicovitch for $\{\mu_n\}$ if it is in the r-seminorm closure of the almost periodic functions (i.e., for $\epsilon > 0$ there is a $g \in \mathrm{AP}(G)$ with $\| |f - g| \|_{r,\{\mu_n\}} < \epsilon$), and denote this class by $B_{r,\{\mu_n\}}$.

We note that if $\|f\|_{r,\{\mu_n\}} = 0$, $f \in B_{r,\{\mu_n\}}$ since the constant function 0 is almost periodic. (In the abelian case, the term "almost periodic functions" can be replaced by "trigonometric polynomials".) Besicovitch classes on the semigroup $\mathbb{Z}^+$ with $\mu_n = \dfrac{\delta_0 + \cdots + \delta_{n-1}}{n}$, where δ_k is the Dirac measure at k, have been similarly defined in **[RN-2]**, **[Ba-O]** and **[B-L]**, while Besicovitch classes on $\mathbb{Z}_d^+$ have been studied in **[J-O]**. We will need the following fact, which is proved in the $\mathbb{Z}^+$ and $\mathbb{Z}_d^+$ cases in **[B-L]** and **[J-O]** respectively.

Theorem 2.1. *$B_{r,\{\mu_n\}} \cap L_\infty(G) = B_{1,\{\mu_n\}} \cap L_\infty(G)$ for all r, $1 \le r < \infty$.*

Before proceeding to the proof of Theorem 2.1, we illustrate its content with the following examples for $G = \mathbb{Z}$.

Let $r > 1$, a a real number such that $3 > a > 3^{\frac{1}{r}}$, and let f be the function defined on $\mathbb{Z}$ by

$$f(k) = \begin{cases} a^n & \text{if } k = 3^n, n = 0, 1, 2, \ldots \\ 0, & \text{otherwise.} \end{cases}$$

Then if $\mu_n = \frac{1}{n}(\delta_0 + \cdots + \delta_{n-1})$,

$$\limsup \int |f(k)| \, d\mu_n = \limsup \frac{\sum_{k=0}^n a^k}{3^n} = \limsup \left(\frac{a}{3}\right)^n \left(\frac{a}{a-1}\right) = 0,$$

while

$$\limsup \int |f(k)|^r \, d\mu_n(k) = \limsup \left(\frac{a^r}{3}\right)^n \left(\frac{a^r}{a^r - 1}\right) = +\infty.$$

Thus, $f \in B_{1,\{\mu_n\}}$, $f \notin B_{r,\{\mu_n\}}$ and f is not bounded. This shows that while we evidently have $B_{r,\{\mu_n\}} \subset B_{1,\{\mu_n\}}$, the containment is strict and that $B_{1,\{\mu_n\}}$ contains unbounded functions which must therefore be not almost periodic.

With the same G and $\{\mu_n\}$ as above, let f be defined by

$$f(k) = \begin{cases} 1, & \text{if } k = 3^n - 2^n + 1,\ 3^n - 2^n + 2,\ \ldots, 3^n \text{ some } n \in \mathbb{Z}^+ \\ 0, & \text{otherwise.} \end{cases}$$

Then

$$\limsup \int |f(k)|\, d\mu_n \le \lim \frac{2^{n+1}}{3^n} = 0,$$

So again $f \in B_{1,\{\mu_n\}}$. f is not almost periodic (or even weakly almost periodic), since in that case we should have uniform convergence in z to zero of $\int f(k+z)\, d\mu_n$ by Lemma 2.2 below. However, for every N there exists $n > N$ and t such that $\int f(k+t)\, d\mu_n = 1$. Therefore there are also bounded elements of $B_{1,\{\mu_n\}}$ that are not weakly almost periodic.

Proof of Theorem 2.1. We first note that Hölder's inequality holds for the $(r, \{\mu_n\})$ seminorms and that the constant function $1 \in B_{r,\{\mu_n\}}$ for every r, $1 \le r < \infty$. Therefore, $B_{r,\{\mu_n\}} \subset B_{1,\{\mu_n\}}$ for all $r \ge 1$.

Now let $f \in B_{1,\{\mu_n\}} \cap L_\infty(G)$, $r > 1$. Then $\forall \epsilon > 0\ \exists\, \ell_\epsilon \in \mathrm{AP}(G)$ satisfying

$$\limsup_n \int |f - \ell_\epsilon|\, d\mu_n < \epsilon.$$

If there is a uniform bound on $|\ell_\epsilon|$, M say, we have $f \in B_{r,\{\mu_n\}}$, since

$$\begin{aligned}\limsup_n \int |f - \ell_\epsilon|^r\, d\mu_n &= \limsup_n \int |f - \ell_\epsilon|\, |f - \ell_\epsilon|^{r-1}\, d\mu_n \\ &\le \limsup_n \Big(\int |f - \ell_\epsilon|\, d\mu_n\Big)(\|f\|_\infty + M)^{r-1}.\end{aligned}$$

In general, write $\ell_\epsilon(t) = R(t)e^{i\theta(t)}$. Noting that the function g on C defined by $g(z) = (|z| \wedge M)e^{i \arg z}$ is uniformly continuous on C and that a uniformly continuous function of an almost periodic function is almost periodic, we have that the function $\tilde{\ell}_\epsilon(t) = (R(t) \wedge \|f\|_\infty)e^{i\theta(t)}$ is almost periodic. Since $\|\tilde{\ell}_\epsilon\|_\infty \le \|f\|_\infty$ and $\|f - \tilde{\ell}_\epsilon\|_{(1,\mu_n)} \le \|f - \ell_\epsilon\|_{(1,\mu_n)}$, Theorem 2.1 is proved. □

We remark that this proof is applicable to the analogous results in **[B-L]** and **[J-O]**, although being a different proof.

Lemma 2.2. *Let G be a locally compact σ-compact amenable group. Then there exists a projection M of* WAP(G) *onto the constants such that for every ergodic sequence $\{\mu_n\}$ we have $\|\mu_n * f - Mf\| \to 0$ for $f \in$* WAP(G). *Furthermore, $M(\delta_t * f) = M(f)$, and $M(f)$ is the unique invariant mean on* WAP(G).

Proof. The existence and uniqueness of an invariant mean on WAP(G) was proved in **[RN-1]** (see also **[Bu**, p. 15**]**). Apply Theorem 2.10 (iv) of **[Li-W]** (the only bounded continuous functions invariant under all the translations are the constants). □

Proposition 2.3. *Let G be a locally compact σ-compact amenable group. For any ergodic sequence $\{\mu_n\}$ we have* WAP$(G) \subset B_{r,\{\mu_n\}}$ *for every $1 \le r < \infty$.*

Proof. Let M be the unique invariant mean on WAP(G). Let $W_0(G) = \{f \in \text{WAP}(G) : M(|f|) = 0\}$. If $f \in W_0(G)$, then by Lemma 2.2

$$\int |f|\, d\mu_n = \mu_n * |f|(e) \to 0.$$

The decomposition WAP(G) = AP$(G) \oplus W_0(G)$ (see **[Bu**, pp. 29–30**]**) yields the proposition for $r = 1$. For $r > 1$, we apply Theorem 2.1. □

Remark. For G abelian and $r = 1$, this result is due to Eberlein **[E]**.

Theorem 2.4. *Let $\{\mu_n\}$ be an ergodic sequence of measures on a locally compact σ-compact amenable group. Then $\lim_{n\to\infty} \int f\, d\mu_n$ exists for any $f \in B_{1,\{\mu_n\}}$.*

Proof. Following Tempelman **[T-3]**, let $w_k \in \text{AP}(G)$ with

$$\lim_{k\to\infty} \limsup_{n\to\infty} \int |f - w_k|\, d\mu_n = 0\,.$$

By Lemma 2.2, $\alpha_k = \lim_{n\to\infty} \int w_k\, d\mu_n$ exists for each k. Now, for $\epsilon > 0$

$$\begin{aligned} |\alpha_k - \alpha_j| &= \lim_n \left| \int (w_k - w_j)\, d\mu_n \right| \\ &\le \limsup_{n\to\infty} \int |f - w_k|\, d\mu_n + \limsup_{n\to\infty} \int |f - w_j|\, d\mu_n < \epsilon \end{aligned}$$

if k and j are large enough. Hence $\alpha = \lim \alpha_k$ exists. Now

$$\left| \int f\, d\mu_n - \alpha \right| \le \int |f - w_k|\, d\mu_n + \left| \int w_k\, d\mu_n - \alpha_k \right| + |\alpha_k - \alpha|$$

proves the desired convergence. □

Remark. The result is proved in **[T-3]** for G abelian and $\{\mu_n\}$ with densities $\{p_n\}$ satisfying $\lim_{n\to\infty} \int |p_n(ts) - p_n(s)|\, d\lambda(s) = 0$, (which implies that $\{\mu_n\}$ converges weakly to invariance).

We now analyse the abelian case, and obtain Fourier expansions (generalizing some of the results obtained for R in **[B]**). We denote the dual group of G by $\hat{G}$. Clearly every character $\chi \in \hat{G}$ is in AP(G). A finite linear combination of characters is called a *trigonometric polynomial.* It is well-known that (for G abelian) every $f \in \text{AP}(G)$ can be uniformly approximated by trigonometric polynomials **[Lo**, §41**]**, **[Bu**, p. 77**]**.

Theorem 2.5. *Let $\{\mu_n\}$ be an ergodic sequence of probabilities on a locally compact σ-compact abelian group G. Then $c_f(\chi) = \lim_{n\to\infty} \int f\bar{\chi}\, d\mu_n$ exists for any $\chi \in \hat{G}$ and any $f \in B_{1,\{\mu_n\}}$.*

Proof. Let $w_k \in \text{AP}(G)$ satisfy $\lim_{k\to\infty} \limsup_{n\to\infty} \int |f - w_k|\, d\mu_n = 0$. Then

$$\lim_{k\to\infty} \limsup_{n\to\infty} \int |f\bar{\chi} - w_k\bar{\chi}|\, d\mu_n = 0\,.$$

Since $w_k \bar{\chi} \in \mathrm{AP}(G)$, we can apply Theorem 2.4 to $f\bar{\chi} \in B_{1,\{\mu_n\}}$. □

Definition. $c_f(\chi)$ are called the *Fourier coefficients* of $f \in B_{1,\{\mu_n\}}$.

Proposition 2.6. *Let $\{\mu_n\}$ be an ergodic sequence of probabilities on a locally compact σ-compact abelian group G. Then:*

(i) *For $f \in \mathrm{WAP}(G)$ we have $c_f(\chi) = M(f\bar{\chi})$ for $\chi \in \hat{G}$. (M is the invariant mean on* WAP(G) *).*

(ii) $M(\chi) = 0$ *for* $1 \neq \chi \in \hat{G}$.

(iii) *For a trigonometric polynomial $f = \Sigma_{j=1}^m a_j \chi_j$ we have: $c_f(\chi_j) = a_j$, $c_f(\chi) = 0$ for $\chi \neq \chi_j$, and $\|f\|_{2,\{\mu_n\}}^2 = \Sigma_{j=1}^m |a_j|^2$.*

Proof. (i) For $f \in \mathrm{WAP}(G)$, also $f\bar{\chi} \in \mathrm{WAP}(G)$, and we apply Lemma 2.2 to $f\bar{\chi}$.

(ii) Let $M(\chi) \neq 0$. Then $\chi(t) = 1$ for any $t \in G$ since

$$0 \neq M(\chi) = M(\delta_t * \chi) = M(\chi(t)\chi) = \chi(t)M(\chi).$$

(iii) Applying (i) to f we obtain $c_f(\chi) = \Sigma_{j=1}^m a_j M(\chi_j \bar{\chi})$. By (ii), $M(\chi_j\bar{\chi}) = 0$ if $\chi \neq \chi_j$. Since $M(1) = 1$, we have $c_f(\chi_k) = a_k$ for $1 \leq k \leq m$. Finally, (ii) yields

$$\|f\|_{2,\{\mu_n\}}^2 = \lim_{n\to\infty} \int |\sum_{j=1}^m a_j \chi_j|^2 \, d\mu_n = \sum_{j=1}^m \sum_{k=1}^m a_j \bar{a}_k M(\chi_j \bar{\chi}_k) = \sum_{j=1}^m |a_j|^2.$$

Remarks. 1. Part (i) shows that for $f \in \mathrm{WAP}(G)$ the Fourier coefficients do not depend on the ergodic sequence.

2. The representation of $f \in \mathrm{WAP}(G)$ as $f = f_1 + f_0$ with $f_1 \in \mathrm{AP}(G)$ and $M(|f_0|) = 0$ **[Bu**, pp. 29–30**]** yields that f and f_1 have the same Fourier coefficients.

Theorem 2.7. *Let $\{\mu_n\}$ be an ergodic sequence of probabilities on a locally compact σ-compact abelian group G, and let $f \in B_{2,\{\mu_n\}}$. Then $c_f(\chi) \neq 0$ only for countably many $\chi \in \hat{G}$, $f = \sum_{\chi\in\hat{G}} c_f(\chi)\chi$ (convergence in $B_{2,\{\mu_n\}}$-seminorm), and $\|f\|_{2,\{\mu_n\}}^2 = \sum_{\chi\in\hat{G}} |c_f(\chi)|^2$.*

Proof. By Proposition 2.6 (iii), the result is true for trigonometric polynomials. Fix $f \in B_{2,\{\mu_n\}}$, and let $\{g_k\}$ be trigonometric polynomials with $\lim_{k\to\infty} \|f - g_k\|_{2,\{\mu_n\}} = 0$. Let $\{\chi_j\}$ be the sequence of characters which appear in some of the g_k. We can rewrite Proposition 2.6 (iii) as $g_k = \sum_j c_{g_k}(\chi_j)\chi_j$ (which in fact has only finitely many non-zero terms), and

$$\|g_k - g_m\|_{2,\{\mu_n\}}^2 = \sum_j |c_{g_k}(\chi_j) - c_{g_m}(\chi_j)|^2.$$

Hence the sequence of vectors $\mathbf{c}^{(k)}$, defined for fixed k by

$$\mathbf{c}^{(k)} = (c_{g_k}(\chi_1), \ldots, c_{g_k}(\chi_j), \ldots),$$

is Cauchy in ℓ_2. Denote the limit sequence by $\mathbf{a} = \{a_j\}$. Since

$$|c_f(\chi) - c_{g_k}(\chi)| \le \|f - g_k\|_{1,\{\mu_n\}} \le \|f - g_k\|_{2,\{\mu_n\}},$$

we have $a_j = c_f(\chi_j)$, and also $c_f(\chi) = 0$ if $\chi \ne \chi_j$ for every j. For $\varepsilon > 0$, fix k with $\|f - g_k\|_{2,\{\mu_n\}} < \varepsilon$ and $\|\mathbf{c}^{(k)} - \mathbf{a}\|_{\ell_2} < \varepsilon$. Take m large so $c_{g_k}(\chi_j) = 0$ for $j > m$. Then

$$\begin{aligned}\|f - \sum_{j=1}^{m} c_f(\chi_j)\chi_j\|_{2,\{\mu_n\}} &\le \|f - g_k\|_{2,\{\mu_n\}} + \|g_k - \sum_{j=1}^{m} c_f(\chi_j)\chi_j\|_{2,\{\mu_n\}} \\ &\le \varepsilon + (\sum_{j=1}^{m} |c_{g_k}(\chi_j) - a_j|^2)^{1/2} < 2\varepsilon.\end{aligned}$$

This proves the seminorm convergence of the Fourier series. Finally,

$$\|f\|^2_{2,\{\mu_n\}} = \lim_k \|g_k\|^2_{2,\{\mu_n\}} = \lim_k \|\mathbf{c}^{(k)}\|^2_{\ell_2} = \|\mathbf{a}\|^2_{\ell_2} = \sum_j |c_f(\chi_j)|^2.$$

Let $\ell_2(\hat{G})$ be the set of complex functions $\mathbf{b}$ defined on $\hat{G}$ such that $\sum_{\chi \in \hat{G}} |\mathbf{b}(\chi)^2| < \infty$ (only countably non-zero terms). The previous theorem shows that for an ergodic sequence $\{\mu_n\}$, the Fourier coefficient function c_f of any $f \in B_{2,\{\mu_n\}}$ is in $\ell_2(\hat{G})$. Using the method of **[B]**, we can show that if $\{A_n\}$ is an increasing Følner sequence with union all of G, and $d\mu_n = \lambda(A_n)^{-1}\mathbf{1}_{A_n} d\lambda$, then any $\mathbf{b} \in \ell_2(\hat{G})$ is the Fourier coefficient function of some $f \in B_{2,\{\mu_n\}}$. We do not know if the same holds for an arbitrary ergodic sequence.

3. Weighted ergodic theorems for representations of abelian groups

If $T(t)$ is a weakly almost periodic representation of G, it was proved in **[Li-W]** that $\int T(t)x\,d\mu_n(t)$ converges strongly for every $x \in X$ if $\{\mu_n\}$ is ergodic. In this section we deal with the problem of *weighted averages*—finding $w(t)$ such that $\int w(t)T(t)x\,d\mu_n(t)$ converges for every $x \in X$, when the ergodic sequence $\{\mu_n\}$ is given.

Theorem 3.1. *Let G be an* LCA *group, and let $w(t) \in$* WAP(G). *Let $T(t)$ be a weakly almost periodic representation of G in a Banach space X. Then there exists a (bounded linear) operator E such that for every ergodic sequence $\{\mu_n\}$,*

$$\lim_{n\to\infty} \| \int w(t)T(t)x\,d\mu_n(t) - Ex\| \to 0.$$

Proof. Let $\chi(t)$ be a character of G. Then $S(t) = \chi(t)T(t)$ is also a bounded continuous representation of G with weakly compact orbits. By **[Li-W**, Theorem 2.10**]** we have

$$X = \{y : S(t)y = y \quad \forall t \in G\} \oplus \operatorname{clm} \bigcup_{t\in G} \big(I - S(t)\big)X, \tag{3.1}$$

and

$$\lim_{n\to\infty} \Big\| \int \chi(t)T(t)x\, d\mu_n(t) - Px \Big\| = 0 \tag{3.2}$$

for every ergodic sequence $\{\mu_n\}$ and $x \in X$, where P is the projection corresponding to (3.1). Hence the theorem is true for every trigonometric polynomial $w(t) = \sum_{i=1}^n a_i \chi_i(t)$ with $\chi_i \in \hat{G}$.

Let $\{w_k\}$ be a Cauchy sequence (in the uniform norm) of trigonometric polynomials, and let $E_k x = \lim_{n\to\infty} \int w_k(t)T(t)x\, d\mu_n(t)$ be the limits we have just shown to exist. Then

$$\|E_k x - E_m x\| \le \|x\|\, \|w_k - w_m\|_\infty \sup_{t\in G} \|T(t)\|$$

shows that $Ex = \lim E_k x$ exists.

Now let $w \in \mathrm{AP}(G)$. Since G is abelian, there exists a sequence $\{w_k\}$ of trigonometric polynomials with $\|w_k - w\|_\infty \to \infty$ (see, e.g., **[Lo**, §41**]**, **[Bu**, p. 77**]**). Hence $Ex = \lim_{k\to\infty} E_k x$ is well defined. Let $\{\mu_n\}$ be an ergodic sequence. Then

$$\begin{aligned}
&\Big\| \int w(t)T(t)x\, d\mu_n(t) - Ex \Big\| \\
&\le \Big\| \int [w(t) - w_k(t)]T(t)x\, d\mu_n(t) \Big\| + \Big\| \int w_k(t)T(t)x\, d\mu_n(t) - E_k x \Big\| + \|E_k - Ex\| \\
&\le \|w - w_k\|_\infty \|x\| \sup_t \|T(t)\| + \|E_k x - Ex\| + \Big\| \int w_k(t)T(t)x\, d\mu_n(t) - E_k x \Big\|
\end{aligned}$$

shows that $\lim_{n\to\infty} \int w(t)T(t)x\, d\mu_n(t) = Ex$ for $x \in X$.

Thus the theorem is true for $w \in \mathrm{AP}(G)$.

By Lemma 2.2, the unique invariant mean on $\mathrm{WAP}(G)$ can be computed as $M(f) = \lim \int f(t)\, d\mu_n(t)$ for any ergodic sequence $\{\mu_n\}$. Let $w(t) \in \mathrm{WAP}(G)$. By Eberlein's theorem **[E]** (see **[Bu**, pp. 29–30**]**), we have $w = w_1 + w_0$, with $w_1 \in \mathrm{AP}(G)$ and $M(|w_0|) = 0$. Since $M(|w_0|) = \lim \int |w_0|\, d\mu_n$, we have

$$\Big| \int w_0(t)T(t)x\, d\mu_n(t) \Big| \le \|x\| \sup_{t\in G} \|T(t)\| \int |w_0|\, d\mu_n \to 0,$$

and the theorem is proved. □

Theorem 3.2. *Let* $\{\mu_n\}$ *be an ergodic sequence on a* σ*-compact* LCA *group, and let* $T(t)$ *be a weakly almost periodic representation of* G *in a Banach space* X. *If* $w \in B_{1,\{\mu_n\}}$, *then* $\int w(t)T(t)x\, d\mu_n(t)$ *converges strongly for every* $x \in X$.

Proof. Since $w \in B_{1,\{\mu_n\}}$, there exists $\{w_k\} \in \mathrm{AP}(G)$ with

$$\lim_{k\to\infty} \limsup_{n\to\infty} \int |w_k - w|\, d\mu_n = 0\,.$$

Fix $x \in X$. Let $y_{k,n} = \int w_k(t)T(t)x\,d\mu_n$, $y_n = \int w(t)T(t)x\,d\mu_n$. By Theorem 3.1, $z_k = \lim_{n\to\infty} y_{k,n}$ exists for each n. It now follows easily that $z = \lim_k z_k$ exists, and $\|y_n - z\| = 0$ (see **[T-3**, Lemma 1]). □

Remarks.. 1. The theorem does not require the boundedness of $w(t)$.
2. The ergodic sequence $\{\mu_n\}$ can consist of singular measures.

In case that $w \in B_{2,\{\mu_n\}}$, we can identify the limit obtained in the previous theorem. In particular, this provides an identification of the limit in Theorem 3.1.

Definition. A vector $0 \neq y \in X$ is an *eigenvector* for the representation $T(t)$ of G if there is a function $\phi(t)$ with $T(t)y = \phi(t)y$ for every $t \in G$.

When $T(t)$ is a bounded continuous representation, ϕ is clearly continuous, and is a character. It will be called an *eigencharacter*. The *eigenspace* of a character χ is the set of vectors $y \in X$ such that $T(t)y = \chi(t)y$ for every $t \in G$ (which is $\{0\}$ if χ is not an eigencharacter).

Theorem 3.3. *Let $\{\mu_n\}$ be an ergodic sequence on a σ-compact* LCA *group, and let $T(t)$ be a weakly almost periodic representation of G in a Banach space X.*

(i) *For $\chi \in \hat{G}$, $P(\chi)x = \lim \int \bar{\chi}(t)T(t)x\,d\mu_n(t)$ is a projection on the eigenspace of χ, with null space* $\operatorname{clm}\bigcup_{t\in G}\big(\chi(t)I - T(t)\big)X$.

(ii) *For $w \in B_{2,\{\mu_n\}}$ with Fourier coefficients $c_w(\chi)$, we have (with strong convergence on both sides)*

$$\lim_{n\to\infty} \int w(t)T(t)x\,d\mu_n(t) = \sum_{\chi\in\hat{G}} c_w(\chi)P(\bar{\chi})x.$$

Proof. (i) is proved in Theorem 3.1 (put $\bar{\chi}$ instead of χ in defining $S(t)$).

(ii): $Ex = \lim_{n\to\infty} \int w(t)T(t)x\,d\mu_n(t)$ exists for every $x \in X$ by the previous theorem. By (i) the result holds for any trigonometric polynomial. Let $\{\chi_j\}$ be the sequence of characters for which $c_j := c_w(\chi_j) \neq 0$ (see Theorem 2.7), and denote $w_k = \sum_{j=1}^k c_j\chi_j$. By Theorem 2.7 $\|w\|^2_{2,\{\mu_n\}} = \sum_{j=1}^\infty |c_j|^2$, and $w = \sum_{j=1}^\infty c_j\chi_j$ (convergence in $B_{2,\{\mu_n\}}$-seminorm). In the following we denote $P(\bar{\chi}_j)$ by P_j. For $\varepsilon > 0$ fix k with $\|w - w_k\|^2_{2,\{\mu_n\}} = \sum_{j=k+1}^\infty |c_j|^2 < \varepsilon^2$. Then

$$\Big\|Ex - \sum_{j=1}^k c_jP_jx\Big\| \le \Big\|Ex - \int w(t)T(t)x\,d\mu_n(t)\Big\| + \Big\|\int [w(t) - w_k(t)]T(t)x\,d\mu_n(t)\Big\| + \Big\|\int w_k(t)T(t)x d\,\mu_n(t) - \sum_{j=1}^k c_jP_jx\Big\|.$$

As $n \to \infty$, the first term tends to 0 by definition, and the last term tends to 0 by the result for trigonometric polynomials. Since

$$\|\int [w(t) - w_k(t)]T(t)x\, d\mu_n(t)\|$$

$$\leq \|x\| \sup_t \|T(t)\| \int |w(t) - w_k(t)|\, d\mu_n(t)$$

$$\leq [\int |w(t) - w_k(t)|^2 d\mu_n(t)]^{1/2}$$

is less than ε for large n by the choice of k, we have $\|Ex - \sum_{j=1}^k c_j P_j x\| < \varepsilon$. □

Remarks. 1. If $T(t)$ is a unitary representation in a Hilbert space, then the projections $P(\chi)$ are orthogonal.

2. The result for $G = \mathbb{Z}$ and T unitary was proved in **[O]**, and the identification of the limit was extended to weighted averages of contractions of a Hilbert space.

It is clear that if $\{A_n\}$ is a Følner sequence, then $d\mu_n = \lambda(A_n)^{-1}\mathbf{1}_{A_n} d\lambda$ is an ergodic sequence. However, for pointwise ergodic theorems for measure preserving group actions more properties are needed (for a discussion, see **[K**, pp. 223–227**]**). Such conditions were first discovered by Tempelman **[T-1]**, **[T-2]** (see also **[K]** and **[T-4]**), in the context of *amenable* groups. Weighted pointwise ergodic theorems for actions of a LCA group, using Besicovitch functions, were proved in **[T-3]**] (a paper which seems to have been unnoticed by more recent authors when dealing with actions of $\mathbb{Z}$ or N).

Definition. A *Tempelman sequence* is a Følner sequence $\{A_n\}$ satisfying the additional hypotheses

i) $A_n \subset A_{n+1}$, $n = 1, 2, \ldots$;

ii) There exists $K_1 < \infty$ with $\dfrac{\lambda^*(A_N A_n)}{\lambda(A_n)} \leq K_1$ for all N;

iii) There exists $K_2 < \infty$ with $\lambda^*(A_n^{-1} A_n) \leq K_2 \lambda(A_n)$, $n = 1, 2, \ldots$.
Here, λ^* denotes outer measure.

If $\{A_n\}$ is a sequence of subsets of G, we will write $B_r(\{A_n\})$ for $B_{r,\{\mu_n\}}$ where the sequence of measures $\{\mu_n\}$ is defined by $d\mu_n = \frac{1}{\lambda(A_n)}\mathbf{1}_{A_n} d\lambda$.

Tempelman's result, not superseded by subsequent papers, is

Theorem 3.4. *Let $\{A_n\}$ be a Tempelman sequence in a σ-compact* LCA *group, and let $\{\theta_t : t \in G\}$ be a measure preserving group action in a σ-finite measure space (Ω, Σ, m). Then for $f \in L_p(m)$, $1 < p < \infty$, we have that $\frac{1}{\lambda(A_n)} \int_{A_n} w(t) f(\theta_t x)\, d\lambda(t)$ converges a.e. for any $w \in B_q(\{A_n\})$, where, as usual, $\frac{1}{p} + \frac{1}{q} = 1$.*

Remarks. 1. Although stated in **[T-3]** for m finite, the proof works for the σ-finite case.

2. Tempelman considers a more general case, of "regular ergodic sequences" $\{\mu_n\}$, but we have stated the theorem as it specializes to our case to avoid unnecessary repetition of definitions.

3. L_p-norm convergence follows from Theorem 3.2 directly, so the limit is in $L_p(m)$.

4. For $1 < p \leq 2$ we have $B_q(\{A_n\}) \subset B_2(\{A_n\})$, and the limit can be identified by Theorem 3.3 (see also **[T-3]**).

Theorem 3.5. *Under the assumptions of Theorem 3.4, for $f \in L_1(m)$ we have:*

(i) $\frac{1}{\lambda(A_n)} \int_{A_n} w(t) f(\theta_t x)\, d\lambda(t)$ *converges a.e. for any* $w \in B_1(\{A_n\}) \cap L_\infty(G)$.

(ii) *The limit function is in* $L_1(m)$.

(iii) *If m is finite, we also have convergence in the L_1-norm.*

(iv) *For any* $\chi \in \hat{G}$, $P(\chi) f = \lim_n \frac{1}{\lambda(A_n)} \int_{A_n} \bar{\chi}(t) f(\theta_t x)\, d\lambda(t)$ *is a projection in* $L_1(m)$.

(v) *For* $w \in B_1(\{A_n\}) \cap L_\infty(G)$ *with Fourier coefficients* $c_w(\chi)$, *and* $f \in L_1(m)$ *we have L_1-norm convergence of* $\sum_{\chi \in \hat{G}} c_w(\chi) P(\bar{\chi}) f$, *and the sum is the limit in* (ii).

Proof. (i) Fix $w \in B_1(\{A_n\}) \cap L_\infty(G)$. Since $w(t)$ is bounded, Tempelman's maximal inequality **[K**, Theorem 4.2, p. 223] yields the maximal inequality for the operators $U_n f(x) = \frac{1}{\lambda(A_n)} \int_{A_n} w(t) f(\theta_t x)\, d\lambda(t)$ defined on $L_1(\Omega, m)$.

By Theorem 2.1, $w \in B_2(\{A_n\})$, so by Theorem 3.4 we have that $U_n f(x)$ converges a.e. for $f \in L_2(\Omega, m) \cap L_1(\Omega, m)$. Almost everywhere convergence now follows from the Banach principle.

(ii) Let $Uf(x) = \lim_{n\to\infty} \frac{1}{\lambda(A_n)} \int_{A_n} w(t) f(\theta_t x)\, d\lambda(t)$, which is defined (a.e.) for $f \in L_1(\Omega, m)$. Clearly U is linear. It is also bounded in L_1, since

$$\begin{aligned}
\int |Uf|\, dm &= \int \lim \frac{1}{\lambda(A_n)} \Big| \int_{A_n} w(t) f(\theta_t x)\, d\lambda(t) \Big|\, dm(x) \\
&\leq \liminf \frac{1}{\lambda(A_n)} \int \Big| \int_{A_n} w(t) f(\theta_t x)\, d\lambda(t) \Big|\, dm(x) \\
&\leq \liminf \frac{1}{\lambda(A_n)} \int \Big[\int_{A_n} |w(t) f(\theta_t x)|\, d\lambda(t) \Big]\, dm(x) \\
&\leq \liminf \|w\|_\infty \frac{1}{\lambda(A_n)} \int_{A_n} \int |f(\theta_t x)|\, dm(x)\, d\lambda(t) \\
&= \|w\|_\infty \|f\|_1 .
\end{aligned}$$

(iii) follows from the $L_2(m)$-norm convergence, since m is finite.

(iv) follows from the computations in (ii) for $w = \bar{\chi}$.

(v) Since $w \in B_2(\{A_n\})$, it follows from Theorem 2.7 that $w_k := \sum_{j=1}^{k} c_w(\chi_j) \chi_j$ converges to w in $B_1(\{A_n\})$-norm. By (iv) we have that (v) holds for trigonometric polynomials. Hence

$$\begin{aligned}\Big\| \sum_{j=1}^{k} c_w(\chi_j) P(\bar{\chi}_j) f - Uf \Big\| &= \Big\| \lim_n \frac{1}{\lambda(A_n)} \int_{A_n} [w_k(t) - w(t)] T(t) f \, d\lambda(t) \Big\| \\ &\leq \|f\|_1 \limsup_n \frac{1}{\lambda(A_n)} \int_{A_n} |w_k(t) - w(t)| \, d\lambda(t) \\ &= \|f\|_1 \, \|w_k - w\|_{B_1(\{A_n\})} \to_{k\to\infty} 0.\end{aligned}$$

Remarks.. 1. Tempelman **[T-3]** proved the theorem only for $w \in \mathrm{AP}(G)$.

2. A similar L_1 convergence result is proved in **[J-O]** for multi-parameter families of commuting Dunford–Schwartz operators (without identification of the limit).

3. If $w \in \mathrm{WAP}(G)$, then the limit function does not depend on the Tempelman sequence: For $1 < p < \infty$, we have norm convergence in Theorem 3.4, with the limit independent of $\{A_n\}$ by Theorem 3.1. Let $p = 1$. If $\{A_n\}$ and $\{A'_n\}$ are Tempelman sequences with limit operators U and U', these operators are bounded in L_1, and agree on the dense subspace $L_2(m) \cap L_1(m)$, so they are equal.

Acknowledgments. (1) Part of this research was carried out during the second author's visit to Ben-Gurion University. He wishes to thank Ben-Gurion University for its generous support and warm hospitality.

(2) The first author's research was partly supported by the Israel Ministry of Science.

(3) This research was completed while both authors visited Northwestern University, with partial support by its Emphasis Year in Probability program.

References

[B] Besicovitch, A. S., Almost Periodic Functions, Cambridge University Press, London, 1932 (reprinted Dover, 1954).

[Ba-O] Baxter, J.R. and Olsen, J. H., Weighted and subsequential ergodic theorems, Canad. J. Math 35 (1983), 145–166.

[B-L] Bellow, A. and Losert, V., The weighted pointwise ergodic theorem and the individual ergodic theorem along subsequences, Trans. Amer. Math. Soc., 288 (1985), 307–345.

[Bu] Burckel, R. B., Weakly Almost Periodic Functions on Semigroups, Gordon and Breach, New York 1970.

[D] Day, M.M., Amenable semi-groups, Illinois J. Math. 1 (1957), 509–544.

[E] Eberlein, W., Abstract ergodic theorems and weak almost periodic functions, Trans. Amer. Math. Soc. 67 (1949), 217–240.

[J-O] Jones, R. L. and Olsen, J., Multiparameter weighted ergodic theorems, Canad. J. Math. 46 (1994), 343–356.

[K] Krengel, U., Ergodic Theorems, de Gruyter Stud. Math. 6, de Gruyter, Berlin-New York 1985.

[Li-W] Lin, M., and Wittmann, R., Ergodic sequences of averages of group representations, Ergodic Theory Dynamical Systems 14 (1994), 181–196.

[Lo] Loomis, L. H., Abstract Harmonic Analysis, Van Nostrand, Princeton, 1953.

[O] Olsen, J., Calculation of the limit in the return times theorem for Dunford–Schwartz operators, in: Proc. Alexandria (Egypt) Conference on Ergodic Theory and its Connections with Harmonic Analysis, K. E. Petersen and I. Salama, eds., London Math. Soc. Lecture Note Ser. 205, pp. 359–367, Cambridge 1995.

[RN-1] Ryll-Nardzewski, C., On fixed points of semi-groups of endomorphisms of linear spaces, Proc. Fifth Berkeley Symp. Math. Stat. Prob. (1965/6) II(1), 55–61.

[RN-2] Ryll-Nardzewski, C., Topics in ergodic theory, in: Proceedings of the Winter School in Probability, Karpacz, Poland, Lecture Notes in Math. 472, pp. 131–156, Springer-Verlag, Berlin 1975.

[T-1] Tempelman, A., Ergodic theorems for general dynamical systems, Doklady Akad. Nauk SSSR 176 (1967), 790–793. English trans. Soviet Math. Dokl. 8 (1967), 1213–1216.

[T-2] Tempelman, A., Ergodic theorems for general dynamical systems, Trans. Moscow Math. Soc. 26 (1972), 94–132.

[T-3] Tempelman, A., Ergodic theorems for amplitude modulated homogeneous random fields, Litovsk. Mat. Sb. 14 (1974), 221–229 (in Russian). English trans. in Lithuanian Math. J. 14 (1974), 698–704.

[T-4] Tempelman, A., Ergodic Theorems for Group Actions: Informational and Thermodynamical Aspects, Kluwer, Dordrecht 1992.

Department of Mathematics and Computer Science
Ben-Gurion University of the Negev
Beer-Sheva, Israel
lin@black.bgu.ac.il

Department of Mathematics
North Dakota State University
Fargo, ND 58105, U.S.A.
Jolsen@Plains.NoDak.edu

A Remark on the Strong Sweeping Out Property

Viktor Losert

Let (X, Σ, m) denote a non-atomic probability space, $U_t: X \to X$ an aperiodic (measurable) flow of measure preserving transformations ($t \in \mathbb{R}$). Let W be a countable subset of $\mathbb{R}_+$. For ν a probability measure on W put

$$\nu_n f(x) = \sum_{w \in W} \nu_n(w) f(U_w x) .$$

The sequence of measures (ν_n) is said to have the *"strong sweeping out property"* if, given $\varepsilon > 0$, there is a set B with $m(B) < \varepsilon$ such that

$$\limsup_{n\to\infty} \nu_n \chi_B(x) = 1 \text{ a.e.} \quad \text{and} \quad \liminf_{n\to\infty} \nu_n \chi_B(x) = 0 \text{ a.e.}$$

There is a general "transfer principle" (**[A]** Theorem 2.2, see also **[C]** § 1 for the discrete case), saying that if this property holds for one flow (U_w), then it holds for any such flow as well.

The sequence (ν_n) of measures supported on W is said to be *dissipative* if for each finite set $A \subset W$ we have $\lim_{n\to\infty} \nu_n(A) = 0$.

Notation. For $A \subseteq \mathbb{R}^n$, $|A|$ denotes its Lebesgue measure, for $A \subseteq X$, χ_A denotes its indicator function, for $x \in \mathbb{R}$, $\{x\} = x - [x] \in [0, 1)$ denotes its fractional part.

Theorem. *Let $W = (w_k)$ be a finite union of real lacunary sequences. Then for any dissipative sequence (ν_n) of probability measures on W the "strong sweeping out property" holds.*

Lemma 1. *Let (w_k) be an increasing sequence of positive real numbers satisfying for some $t \in \mathbb{N}$, $\lambda > 1$ the estimate $w_{k+t}/w_k \geq \lambda$ for all $k \in \mathbb{N}$, (i.e. (w_k) is a union of t lacunary sequences). Choose a multiple h of t such that $\lambda^{h/t} \geq 2h + 2$ and define $\varepsilon = \frac{1}{4} \cdot \frac{1}{1+h\lambda^{2h}}$. Then, given a (non-degenerate) closed real interval I, there exists $k_0 \in \mathbb{N}$ such that for any given $c_k \in \mathbb{R}$ one can find $\theta \in I$ with $\{c_k + w_k\theta\} \in [\varepsilon, 1 - \varepsilon]$ for all $k \geq k_0$. (If $|I| \geq (1 - 2\varepsilon)/(w_1\lambda^{2h})$, one can take $k_0 = 1$).*

Proof. By adding some terms, if necessary, we may assume that $w_{k+1}/w_k < \lambda^2$ for all k. By dropping finitely many (initial) terms or reducing I, we may assume that $w_1\lambda^{2h} \cdot |I| = 1 - 2\varepsilon$. We will show below that there exists a closed subinterval I_1 of I with $w_{h+1}\lambda^{2h} \cdot |I_1| = 1 - 2\varepsilon$, such that $\{c_k + w_k\theta\} \in [\varepsilon, 1 - \varepsilon]$ for all $\theta \in I_1$ and $k = 1, \ldots, h$. Then we can iterate the construction, replacing next I by I_1 and dropping $w_1, \ldots, w_h$ from the sequence, and so on. We get a decreasing sequence of closed intervals $I_1, I_2, \ldots$ and any θ in the intersection has the desired property. I_1 is found as follows: for $1 \leq k \leq h$ put $G_k = \{x \in I: \{c_k + w_k x\} \notin [\varepsilon, 1-\varepsilon]\}$. Since by assumption $w_h \leq w_1 \lambda^{2h}$, we have $|w_k \cdot I| \leq |w_h \cdot I| \leq 1 - 2\varepsilon$, and it follows that $c_k + w_k G_k$ is an interval of length at most 2ε (observe that $(-\varepsilon, \varepsilon) + \mathbb{Z}$ consists of open intervals of distance $1 - 2\varepsilon$). Hence, G_k is an interval of length at most $2\varepsilon / w_k \leq 2\varepsilon / w_1$, relatively open in I. Put $G = \bigcup_{k=1}^{h} G_k$. It follows that $|G| \leq 2\varepsilon h / w_1$ and that $I \setminus G$ consists of at most $h+1$ closed intervals. Therefore, one of them will have length at least $|I \setminus G|/(h+1)$. If we can show that this is not smaller than $\frac{1-2\varepsilon}{w_{h+1}\lambda^{2h}}$, the proof will be finished. Now, $|I \setminus G| \geq \frac{1-2\varepsilon}{w_1 \lambda^{2h}} - \frac{2\varepsilon h}{w_1}$. Hence $w_1 \lambda^{2h} \cdot |I \setminus G| \geq 1 - 2\varepsilon - 2\varepsilon\, h\, \lambda^{2h} = 1 - \frac{1}{2} = \frac{1}{2}$. Since $\frac{1}{w_{h+1}} \leq \frac{1}{\lambda^{h/t} w_1} \leq \frac{1}{2(h+1) w_1}$, our claim follows. □

Remark. This is just a (slightly simplified and adapted) variant of the proofs in **[M]** and **[P]** on lacunary sequences. In the language of **[A]** this means that any finite union of lacunary sequences has the property $C(\alpha)$ with $\alpha = \varepsilon$ as above. I did not try to optimize the constant ε, e.g. in the case of lacunary sequences (i.e. $t = 1$) it is easy to see that for $\lambda > 2$, one can take $\varepsilon = \frac{1}{2} - \frac{1}{\lambda}$ (see e.g. **[A]** § 4). A special case of this result can be traced back further to **[E]**, proof of Theorem 14.

For the remaining part of the proof fix (ν_n) as described in the theorem and $\varepsilon > 0$ as in Lemma 1. Put $\delta = \frac{\varepsilon}{3}$ and $A = (\delta, 1-\delta)$. For $N \in \mathbb{N}$ and $(x_s), (\theta_s) \in \mathbb{R}^N$ write

$$S(\nu, N, (x_s), (\theta_s) = \sum_k \nu(w_k) \prod_{s=1}^{N} \chi_A(\{x_s + \theta_s w_k\}).$$

Lemma 2. *Fix $N \in \mathbb{N}$ and let I be a (non-degenerate) N-dimensional interval, $y_s \in \mathbb{R}$ $(s = 1, \ldots, N)$, $\gamma > 0$, $n_0 \in \mathbb{N}$. Then there exists $n \in \mathbb{N}$ such that $n > n_0$ and a (non-degenerate) subinterval J of I such that*

$$S(\nu_n, N, (x_s), (\theta_s)) > 1 - \gamma \text{ for all } (\theta_s) \in J \text{ and all } |x_s - y_s| < \delta.$$

Proof. By Lemma 1 (applied to each coordinate) there exists $(\theta_s) \in I$ and $k_0 \in \mathbb{N}$ such that $\{y_s + \theta_s w_k\} \in [\varepsilon, 1-\varepsilon] \subseteq (2\delta, 1-2\delta)$ for $k \geq k_0$, $s = 1, \ldots, N$. Since (ν_n) is dissipative, there exists $n \in \mathbb{N}$ $(n > n_0)$ such that $\nu_n(\{w_k : k < k_0\}) < \gamma/2$. Then there exists k_1 such that $\nu_n(\{w_k : k > k_1\}) < \gamma/2$. By continuity, one has even a subinterval J of I such that $\{y_s + \theta_s w_k\} \in (2\delta, 1-2\delta)$ for all $(\theta_s) \in J$, $k_0 \leq k \leq k_1$. Then $\{x_s + \theta_s w_k\} \in (\delta, 1-\delta)$ for $|x_s - y_s| < \delta$ and our claim follows, since $\sum_{k=k_0}^{k_1} \nu(w_k) > 1 - \gamma$. □

Lemma 3. *Let N, I, n_0, γ be as in Lemma 2. Then there exists a finite subset F of $\{n \in \mathbb{N} : n > n_0\}$ and a (non-degenerate) subinterval J of I such that*

$$\sup_{n \in F} S(\nu_n, N, (x_s), (\theta_s)) > 1 - \gamma \text{ for all } (\theta_s) \in J \text{ and all } (x_s) \in \mathbb{R}^N.$$

Proof. It is clearly sufficient to consider $x_s \in [0, 1]$. Choose $p \in \mathbb{N}$ such that $\frac{1}{p} < \delta$. Consider $y_s = \frac{r_s}{p}$ with $r_s \in \{0, \ldots, p-1\}$. Now apply Lemma 2 successively with all possible choices of such (y_s). □

Proof of the Theorem. We apply Lemma 3 sucessively for $N = 1, 2, \ldots$ with $\gamma = \frac{1}{N}$ to get closed intervals I_N $(= J)$ such that $I_1 \subseteq [0, 1], I_{N+1} \subseteq I_N \times [0, 1]$. Choose $(\theta_s)_{s=1}^{\infty}$ such that $(\theta_s)_{s=1}^{N} \in I_N$ for all N. Then

$$\limsup_{n\to\infty} S(\nu_n, N, (x_s), (\theta_s)) = 1 \text{ for all } N \in \mathbb{N} \text{ and all } (x_s) \in \mathbb{R}^N.$$

Consider $X = [0, 1]^{\mathbb{N}}$ with product measure and $U_w(x_s) = (\{x_s + w\,\theta_s\})$. Then for each $N \in \mathbb{N}$, $A_N = \{(x_s) \in X : x_s \in A \text{ for } s = 1, \ldots, N\}$ has the property that $|A_N| = (1-2\delta)^N$ and $\limsup_{n\to\infty} \nu_n(\chi_{A_N})(x_s) = 1$ everywhere. It follows from [*J*] Th. 1.3 and the transfer principle mentioned at the beginning that (ν_n) has the strong sweeping out property. □

Remark. 1) Clearly, a corresponding result holds in the discrete case, i.e. $W \subseteq \mathbb{N}$ and instead of a flow one has the powers τ^n of a single aperiodic, measure preserving transformation.

2) In **[B]** it was shown that the individual ergodic theorem with respect to Cesàro averages along lacunary sequences is not true for indicator functions of subintervals ($X = [0, 1]$, $\tau(x) = \{x + \theta\}$). In a one-dimensional setting, the construction that is used in the proof of the Theorem here was found in summer 1985. I used it to show that there exists θ for which the upper limit of the Cesàro averages is 1 a.e. (but, in general, one could not make the interval arbitrarily small, i.e. one had a property similar to "δ-sweeping out" of **[A]**). Combined with the results of Erdös–Taylor (**[E]**, Theorem 8), it resulted also that super-lacunary sequences (i.e. $w_{k+1}/w_k \to \infty$) are "strongly sweeping out". But only after this was generalized to arbitrary lacunary sequences in **[A]**, Cor. 1.11, I found out quite recently that the original technique can be generalized by considering product spaces (partly motivated by the approach in **[A]** Proposition 3.1).

References

[A] M. Akcoglu, A. Bellow, R. L. Jones, V. Losert, K. Reinhold-Larsson, M. Wierdl, The strong sweeping out property for lacunary sequences, Riemann sums convolution powers, and related matters, Ergodic Theory Dynam. Systems, to appear.

[B] A. Bellow, Sur la structure des suites Mauvaises Universelles en theorie Ergodique, C. R. Acad. Sci. Paris Sér. I Math. 294 (1982), 55–58.

[C] J.-P. Conze, Convergence des moyennes ergodiques pour des sous-suites, Bull. Soc. Math. France 35 (1973), 7–15.

[E] P. Erdös, S. J. Taylor, On the set of points of convergence of a lacunary trigonometric series, and the equidistribution properties of related sequences, Proc. London Math. Soc. (3) 7 (1957), 598–615.

[J] A. del Junco, J. Rosenblatt, Counter examples in ergodic theory and number theory, Math. Ann. 245 (1979), 185–197.

[M] B. de Mathan, Numbers contravening a condition in density modulo 1, Acta Math. Hungar. 36 (1980), 237–241.

[P] A. D. Pollington, On the density of sequence $\{n_k\xi\}$, Illinois J. Math. 23 (1979), 511–515.

Institut für Mathematik der Universität Wien
Strudlhofgasse 4,
A-1090 Wien, Austria

Ergodic Invertible Liftings

Dorothy Maharam

Abstract. A relatively simple construction is given to show that a measure-preserving ergodic tranformation on a Borel probability measure space can be lifted to an invertible ergodic measure-preserving transformation on a larger space.

1. Introduction

The present paper is a complement to **[3]**, in which a relatively simple construction was used to lift (under mild hypotheses) a given measurable surjection T, of a measure space S, to an invertible one, θ, on a larger space—a result originally due (with a complicated construction) to Rokhlin **[4]** (see also Silva **[5]**). However, unlike Rokhlin's, the simpler construction did not in general preserve ergodicity (or entropy). Here we show that, under somewhat stronger (but still reasonable) hypotheses, a modification of the simple construction in **[3]** produces an ergodic lifting when the given transformation T is ergodic. The main extra hypothesis we make is that T is finite-measure-preserving (an assumption made in **[4]** and **[5]**), and (as in **[4]** and **[5]**) we obtain a finite-measure-preserving ergodic invertible lifting Θ. Without the assumption that the ergodic transformation T preserves measure, Silva **[5]** has also treated the situation in which T satisfies a requirement ("μ-recurrent") a little stronger than "conservative," obtaining a conservative ergodic invertible lifting.

The present construction is simpler than that in **[4]** and **[5]**; essentially it consists, as in **[3]**, of multiplying S by the unit interval I. Here, however, we regard I as its measure-theoretic equivalent $P = \prod\{I_n : n \in \mathbb{Z}\}$, where each I_n is a copy of I, and combine the invertible lifting θ to $S \times I_0$ of T, given by **[3]**, with a shift on P, to produce an ergodic measure-preserving lifting Θ of T to $S \times P$. (This method, however, does not lend itself to preserving entropy, which was one of the achievements of **[4]** and **[5]**. But in many cases of interest the entropy of T will be infinite, and then so is that of Θ.)

The extra assumptions on T, over and above those in **[3]**, are:

(a) T is ergodic and preserves a finite measure μ (as in **[4]**), and

(b) μ has no atoms.

In fact, if μ has one or more atoms, the other assumptions imply that, apart from a null set, S consists of a finite number of atoms permuted cyclically by T.

In §5 we show that, in a significant special case, the present construction can be simplified further; the invertible lifting θ of **[3]** is not needed, and we can combine T

with a shift on a product of unit intervals directly. The detailed statement and proof are in §5 below.

Finally, the following questions arise naturally from the present work. First, if we drop the condition that T is measure-preserving (assuming T merely measurable and non-singular) the construction in §4 still applies unchanged to give an invertible lifting Θ of T. If now T is ergodic and conservative, need Θ be ergodic and conservative?

Second, if we allow the measure μ to be infinite (but σ-finite)—or (in effect) more generally if (as in **[3]**) we drop the condition that μ is preserved by T—the case in which μ has atoms is no longer trivial and deserves investigation. In particular, does every transformation on a countable (purely atomic) measure space admit an invertible lifting to a countable space ? preserving ergodicity ? and conservatism ?

2. Notation

The closed unit interval $[0, 1]$ is denoted by I, the Hilbert cube $\prod\{I_n : n \in \mathbb{Z}\}$ by P. Lebesgue measure on I is λ; product Lebesgue measure on P is $\bar{\lambda}$. We restrict attention to Borel sets throughout. The family of Borel subsets of a space X is written $\mathcal{B}(X)$. The identity map on a set X is i_X.

Since the transformations considered need not be one-to-one, some care is needed in formulating the basic definitions. We suppose given: a Polish space S, with a (Borel) probability measure μ defined on $\mathcal{B}(S)$, and a (Borel) measurable surjection T of S onto S, preserving μ; that is, for all $B \in \mathcal{B}(S)$, $T^{-1}(B) \in \mathcal{B}(S)$ and $\mu(T^{-1}(B)) = \mu(B)$. (It does not follow that $\mu(T(B)) = \mu(B)$, and $T(B)$ need not be Borel, though it will be analytic.) We say that a Borel set B is "invariant" provided $T^{-1}(B) = B$; this implies, but is not equivalent to, $T(B) = B$. We say B is "almost invariant" if the symmetric difference $T^{-1}(B)\Delta B$ has measure 0. We assume throughout that T is ergodic: that is, if B is invariant then $\mu(B) = 0$ or 1. Here "invariant" can be replaced by "almost invariant," since it is easily seen that if B is almost invariant it differs by a null set from an invariant set (*e.g.*, $\limsup_{n\to\infty} T^{-n}(B)$).

Theorem 1. *There exists a Borel set* $V \subset S$, *satisfying* $\mu(S \setminus V) = 0$ *and* $T(V) = V$, *such that the restriction* $T|V$ *has a lifting to an ergodic measure-preserving Borel bijection* Θ *("point-isometry") of* $(V \times P, \mathcal{B}(V \times P), \mu \times \bar{\lambda})$ *onto itself. Further, this lifting can be made "measure-preserving on fibres", in a sense explained at the end of* §2 *below.*

Here the statement that $\Theta : V \times P \to V \times P$ is a lifting of $T|V$ means that the diagram

$$\begin{array}{ccc} V \times P & \overset{\Theta}{\rightleftarrows} & V \times P \\ \pi\downarrow & & \downarrow\pi \\ V & \xrightarrow{\tau} & V \end{array}$$

commutes, where τ denotes $T|V$ and π denotes the projection map.

The first step in the proof is to apply **[3**, Th. 2], noting that the hypothesis of strong measurability of T is satisfied here because of **[3**, Prop. 2.4] and the fact that T is measure-preserving. This theorem provides the set V that we require here. It ensures $V \in \mathcal{B}(S)$, $T(V) = V$ and $\mu(S \setminus V) = 0$.[1] Further (*cf.* **[3**, 6.5]) V can be identified with the set M of an "adjusted planar model disintegration," induced by $\tau = T|V$—that is, the disintegration of $\mathcal{B}(V)$ with respect to $\tau^{-1}(\mathcal{B}(V))$. And V here is isomorphic to the base V' of this disintegration, under an isomorphism $\rho : V' \to V$, and we have $\tau = \rho \circ \pi_M$ where π_M is the map projecting M to the first coordinate. (Though $M = V$ and $V = V'$ in a sense, it is convenient to distinguish the roles played by V by different symbols.)

We use the notation $v' = \rho^{-1}(v)$ (where $v \in V = M$), so that $v' \in V'$, and define the "v-slice" of M to be

$$\tau^{-1}(v) = (\pi_M)^{-1}(v') = M_{v'},$$

the "M-slice over v'." The disintegration produces a Borel measure $\nu_{v'}$ on each such slice, such that, for all $A \in \mathcal{B}(V)$,

$$\mu(A) = \int_{V'} \nu_{v'}(A \cap M_{v'})\, d\mu'(v')$$

where μ' is the measure on $\mathcal{B}(V')$ given by

$$\mu'(E) = \mu(\pi_M^{-1}(E)).$$

The present decomposition has some additional properties. First, the "garbage set" G is empty and each $\nu_{v'}$ is a probability measure **[3**, Lemma 6.1]. Second, because τ is measure-preserving, ρ also preserves measure in the sense that, for all $E \in \mathcal{B}(V')$, $\mu'\left(\rho^{-1}(E)\right) = \mu(E)$. Because $\mu(V) = 1$ we have $\mu'(V') = 1$ also.

It is convenient to work with v $(= \rho(v'))$ rather than v'; we denote the measure $\nu_{v'}$, where $v' = \rho^{-1}(v)$, by ν^v. This is, of course, a Borel measure on $M_{v'} = \tau^{-1}(v)$. In this notation, because of the measure-preserving property of ρ, we have

$$\mu(A) = \int_V \nu^v(A \cap \tau^{-1}(v))\, d\mu(v).$$

We can now state the final "measure-preserving on fibres" property in Theorem 1 above: *for each* $v \in V$, Θ *takes* $\tau^{-1}(v) \times P$ *with measure* $\nu^v \times \bar{\lambda}$, *to* $\{v\} \times P$ *with measure* $\bar{\lambda}$.

1 In **[3**, p. 243, line 7] it was incorrectly stated that the null set $S \setminus V$ is invariant.

3. The first-stage lifting

In [3, Th. 2] an invertible (but not in general ergodic) lifting θ of τ $(= T|V)$ was constructed, constituting (since now T is measure-preserving) a measure-preserving Borel isomorphism of $V \times I$, with measure $\mu \times \lambda$, onto itself, making the diagram

$$\begin{array}{ccc} V \times I & \overset{\theta}{\underset{\leftarrow - -}{\longrightarrow}} & V \times I \\ \varpi \big\downarrow & & \big\downarrow \varpi \\ V & \underset{\tau}{\longrightarrow} & V \end{array}$$

commute; and further θ is "measure-preserving on fibres" in the sense that θ maps each $\tau^{-1}(v) \times I$, with measure $\nu^v \times \lambda$, isometrically onto $\{v\} \times I$, with measure λ.[2] As the diagram shows, the first coordinate of $\theta^{-1}(v, t)$ (where $v \in V$ and $t \in I$) belongs to $\tau^{-1}(v)$; we denote it by $\sigma(v, t)$. Also, the first coordinate of $\theta(w, y)$ (where $w \in V$ and $y \in I$) is $\tau(w)$; we denote the second coordinate by $\xi_w(y)$. For fixed w, ξ_w is a one-to-one map of I into I (because θ is one-to-one). (It is not *onto* I, in general.) We observe:

For fixed $v \in V$, the map $(w, y) \mapsto \xi_w(y)$ $(w \in \tau^{-1}(v),\ y \in I)$ is a measure-preserving Borel isomorphism of $\tau^{-1}(v) \times I$, with measure $\nu^v \times \lambda$, onto I with measure λ. Its inverse is the map $z \mapsto (\sigma(v, z), \xi^{-1}_{\sigma(v,z)}(z))$ $(z \in I)$. (1)

For these maps (if we identify I with $\{v\} \times I$) are just the restrictions of θ and θ^{-1} to $\tau^{-1}(v) \times I$ and $\{v\} \times I$ respectively.

The foregoing notation, restated for later reference, gives:

$$\theta(w, y) = (\tau(w), \xi_w(y)) \quad (w \in V,\ y \in I), \tag{2}$$

$$\theta^{-1}(v, x) = (\sigma(v, x), \xi_w^{-1}(x)) \quad (v \in V,\ x \in I,\ w = \sigma(v, x)). \tag{3}$$

Since θ is a measure-preserving Borel isomorphism (of $V \times I$ onto itself), the same is true for θ^n $(n \in \mathbb{Z})$. We further note:

$$\text{if } E \in \mathcal{B}(V),\ \theta^{-n}(E \times I) = \tau^{-n}(E) \times I \quad (n = 1, 2, \ldots). \tag{4}$$

4. The ergodic lifting

The underlying space will be $V \times P$, where $P = I^{\mathbb{Z}} = \prod_{n=-\infty}^{\infty} I_n$, with product measure $\mu \times \bar{\lambda}$, as in §2. We often regard $V \times P$ as $(V \times I_0) \times \prod_{n \neq 0} I_n$, and we write θ_0 for the map $\theta : V \times I \to V \times I$ with I identified with I_0. This map will be combined with the "backward shift" S on P, as follows.

---

2 I take this opportunity to correct some misprints in [3]: (a) at the end of the statement of [3, Th. 1, p. 231], the symbol ϖ^{-1} should be deleted; (b) in the diagrams on ibid. p. 243, I^{-1} should be I^-; and (c) in the second of these diagrams, the symbols ϖ and ϖ' have been interchanged.

Let $\bar{p} = (\ldots, p_{-1}, p_0, p_1, \ldots) \in P$, and put $S(\bar{p}) = \bar{q}$ where $q_n = p_{n+1}$ for all $n \in \mathbb{Z}$. Define $S^* : V \times P \to V \times P$ to be $i_V \times S$; that is,

$$S^*(v, \bar{p}) = (v, S(\bar{p})). \tag{1}$$

Define $\Phi_0 : V \times P \to V \times P$ by

$$\Phi_0 = \left(\theta_0^{-1} \text{ on } V \times I_0\right) \times \left(\text{identity on } \prod_{n \neq 0} I_n\right), \tag{2}$$

and put

$$\Phi = S^* \circ \Phi_0. \tag{3}$$

This is a measure-preserving Borel isomorphism of $V \times P$ onto $V \times P$, since the same is true of both S^* and Φ_0. Explicitly we have from 3.3, on writing $w = \sigma(v, p_0)$,

$$\Phi(v, \bar{p}) = (w, \ldots, p_{-1}, \xi_w^{-1}(p_0), p_1, p_2, p_3, \ldots) \tag{4}$$

where $\xi_w^{-1}(p_0)$ is the (-1)-place coordinate in P, and each p_n $(n \neq 0)$ is in the $(n-1)$st place.

We define $\Theta = \Phi^{-1} = (\Phi_0^{-1}) \circ (S^*)^{-1}$, and show that Θ has the desired properties. By construction, it is invertible and is a measure-preserving Borel isomorphism.

From 3.2 and 3.1, the Borel isomorphism θ_0 maps each $\tau^{-1}(v) \times I_0$ $(v \in V)$ onto $\{v\} \times I_0$ (a restatement of the fact that θ is a lifting of τ), taking the measure $\nu^v \times \lambda$ to λ. Thus (from 2 above) the Borel isomorphism Φ_0 maps each $\{v\} \times P$ onto $\tau^{-1}(v) \times P$, taking the measure $\bar{\lambda}$ to $\nu^v \times \bar{\lambda}$. Since the map $(S^*)^{-1}$ does not affect this, we see that Θ maps each $\tau^{-1}(v) \times P$ onto $\{v\} \times P$, taking the measure $\nu^v \times \bar{\lambda}$ to $\bar{\lambda}$, as promised at the end of §2.

In particular, the first coordinate of $\Theta(v, \bar{p})$ (where $v \in V$ and $\bar{p} \in P$) is $\tau(v)$, which verifies that Θ is indeed a lifting of $\tau = T|V$, as in Theorem 1 (§2). (It is not, in general, a lifting of θ; if it were, θ would itself be ergodic.)

All that remains is to show that Θ is ergodic—or, what comes to the same (since Θ is a measure-preserving bijection), that its inverse Φ is ergodic; and it is more convenient to work with Φ. The foregoing has shown

$$\Phi(\{v\} \times P) = \tau^{-1}(v) \times P \quad (v \in V),$$

from which it follows that

$$\Phi(E \times P) = \tau^{-1}(E) \times P \quad (E \subset V),$$

and hence that, for $n = 1, 2, \ldots$ and $E \subset V$,

$$\Phi^n(E \times P) = \tau^{-n}(E) \times P. \tag{5}$$

To prove ergodicity, we first consider the effect of Φ on certain "cylinder sets." For k, $\ell \in \mathbb{Z}$ with $k \leq \ell$, write $P(k, \ell)$ for the sub-product $\prod\{I_n : k \leq n \leq \ell\}$ of P, and $Q(k, \ell)$

for the "complementary" sub-product $\prod\{I_n : n \notin [k, \ell]\}$. We claim:

$$\begin{aligned}&\text{Given } H \in \mathcal{B}(V \times P(k,\ell)), \text{ there is some } K \in \mathcal{B}(V \times P(k-1,\ell-1))\\ &\text{such that } \Phi(H \times Q(k,\ell)) = K \times (k-1,\ell-1).\end{aligned} \tag{6}$$

It suffices to prove 6 when H is replaced by a singleton, say $\{h\}$; for if it is known that $\Phi(\{h\} \times Q(k,\ell)) = K_h \times Q(k-1,\ell-1)$ for some $K_h \subset V \times P(k-1,\ell-1)$ whenever $h \in V \times P(k,\ell)$, we put $K = \bigcup\{K_h : h \in H\}$ to obtain 6. (The requirement that K is Borel is satisfied because Φ preserves Borel sets.)

It further suffices to deal with the case $k = \ell$. For suppose it has been established that

$$\begin{aligned}&\text{for all } j \in \mathbb{Z} \text{ and } (v, p_j) \in V \times I_j, \text{ there is } A_{j-1} \subset V \times I_{j-1} \text{ such that}\\ &\Phi((v, p_j) \times Q(j,j)) = A_{j-1} \times Q(j-1, j-1).\end{aligned} \tag{7}$$

Then, given $h = (v, p_k, p_{k+1}, \ldots, p_\ell) \in V \times P(k,\ell)$, the desired K_h is given by $\bigcap\{A_{j-1} : k \leq j \leq \ell\}$.

To prove 7, consider two cases.

If $j \neq 0$, the left side of 7 can be written

$$\begin{aligned}&\Phi\big(\{v\} \times I_0 \times \{p_j\} \times \prod\{I_n : n \neq 0, j\}\big)\\ &\quad = S^*\big(\tau^{-1}(v) \times I_0 \times \{p_j\} \times \prod\{I_n : n \neq 0, j\}\big)\\ &\quad = \tau^{-1}(v) \times I_{-1} \times \{p_j \text{ in } (j-1)\text{st place}\} \times \prod\{I_n : n \neq -1, j-1\},\end{aligned}$$

of the desired form.

In the remaining case, $j = 0$ and the left side of 7 is

$$\begin{aligned}\Phi\big((v, p_0) \times \prod_{n\neq 0} I_n\big) &= S^*\big(\theta_0^{-1}(v, p_0) \times \prod_{n\neq 0} I_n\big)\\ &= S^*\big(\sigma^{-1}(v, p_0), \xi_w^{-1}(p_0) \times \prod_{n\neq 0} I_n\big) \text{ from 3.3}\\ (\text{where } w = \sigma(v, p_0)) \quad &= \{\sigma^{-1}(v, p_0)\} \times \{\xi_w^{-1}(p_0) \text{ in } (-1)\text{st place}\} \times \prod_{n\neq -1} I_n,\end{aligned}$$

again of the required form.

From 6 and induction over r, we obtain that, for $r = 1, 2, \ldots,$

$$\begin{aligned}&\text{given } H \in \mathcal{B}(V \times P(k,\ell)) \text{ there is } K_r \in \mathcal{B}(V \times P(k-r,\ell-r)) \text{ such}\\ &\text{that } \Phi^r(H \times Q(k,\ell)) = K_r \times Q(k-r,\ell-r).\end{aligned} \tag{8}$$

Next we need the following lemma, essentially known:

Lemma. *Given $A \in \mathcal{B}(V \times P)$ and $\epsilon > 0$, there exist a positive integer n and a set $H \in \mathcal{B}(V \times P(-n,n))$ such that the symmetric difference $A \Delta (H \times Q(-n,n))$ has $\mu \times \bar{\lambda}$-measure less than ϵ.*

Proof. Standard measure theory.

Now suppose, for a contradiction, that there is a nontrivial set $A \in \mathcal{B}(V \times P)$, invariant under Φ. Thus $A = \Phi^n(A)$ for all $n \in \mathbb{Z}$, and $0 < (\mu \times \bar{\lambda})(A) < 1$. Write

$$(\mu \times \bar{\lambda})(A) = a. \tag{9}$$

Fixing $\epsilon > 0$ for the present, we apply the Lemma to provide a positive integer n and a Borel subset H of $V \times P(-n, n)$ such that

$$(\mu \times \bar{\lambda})(A \Delta (H \times Q(-n, n))) = \eta^2 \text{ (say) } < \min\big(1, (\epsilon/7)^2\big). \tag{10}$$

For $k = 1, 2, \ldots$, write $\Phi^k(H \times Q(-n, n)) = H^k$. Because $\Phi^k(A) = A$ we have $(\mu \times \bar{\lambda})(A \Delta H^k) = \eta^2$; *a fortiori*,

$$(\mu \times \bar{\lambda})(A \setminus H^k) \leq \eta^2. \tag{11}$$

Now put $r = 2n + 2$, and consider the sets H^0, H^r. From 9 and 11 we have

$$(\mu \times \bar{\lambda})(H^0 \cap H^r) \geq a - 2\eta^2. \tag{12}$$

We use the notation: if $E \subset V \times P$, E_v denotes $\{p \in P : (v, p) \in E\}$. Define (for $k = 0$ or r) $M^k = \{v \in V : \bar{\lambda}(A \Delta H^k)_v \geq \eta\}$. By Fubini's theorem,

$$\int_V \bar{\lambda}(A \Delta H^k)_v \, d\mu(v) = (\mu \times \bar{\lambda})(A \Delta H^k) = \eta^2.$$

Hence $\mu(M^k) \leq \eta$. For all $v \in V \setminus (M^0 \cup M^r)$ we have $\bar{\lambda}(H^0_v) < \bar{\lambda}(A_v) + \eta$ and $\bar{\lambda}\left(H^r_v\right) < \bar{\lambda}\,(A_v) + \eta$. We also have

$$\bar{\lambda}(H^0_v \cap H^r_v) = \bar{\lambda}(H^0_v)\bar{\lambda}(H^r_v) \quad (v \in V), \tag{13}$$

because the sets H^0_v, H^r_v are "cylinders" on disjoint sets of co-ordinates; more precisely, $H^0_v =$ (subset of $P(-n, n)) \times Q(-n, n)$ and $H^r_v =$ (subset of $P(-n-r, n-r)) \times Q(-n-r, n-r)$, from 8. Thus $\bar{\lambda}(H^0 \cap H^r)_v < (\bar{\lambda}(A_v) + \eta)^2$ on $V \setminus \big(M^0 \cup M^r\big)$, showing (in view of 12) that

$$a - 2\eta^2 \leq (\mu \times \bar{\lambda})(H^0 \cap H^r) \leq \int_V (\bar{\lambda}(A_v) + \eta)^2 \, d\mu(v) + 2\eta.$$

Thus

$$\begin{aligned} \int_V \bar{\lambda}(A_v) d\mu(v) = a &\leq \int_V (\bar{\lambda}(A_v))^2 d\mu(v) + 4\eta + 3\eta^2 \\ &< \int_V (\bar{\lambda}(A_v))^2 d\mu(v) + \epsilon. \end{aligned}$$

Since $\epsilon > 0$ was arbitrary, this proves

$$\int_V \bar{\lambda}(A_v)(1 - \bar{\lambda}(A_v)) \, d\mu(v) \leq 0.$$

The integrand (being non-negative) must therefore be 0 for μ-almost all v. Put $E = \{v \in V : \bar{\lambda}(A_v) = 1\}$; thus, for λ-almost all $v \in V \setminus E$, $\bar{\lambda}\,(A_v) = 0$. We show

$$(\mu \times \bar{\lambda})((E \times P) \Delta A) = 0. \tag{14}$$

For, by Fubini's theorem again, the left side of 14 is

$$\int_V \bar{\lambda}((E \times P) \setminus A)_v d\mu(v) + \int_V \bar{\lambda}(A \setminus (E \times P))_v d\mu(v).$$

We split these integrals into $\int_E$ and $\int_{V \setminus E}$, and easily see that all 4 resulting integrals are zero because their integrands vanish almost everywhere.

Since Φ is measure-preserving and $\Phi^n(A) = A$, it follows from 14 that, for $n = 1, 2, \ldots$,

$$(\mu \times \bar{\lambda})(\Phi^n(E \times P)\Delta A) = 0.$$

From 5 this says $(\mu \times \bar{\lambda})((\tau^{-n}(E) \times P)\Delta A) = 0$, so that $(\mu \times \bar{\lambda})(\tau^{-n}(E) \times P) = (\mu \times \bar{\lambda})(A) = a$, and therefore $\mu(\tau^{-n}(E)) = a$.

Write $\bar{E} = \limsup_{n\to\infty} \tau^{-n}(E) = \bigcap_{n\geq 1} \bigcup_{k\geq 0} \tau^{-(n+k)}(E)$. It is a routine matter to deduce that $\tau^{-1}(\bar{E}) = \bar{E}$ and $\mu(\bar{E}) = a$. But this contradicts the assumption that τ (or equivalently T) is ergodic; and the theorem is proved.

5. A simpler construction in a special case

In some cases of interest, the measure space $(S, \mathcal{B}(S), \mu)$ and transformation T of §2 will have the further property that the measure algebra $\mathcal{E} = \mathcal{B}(S)/\mu$ (of Borel sets modulo null sets) is homogeneous of order $\aleph_0$ over the subalgebra $\mathcal{F} = T^{-1}(\mathcal{B}(S))/\mu$ of inverse elements (*cf.* **[1**, p. 301**]**). This is equivalent here (see **[2**, p. 144**]**) to saying that $\mathcal{E}$ has no non-zero elements of order 0 over $\mathcal{F}$—that is, there is no $B \in \mathcal{B}(S)$ of positive μ whose Borel subsets can all be written (modulo null sets) as $B \cap T^{-1}(C)$ where $C \in \mathcal{B}(S)$. Equivalently, there are no indecomposable elements (*cf.* **[1**, p. 295**]**); that is, given disjoint non-null Borel sets B_1, B_2 in S, there is always some inverse set $T^{-1}(C)$ ($C \in \mathcal{B}(S)$) that (mod null sets) contains one of B_1, B_2, and not the other. It is also equivalent to saying that, in the disintegration of $\mathcal{B}(S)$ over $T^{-1}(\mathcal{B}(S))$ there are (almost) no slice-measure atoms.

In this significant special case, there is a simpler construction lifting T to an ergodic invertible measure-preserving transformation, preserving fibre-measures, as in Theorem 1. Here we do not need the "first stage" invertible lifting θ of **[3]**; instead we combine T (or rather $\tau = T|V$) directly with a shift on $\prod_{n>1} I_n$, as follows.

We use the same set V as before, and again identify V with an "adjusted planar model" M in which the base V' is isomorphic to V under the measure-preserving isomorphism $\rho : V' \to V$, with $T|V = \tau = \rho \circ \pi_M$. Here (modulo null sets) $\mathcal{B}(V)$ is homogeneous of order $\aleph_0$ over the field of Borel cylinder sets on V', so that the slice-measures ν^v ($= \nu_{v'}$ where $v' = \rho^{-1}(v)$) have no atoms. Thus V ($= M$) becomes the unit square $I_1 \times I_2$, with $\mu =$ planar Lebesgue measure, I_2 (with measure $\mu' = \lambda$) corresponding to V', and all the "slice-measures" reducing to Lebesgue measure λ on I_1.

For $v \in V$, writing $v = (t_1, t_2) \in I_1 \times I_2$, we have

$$\tau^{-1}(v) = \rho^{-1}(t_1, t_2) \times I_2 \quad \text{and} \quad \tau(v) = \tau(t_1, t_2) = \rho(t_1) \in I_1 \times I_2.$$

We put $\rho(t_1) = (\rho_1(t_1), \rho_2(t_1))$.

The underlying space of the present lifting will be

$$V^* = V \times I^{\aleph_0} = (I_1 \times I_2) \times P'$$

where $P' = \prod\{I_n : n \geq 3\}$. We use Lebesgue measure on each I_n, and the appropriate product Lebesgue measures on the products.

Define $\tau^* : V^* \to V^*$ to be the composition of $\rho : I_1 \to I_1 \times I_2$ and the shift $I_2 \times P' \to P'$; that is,

$$\tau^*(t_1, t_2, \ldots, t_n, \ldots) = (\rho_1(t_1), \rho_2(t_1), t_2, \ldots, t_{n-1}, \ldots)$$

(with t_{n-1} in the nth place when $n \geq 3$).

Thus τ^* is a bijection, with inverse given by

$$(\tau^*)^{-1}(s_1, s_2, \ldots, s_n, \ldots) = (\rho^{-1}(s_1, s_2), s_3, \ldots, s_{n+1}, \ldots).$$

Clearly τ^* is measure-preserving and measure-preserving on fibres, since both ρ and the shift preserve these measures. And, on writing π_{12} for the projection of $\prod\{I_n : n \geq 1\}$ onto $I_1 \times I_2$, we have that the diagram

$$\begin{array}{ccc} V^* & \xrightarrow{\tau^*} & V^* \\ \pi_{12}\downarrow & & \downarrow\pi_{12} \\ V & \xrightarrow[\tau]{} & V \end{array}$$

commutes, verifying that τ^* is a lifting of τ. All that remains is to prove τ^* ergodic.

We use the following notation. For $H_n \in \mathcal{B}(I_1 \times \cdots \times I_n)$, the set $H_n \times \prod\{I_i : i > n\}$ is denoted by H_n^*, and we write $\mathcal{B}^*(I_1 \times \cdots \times I_n) = \{H_n^* : H_n \in \mathcal{B}(I_1 \times \cdots \times I_n)\}$. Since $\tau^{-1}(\mathcal{B}(I_1 \times I_2)) = \mathcal{B}(I_1) \times I_2$ it readily follows that

$$\text{if } H_1 \in \mathcal{B}(I_1) \text{ then } (\tau^*)^{-1}(H_1^*) \in \mathcal{B}^*(I_1), \tag{1}$$

and that, if $n \geq 2$ and $H_n \in \mathcal{B}(I_1 \times \cdots \times I_n)$ then

$$(\tau^*)^{-1}(H_n^*) \in \mathcal{B}^*(I_1 \times \cdots \times I_{n-1}). \tag{2}$$

From 1 and 2, by induction over k, we get:

$$\text{If } n > k \geq 1,\ (\tau^*)^{-k}(\mathcal{B}^*(I_1 \times \cdots \times I_n)) \subset \mathcal{B}^*(I_1 \times I_2 \times \cdots \times I_{n-k}) \tag{3}$$

and (the case $n = k$)

$$\text{if } k \geq 1,\ (\tau^*)^{-k}(\mathcal{B}^*(I_1 \times \cdots \times I_k)) \subset \mathcal{B}^*(I_1). \tag{4}$$

Now suppose, for a contradiction, that τ^* is not ergodic; thus there is some $A^* \in \mathcal{B}(I_1 \times I_2 \times \cdots)$ such that $(\tau^*)^{-1}(A^*) = A^*$ and $0 < \lambda^*(A^*) < 1$, where λ^* is product Lebesgue measure. Fixing $\epsilon > 0$ for the present, we obtain a positive integer n and a set $H_n \in \mathcal{B}(I_1 \times \cdots \times I_n)$ such that $\lambda^*(A^* \Delta H_n^*) < \epsilon$. (Compare the Lemma in §4.) Put $K^* = (\tau^*)^{-n}(H_n^*)$. By 4, $K^* \in \mathcal{B}^*(I_1)$ and (because τ^* is a measure-preserving bijection)

$$\lambda^*(A^* \Delta K^*) = \lambda^*(A^* \Delta H_n^*) < \epsilon.$$

Here K^* is of the form $K_1(\epsilon) \times \prod\{I_n : n \geq 2\}$ for some $K_1(\epsilon) \in \mathcal{B}(I_1)$.

Now we let ϵ vary over the sequence (say) $\epsilon_i = 2^{-i}$ $(i = 1, 2, \ldots)$, and put $L_1 = \limsup_{i\to\infty} K_1(\epsilon_i) = \bigcap_{i\geq 1} \bigcup_{j\geq 0} K_1(\epsilon_{i+j}) \in \mathcal{B}(I_1)$. It is easy to see that $\limsup_{i\to\infty}(K_1(\epsilon_i)^*) = L_1^*$ and that $\lambda^*(A^* \Delta L_1^*) = 0$, so that L_1^* is almost invariant under τ^*. Here $L_1^* = L_1 \times \prod_{n=2}^{\infty} I_n$. It follows that $L_1 \times I_2$ is almost invariant under τ. But $\mu(L_1 \times I_2) = \lambda^*(L_1^*) = \lambda^*(A^*)$, and this contradicts the assumption that τ is ergodic.

References

[1] D. Maharam, The representation of abstract measure functions, Trans. Amer. Math. Soc. 65 (1949), 279–330.

[2] D. Maharam, Decompositions of measure algebras and spaces, Trans. Amer. Math. Soc. 69 (1950), 142–160.

[3] D. Maharam, Invertible liftings of measurable transformations, Israel J. Math. 73 (1991), 225–245.

[4] V. A. Rokhlin, Exact endomorphisms of a Lebesgue space, Amer. Math. Soc. Transl. (2) 39, (1964) 1–36.

[5] C. E. Silva, On μ-recurrent nonsingular endomorphisms, Israel J. Math. 61 (1988), 1–13.

Department of Mathematics
Northeastern University
Boston, MA 02115, U.S.A.

The Uniform Ergodic Theorem for Dynamical Systems

Goran Peškir and Michel Weber*

Abstract. Necessary and sufficient conditions are given for the uniform convergence over an arbitrary index set in von Neumann's mean and Birkhoff's pointwise ergodic theorem. Three different types of conditions already known from probability theory are investigated. Firstly it is shown that the property of being eventually totally bounded in the mean is necessary and sufficient. This condition involves as a particular case Blum–DeHardt's theorem which offers the best known sufficient condition for the uniform law of large numbers in the independent case. Secondly it is shown that eventual tightness is necessary and sufficient. In this way a link with a weak convergence is obtained. Finally it is shown that the existence of some particular totally bounded pseudo-metrics is necessary and sufficient. The conditions derived are of Lipschitz type, while the method of proof relies upon a result of independent interest, called the uniform ergodic lemma. This result considerably extends Hopf–Yosida–Kakutani's maximal ergodic lemma to a form more suitable for examinations of the uniform convergence under consideration. From this lemma an inequality is also derived which extends the classical (weak) maximal ergodic inequality to the uniform case. In addition, a uniform approximation by means of a dense family of maps satisfying the uniform ergodic theorem in a trivial way is investigated, and a particular result of this type is established. This approach is in the spirit of the classical Hilbert space method for the mean ergodic theorem of von Neumann, and therefore from the ergodic theory point of view it could be seen as the natural one. After this, a simple characterization is obtained for the uniform convergence of moving averages. Finally, a counter-example is constructed for a symmetrization inequality in the stationary ergodic case. This inequality is known to be of vital importance to support the Vapnik–Chervonenkis random entropy approach in the independent case. Further developments in this direction are indicated.

1. Introduction

What is nowadays called the *uniform ergodic theorem* goes back to 1941 when Yosida and Kakutani published the study **[52]**. In this study they obtained conditions[1] on a bounded linear operator T in a Banach space B which are sufficient to provide that the averages

---

* Research partially supported by Danish Natural Science Research Council and partially by Danish Research Academy

1 If T is power bounded and quasi–compact, then (1.1) holds true (see **[25]** p. 86–92).

$n^{-1}\sum_{j=0}^{n-1} T^j$ converge in the uniform operator topology to an operator P in B:

$$\left\| \frac{1}{n} \sum_{j=0}^{n-1} T^j - P \right\| \to 0 \tag{1.1}$$

as $n \to \infty$. In this context it should be recalled that the operator T satisfying (1.1) is called *uniformly ergodic* (see **[25]** p. 86). Their work generalizes previously established results of Fréchet and Visser, and in particular of Krylov and Bogolioubov whose concept of *compact operator* is shown to be of vital importance in this direction. Here we find it convenient to recall the words of Yosida and Kakutani which describe how the uniform ergodic theorem has been discovered. They wrote (see **[52]** p. 195):

> ... This condition[2] was investigated by J. L. Doob[3] without being noticed that the linear operation T becomes weakly completely continuous[4] under this condition. We have once[5] treated this case of the condition of J. L. Doob as an application of the mean ergodic theorem. But, on looking precisely into the detail of the fact, we have found that, in this case, the linear operation T satisfies even the (in some sense stronger) condition[6] of N. Kryloff–N. Bogolioboff, and that the uniform ergodic theorem is true in this case. (Indeed, the condition of N. Kryloff–N. Bogolioboff follows from the condition[7] of W. Doeblin, which is weaker than that of J. L. Doob.)
> ...

In the rest of the Yosida–Kakutani study the result has been well applied to some problems in the theory of Markov processes. A number of studies have followed. Among those we point out **[1]**, **[2]**, **[4]**, **[5]**, **[14]**, **[19]**, **[23]**[8], **[24]**, **[30]**[9], **[31]**, **[32]**[10], **[33]**, **[34]**, **[43]**, as well as the presentation of the Yosida–Kakutani result in **[13]** (p. 708–717) and **[25]** (p. 86–94) where further references can be found. Of the particular Banach spaces that were under consideration, it seems that the case of the Banach spaces $C(S)$ and $L_1(S, \Sigma, \mu)$ in the notation from **[13]** has been studied in more detail (see **[1]**, **[2]**,

---

2 A uniform integrability condition which characterizes weak compactness of the (integral) operator under consideration.

3 It has been done in **[10]**.

4 It means *weakly compact* in today's language.

5 It was in **[50]**.

6 It is *quasi-compactness* (in today's language) which was introduced in **[26]** and **[27]**.

7 There exist $d \geq 1$ and $\varepsilon, \delta > 0$ such that $\mu(A) < \delta$ implies $P^{(d)}(t, A) \leq 1 - \varepsilon$ for all t. (Here $P(t, A)$ is a transition probability.) For more details see Doeblin's paper **[9]**.

8 If T is a conservative and ergodic positive contraction of L^1, then it is uniformly ergodic if and only if it is quasi-compact.

9 If $\|T^n\|/n \to 0$, then T is uniformly ergodic if and only if $(I - T)B$ is closed.

10 If T is a positive operator on a Banach lattice with $\|T^n\|/n \to 0$ then T is quasi-compact if (and only if) the averages of its iterates converge uniformly to a finite-dimensional projection.

[4], **[13]**, **[23]**, **[34]**). Less seems to be known in this context about some of the other spaces.

On the other hand in probability theory nowadays, we have the fundamental *Glivenko–Cantelli theorem* which goes back to 1933 when papers **[6]** and **[18]** were published. A large number of studies were followed. As general references we point out **[11]**, **[15]**, **[41]** and **[42]** where additional references can be found. In the recent years various extensions of the classical Glivenko–Cantelli theorem have appeared. They could be commonly called *uniform laws of large numbers*. The main problem under considerations in this context, roughly speaking, may be stated as follows. Given a sequence of independent and identically distributed random variables $\{\xi_j \mid j \geq 1\}$ with values in a space S, and a map f from $S \times \Theta$ into $\mathbb{R}$, determine conditions under which the uniform convergence is valid:

$$\sup_{\theta \in \Theta} \Big| \frac{1}{n} \sum_{j=1}^{n} f(\xi_j, \theta) - M(\theta) \Big| \to 0 \tag{1.2}$$

as $n \to \infty$, where $M(\theta)$ is the average of $f(\xi_1, \theta)$ for $\theta \in \Theta$, and the convergence in (1.2) is one of the standard probabilistic ones. It seems apparent at the present time that two distinct approaches towards solution for this problem have emerged. The first one offers Lipschitz type conditions and relies upon *Blum–DeHardt's law of large numbers* that is established in **[3]** and **[7]**. This law offers presently the best known sufficient condition for (1.2) to be valid. Subsequently, a necessary and sufficient condition for (1.2), called *eventually total boundedness in the mean*, is obtained in **[20]**, and although not explicitly recorded there, this condition involves Blum–DeHardt's law of large numbers as a particular case. That result has been recently shown to be valid in the stationary ergodic case as well (see **[37]**). This fact clearly indicates that further extensions of this approach into ergodic theory might be possible. (For a related work we refer the reader to **[21]**, and in particular to **[36]** where it is shown that *Hardy's regular convergence* of certain means is necessary and sufficient for the most useful consequence of (1.2) in statistics.) The second approach has been developed by *Vapnik and Chervonenkis* through the fundamental studies **[45]** and **[46]**. It offers combinatorial conditions in terms of random entropy numbers. In this direction we point out papers **[17]** and **[53]** as well as the presentation in **[29]** (p. 394–420). Let us also mention that somewhat different necessary and sufficient conditions for (1.2) are obtained in **[44]**.

It is the purpose of the present paper to point out the closeness of problems (1.1) and (1.2), and to obtain their unification through a single approach. Let us for this reason look more closely at (1.1) and (1.2) itself. First it should be noted that (1.1) does not provide any application to (1.2). Moreover, if $B = L^1(\mu)$ and T is the composition with a measure-preserving transformation, then T satisfies (1.1) if and only if T is periodic with bounded period. Similarly, it seems that having any particular version of (1.2), the best we can deduce in the direction of (1.1) is the mean ergodic theorem, that is the pointwise convergence on B of the averages from (1.1). This should follow from uniform integrability of averages along the lines of **[11]** (p. 42–43), at least for T being the composition with a stationary ergodic shift in a countable product of copies of a separable Banach space, and with Θ being a norming subset of the unit ball in the dual space (see also **[11]** p. 3). Therefore it seems to be of interest to reconsider the uniform ergodic theorem (1.1) as follows. We suppose that T and P are maps on some space B (which is (are) to be specified from the character of the objects being involved).

Prime Problem. *Determine conditions which are (necessary and) sufficient for the convergence to be valid:*

$$\frac{1}{n}\sum_{j=0}^{n-1} T^j(f_\theta) - P(f_\theta) \to 0 \quad \textit{uniformly in} \quad \theta \in \Theta$$

as $n \to \infty$, *where* f_θ *are with values in* B *for* $\theta \in \Theta$.

Dual Problem. *Determine conditions which are (necessary and) sufficient for the convergence to be valid:*

$$\frac{1}{n}\sum_{j=0}^{n-1} f_\theta(T^j) - f_\theta(P) \to 0 \quad \textit{uniformly in} \quad \theta \in \Theta$$

as $n \to \infty$, *where* f_θ *are with arguments in* B *for* $\theta \in \Theta$.

These two problems seem to be of different nature, but each is interesting from its own standpoint. For instance, taking for Θ in the prime problem the unit ball in a Banach space B and norm convergence, we obtain the uniform ergodic theorem (1.1). Similarly, taking for Θ in the dual problem the unit ball in its dual space B^* and weak*-convergence, we obtain the mean ergodic theorem for T in B. It should be noted here that in the convergence statements in the prime problem and in the dual problem we did not specify which convergence takes the place. We take this liberty to clarify the main ideas of unification, and at the same time to leave various possibilities open. For instance, if B is a function space, then the expression on the left-hand side in the prime problem could define a function with values in a normed space. Thus the question as to which convergence takes the place in the prime problem depends on the norm we consider, due to the fact that in this case instead of the function norm we can consider a norm in the range space of the function space as well. The same remark might be directed to the dual problem. Moreover, replacing the unit ball of a Banach space B in the prime problem by a smaller family of vectors (that may be of some interest) allows the uniform ergodic theorem (1.1) to become weaker and at the same time more easily established (see Example 4.6). In this way the anomaly with measure-preserving transformations stated above could be avoided. Finally, it is obvious that these two problems become mutually equivalent in the case where the operator T is induced by a measure-preserving transformation (at least when this transformation is invertible), and therefore they involve (1.2) as a particular case. This is precisely the case which is mainly considered and investigated in the paper.

We shall say just a few words about the organization of the paper, and rather concentrate to the results in a straightforward way. In the next section we prove a result which is of a considerable interest in itself. We call it the *uniform ergodic lemma* (for positive contractions), and to the best of our knowledge this sort of result has not been studied or recorded previously. Using this result in the third section we offer solutions for the prime problem and dual problem in the case where the operator T is induced by a measure-preserving transformation. In this context it should be recalled that the pointwise ergodic theorem of Birkhoff is proved by using the maximal ergodic lemma, while now the uniform ergodic theorem for dynamical systems is proved by using the uniform ergodic lemma. We think that this fact is by itself of theoretical interest. In the last section we present another approach towards the uniform ergodic theorem for dynamical systems (which is in the spirit of the classical Hilbert space method for the mean ergodic theorem of von Neumann) and various examples.

2. The uniform ergodic lemma

The proof of the forthcoming uniform ergodic theorem in the next section relies upon a fact of independent interest that is presented in Theorem 2.1 below. This result is instructive to be compared with Hopf–Yosida–Kakutani's maximal ergodic lemma (see **[25]** p. 8). In Corollary 2.2 below we present its consequence which should be compared with the maximal ergodic inequality (see **[16]** p. 24, 29). We begin by introducing the notation and recalling some facts on the non-measurable calculus needed in the sequel.

Given a linear operator T in $L^p(\mu)$ and $f \in L^p(\mu)$ for some $1 \le p < \infty$, we denote:

$$S_n(f) = \sum_{j=0}^{n-1} T^j f \tag{2.1}$$

$$M_n(f) = \max_{1 \le j \le n} S_j(f) \tag{2.2}$$

$$R_n(f) = \max_{1 \le j \le n} \frac{S_j(f)}{j} \tag{2.3}$$

for $n \ge 1$.

We shall restrict our attention to the case where the underlying measure space $(X, \mathcal{A}, \mu)$ is σ-finite, but we remark that further extensions are possible. The symbols μ^* and μ_* denote the *outer μ-measure* and the *inner μ-measure* respectively. The *upper μ-integral* of an arbitrary function f from X into $\bar{\mathbb{R}}$ is defined as follows $\int^* f\, d\mu = \inf\{\int g\, d\mu \mid g \in L^1(\mu),\ f \le g\}$, with the convention $\inf \emptyset = +\infty$. The *lower μ-integral* of an arbitrary function f from X into $\bar{\mathbb{R}}$ is defined as follows $\int_* f\, d\mu = \sup\{\int g\, d\mu \mid g \in L^1(\mu),\ g \le f\}$, with the convention $\sup \emptyset = -\infty$. We denote by f^* the *upper μ-envelope* of f. This means that f^* is an $\mathcal{A}$-measurable function from X into $\bar{\mathbb{R}}$ satisfying $f \le f^*$ and if g is another $\mathcal{A}$-measurable function from X into $\bar{\mathbb{R}}$ satisfying $f \le g$ μ-a.e., then $f^* \le g$ μ-a.e. We denote by f_* the *lower μ-envelope* of f. This means that f_* is an $\mathcal{A}$-measurable function from X into $\bar{\mathbb{R}}$ satisfying $f_* \le f$ and if g is another $\mathcal{A}$-measurable function from X into $\bar{\mathbb{R}}$ satisfying $g \le f$ μ-a.e., then $g \le f_*$ μ-a.e. It should be noted that such envelopes exist under the assumption that μ is σ-finite. It is well-known that we have $\int^* f\,d\mu = \int f^*\,d\mu$, whenever the integral on the right-hand side exists in $\bar{\mathbb{R}}$, and $\int^* f\,d\mu = \infty$ otherwise. Similarly we have $\int_* f\,d\mu = \int f_*\,d\mu$, whenever the integral on the right-hand side exists in $\bar{\mathbb{R}}$, and $\int_* f\,d\mu = -\infty$ otherwise. For more information in this direction we refer to **[35]**. To conclude the preliminary part of the section, we clarify that $\int_A^* f\,d\mu$ stands for $\int^* f \cdot 1_A\,d\mu$, whenever $A \subset X$ is an arbitrary set and $f\colon X \to \bar{\mathbb{R}}$ is an arbitrary function.

Theorem 2.1 (The uniform ergodic lemma). *Let T be a positive contraction in $L^1(\mu)$, and let $\{f_\theta \mid \theta \in \Theta\}$ be a family of functions from $L^1(\mu)$. Let us denote*

$$A_n = \{\sup_{\theta\in\Theta} M_n(f_\theta) > 0\}$$

and

$$B_n = \{(\sup_{\theta\in\Theta} M_n(f_\theta))^* > 0\}$$

for all $n \geq 1$. Then we have:

$$\int_{A_n}^* \sup_{\theta\in\Theta} f_\theta\,d\mu \geq 0 \tag{2.4}$$

$$\int_{B_n}^* \sup_{\theta\in\Theta} f_\theta\,d\mu \geq 0 \tag{2.5}$$

for all $n \geq 1$.

Proof. We shall first assume that $\sup_{\theta\in\Theta} f_\theta \leq g$ for some $g \in L^1(\mu)$. In this case we have $(\sup_{\theta\in\Theta} f_\theta)^* \in L^1(\mu)$. Let $n \geq 1$ be given and fixed. By monotonicity of T we get:

$$S_j(f_\theta) = f_\theta + TS_{j-1}(f_\theta) \leq f_\theta + T((M_n(f_\theta))^+) \tag{2.6}$$

for all $j = 2, \ldots, n+1$, and all $\theta \in \Theta$. Moreover, since $T((M_n(f_\theta))^+) \geq 0$ for all $\theta \in \Theta$, we see that (2.6) is valid for $j = 1$ as well. Hence we find:

$$M_n(f_\theta) \leq f_\theta + T((M_n(f_\theta))^+)$$

for all $\theta \in \Theta$. Taking supremum over all $\theta \in \Theta$ we obtain:

$$\sup_{\theta\in\Theta} M_n(f_\theta) \leq \sup_{\theta\in\Theta} f_\theta + \sup_{\theta\in\Theta} T((M_n(f_\theta))^+). \tag{2.7}$$

Since $f_\theta \leq g$ for all $\theta \in \Theta$, then we have $\sup_{\theta\in\Theta}(M_n(f_\theta))^+ \leq (M_n(g))^+$. Hence we see that $(\sup_{\theta\in\Theta}(M_n(f_\theta))^+)^* \in L^1(\mu)$. Therefore by (2.7) and monotonicity of T we get:

$$\sup_{\theta\in\Theta} M_n(f_\theta) \leq \sup_{\theta\in\Theta} f_\theta + T((\sup_{\theta\in\Theta}(M_n(f_\theta))^+)^*). \tag{2.8}$$

Multiplying both sides by 1_{A_n} we obtain:

$$\begin{aligned}\sup_{\theta\in\Theta}(M_n(f_\theta))^+ &= (\sup_{\theta\in\Theta} M_n(f_\theta))^+\\ &= \sup_{\theta\in\Theta} M_n(f_\theta)\cdot 1_{A_n}\\ &\le \sup_{\theta\in\Theta} f_\theta \cdot 1_{A_n} + T((\sup_{\theta\in\Theta}(M_n(f_\theta))^+)^*)\cdot 1_{A_n}\\ &\le \sup_{\theta\in\Theta} f_\theta \cdot 1_{A_n} + T((\sup_{\theta\in\Theta}(M_n(f_\theta))^+)^*).\end{aligned}$$

Integrating both sides we get:

$$\int^* \sup_{\theta\in\Theta}(M_n(f_\theta))^+\,d\mu \le \int^*_{A_n} \sup_{\theta\in\Theta} f_\theta\,d\mu + \int T((\sup_{\theta\in\Theta}(M_n(f_\theta))^+)^*)\,d\mu.$$

Finally we may conclude:

$$\int^* \sup_{\theta\in\Theta}(M_n(f_\theta))^+ d\mu - \int T((\sup_{\theta\in\Theta}(M_n(f_\theta))^+)^*)\,d\mu \le \int^*_{A_n} \sup_{\theta\in\Theta} f_\theta\,d\mu$$

and the proof of (2.4) follows from contractibility of T. In addition from (2.8) we get:

$$(\sup_{\theta\in\Theta} M_n(f_\theta))^* \le (\sup_{\theta\in\Theta} f_\theta)^* + T((\sup_{\theta\in\Theta}(M_n(f_\theta))^+)^*).$$

Multiplying both sides by 1_{B_n} we obtain:

$$\begin{aligned}(\sup_{\theta\in\Theta} M_n(f_\theta))^+)^* &= ((\sup_{\theta\in\Theta} M_n(f_\theta))^+)^*\\ &= ((\sup_{\theta\in\Theta} M_n(f_\theta))^*)^+\\ &= (\sup_{\theta\in\Theta} M_n(f_\theta))^*\cdot 1_{B_n}\\ &\le (\sup_{\theta\in\Theta} f_\theta)^*\cdot 1_{B_n} + T((\sup_{\theta\in\Theta}(M_n(f_\theta))^+)^*)\cdot 1_{B_n}\\ &\le (\sup_{\theta\in\Theta} f_\theta)^*\cdot 1_{B_n} + T((\sup_{\theta\in\Theta}(M_n(f_\theta))^+)^*).\end{aligned}$$

Integrating both sides we get:

$$\int^* \sup_{\theta\in\Theta}(M_n(f_\theta))^+\,d\mu \le \int^*_{B_n} \sup_{\theta\in\Theta} f_\theta\,d\mu + \int T((\sup_{\theta\in\Theta}(M_n(f_\theta))^+)^*)\,d\mu.$$

Finally, as above, we may conclude:

$$\int^* \sup_{\theta\in\Theta}(M_n(f_\theta))^+ d\mu - \int T((\sup_{\theta\in\Theta}(M_n(f_\theta))^+)^*)\,d\mu \le \int^*_{B_n} \sup_{\theta\in\Theta} f_\theta\,d\mu$$

and the proof of (2.5) follows from contractibility of T.

Next suppose that there is no $g \in L^1(\mu)$ satisfying $\sup_{\theta\in\Theta} f_\theta \le g$. In this case we have $\int^* \sup_{\theta\in\Theta} f_\theta\,d\mu = +\infty$. Let $n \ge 1$ be given and fixed, then by subadditivity we get:

$$\int^* \sup_{\theta\in\Theta} f_\theta \, d\mu \le \int^*_{C_n} \sup_{\theta\in\Theta} f_\theta \, d\mu + \int^*_{C_n^c} \sup_{\theta\in\Theta} f_\theta \, d\mu$$

with C_n equal to either A_n or B_n. However on the set C_n^c in both cases we evidently have $\sup_{\theta\in\Theta} f_\theta \le \sup_{\theta\in\Theta} M_n(f_\theta) \le 0$. Therefore $\int^*_{C_n} \sup_{\theta\in\Theta} f_\theta \, d\mu = +\infty$, and the proof of theorem is complete. □

Corollary 2.2 (The uniform ergodic inequality). *Under the hypotheses of Theorem 2.1 suppose moreover that μ is finite, and that $T(\mathbf{1}) = \mathbf{1}$. Let us denote*

$$A_{n,\alpha} = \{\sup_{\theta\in\Theta} R_n(f_\theta) > \alpha\}$$

and

$$B_{n,\alpha} = \{(\sup_{\theta\in\Theta} R_n(f_\theta))^* > \alpha\}$$

for all $n \ge 1$ and all $\alpha > 0$. Then we have:

$$\int^*_{A_{n,\alpha}} \sup_{\theta\in\Theta} (f_\theta - \alpha) \, d\mu \ge 0 \tag{2.9}$$

$$\mu_*\{\sup_{\theta\in\Theta} R_n(f_\theta) > \alpha\} \le \frac{1}{\alpha} \int^*_{A_{n,\alpha}} \sup_{\theta\in\Theta} f_\theta \, d\mu \tag{2.10}$$

$$\int^*_{B_{n,\alpha}} \sup_{\theta\in\Theta} (f_\theta - \alpha) \, d\mu \ge 0 \tag{2.11}$$

$$\mu^*\{\sup_{\theta\in\Theta} R_n(f_\theta) > \alpha\} \le \frac{1}{\alpha} \int^*_{B_{n,\alpha}} \sup_{\theta\in\Theta} f_\theta \, d\mu \tag{2.12}$$

for all $n \ge 1$ and all $\alpha > 0$.

Proof. Let $n \ge 1$ and $\alpha > 0$ be given and fixed. Consider $g_\theta = f_\theta - \alpha$ for $\theta \in \Theta$. Since $T(\mathbf{1}) = \mathbf{1}$, it is easily verified that $A_{n,\alpha} = \{\sup_{\theta\in\Theta} M_n(g_\theta) > 0\}$ and therefore $B_{n,\alpha} = \{(\sup_{\theta\in\Theta} M_n(g_\theta))^* > 0\}$. Hence by subadditivity and Theorem 2.1 we obtain:

$$\int^*_{C_{n,\alpha}} \sup_{\theta\in\Theta} f_\theta \, d\mu + \int^*_{C_{n,\alpha}} (-\alpha) \, d\mu \ge \int^*_{C_{n,\alpha}} \sup_{\theta\in\Theta} g_\theta \, d\mu \ge 0$$

with $C_{n,\alpha}$ equal to either $A_{n,\alpha}$ or $B_{n,\alpha}$. Thus the proof follows straightforward from the facts that $\int^*_{A_{n,\alpha}} (-\alpha) \, d\mu = -\alpha \cdot \mu_*(A_{n,\alpha})$ and $\int^*_{B_{n,\alpha}} (-\alpha) \, d\mu = -\alpha \cdot \mu_*(B_{n,\alpha})$. □

Remark 2.3. Under the hypotheses of Theorem 2.1 and Corollary 2.2 respectively let us suppose that $(\Theta, \mathcal{B})$ is a *Blackwell space*[11] (see **[22]**). We remark that any *analytic*

11 A measurable space $(\Theta, \mathcal{B})$ is called Blackwell, if $f(\Theta)$ is analytic (see below) for every measurable $f: \Theta \to \mathbb{R}$.

space[12], and thus any *polish space*[13], is a Blackwell space. Suppose moreover that the map $(x, \theta) \mapsto f_\theta(x)$ from $X \times \Theta$ into $\bar{\mathbb{R}}$ is measurable with respect to the product σ-algebra $\mathcal{A} \times \mathcal{B}$ and Borel σ-algebra $\mathcal{B}(\bar{\mathbb{R}})$. Then by the *projection theorem*[14] (see **[22]**) we may conclude that the map $x \mapsto \sup_{\theta \in \Theta} f_\theta(x)$ is *universally $\mathcal{A}$-measurable* from X into $\bar{\mathbb{R}}$ (μ-measurable with respect to any measure μ on $\mathcal{A}$). In this way all upper μ-integrals (outer and inner μ-measures) in Theorem 2.1 and Corollary 2.2 become the ordinary ones (without stars). Moreover, then both Theorem 2.1 and Corollary 2.2 extend to the case where supremum in the definitions of functions M and R is taken over all integers. We leave the formulation of these facts and remaining details to the reader.

3. The uniform ergodic theorem for dynamical systems

The main object under our consideration in this section is a given *ergodic* dynamical system $(X, \mathcal{B}, \mu, T)$. We clarify that $(X, \mathcal{B}, \mu)$ is supposed to be a probability space, and T is an ergodic measure-preserving transformation of X. In addition, we assume that a parameterized family $\mathcal{F} = \{f_\theta \mid \theta \in \Theta\}$ of measurable maps from X into $\mathbb{R}$ is given, such that the condition is satisfied:

$$\int^* \sup_{\theta \in \Theta} |f_\theta| \, d\mu < \infty. \tag{3.1}$$

In particular, we see that f_θ belongs to $L^1(\mu)$ for all $\theta \in \Theta$, and therefore from Birkhoff's pointwise ergodic theorem (see **[25]** p. 9) we have:

$$\frac{1}{n} \sum_{j=0}^{n-1} T^j(f_\theta) \to M(\theta) \quad \mu\text{-a.s.} \tag{3.2}$$

as $n \to \infty$, for all $\theta \in \Theta$. The limit function M is called the *μ-mean function* of $\mathcal{F}$, and we have $M(\theta) = \int f_\theta \, d\mu$ for all $\theta \in \Theta$. Hence we see that M belongs to $B(\Theta)$ whenever (3.1) is satisfied, where $B(\Theta)$ denotes the set of all bounded real valued functions on Θ. It is the purpose of this section to present a solution for the *prime and dual problem* from Section 1 that now may be restated as follows. Determine conditions which are necessary and sufficient for the uniform convergence to be valid:

$$\sup_{\theta \in \Theta} \Big| \frac{1}{n} \sum_{j=0}^{n-1} T^j(f_\theta) - M(\theta) \Big| \to 0 \tag{3.3}$$

---

12 An analytic space is a Hausdorff space which is a continuous image of a Polish space.

13 A Polish space is a separable topological space that can be metrized by means of a complete metric.

14 Let $(X, \mathcal{A})$ be a measurable space, and let $(\Theta, \mathcal{B})$ be a Blackwell space. If π_X denotes the projection of $X \times \Theta$ onto X, then $\pi_X(C)$ is universally $\mathcal{A}$-measurable for every $C \in \mathcal{A} \times \mathcal{B}$. (This is a version of the projection theorem that suffices for our remark above. We remark that its further extensions and generalizations are available (see **[22]**).)

as $n \to \infty$. More precisely, we shall consider three kinds of convergence notions in (3.3), namely (a.s.)*-convergence, $(L^1)^*$-convergence, and (μ^*)-convergence. Given a sequence of arbitrary maps $\{Z_n \mid n \geq 1\}$ from X into $\mathbb{R}$, we recall that $Z_n \to 0$ (a.s.)* if $|Z_n|^* \to 0$ μ-a.s., that $Z_n \to 0$ $(L^1)^*$ if $|Z_n|^* \to 0$ in μ-mean, and that $Z_n \to 0$ (μ^*) if $|Z_n|^* \to 0$ in μ-measure. For more information in this direction we shall refer the reader to **[36]**. It should be noted that if all maps under consideration are measurable, then the convergence notions stated above coincide with the notion of μ-a.s. convergence, convergence in μ-mean, and convergence in μ-measure, respectively. This could be achieved in quite a general setting as it was already described in Remark 2.3. In order to keep generality and be able to handle measurability problems appearing in the sequel, we shall often assume that the transformation T is μ-perfect. This means that for every $B \in \mathcal{B}$ there exists $C \in \mathcal{B}$ satisfying $C \subset T(B)$ and $\mu(B \setminus T^{-1}(C)) = 0$. We can require equivalently that $\mu^*(T^{-1}(C)) = 1$ whenever C is a subset of X satisfying $\mu^*(C) = 1$ (see **[35]**). Under this assumption we have (see **[35]**):

$$(\psi \circ T)^* = \psi^* \circ T \tag{3.4}$$

$$\int^* \psi \circ T \, d\mu = \int^* \psi \, d\mu \tag{3.5}$$

whenever $\psi : X \to \bar{\mathbb{R}}$ is an arbitrary map. Dynamical system $(X, \mathcal{B}, \mu, T)$ is said to be *perfect*, if T is μ-perfect. The best known sufficient condition for the μ-perfectness of T could be:

$$T(B) \in \mathcal{B}_\mu \quad \text{for all} \quad B \in \mathcal{B} \tag{3.6}$$

where $\mathcal{B}_\mu$ denotes the μ-completion of $\mathcal{B}$. This condition is by the image theorem satisfied whenever $(X, \mathcal{B})$ is a countably separated Blackwell space (see **[22]**). Moreover, if $(X, \mathcal{B})$ equals to the countable product $(S^{\mathbb{N}}, \mathcal{A}^{\mathbb{N}})$ of copies of a measurable space $(S, \mathcal{A})$, then by the projection theorem we see that condition (3.6) is satisfied whenever $(S, \mathcal{A})$ is a Blackwell space (see **[22]**). We remark again that any analytic space is a Blackwell space, and that any analytic metric space, and thus any polish space, is a countably separated Blackwell space. To conclude the preliminary part of this section we remark that from the point of view of two distinct approaches towards solution for (3.3) described in Section 1, our approach below offers Lipschitz type conditions. For the random entropy combinatorial approach we shall refer the reader to **[39]**. Finally, it should be mentioned that our conditions on the dynamical system $(X, \mathcal{B}, \mu, T)$ and family $\mathcal{F} = \{f_\theta \mid \theta \in \Theta\}$ imposed above seem to be as optimal as possible from many points of view. However, we shall not present the details in this direction here, and for more detailed discussion on this point we refer the reader to **[37]**. We pass to the prime and dual problem (3.3) itself. For this we shall say that the family $\mathcal{F} = \{f_\theta \mid \theta \in \Theta\}$ is *eventually totally bounded in μ-mean* with respect to T, if the following condition is satisfied (see **[20]** and **[37]**):

For every $\varepsilon > 0$ there exists $\gamma_\varepsilon \in \Gamma(\Theta)$ such that:

$$\inf_{n\geq 1} \frac{1}{n} \int^* \sup_{\theta',\theta''\in A} \Big|\sum_{j=0}^{n-1} T^j(f_{\theta'} - f_{\theta''})\Big|\, d\mu < \varepsilon \tag{3.7}$$

for all $A \in \gamma_\varepsilon$. Here and in the sequel $\Gamma(\Theta)$ denotes the family of all finite covers of Θ. We recall that a finite cover of Θ is any family $\gamma = \{A_1, \dots, A_n\}$ of non-empty subsets of Θ satisfying $\Theta = \bigcup_{j=1}^n A_j$ with $n \geq 1$. It is our next aim to show that under (3.1), condition (3.7) is equivalent to the uniform convergence in (3.3), with respect to any of the convergence notions stated above. It should be noted that our method below in essence relies upon the uniform ergodic lemma from the previous section. The first result may be stated as follows.

Theorem 3.1. *Let $(X, \mathcal{B}, \mu, T)$ be a perfect ergodic dynamical system, and let $\mathcal{F} = \{f_\theta \mid \theta \in \Theta\}$ be a parameterized family of measurable maps from X into $\mathbb{R}$ satisfying* (3.1). *If $\mathcal{F}$ is eventually totally bounded in μ-mean with respect to T, then we have:*

$$\sup_{\theta\in\Theta} \Big|\frac{1}{n}\sum_{j=0}^{n-1} T^j(f_\theta) - M(\theta)\Big| \to 0 \quad (a.s)^* \text{ and } (L^1)^* \tag{3.8}$$

as $n \to \infty$, where M is the μ-mean function of $\mathcal{F}$.

Proof. Let $\varepsilon > 0$ be given and fixed, then by our assumption there exists $\gamma_\varepsilon \in \Gamma(\Theta)$ such that:

$$\inf_{n\geq 1} \frac{1}{n} \int^* \sup_{\theta',\theta''\in A} \Big|\sum_{j=0}^{n-1} T^j(f_{\theta'} - f_{\theta''})\Big|\, d\mu < \varepsilon \tag{3.9}$$

for all $A \in \gamma_\varepsilon$. Since $M \in B(\Theta)$, then there is no restriction to assume that we also have:

$$\sup_{\theta',\theta''\in A} |M(\theta') - M(\theta'')| < \varepsilon \tag{3.10}$$

for all $A \in \gamma_\varepsilon$. In addition choose a point $\theta_A \in A$ for every $A \in \gamma_\varepsilon$. From (3.10) we get:

$$\begin{aligned}
\sup_{\theta\in\Theta} \Big|\frac{1}{n}\sum_{j=0}^{n-1} T^j(f_\theta) - M(\theta)\Big| &= \max_{A\in\gamma_\varepsilon} \sup_{\theta\in A} \Big|\frac{1}{n}\sum_{j=0}^{n-1} T^j(f_\theta) - M(\theta)\Big| \\
&\leq \max_{A\in\gamma_\varepsilon} \Big(\sup_{\theta',\theta''\in A} \Big|\sum_{j=0}^{n-1} T^j(f_{\theta'} - f_{\theta''})\Big| + \Big|\frac{1}{n}\sum_{j=0}^{n-1} T^j(f_{\theta_A}) - M(\theta_A)\Big| \\
&\qquad + \sup_{\theta',\theta''\in A} |M(\theta') - M(\theta'')| \Big) \\
&\leq \max_{A\in\gamma_\varepsilon} \sup_{\theta',\theta''\in A} \Big|\sum_{j=0}^{n-1} T^j(f_{\theta'} - f_{\theta''})\Big| + \max_{A\in\gamma_\varepsilon} \Big|\frac{1}{n}\sum_{j=0}^{n-1} T^j(f_{\theta_A}) - M(\theta_A)\Big| + \varepsilon
\end{aligned}$$

for all $n \geq 1$. Hence by (3.2) we easily obtain:

$$\limsup_{n\to\infty} \Big(\sup_{\theta\in\Theta} \Big|\frac{1}{n}\sum_{j=0}^{n-1} T^j(f_\theta) - M(\theta)\Big|\Big)^*$$

$$\leq \max_{A\in\gamma_\varepsilon} \limsup_{n\to\infty} \Big(\sup_{\theta',\theta''\in A} \Big|\frac{1}{n}\sum_{j=0}^{n-1} T^j(f_{\theta'} - f_{\theta''})\Big|\Big)^* + \varepsilon \quad \mu\text{-a.s.}$$

From this inequality and (3.9) we see that(a.s.)*-convergence in (3.8) will be established as soon as we obtain the inequality:

$$\limsup_{n\to\infty} \Big(\sup_{\theta',\theta''\in A} \Big|\frac{1}{n}\sum_{j=0}^{n-1} T^j(f_{\theta'} - f_{\theta''})\Big|\Big)^* \leq \inf_{n\geq 1} \int^* \sup_{\theta',\theta''\in A} \Big|\frac{1}{n}\sum_{j=0}^{n-1} T^j(f_{\theta'} - f_{\theta''})\Big|\, d\mu \tag{3.11}$$

for all $A \subset \Theta$. We leave this inequality to be established with some facts of independent interest in the next proposition. We proceed with the $(L^1)^*$-convergence in (3.8). From (3.4) we get:

$$\Big(\sup_{\theta\in\Theta} \Big|\frac{1}{n}\sum_{j=0}^{n-1} T^j(f_\theta) - M(\theta)\Big|\Big)^* \leq \frac{1}{n}\sum_{j=0}^{n-1}\Big(\sup_{\theta\in\Theta} |T^j(f_\theta)|\Big)^* + \sup_{\theta\in\Theta} |M(\theta)|$$

$$= \frac{1}{n}\sum_{j=0}^{n-1} T^j\Big(\big(\sup_{\theta\in\Theta} |f_\theta|\big)^*\Big) + \sup_{\theta\in\Theta} |M(\theta)|$$

for all $n \geq 1$. Hence we see that the sequence on the left-hand side is uniformly integrable. Thus $(L^1)^*$-convergence in (3.8) follows from (a.s.)*-convergence, and the proof is complete. □

Proposition 3.2. *Under the hypotheses of Theorem* 3.1 *we have:*

$$\limsup_{n\to\infty} \Big(\sup_{\theta',\theta''\in A} \Big|\frac{1}{n}\sum_{j=0}^{n-1} T^j(f_{\theta'} - f_{\theta''})\Big|\Big)^* = C \quad \mu\text{-a.s.} \tag{3.12}$$

$$C \leq \inf_{n\geq 1} \int^* \sup_{\theta',\theta''\in A} \Big|\frac{1}{n}\sum_{j=0}^{n-1} T^j(f_{\theta'} - f_{\theta''})\Big|\, d\mu \tag{3.13}$$

$$\inf_{n\geq 1} \int^* \sup_{\theta',\theta''\in A} \Big|\frac{1}{n}\sum_{j=0}^{n-1} T^j(f_{\theta'} - f_{\theta''})\Big|\, d\mu = \limsup_{n\to\infty} \int^* \sup_{\theta',\theta''\in A} \Big|\frac{1}{n}\sum_{j=0}^{n-1} T^j(f_{\theta'} - f_{\theta''})\Big|\, d\mu \tag{3.14}$$

where A is an arbitrary subset of Θ, and C is a real number depending on A.

First Proof. We could notice that if T is μ-perfect, then T^{j-1} is μ-perfect for all $j \geq 1$. This fact easily implies that the map $x \mapsto (T^0(x), T^1(x), \dots)$ is μ-perfect as a map from $(X, \mathcal{B})$ into $(X^{\mathbb{N}}, \mathcal{B}^{\mathbb{N}})$. Hence we see that if T is assumed to be μ-perfect, then the hypotheses of Theorem 1 and Proposition 2 from Section 3 in **[37]** are satisfied with $\xi_j = T^{j-1}$ for $j \geq 1$. The proof then could be carried out in exactly the same way as the proof of Proposition 2 from Section 3 in **[37]**. This completes the first proof.

Second Proof. We could deduce the result of Proposition 3.2 by using the subadditive ergodic theorem of Kingman (see **[12]** p. 292–299). Indeed, let us in the notation of Proposition 3.2 put:

$$g_n = \Big(\sup_{\theta',\theta'' \in A} \Big| \sum_{j=0}^{n-1} T^j (f_{\theta'} - f_{\theta''} \Big| \Big)^*$$

for all $n \geq 1$, and let us assume that T is μ-perfect. Then it is quite easily verified that the sequence $\{g_n \mid n \geq 1\}$ is subadditive in $L^1(\mu)$. Therefore (3.12), (3.13) and (3.14) in Proposition 3.2 follow from Kingman's theorem in **[12]**. Moreover, even though it is irrelevant for our purposes, it should be noted that in this way we obtain immediately a more precise information. Namely, we have equality in (3.13) with the limit in (3.12) instead of the limit superior. (Since this follows from either the first proof or the third proof given below, together with the result of Theorem 3.1, we did not state the result of Proposition 3.2 in this stronger form. We have rather chosen the form which appears to be essential in the process of proving Theorem 3.1 by the method presented. For this one should recall (3.11) above.) Finally, it might be instructive in this context to recall the classical fact that a subadditive sequence of real numbers $\{\gamma_n \mid n \geq 1\}$ converges to $\inf_{n \geq 1} \gamma_n / n$ as $n \to \infty$. This completes the second proof.

Third Proof. We now give a proof of Proposition 3.2, based on our uniform ergodic lemma (see Theorem 2.1). Although this proof yields somewhat less (in comparison with the second proof), we include it for the sake of keeping the analogy with Birkhoff's theorem, with the hope that this method will find other applications.

Let $A \subset \Theta$ be given and fixed. Let us denote:

$$\Psi_n(x) = \sup_{\theta',\theta'' \in A} \Big| \frac{1}{n} \sum_{j=0}^{n-1} T^j (f_{\theta'} - f_{\theta''})(x) \Big|$$

for all $x \in X$ and all $n \geq 1$. Then the proof might be carried out as follows.

(3.12): Since T is ergodic, then it is enough to show that:

$$\limsup_{n \to \infty} \Psi^* \circ T = \limsup_{n \to \infty} \Psi_n^* \quad \mu\text{-a.s.} \tag{3.15}$$

For this it should be noted that we have:

$$\psi_N \circ T = \sup_{\theta',\theta'' \in A} \Big| \frac{n+1}{n} \frac{1}{n+1} \sum_{j=0}^{n-1} T^j (f_{\theta'} - f_{\theta''}) - \frac{1}{n} (f_{\theta'} - f_{\theta''}) \Big|$$

for all $n \geq 1$. Hence we easily find:

$$\frac{n+1}{n}\Psi_{n+1} - \frac{2}{n}\sup_{\theta\in\Theta}|f_\theta| \leq \Psi_n \circ T \leq \frac{n+1}{n}\Psi_{n+1} + \frac{2}{n}\sup_{\theta\in\Theta}|f_\theta|$$

for all $n \geq 1$. Taking upper μ-envelopes on both sides we obtain:

$$\frac{n+1}{n}\Psi_{n+1}^* - \frac{2}{n}\big(\sup_{\theta\in\Theta}|f_\theta|\big)^* \leq \big(\Psi_n \circ T\big)^* \leq \frac{n+1}{n}\Psi_{n+1}^* + \frac{2}{n}\big(\sup_{\theta\in\Theta}|f_\theta|\big)^*$$

for all $n \geq 1$. Letting $n \to \infty$, and using (3.1) and (3.4) we obtain (3.15), and the proof of (3.12) is complete.

(3.13): Here we essentially use the uniform ergodic lemma from the last section. For this we denote $S_n(f) = \sum_{j=0}^{n-1} T^j f$ and $M_n(f) = \max_{1\leq j\leq n} S_j(f)$, whenever $f \in L^1(\mu)$ and $n \geq 1$. Let us for fixed $N, m \geq 1$ and $\varepsilon > 0$ consider a set as follows:

$$B_{N,m,\varepsilon} = \big\{\big(\sup_{\theta',\theta''\in A} M_m\big(|S_N(f_{\theta'} - f_{\theta''})/N| - (C-\varepsilon)\big)\big)^* > 0\big\}.$$

Then by Theorem 2.1 we may conclude:

$$\int^*_{B_{N,m,\varepsilon}} \sup_{\theta',\theta''\in A} (|S_N(f_{\theta'} - f_{\theta''})/N| - (C-\varepsilon))\, d\mu \geq 0.$$

Hence by subadditivity of upper μ-integral we obtain:

$$\int^*_{B_{N,m,\varepsilon}} \sup_{\theta',\theta''\in A} |S_N(f_{\theta'} - f_{\theta''})/N|\, d\mu \geq (C-\varepsilon)\mu(B_{N,m,\varepsilon}).$$

Therefore by monotone convergence theorem for upper μ-integral (see **[35]**) in order to complete the proof of (3.13) it is enough to show that $\mu(B_{N,m,\varepsilon}) \uparrow 1$ as $m \to \infty$ with $N \geq 1$ and $\varepsilon > 0$ being fixed. We shall establish this fact by proving the following inequality:

$$\begin{aligned}\big(\sup_{\theta',\theta''\in A} |S_n(f_{\theta'} - f_{\theta''})/n|\big)^* \leq &\big(\sup_{\theta',\theta''\in A}\sup_{m\geq 1} S_m\big(|S_N(f_{\theta'} - f_{\theta''})/N|\big)\big)^* \\ &+ \big(\frac{1}{n}\cdot\sup_{\theta',\theta''\in A}\big|\sum_{N[n/N]\leq j<n} T^j(f_{\theta'} - f_{\theta''})\big|\big)^*\end{aligned} \tag{3.16}$$

being valid for all $n \geq N$. For this it should be noted that we have:

$$|S_n(f_{\theta'} - f_{\theta''})/n| \leq |S_{N[n/N]}(f_{\theta'} - f_{\theta''})/n| + \frac{1}{n}\big|\sum_{N[n/N]\leq j\leq n} T^j(f_{\theta'} - f_{\theta''})\big| \tag{3.17}$$

being valid for all $n \geq N$ and all $\theta', \theta'' \in A$. Moreover, given $n \geq N$, taking $m \geq 1$ large enough we get:

$$
\begin{aligned}
S_m\big(|S_N(f_{\theta'} - f_{\theta''})|\big) &= \Big|\sum_{j=0}^{N-1} T^j(f_{\theta'} - f_{\theta''})\Big| + \Big|\sum_{j=1}^{N} T^j(f_{\theta'} - f_{\theta''})| + \cdots \\
&\qquad \cdots + \Big|\sum_{j=N}^{2N-1} T^j(f_{\theta'} - f_{\theta''})\Big| + \cdots \\
&\geq \Big|\sum_{j=0}^{N-1} T^j(f_{\theta'} - f_{\theta''})\Big| + \Big|\sum_{j=N}^{2N-1} T^j(f_{\theta'} - f_{\theta''})\Big| + \cdots \\
&\qquad \cdots + \Big|\sum_{j=N([n/N]-1)}^{N[n/N]-1} T^j(f_{\theta'} - f_{\theta''})\Big| \\
&\geq |S_{N[n/N]}(f_{\theta'} - f_{\theta''})|
\end{aligned}
$$

for all $\theta', \theta'' \in A$. Hence we obtain:

$$
|S_{N[n/N]}(f_{\theta'} - f_{\theta''})/n| \leq S_m\big(|S_N(f_{\theta'} - f_{\theta''})/n|\big) \leq S_m\big(|S_N(f_{\theta'} - f_{\theta''})/N|\big) \tag{3.18}
$$

with $n, N, m \geq 1$ as above. Thus (3.16) follows straightforward by (3.17) and (3.18). In addition, for the last term in (3.16) we have:

$$
\begin{aligned}
\Big(\frac{1}{n} \cdot \sup_{\theta',\theta'' \in A} \Big| \sum_{N[n/N] \leq j < n} T^j(f_{\theta'} - f_{\theta''})\Big|\Big)^* &\leq \frac{2}{n} \sum_{N[n/N] \leq j < n} T^j\big((\sup_{\theta \in A} |f_\theta|)^*\big) \\
&= 2\Big(\frac{1}{n} S_n\big((\sup_{\theta \in A} |f_\theta|)^*\big) - \frac{N[n/N]}{n} \cdot \frac{1}{N[n/N]} S_{N[n/N]}\big((\sup_{\theta \in A} |f_\theta|)^*\big)\Big)
\end{aligned}
$$

for all $n \geq N$. Since $N[n/N]/n \to 1$ as $n \to \infty$, then by (3.1) and Birkhoff's theorem, the right-hand side tends to zero μ-a.s. as $n \to \infty$. Letting $n \to \infty$ in (3.16) therefore we obtain:

$$
\limsup_{n\to\infty}\Big(\sup_{\theta',\theta'' \in A} |S_n(f_{\theta'} - f_{\theta''})/n|\Big)^* \leq \Big(\sup_{\theta',\theta'' \in A} \sup_{m \geq 1} S_m\big(|S_N(f_{\theta'} - f_{\theta''})/N|\big)\Big)^* \tag{3.19}
$$

for all $N \geq 1$. Finally, from (3.12) and (3.19) we get:

$$
\begin{aligned}
\lim_{m\to\infty} \mu(B_{N,m,\varepsilon}) &= \lim_{m\to\infty} \mu^*\Big\{\sup_{\theta',\theta'' \in A} M_m(|S_N(f_{\theta'} - f_{\theta''})/N|) > (C - \varepsilon)\Big\} \\
&= \mu^*\Big\{\sup_{\theta',\theta'' \in A} \sup_{m \geq 1} S_m(|S_N(f_{\theta'} - f_{\theta''})/N|) > (C - \varepsilon)\Big\} \\
&= \mu\Big\{\Big(\sup_{\theta',\theta'' \in A} \sup_{m \geq 1} S_m(|S_N(f_{\theta'} - f_{\theta''})/N|)\Big)^* > (C - \varepsilon)\Big\} \\
&\geq \mu\Big\{\limsup_{n\to\infty}\Big(\sup_{\theta',\theta'' \in A} |S_n(f_{\theta'} - f_{\theta''})/n|\Big)^* > (C - \varepsilon)\Big\} = 1.
\end{aligned}
$$

This fact completes the proof of (3.13).

(3.14): It should be noted that we have:

$$\Big(\sup_{\theta',\theta'' \in A} \Big| \frac{1}{n} \sum_{j=0}^{n-1} T^j (f_{\theta'} - f_{\theta''}) \Big| \Big)^* \leq \frac{2}{n} \sum_{j=0}^{n-1} T^j \big((\sup_{\theta \in A} |f_\theta|)^* \big)$$

for all $n \geq 1$. Therefore by (3.1) it follows that the sequence on the left-hand side is uniformly integrable. Thus by Fatou's lemma we can conclude:

$$\limsup_{n \to \infty} \int^* \sup_{\theta',\theta'' \in A} \Big| \frac{1}{n} \sum_{j=0}^{n-1} T^j (f_{\theta'} - f_{\theta''}) \Big| \, d\mu \leq C.$$

Hence (3.14) follows straightforward by (3.13). This fact completes the proof. □

Remark 3.3. Under the hypotheses of Theorem 3.1 and Proposition 3.2 it is easily verified that in the case where the map $x \mapsto \sup_{\theta',\theta'' \in A} | \sum_{j=0}^{n-1} T^j (f_{\theta'} - f_{\theta''})(x)|$ is μ-measurable as a map from X into $\mathbb{R}$ for all $n \geq 1$ and given $A \subset \Theta$, the assumption of μ-perfectness of T is not needed for their conclusions to remain valid.

Remark 3.4. It should be noted that putting $n = 1$ in the definition of eventual total boundedness in μ-mean, we obtain an extension of the Blum–DeHardt sufficient condition, which is the best known for the uniform law of large numbers, at least in the independent case. There are examples showing that a parameterized family could be eventually totally bounded in μ-mean with respect to a measure preserving transformation, though it does not satisfy the Blum–DeHardt condition (see **[11]** p. 41). It is not a surprise indeed, since we shall see in the next theorem that the eventual total boundedness in μ-mean property characterizes the uniform ergodic theorems, and thus the uniform laws of large numbers as well. In this way we may conclude that we have obtained a characterization of the uniform ergodic theorem which contains the best known sufficient condition as a particular case. Finally, let us mention a well-known and interesting fact that Blum–DeHardt's law of large numbers contains the classical Mourier's law of large numbers (see **[11]** p. 43).

In order to state the next theorem we clarify that $B(\Theta)$ denotes the Banach space of all bounded real valued functions on Θ with respect to the sup-norm. We denote by $C(B(\Theta))$ the set of all bounded continuous functions from $B(\Theta)$ into $\mathbb{R}$, while by $\mathcal{K}(B(\Theta))$ we denote the family of all compact subsets of $B(\Theta)$. A pseudo-metric d on a set Θ is said to be *totally bounded*, if Θ can be covered by finitely many d-balls of any given radius $\varepsilon > 0$. A pseudo-metric d on a set Θ is said to be a *ultra pseudo-metric*, if it satisfies $d(\theta_1, \theta_2) \leq d(\theta_1, \theta_3) \vee d(\theta_3, \theta_2)$ for all $\theta_1, \theta_2, \theta_3 \in \Theta$.

Theorem 3.5. *Let $(X, \mathcal{B}, \mu, T)$ be a perfect ergodic dynamical system, and let $\mathcal{F} = \{f_\theta \mid \theta \in \Theta\}$ be a parameterized family of measurable maps from X into $\mathbb{R}$ satisfying* (3.1). *Let M be the μ-mean function of $\mathcal{F}$, then the following eight statements are equivalent:*

(3.20) *The family $\mathcal{F}$ is eventually totally bounded in μ-mean with respect to T.*

(3.21) $\sup_{\theta \in \Theta} | \frac{1}{n} \sum_{j=0}^{n-1} T^j (f_\theta) - M(\theta)| \to 0 \quad (a.s.)^*$.

(3.22) $\sup_{\theta\in\Theta} |\frac{1}{n}\sum_{j=0}^{n-1} T^j(f_\theta) - M(\theta)| \to 0 \quad (L^1)^*$.

(3.23) $\sup_{\theta\in\Theta} |\frac{1}{n}\sum_{j=0}^{n-1} T^j(f_\theta) - M(\theta)| \to 0 \quad (\mu^*)$.

(3.24) $\frac{1}{n}\sum_{j=0}^{n-1} T^j(f_\theta) \to M$ *weakly in* $B(\Theta)$*:*

$$\lim_{n\to\infty} \int^* F\Big(\frac{1}{n}\sum_{j=0}^{n-1} T^j(f)\Big) d\mu = F(M)$$

for all $F \in C(B(\Theta))$.

(3.25) *The sequence* $\{\frac{1}{n}\sum_{j=0}^{n-1} T^j(f) \mid n \geq 1\}$ *is eventually tight in* $B(\Theta)$*:*

$$\limsup_{n\to\infty} \int^* F\Big(\frac{1}{n}\sum_{j=0}^{n-1} T^j(f)\Big)\, d\mu \leq \varepsilon$$

for some $K_\varepsilon \in \mathcal{K}(B(\Theta))$, *and for all* $F \in C(B(\Theta))$ *satisfying* $0 \leq F \leq 1_{B(\Theta)\setminus K_\varepsilon}$, *whenever* $\varepsilon > 0$.

(3.26) *There exists a totally bounded ultra pseudo-metric d on* Θ *such that the condition is satisfied:*

$$\liminf_{r\downarrow 0}\, \inf_{n\geq 1} \int^* \sup_{d(\theta',\theta'')<r} \Big|\frac{1}{n}\sum_{j=0}^{n-1} T^j(f_{\theta'} - f_{\theta''})\Big|\, d\mu = 0.$$

(3.27) *For every* $\varepsilon > 0$ *there exist a totally bounded pseudo-metric* d_ε *on* Θ *and* $r_\varepsilon > 0$ *such that:*

$$\inf_{n\geq 1} \mu^*\Big\{ \sup_{d_\varepsilon(\theta,\theta')<\varepsilon} \Big|\frac{1}{n}\sum_{j=0}^{n-1} T^j(f_\theta - f_{\theta'})\Big| > \varepsilon\Big\} < \varepsilon$$

for all $\theta \in \Theta$.

Proof. Since T is μ-perfect, then T^{j-1} is μ-perfect for all $j \geq 1$. This fact as before easily implies that the map $x \mapsto (T^0(x), T^1(x), \dots)$ is μ-perfect as a map from $(X, \mathcal{B})$ into $(X^{\mathbb{N}}, \mathcal{B}^{\mathbb{N}})$. Putting $\xi_j = T^{j-1}$ for $j \geq 1$ we see that the hypotheses of Theorem 4, Corollary 5, Theorem 7, Corollary 8 and Theorem 10 from Section 3 in **[37]** are satisfied. The result therefore follows from Corollary 5, Theorem 7, Corollary 8 and Theorem 10 from Section 3 in **[37]**. It should be noted that a characterization of compact sets in the Banach space $B(\Theta)$ is used (see **[13]** p. 260) for sufficiency of (3.25). It is also not difficult to verify (by using uniform integrability) that (3.23) implies (3.20). These facts complete the proof. □

4. Examples and complements

We continue as in the preceding section by considering a given dynamical system $(X, \mathcal{B}, \mu, T)$ and a parameterized family $\mathcal{F} = \{f_\theta \mid \theta \in \Theta\}$ of measurable maps from X into $\mathbb{R}$. First we discuss a necessary condition for the prime and dual problem as follows. Suppose that (3.3) is satisfied with respect to (a.s.)*-convergence. We could then question does this fact imply validity of condition (3.1). The answer is negative in general, and a more detailed discussion on this point with references can be found in **[37]**. Here we show that a weaker condition still remains true. The result may be stated as follows.

Proposition 4.1. *Let $(X, \mathcal{B}, \mu, T)$ be a perfect dynamical system, and let $\mathcal{F} = \{f_\theta \mid \theta \in \Theta\}$ be a parameterized family of measurable maps from X into $\mathbb{R}$. Suppose that:*

$$\limsup_{n\to\infty} \Big(\sup_{\theta\in\Theta} \Big|\frac{1}{n}\sum_{j=0}^{n-1} T^j(f_\theta)\Big|\Big)^* < \infty \quad \mu\text{-a.s.} \tag{4.1}$$

Then we have:

$$\big(\sup_{\theta\in\Theta} |f_\theta|\big)^* < \infty \quad \mu\text{-a.s.} \tag{4.2}$$

Proof. We show that under the primary hypotheses stated above we have:

$$\mu\Big\{\limsup_{n\to\infty}\Big(\sup_{\theta\in\Theta}\Big|\frac{1}{n}\sum_{j=0}^{n-1} T^j(f_\theta)\Big|\Big)^* < \infty\Big\} \le \mu\Big\{\big(\sup_{\theta\in\Theta}|f_\theta|\big)^* < \infty\Big\}. \tag{4.3}$$

For this it should be noted that we have:

$$T^n(f_\theta) = (n+1)\cdot \frac{1}{n+1}\sum_{j=0}^{n} T^j(f_\theta) - n\cdot\frac{1}{n}\sum_{j=0}^{n-1} T^j(f_\theta)$$

for all $n \ge 1$ and all $\theta \in \Theta$. Hence we get:

$$\big(\sup_{\theta\in\Theta}|T^n(f_\theta)|\big)^* \le (n+1)\cdot\Big(\sup_{\theta\in\Theta}\Big|\frac{1}{n+1}\sum_{j=0}^{n} T^j(f_\theta)\Big|\Big)^* + n\cdot\Big(\sup_{\theta\in\Theta}\Big|\frac{1}{n}\sum_{j=0}^{n-1} T^j(f_\theta)\Big|\Big)^*$$

for all $n \geq 1$. From this and the μ-perfectness of T we may easily conclude:

$$\mu\{\limsup_{n\to\infty}\big(\sup_{\theta\in\Theta}\big|\frac{1}{n}\sum_{j=0}^{n-1}T^j(f_\theta)\big|\big)^* < \infty\} = \mu\Big(\bigcup_{M=1}^{\infty}\bigcup_{N=1}^{\infty}\bigcap_{n=N}^{\infty}\{\big(\sup_{\theta\in\Theta}\big|\frac{1}{n}\sum_{j=0}^{n-1}T^j(f_\theta)\big|\big)^* \leq M\}\Big)$$

$$\leq \mu\Big(\bigcup_{M=1}^{\infty}\bigcup_{N=1}^{\infty}\bigcap_{n=N}^{\infty}\{\big(\sup_{\theta\in\Theta}|T^n(f_\theta)|\big)^* \leq (2n+1)M\}\Big)$$

$$\leq \mu\Big(\bigcup_{N=1}^{\infty}\bigcap_{n=N}^{\infty}\{\big(\sup_{\theta\in\Theta}|T^n(f_\theta)|\big)^* < \infty\}\Big)$$

$$\leq \liminf_{n\to\infty}\mu\{\big(\sup_{\theta\in\Theta}|T^n(f_\theta)|\big)^* < \infty\} = \mu\{\big(\sup_{\theta\in\Theta}|f_\theta|\big)^* < \infty\}.$$

Actually, in the last step we use the fact that T^n is μ-perfect for every $n \geq 1$. Thus (4.3) is satisfied, and (4.2) follows straightforward by (4.1). These facts complete the proof. □

Corollary 4.2. *Under the hypotheses of Proposition* 4.1 *suppose moreover that* $\mathcal{F} \subset L^1(\mu)$ *and that* (3.3) *is satisfied with respect to either (a.s.)*-convergence,* $(L^1)^*$*-convergence, or* (μ^*)*-convergence, where* M *is the* μ*-mean function of* $\mathcal{F}$ *satisfying* $\sup_{\theta\in\Theta}|M(\theta)| < \infty$. *Then* (4.2) *in Proposition* 4.1 *is satisfied.*

Proof. It follows easily from Proposition 4.1 by using the inequality:

$$\limsup_{n\to\infty}\big(\sup_{\theta\in\Theta}\big|\frac{1}{n}\sum_{j=0}^{n-1}T^j(f_\theta)\big|\big)^* \leq \limsup_{n\to\infty}\big(\sup_{\theta\in\Theta}\big|\frac{1}{n}\sum_{j=0}^{n-1}T^j(f_\theta) - M(\theta)\big|\big)^* + \sup_{\theta\in\Theta}|M(\theta)|.$$

This completes the proof. □

Next we investigate a uniform approximation in the L^1-sense by means of a dense family from $L^1(\mu)$ which satisfies the uniform ergodic theorem in a trivial way. This approach is in the spirit of the classical Hilbert space method for the mean ergodic theorem of von Neumann, and therefore from the ergodic theory point of view it could be seen as the natural one. We consider as before a given ergodic dynamical system $(X, \mathcal{B}, \mu, T)$ and a parameterized family $\mathcal{F} = \{f_\theta \mid \theta \in \Theta\}$ of measurable maps from X into $\mathbb{R}$. We recall that $L^2(\mu)$ is a Hilbert space which is dense in $L^1(\mu)$, and for which by the general Hilbert space theory we have (see **[25]** p. 4):

$$L^2(\mu) = G \oplus cl(H)$$

where $G = \{g \in L^2(\mu) \mid Tg = g\}$ and $H = \{h - Th \mid h \in L^\infty(\mu)\}$. It is instructive to observe in this context that G consists of constants, since T is assumed to be ergodic. In addition, suppose that maps $g_\theta = h_\theta - Th_\theta \in H$ for $\theta \in \Theta$ are given, such that

$\big(\sup_{\theta\in\Theta} |h_\theta|\big)^* \in L^1(\mu)$. Then by Birkhoff's pointwise ergodic theorem we have:

$$\begin{aligned}\Big(\sup_{\theta\in\Theta}\Big|\frac{1}{n}\sum_{j=0}^{n-1} T^j(g_\theta)\Big|\Big)^* &= \Big(\sup_{\theta\in\Theta}\Big|\frac{1}{n}(h_\theta - T^n(h_\theta))\Big|\Big)^* \\ &\le \frac{1}{n}\big(\sup_{\theta\in\Theta}|h_\theta|\big)^* + \frac{1}{n}T^n\big(\big(\sup_{\theta\in\Theta}|h_\theta|\big)^*\big) \to 0 \quad \mu\text{-a.s.}\end{aligned} \tag{4.4}$$

as $n \to \infty$. In this way we obtain a natural family of maps satisfying the uniform ergodic theorem in a trivial way. Moreover, let us put $\Phi^\infty(H)$ to denote the class of all families $Z = \{h_\theta - Th_\theta \mid \theta \in \Theta\} \subset H$ with $\big(\sup_{\theta\in\Theta} |h_\theta|\big)^* \in L^1(\mu)$. Then every g_θ from $Z \in \Phi^\infty(H)$ for $\theta \in \Theta$ satisfies (4.4), and the class $\Phi^\infty(H)$ could be enlarged to satisfy the uniform ergodic theorem as follows.

Theorem 4.3. *Let $(X, \mathcal{B}, \mu, T)$ be a perfect ergodic dynamical system, and let $\mathcal{F} = \{f_\theta \mid \theta \in \Theta\}$ a parameterized family of measurable maps from X into $\mathbb{R}$ satisfying $\int f_\theta d\mu = 0$ for all $\theta \in \Theta$. Suppose that the following condition is satisfied:*

(4.5) *For every $\varepsilon > 0$, there exists $Z_\varepsilon = \{g_{\theta,\varepsilon} \mid \theta \in \Theta\} \in \Phi^\infty(H)$ satisfying:*

$$\int^* \sup_{\theta\in\Theta} |f_\theta - g_{\theta,\varepsilon}|\, d\mu < \varepsilon.$$

Then we have:

$$\sup_{\theta\in\Theta}\Big|\frac{1}{n}\sum_{j=0}^{n-1} T^j(f_\theta)\Big| \to 0 \quad (a.s.)^* \ \&\ (L^1)^* \tag{4.6}$$

as $n \to \infty$.

Proof. Let $\varepsilon > 0$ be given, then there exists $Z_\varepsilon = \{g_{\theta,\varepsilon} \mid \theta \in \Theta\} \in \Phi^\infty(H)$ the inequality in (4.5). Hence by Birkhoff's theorem and (4.4) we get:

$$\begin{aligned}&\limsup_{n\to\infty}\Big(\sup_{\theta\in\Theta}\Big|\frac{1}{n}\sum_{j=0}^{n-1} T^j(f_\theta)\Big|\Big)^* \\ &\le \limsup_{n\to\infty}\Big(\sup_{\theta\in\Theta}\Big|\frac{1}{n}\sum_{j=0}^{n-1} T^j(f_\theta - g_{\theta,\varepsilon})\Big|\Big)^* + \limsup_{n\to\infty}\Big(\sup_{\theta\in\Theta}\Big|\frac{1}{n}\sum_{j=0}^{n-1} T^j(g_{\theta,\varepsilon})\Big|\Big)^* \\ &\le \lim_{n\to\infty}\frac{1}{n}\sum_{j=0}^{n-1} T^j\big(\big(\sup_{\theta\in\Theta}|f_\theta - g_{\theta,\varepsilon}|\big)^*\big) \\ &= \int^* \sup_{\theta\in\Theta}|f_\theta - g_{\theta,\varepsilon}|\, d\mu < \varepsilon.\end{aligned}$$

This fact establishes (a.s.)*-convergence in (4.6). Since $(\sup_{\theta\in\Theta} |g_{\theta,1}|)^* \in L^1(\mu)$, then it is easily seen from (4.5) (with $\varepsilon = 1$) that condition (3.1) is fulfilled. Since we have

$$\big(\sup_{\theta\in\Theta}\big|\frac{1}{n}\sum_{j=0}^{n-1}T^j(f_\theta)\big|\big)^* \leq \frac{1}{n}\sum_{j=0}^{n-1}T^j\big(\big(\sup_{\theta\in\Theta}|f_\theta|\big)^*\big)$$

for all $n \geq 1$, then by (3.1) we see that $(L^1)^*$-convergence in (4.6) follows from (4.5) and $(a.s.)^*$-convergence in (4.6) by uniform integrability. This fact completes the proof. □

We pass to a question of independent interest. It concerns a uniform convergence of *moving averages*, and its characterization obtained by means of *Banach limits*. We recall that a linear functional L on l_∞ is called a *Banach limit*, if the following three conditions are satisfied:

(4.7) $L(x) \geq 0$ whenever $x \geq 0$, meaning that all coordinates of x are non–negative.

(4.8) $L(x_1, x_2, x_3, \ldots) = L(x_2, x_3, \ldots)$ for all $(x_1, x_2, x_3, \ldots) \in l_\infty$.

(4.9) $L(\mathbf{1}) = 1$ where $\mathbf{1} = (1, 1, 1, \ldots)$.

It is well-known that Banach limits exist. Moreover if L is a Banach limit, then we have:

(4.10) $\|L\| = 1$

(4.11) $\liminf_{n\to\infty} x_n \leq L(x) \leq \limsup_{n\to\infty} x_n$.

In addition we have:

(4.12) $\sup\{L(x) \mid L \text{ is a Banach limit }\} = \limsup_{n\to\infty} \sup_{i\geq 1} \frac{1}{n}\sum_{j=0}^{n-1} x_{j+i}$ for all $x = (x_1, x_2, \ldots) \in l_\infty$. Finally, if $x = (x_1, x_2, \ldots) \in l_\infty$ and $m \in \mathbb{R}$, then $L(x) = m$ for every Banach limit L, if and only if we have:

$$\lim_{n\to\infty} \sup_{i\geq 1}\big|\frac{1}{n}\sum_{j=0}^{n-1} x_{j+i} - m\big| = 0. \tag{4.13}$$

For proof of all these facts we shall refer the reader to **[25]** (p. 135–136). However, the last one seems to be a weak characterization of (4.13), since there could be too many Banach limits for verification. A more convenient characterization of (4.13) may be stated as follows.

Proposition 4.4. *Let $x = (x_1, x_2, \ldots)$ be an arbitrary element of l_∞, and let m be a real number. Then* (4.13) *is valid, if and only if the following two conditions are satisfied:*

(4.14) $\lim_{n\to\infty} \frac{1}{n}\sum_{j=1}^{n} x_j = m$.

(4.15) *The set* $\big\{\big(n^{-1}\sum_{j=0}^{n-1} x_{j+i}\big)_{n\geq 1} \mid i \geq 1\big\}$ *is totally bounded in* l_∞.

Proof. It follows easily by definition that (4.13) implies (4.14) and (4.15). Conversely, it should be noted that (4.14) is equivalent to:

$$\lim_{n\to\infty} \frac{1}{n}\sum_{j=1}^{n} x_{j+i} = m \tag{4.16}$$

being valid for all $i \geq 1$. In addition, let $\varepsilon > 0$ be given and fixed. Then by (4.15) there exist $i_1, \ldots, i_p \geq 1$ such that:

$$\min_{1 \leq k \leq p} \sup_{n \geq 1} \Big| \frac{1}{n} \sum_{j=0}^{n-1} x_{j+i} - \frac{1}{n} \sum_{j=0}^{n-1} x_{j+i_k} \Big| < \varepsilon$$

for all $i \geq 1$. Moreover, by (4.16) we can choose $n \geq N_\varepsilon$ large enough to have:

$$\Big| \frac{1}{n} \sum_{j=0}^{n-1} x_{j+i_k} - m \Big| < \varepsilon$$

for all $k = 1, \ldots, p$. Hence by the triangle inequality we easily obtain:

$$\Big| \frac{1}{n} \sum_{j=0}^{n-1} x_{j+i} - m \Big| < 2\varepsilon$$

for all $n \geq N_\varepsilon$ and all $i \geq 1$. This fact completes the proof. □

Remark 4.5. One could think that condition (4.15) in Proposition 4.4 may be replaced by the following condition:

(4.15′) The set $\big\{(x_{n-1+i})_{n \geq 1} \mid i \geq 1\big\}$ is totally bounded in l_∞.

However, this is not true. Indeed, even though (4.15′) implies (4.15), and therefore (4.14)+(4.15′) implies (4.13), we could have (4.13) without (4.15') being valid. For example, take x to be $(1, -1, 0, 1, -1, 0, 0, 1, -1, 0, 0, 0, 1, -1, \ldots)$. Then (4.13) holds with $m = 0$, but we have $\|(x_{n-1+i_1})_{n \geq 1} - (x_{n-1+i_2})_{n \geq 1}\|_\infty \geq 1$ whenever $i_1 \neq i_2$.

In the next example we show that replacing the unit ball of a Banach space in the prime problem from Section 1 by a smaller family of vectors allows the uniform ergodic theorem (1.1) to become weaker and more easily established (but still of a considerable interest). Our emphasis is rather on the interpretation of the result than on the generality in which it is stated.

Example 4.6 (A uniform ergodic theorem over the orbit). Consider the d-dimensional torus $X = [0, 1[^d$ equipped with the d-dimensional Lebesgue measure $\mu = \lambda_d$ for some $d \geq 1$. Then X is a compact group with respect to the coordinatewise addition mod 1. Take a point $\alpha = (\alpha_1, \ldots, \alpha_d)$ in X such that $\alpha_1, \ldots, \alpha_d$ and 1 are integrally independent (if $\sum_{i=1}^d k_i \alpha_i$ is an integer for some integers $k_1, \ldots, k_d$, then $k_1 = \ldots = k_d = 0$). Under this condition the translation $T(x) = x + \alpha$ is an ergodic measure-preserving transformation of X, and the orbit $\mathcal{O}(x) = \{T^j(x) \mid j \geq 0\}$ is dense in X for every $x \in X$ (see **[25]** p. 12). Let $f \colon X \to \mathbb{R}$ be a continuous function, and let $\mathcal{O}(f) = \{T^j(f) \mid j \geq 0\}$ be the orbit of f under T. Then the prime problem from Section 1 may be restated to a problem of the uniform convergence as follows:

$$\sup_{g \in \mathcal{O}(f)} \Big| \frac{1}{n} \sum_{j=0}^{n-1} T^j(g) - \int f \, d\mu \Big| \to 0 \tag{4.17}$$

as $n \to \infty$. It is immediate that (4.17) is equivalent to the following statement:

$$\sup_{i \geq 0} \Big| \frac{1}{n} \sum_{j=0}^{n-1} T^{i+j}(f) - \int f \, d\mu \Big| \to 0 \tag{4.18}$$

as $n \to \infty$. Finally, we obtain by the continuity of f and denseness of $\mathcal{O}(x)$ for every $x \in X$, that (4.18) is equivalent to the following statement:

$$\sup_{x \in X} \Big| \frac{1}{n} \sum_{j=0}^{n-1} T^{j}(f)(x) - \int f \, d\mu \Big| \to 0 \tag{4.19}$$

as $n \to \infty$. However, this is precisely the statement of *Weyl's theorem* on uniform distribution mod 1 (see **[25]** p. 13). This establishes the uniform ergodic theorem (4.17).

In fact, the equivalence of (4.17)-(4.19) extends to any Markov operator T in $C(X)$ with X being a compact Hausdorff space. For this, it should be noted that the pointwise convergence in (4.18) with $i = 0$ means weak convergence in $C(X)$, and implies the strong convergence (see **[25]** p. 72) which is precisely (4.19). For more information in this direction see **[25]** (p. 12) and **[40]** (p. 137).

So far we have considered a Blum–DeHardt approach towards solution for the prime and dual problem by offering Lipschitz type conditions. The next lines concern the Vapnik–Chervonenkis random entropy approach. We begin in this direction by displaying a cornerstone for this type of results in the independent case. For this consider a sequence of independent and identically distributed random variables $\{\xi_j \mid j \geq 1\}$ defined on a probability space $(\Omega, \mathcal{F}, P)$ with values in a measurable space $(S, \mathcal{A})$. Let $f : S \times \Theta \to \mathbb{R}$ be a given map, such that $M(\theta) = \int f(\xi_1, \theta) dP$ is well-defined. Let $\{\varepsilon_j \mid j \geq 1\}$ be a Bernoulli sequence independent of $\{\xi_j \mid j \geq 1\}$. Then passing over measurability problems it is well-known that we have:

$$E \sup_{\theta \in \Theta} \Big| \frac{1}{n} \sum_{j=1}^{n} f(\xi_j, \theta) - M(\theta) \Big| \leq 2 \cdot E \sup_{\theta \in \Theta} \Big| \frac{1}{n} \sum_{j=1}^{n} \varepsilon_j \cdot f(\xi_j, \theta) \Big| \tag{4.20}$$

for all $n \geq 1$. It turns out that this symmetrization inequality plays a vitally important role in the Vapnik–Chervonenkis random entropy approach. Moreover, it might be easily verified that having this inequality in the case where the sequence $\{\xi_j \mid j \geq 1\}$ is only assumed to be stationary and ergodic, the basic random entropy result of Vapnik and Chervonenkis could be established in exactly the same way as in the independent case (see for instance the proof in **[53]**). However, our next example shows that this is not possible in general.

Example 4.7. We show that inequality (4.20) may fail if the sequence $\{\xi_j \mid j \geq 1\}$ is only assumed to be stationary and ergodic. For this we shall consider a simple case where Θ consists of a single point, and where f equals to the identity map on the real line. The sequence $\{\xi_j \mid j \geq 1\}$ itself is for a moment only assumed to be stationary and Gaussian.

Thus our question reduces to verify the following inequality:

$$E\Big|\frac{1}{n}\sum_{j=1}^{n}\xi_j\Big| \le C \cdot E\Big|\frac{1}{n}\sum_{j=1}^{n}\varepsilon_j \cdot \xi_j\Big| \tag{4.21}$$

for all $n \ge 1$ some constant $C > 0$, where $\{\varepsilon_j \mid j \ge 1\}$ is a Bernoulli sequence independent of $\{\xi_j \mid j \ge 1\}$. Let us first consider the right-hand side of this inequality. For this denote by $\|\cdot\|_{\Psi_2}$ the Orlicz norm induced by the function $\Psi_2(x) = \exp(x^2) - 1$ for $x \in \mathbb{R}$. Then it is easily verified that we have $\|X\|_1 \le 6/5\|X\|_{\Psi_2}$ whenever X is a random variable. Hence by Kahane–Khintchine's inequality for $\|\cdot\|_{\Psi_2}$ and Jensen's inequality we get (see **[38]**):

$$\begin{aligned} E\Big|\frac{1}{n}\sum_{j=1}^{n}\varepsilon_j \cdot \xi_j\Big| &= E_\xi E_\varepsilon\Big|\frac{1}{n}\sum_{j=1}^{n}\varepsilon_j \cdot \xi_j\Big| \\ &\le \frac{6}{5}\cdot\frac{1}{n}\cdot E_\xi\Big\|\sum_{j=1}^{n}\varepsilon_j \cdot \xi_j\Big\|_{\Psi_{2,\varepsilon}} \\ &\le \frac{6}{5}\cdot\sqrt{\frac{8}{3}}\cdot\frac{1}{n}\cdot E_\xi\Big(\sum_{j=1}^{n}|\xi_j|^2\Big)^{1/2} \\ &\le 2\cdot\frac{1}{n}\sqrt{nE|\xi_1|^2} \\ &= \frac{2}{\sqrt{n}}\cdot\sqrt{E|\xi_1|^2} \end{aligned} \tag{4.22}$$

for all $n \ge 1$. On the other hand since $n^{-1}\sum_{j=1}^{n}\xi_j$ is Gaussian, then we have:

$$\Big(E\Big|\frac{1}{n}\sum_{j=1}^{n}\xi_j\Big|^2\Big)^{1/2} \le D \cdot E\Big|\frac{1}{n}\sum_{j=1}^{n}\xi_j\Big| \tag{4.23}$$

for some constant $D > 0$ and all $n \ge 1$. Inserting (4.22) and (4.23) into (4.21) we obtain:

$$E\Big|\sum_{j=1}^{n}\xi_j\Big|^2 \le G \cdot n \cdot E|\xi_1|^2 \tag{4.24}$$

for all $n \ge 1$ with $G = \sqrt{2CD}$. Thus it is enough to show that (4.24) may fail in general. Since $\{\xi_j \mid j \ge 1\}$ is stationary, then we have $E(\xi_i\xi_j) = R(i-j)$ for all $i, j \ge 1$. Moreover, it is easily verified that the left-hand side in (4.24) may be written as follows:

$$E\Big|\sum_{j=1}^{n}\xi_j\Big|^2 = \sum_{i=1}^{n}\sum_{j=1}^{n}E(\xi_i\xi_j) = \sum_{i=1}^{n}\sum_{j=1}^{n}R(i-j) = nR(0) + \sum_{k=1}^{n-1}2(n-k)R(k) \tag{4.25}$$

for all $n \ge 1$. Let us in addition consider a particular case by putting $R(k) = 1/(k+1)$ for all $k \ge 0$. Then R is a decreasing convex function, and therefore by Polya's theorem it could be the covariance function of a centered stationary Gaussian sequence $\{\xi_j \mid j \ge 1\}$.

Moreover, by Maruyama's theorem this sequence is strongly mixing, and thus ergodic as well. Finally, from (4.25) we may easily obtain:

$$E\Big|\sum_{j=1}^{n} \xi_j\Big|^2 = n + 2\sum_{k=1}^{n-1} \frac{n-k}{k+1} = 2(n+1)\sum_{k=1}^{n-1} \frac{1}{k+1} - n + 2$$
$$\geq 2(n+1)\int_1^n \frac{1}{x+1}dx - n + 2 = 2(n+1)\log(n+1) - (1+\log 4)(n+1) + 3$$
$$\geq n \log n$$

for all $n \geq 1$. This inequality contradicts (4.24), and therefore (4.21) is false in this case as well.

Remark 4.8. It could be instructive to observe that in order that inequality (4.24) from Example 4.7 holds we should have $\| \sum_{j=1}^{n} \xi_j \|_2 = O(\sqrt{n})$ as $n \to \infty$. This fact is known to be valid if the sequence $\{\xi_j \mid j \geq 1\}$ is mixing enough (see for instance **[8]**). It could indicate that the symmetrization method is intimately related with a mixing property. Moreover, even though Example 4.7 shows that the symmetrization inequality (4.20) does not extend to the stationary ergodic case, it does not contradict the most useful consequence of (4.20) in this context. Namely, for Vapnik–Chervonenkis type results in the notation of (4.20), it seems quite enough to have an asymptotic symmetrization inequality as follows:

$$E\Big(\sup_{\theta\in\Theta}\Big|\frac{1}{n}\sum_{j=1}^{n} \varepsilon_j \cdot f(\xi_j,\theta)\Big|\Big) \to 0 \Rightarrow E\Big(\sup_{\theta\in\Theta}\Big|\frac{1}{n}\sum_{j=1}^{n} f(\xi_j,\theta) - M(\theta)\Big|\Big) \to 0$$

as $n \to \infty$. Our Example 4.7 shows that these convergencies could have different speeds. These observations may leave some open space for further developments (see **[39]**).

Acknowledgment. The authors thank J. Hoffmann-Jørgensen and S. E. Graversen for useful discussions.

References

[1] Atalla, R. E., On uniform ergodic theorems for Markov operators on $C(X)$, Rocky Mountain J. Math. 14 (1984), 451–456.

[2] Atalla, R. E., Correction to "On uniform ergodic theorems for Markov operators on $C(X)$", Rocky Mountain J. Math. 15 (1985), 763.

[3] Blum, J. R., On the convergence of empiric distribution functions, Ann. Math. Statist. 26 (1955), 527–529.

[4] Brunel, A. and Revuz, D., Quelques applications probabilistes de la quasi-compacité, Ann. Inst. H. Poincaré Probab. Statist. 10 (1974), 301–337.

[5] Burke, G., A uniform ergodic theorem, Ann. Math. Statist. 36 (1965), 1853–1858.

[6] Cantelli, F. P., Sulla determinazione empirica delle leggi di probabilità, Giorn. Ist. Ital. Attuari 4 (1933), 421–424.

[7] DeHardt, J., Generalizations of the Glivenko–Cantelli theorem, Ann. Math. Statist. 42 (1971), 2050–2055.

[8] Dehling, H., Limit theorems for sums of weakly dependent Banach space valued random variables, Z. Wahrscheinlichkeitstheorie verw. Geb. 63 (1983), 393–432.

[9] Doeblin, W., Sur les propriétés asymptotiques de mouvements régis par certains types de chaines simples, Bull. Soc. Math. Roumaine 39-1 (1937), 57–115 and 39-2 (1938), 3–61.

[10] Doob, J. L., Stochastic processes with an integral valued parameter, Trans. Amer. Math. Soc. 44 (1938), 87–150.

[11] Dudley, R. M., A course on empirical processes, in: Ecole d'Eté de Probabilitiés de Saint-Flour XII-1982, Lecture Notes in Math. 1097, pp. 1–142, Springer-Verlag, Berlin-Heidelberg-New York 1984.

[12] Dudley, R. M., Real analysis and probability, Wadsworth, Inc., Belmont, California 94002 (1989).

[13] Dunford, N. and Schwartz, J. T., Linear Operators, Part I: General Theory, Interscience Publ. Inc., New York (1958).

[14] Eberlein, W. F., Abstract ergodic theorems and weak almost periodic functions, Trans. Amer. Math. Soc. 67 (1949), 217–240.

[15] Gaenssler, P., Empirical Processes, IMS Lecture Notes – Monograph Series 3 (1983).

[16] Garsia, A. M., Topics in Almost Everywhere Convergence, Lectures in Advanced Mathematics, Markham Publishing Company (1970).

[17] Giné, J. and Zinn, J., Some limit theorems for empirical processes, Ann. Probab. 12 (1984), 929–989.

[18] Glivenko, V. I., Sulla determinazione empirica delle leggi di probabilità, Giorn. Ist. Ital. Attuari 4 (1933), 92–99.

[19] Groch, U., Uniform ergodic theorems for identity preserving Schwartz maps on W^* algebras, J. Operator Theory 11 (1984), 395–404.

[20] Hoffmann-Jørgensen, J., Necessary and sufficient conditions for the uniform law of large numbers, in: Probability in Banach Spaces V, Proc. Conf. Medford, USA, 1984, Lecture Notes in Math. 1153, pp. 258–272, Springer-Verlag, Berlin-Heidelberg 1985.

[21] Hoffmann-Jørgensen, J., Uniform convergence of martingales, in: Probability in Banach spaces VII, Proc. Conf. Oberwolfach, Germany 1988, Progress in Probability 21, pp. 127–137, Birkhäuser, Boston 1988.

[22] Hoffmann-Jørgensen, J., The general monotone convergence theorem and measurable selections, Institute of Mathematics, University of Aarhus, Preprint Series 13 (1991), 37 pp.

[23] Horowitz, S., Transition probabilities and contractions of L_∞, Z. Wahrscheinlickeitstheorie verw. Geb. 24 (1972), 263–274.

[24] Jajte, R., A few remarks on the almost uniform ergodic theorems in von Neumann algebras, in: Probability theory and vector spaces III, Proc. Conf. Lublin, Poland 1983, Lecture Notes in Math. 1080, pp. 120–143, Springer-Verlag, Berlin-Heidelberg-New York 1984.

[25] Krengel, U., Ergodic Theorems, de Gruyter Stud. Math. 6, Walter de Gruyter & Co., Berlin-New York 1985.

[26] Kryloff, N. and Bogolioùboff, N., Sur les propriétés en chaine, C. R. Paris 204 (1937), 1386–1388.

[27] Kryloff, N. and Bogolioùboff, N., Les propriétés ergodiques des suites des probabilités en chaine, C. R. Paris 204 (1937), 1454–1456.

[28] Ladoucer, S. and Weber, M., Speed of convergence of the mean average operator for quasi-compact operators, Preprint (1992).

[29] Ledoux, M. and Talagrand, M., Probability in Banach spaces (Isoperimetry and Processes), Springer-Verlag, Berlin-Heidelberg-New York 1991.

[30] Lin, M., On the uniform ergodic theorem, Proc. Amer. Math. Soc. 43 (1974), 337–340.

[31] Lin, M., On the uniform ergodic theorem II, Proc. Amer. Math. Soc. 46 (1974), 217–225.

[32] Lin, M., Quasi–compactness and uniform ergodicity of positive operators, Israel J. Math. 29 (1978), 309–311.

[33] Lloyd, S. P., On the uniform ergodic theorem of Lin, Proc. Amer. Math. Soc. 83 (1981), 710–714.

[34] Lotz, H. P., Uniform ergodic theorems for Markov operators on $C(X)$, Math. Z. 178 (1981), 145–156.

[35] Peškir, G., Perfect measures and maps, Institute of Mathematics, University of Aarhus, Preprint Series 26 (1991), 34 pp.

[36] Peškir, G., Uniform convergence of reversed martingales, Institute of Mathematics, University of Aarhus, Preprint Series 21 (1992), 27 pp., J. Theoret. Probab. 8 (1995), 387–415.

[37] Peškir, G. and Weber, M., Necessary and sufficient conditions for the uniform law of large numbers in the stationary case, Institute of Mathematics, University of Aarhus, Preprint Series 27 (1992), 26 pp., Proc. Conf. Functional Anal. IV (Dubrovnik 1993), Various Publ. Ser. 43 (1994), 165–190.

[38] Peškir, G., Best constants in Kahane–Khintchine inequalities in Orlicz spaces, Institute of Mathematics, University of Aarhus, Preprint Series 10 (1992), 42 pp., J. Multivariate Anal. 45 (1993), 183–216.

[39] Peškir, G. and Yukich, J. E., Uniform ergodic theorems for dynamical systems under VC entropy conditions, Institute of Mathematics, University of Aarhus, Preprint Series 15 (1993), 25. pp., Proc. Probab. Banach Spaces IX (Sandbjerg 1993), Birkhäuser, Boston 1994, 104–127.

[40] Petersen, K., Ergodic Theory, Cambridge University Press, Cambridge 1983.

[41] Pollard, D., Convergence of Stochastic Processes, Springer-Verlag, New York 1984.

[42] Pollard, D., Empirical processes: Theory and Applications, NSF–CBMS Regional Conference Series in Probability and Statistics 2 (1990).

[43] Shaw, S. Y., Uniform ergodic theorems for locally integrable semigroups and pseudo–resolvents, Proc. Amer. Math. Soc. 98 (1986), 61–67.

[44] Talagrand, M., The Glivenko–Cantelli theorem, Ann. Probab. 15 (1987), 837–870.

[45] Vapnik, V. N. and Chervonenkis, A. Ya., On the uniform convergence of relative frequencies of events to their probabilities, Theory Probab. Appl. 16 (1971), 264–280.

[46] Vapnik, V. N. and Chervonenkis, A. Ya., Necessary and sufficient conditions for the uniform convergence of means to their expectations, Theory Probab. Appl. 26 (1981), 532–553.

[47] Weber, M., Une version fonctionelle du théorème ergodique ponctuel, C. R. Acad. Sci. Paris Sér. I Math. 311 (1990), 131–133.

[48] Weber, M., GC sets, Stein's elements and matrix summation methods, Prepublication IRMA 027 (1993), 96 pp.

[49] Weber, M., Coupling of the GB set property for ergodic averages, J. Theoret. Probab. 9 (1996), 105–112.

[50] Yosida, K. and Kakutani, S., Application of mean ergodic theorem to the problem of Markoff's process, Proc. Imp. Acad. Japan 14 (1938), 333–339.

[51] Yosida, K., Operator-theoretical treatment of Markoff's process, Proc. Imp. Acad. Japan 14 (1938), 363–367; 15 (1939), 127–130.

[52] Yosida, K. and Kakutani, S., Operator-theoretical treatment of Markoff's process and mean ergodic theorem, Ann. of Math. 42 (1941), 188–228.

[53] Yukich, J. E., Sufficient conditions for the uniform convergence of means to their expectations, Sankhya, Ser. A 47 (1985), 203–208.

Institute of Mathematics
University of Aarhus
Ny Munkegade, 8000 Aarhus
Denmark

and

Department of Mathematics
University of Zagreb
Bijenička 13, 41000 Zagreb
Croatia

I.R.M.A.
Université Louis Pasteur et CNRS
7, rue René Descartes
67084 Strasbourg
France

Pointwise Fourier Inversion and Related Jacobi Polynomial Expansions

Mark A. Pinsky[1]

O. Introduction

This report consists of two parts. In the first part we describe results, proved elsewhere, which give necessary and sufficient conditions for the convergence at a preassigned point of the spherical partial sums (resp. partial integrals) of the Fourier series (resp. Fourier integral) in the class of piecewise smooth functions on Euclidean space. These are also formulated for orthogonal series expansions in the eigenfunctions of the Laplace operator of a rotationally invariant Riemannian manifold. The conditions are vacuous in two dimensions, but yield non-trivial examples of divergent Fourier series/integrals in higher dimensions, with respect to a preassigned point. In the second part we extend the results to the expansion in terms of Jacobi polynomials, which generalizes the spherical harmonic expansion. The proof is self-contained and is presented here in its entirety.

I. Pointwise Fourier inversion

I.0 Definitions. A function $f \in L^1(\mathbb{R}^n)$ is *piecewise smooth* with respect to $x \in \mathbb{R}^n$ if there exists a finite set $0 = r_0(x) < r_1(x) < \cdots < r_K(x)$ so that the mean value

$$r \to \bar{f}_x(r) \equiv \int_{S^{d-1}} f(x + r\omega)\, d\omega$$

is smooth on each subinterval (r_{i-1}, r_i) and identically zero for $r \geq r_K(x)$.

The *smoothness index* of f at x is the integer defined by

$$j(f;x) = \max\{j : r \to \bar{f}_x(r) \text{ is differentiable of class } C^j\}$$

If $\bar{f}_x$ is discontinuous, we set $j(f;x) = -1$.

In general we have $j(f;x) \geq -1$ for any piecewise smooth function. For example $j(f;x) = 0$ signifies continuity with a discontinuous derivative, $j(f;x) = +1$ signifies a continuous first derivative only.

1 Research sponsored by National Science Foundation.

I.1 Convergence of Fourier transforms. The Fourier transform and spherical partial sum of an integrable function are defined by

$$\hat{f}(\mu) = \frac{1}{(2\pi)^n} \int_{\mathbb{R}^n} e^{-i\langle \mu, x\rangle} f(x)\, dx \quad \text{and} \quad f_M(x) = \int_{|\mu|\le M} e^{i\langle \mu, x\rangle} \hat{f}(\mu)\, d\mu.$$

The following basic result describes the convergence of the Fourier transform in terms of the smoothness index.

Theorem 1. *Suppose that f is piecewise smooth w.r.t. x.*

If $n = 1, 2$ then $\lim_{M\uparrow\infty} f_M(x) = \bar{f}_x(0+0)$.

If $n \ge 3$ then $\lim_{M\uparrow\infty} f_M(x)$ exists iff $r \to \bar{f}_x(r)$ has $[\frac{n-3}{2}]$ continuous derivatives, i.e. $j(f;x) \ge [(n-3)/2]$; in this case $\lim_{M\uparrow\infty} f_M(x) = \bar{f}_x(0+0)$. Otherwise we have

$$-\infty < \liminf_{M\uparrow\infty} M^{-\nu}[f_M(x) - \bar{f}_x(0+0)] < \limsup_{M\uparrow\infty} M^{-\nu}[f_M(x) - \bar{f}_x(0+0)] < \infty$$

where $\nu = \frac{n-5}{2} - j(f;x)$ and $j(f;x)$ is the smoothness index defined above.

The proof [P1] depends on the integral representation of the spherical partial sum

$$f_M(x) = \int_0^{r_K(x)} D_n^M(r) \bar{f}_x(r)\, dr$$

where the Dirichlet kernel $D_n^M(r)$ is expressed in terms of the Bessel function $J_{\frac{n}{2}}(Mr)$. Integration by parts k times reduces this integral to a sum of k boundary terms and a spherical partial sum in $n-2k$ dimensions, which can be handled by theorems of one or two-dimensional Fourier analysis if $k = 1 + [\frac{n-3}{2}]$.

The explicit analysis of the boundary terms leads to the conditions of the theorem.

I.2 Convergence of Fourier series of periodic functions. The n-dimensional torus is $\mathbb{T}^n = \{x = (x_1, \ldots, x_n) : -\pi \le x_i \le \pi\}$. The Fourier coefficients and spherical partial sum of an integrable function $f(x)$, $x \in \mathbb{T}^n$ are

$$A_m = \frac{1}{(2\pi)^n} \int_{\mathbb{T}^n} f(x) e^{-i\langle m, x\rangle}\, dx, \qquad m \in \mathbb{Z}^n$$

$$f_M(x) = \sum_{|m|\le M} A_m e^{i\langle m, x\rangle}, \qquad x \in \mathbb{R}^n$$

The spherical mean value is $\bar{f}_x(r) = \frac{1}{\omega_{n-1}} \int_{S^{n-1}} f(x + r\omega)\, d\omega$, where we extend the function f periodically to $\mathbb{R}^n$.

Definition. $f \in L^1(\mathbb{T}^n)$ is a *piecewise smooth radial function with respect to $x \in \mathbb{T}^n$* if $r \to \bar{f}_x(r)$ is piecewise C^∞ and the periodic extension satisfies $f(x + r\omega) = \bar{f}_x(r)$, $0 \le r \le \pi$, $\omega \in \mathbb{S}^{n-1}$ and $f(x + r\omega) = 0$ if $r > \pi$, $r\omega \in \mathbb{T}^n$.

In two dimensions the proof of convergence for a piecewise smooth radial function can de accomplished by appealing to a theorem of Hardy and Landau [HL]. Otherwise we proceed as follows; for a radial function the Fourier coefficient is the restriction of a smooth radial function: $A_m = A(|m|)$ where the asymptotic behavior of $A(r)$, $A'(r)$, $r \uparrow \infty$ is determined explicitly. The spherical partial sum is expressed as a Stieltjes integral:

$$f_M(0) := \sum_{|m|\leq M} A_m = \int_0^R A(r)\,dN(r)$$

where $N(r)$ is the counting function of lattice points, satisfying *Landau's estimate*

$$N(r) := \sum_{|m|\leq r} 1 = c_n r^n + O(r^{n-2+\frac{2}{n+1}}).$$

This is used to write

$$f_M(0) - \int_0^M A(r) c_n\, n r^{n-1}\,dr = A(M)[N(M) - c_m M^n] + \int_0^M A'(r)[N(r) - c_n r^n]\,dr$$

and leads to the following theorem[P2], which extends the results in [PST].

Theorem 2. *Suppose that f is piecewise smooth and radial with respect to $x \in \mathbb{T}^n$. If $n = 1, 2$ the Fourier series converges, and*

$$\lim_{M\to\infty} f_M(x) = \bar{f}_x(0+0).$$

If $n \geq 3$ the Fourier series converges iff the function $r \to \bar{f}_x(r)$ has at least $[\frac{n-3}{2}]$ continuous derivatives.

I.3 Fourier series with boundary conditions. The methods of the preceding section can be used to study the expansion of a radial function in an eigenfunction series in the cube $-\pi \leq x_i \leq \pi$:

$$f(x) \sim \sum_{m\in Z^n} A_m \phi_m(x)$$

where the eigenfunctions ϕ_m satisfy $\Delta\phi + \lambda\phi = 0$ in the cube and the Dirichlet condition that $\phi = 0$ on the boundary. (The corresponding problem with Neumann or mixed Dirichlet–Neumann boundary conditions can be dealt with by suitably modifications.)

Separation-of-variables produces the eigenfunctions in the form

$$\phi_{m_1,\ldots,m_n}(x_1,\ldots,x_n) = \prod_{i=1}^n \left(\cos\frac{m_i x_i}{2}\sin\frac{m_i\pi}{2} + \sin\frac{m_i x_i}{2}\cos\frac{m_i\pi}{2}\right).$$

The Fourier coefficients of a radial function $f(x) = F(|x|)$ are computed as

$$A_{m_1,\ldots,m_n} = \pi^{-n}\int_{|x|\leq\pi} F(|x|)\,\phi_{m_1,\ldots,m_n}(x_1,\ldots,x_n)\,dx.$$

At the center of the cube we have $\sum_m A_m\phi_m(0) = \sum_{m_i \text{odd}} A_m \prod_{i=1}^n \sin\frac{m_i\pi}{2}$. But for m_i odd, $A_m = \pi^{-n}\int_{|x|\leq\pi} F(|x|)\prod_{i=1}^n \cos\frac{m_i x_i}{2}\sin\frac{m_i\pi}{2}\,dx$ so that $\sum_m A_m\phi_m(0) = \pi^{-n}\sum_m \int_{|x|\leq\pi} F(|x|)\prod_{i=1}^n \cos\frac{m_i x_i}{2}\sin^2\frac{m_i\pi}{2}\,dx$.

To complete the analysis of the Fourier expansion, we need the asymptotic behavior of the number of *positive* lattice points, defined as

$$N_n^+(\lambda) := \sum_{m_i>0, |m|\leq\lambda} 1.$$

Proposition. *When* $\lambda \to \infty$, $N_n^+(\lambda) = e_n\lambda^n - f_n\lambda^{n-1} + O(\lambda^{n-2+\frac{2}{n+1}})$, *for suitable positive constants* e_n, f_n.

A corresponding analysis leads to the following identical result:

Theorem 3. *Supppose that* $f(x) = F(|x|)$ *where* F *is piecewise smooth on* $[0, \pi]$ *and zero elsewhere. Then the series expansion* $\sum_m A_m\phi_m(0)$ *converges iff* F *is of class* $[\frac{n-3}{2}]$.

I.4 Spherical harmonic expansions. The expansion of an L^2 function on the sphere $\mathbb{S}^n$ is written

$$f \sim \sum_{m=0}^{\infty} Y_m^{(n)}$$

where $Y_m^{(n)} \in \mathcal{H}_m$, the space of spherical harmonics of degree m with respect to the sphere $\mathbb{S}^n$. The partial sum $f_M := \sum_{0\leq m\leq M} Y_m^{(n)}$ is the orthogonal projection of f onto the linear subspace of spherical harmonics of degree $\leq M$. The spherical harmonics of degree m are $\mathrm{SO}(n+1)$-invariantly defined as homogeneous polynomials of degree m which satisfy Laplace's equation $\Delta_{\mathbb{R}^{n+1}} Y = 0$. In order to compute $f_M(Q)$, we may take a basis of spherical harmonics with respect to the axis $Q \in \mathbb{S}^n$. Without loss of generality we will suppose that $Q = (1, 0, \ldots, 0)$ is a fixed origin from which distance on the sphere will be measured. A system of geodesic polar coordinates is written (r, θ), where $0 \leq r \leq \pi$ and $\theta \in \mathbb{S}^{n-1}$. With this notation the metric of $\mathbb{S}^n$ is written

$$ds^2 = dr^2 + \sin^2 r\, d\theta^2$$

and the Laplacian is written

$$\Delta_{\mathbb{S}^n} = \frac{d^2}{dr^2} + (n-1)\cot r \frac{d}{dr} + \frac{1}{\sin^2 r}\Delta_{\mathbb{S}^{n-1}}.$$

An n-dimensional spherical harmonic is written

$$Y_m^{(n)}(r, \theta) = \phi(r)\psi(\theta)$$

where the radial factor ϕ is a solution of the ordinary differential equation

$$\phi'' + (n-1)\cot r\,\phi' + (\lambda - \frac{\beta}{\sin^2 r})\phi = 0, \qquad \lambda = |\boldsymbol{m}|(|\boldsymbol{m}| + n - 1)$$

and $\psi(\theta)$ is a spherical harmonic on $\mathbb{S}^{n-1}$, solution of

$$\Delta_{\mathbb{S}^{n-1}}\psi + \beta\psi = 0.$$

An integrable function $f(Q)$, $Q \in \mathbb{S}^n$ is said to be *piecewise smooth* if $\forall Q_0 \in \mathbb{S}^n$ the spherical mean value

$$(0, \pi) \ni r \to \bar{f}_{Q_0}(r) = \frac{1}{\omega_{n-1}} \int_{\{Q:d(Q,Q_0)=r\}} f(Q)\, dS(Q)$$

is piecewise C^∞ on the interval $0 \le r \le \pi$. The Fourier expansion of a piecewise smooth function is written

$$f \sim \sum_{k,m} A_{k,m} \phi_{k,m}(r) \psi_m(\theta)$$

where

$$A_{k,m} = \frac{\int_{\mathbb{S}^{n-1}} \int_0^\pi f(r,\theta) \phi_{k,m}(r) \psi_m(\theta) \sin^{n-1} r\, dr\, d\omega}{\int_{\mathbb{S}^{n-1}} \int_0^\pi \phi_{k,m}(r)^2 \psi_m(\theta)^2 \sin^{n-1} r\, dr\, d\omega}.$$

From the above remarks, we have the series at $r = 0$ in the form

$$f(0) \sim \sum_{k \ge 1} A_{k,0} \phi_{k,0}(0)$$

and the f-dependence of the Fourier coefficient is given by

$$\int_{\mathbb{S}^{n-1}} \int_0^\pi f(r,\theta) \phi_{k,0}(r) \sin^{n-1} r\, dr\, d\omega = \omega_{n-1} \int_0^\pi \bar{f}_0(r) \phi_{k,0}(r) \sin^{n-1} r\, dr.$$

This shows that *the series for $f(0)$ depends only on the spherical mean value $\bar{f}_0(r)$*, $0 \le r \le \pi$. Therefore we may restrict attention to radial functions and obtain the following result [P2].

Theorem 4. *Let $f(Q), Q \in \mathbb{S}^n$ be a piecewise smooth function. Then the spherical harmonic expansion converges at $Q_0 \in \mathbb{S}^n$ iff the spherical mean $r \to \bar{f}_{Q_0}(r)$ has $[\frac{n-3}{2}]$ continuous derivatives.*

I.5 Dirichlet eigenfunction expansions. In the previous sections we presented results on the convergence of series expansions of functions on complete manifolds without boundary, where we have the full group of translations, allowing any point as the origin. When we pass to manifolds with boundary, we have a privileged origin, where we can study the convergence/divergence of the expansions. To motivate the exposition, we first present the case of a Euclidean ball with Dirichlet boundary conditions, where the analysis is explicitly described by Bessel functions. Previous studies of this case with graphical illistrations are contained in [GP1, GP2]

I.5a Analysis on $\mathbb{R}^n$: Fourier–Bessel series. We consider the Euclidean space $\mathbb{R}^n$ with a ball of radius $a > 0$:

$$B_a = \{x : 0 \le |x| < a\}.$$

The n-dimensional Laplace operator is $\Delta = \sum_{i=1}^{n} \frac{\partial^2}{\partial x_i^2}$. Let $\phi = \phi_{k,\boldsymbol{m}}(x)$ denote the eigenfunctions of the n-dimensional Laplace operator in the ball:

$$\Delta\phi + \lambda\phi = 0, \qquad 0 \leq |\boldsymbol{x}| < a$$

$$\phi(x) = 0, \qquad |\boldsymbol{x}| = a.$$

These may be computed in terms of Bessel functions and spherical harmonics by the technique of separation of variables according to

$$\phi_{k,\boldsymbol{m}}(\boldsymbol{x}) = r^{|\boldsymbol{m}|} \frac{J_{|\boldsymbol{m}|+\frac{1}{2}(n-2)}(r\sqrt{\lambda})}{r^{|\boldsymbol{m}|+\frac{1}{2}(n-2)}} Y_{\boldsymbol{m}}(\omega), \quad r = |\boldsymbol{x}|,\ \omega = \boldsymbol{x}/|\boldsymbol{x}|, \quad \lambda = [\frac{x_k^{(m,n)}}{a}]^2.$$

Here the first factor denotes the standard Bessel function and the second factor is a *spherical harmonic of degree* $|\boldsymbol{m}|$, a solution of the equation $\Delta_{\mathbb{S}^{n-1}} Y + |\boldsymbol{m}|(|\boldsymbol{m}|+n-2)Y = 0$, where $\Delta_{\mathbb{S}^{n-1}}$ is the spherical Laplace operator. The quantity $x_k^{(m,n)}$ denotes the kth zero of the Bessel function $J_{|\boldsymbol{m}|+\frac{1}{2}(n-2)}$.

The index $k = 1, 2, \ldots$ while the index $\boldsymbol{m}$ takes values in a finite set which labels the spherical harmonics of degree $|\boldsymbol{m}| = 0, 1, 2 \ldots$. These functions are orthogonal with respect to the Euclidean volume element in the ball B_a, which, when written in polar coordinates is

$$\begin{aligned}
\langle \phi_{k_1,\boldsymbol{m}_1}, \phi_{k_2,\boldsymbol{m}_2} \rangle &= \int_{B_a} \phi_{k_1,\boldsymbol{m}_1}(\boldsymbol{x}) \phi_{k_2,\boldsymbol{m}_2}(\boldsymbol{x}) d\boldsymbol{x} \\
&= \int_0^a \int_{\mathbb{S}^{n-1}} \phi_{k_1,\boldsymbol{m}_1}(\boldsymbol{x}) \phi_{k_2,\boldsymbol{m}_2}(\boldsymbol{x}) r^{n-1}\, dr\, d\omega = 0, \quad (k_1, \boldsymbol{m}_1) \neq (k_2, \boldsymbol{m}_2).
\end{aligned}$$

The Fourier–Bessel coefficients of an integrable function are defined by

$$A_{k,\boldsymbol{m}} = \frac{\int_{B_a} f(\boldsymbol{x}) \phi_{k,\boldsymbol{m}}(\boldsymbol{x}) d\boldsymbol{x}}{\int_{B_a} \phi_{k,\boldsymbol{m}}(\boldsymbol{x})^2 d\boldsymbol{x}}$$

Theorem 5a. *Let* $f(\boldsymbol{x})$, $\boldsymbol{x} \in B_a$ *be an integrable function for which the mean value*

$$r \to \bar{f}(r) := \frac{\int_{\mathbb{S}^{n-1}} f(r\omega) d\omega}{\int_{\mathbb{S}^{n-1}} d\omega}$$

is piecewise C^∞.

i) *If* $n = 1, 2$, *then* $\lim_{M\to\infty} \sum_{\lambda_k \leq M} A_{k,\boldsymbol{m}} \phi_{k,m}(0)$ *exists.*

ii) *If* $n \geq 3$ *then* $\lim_{M\to\infty} \sum_{\lambda_k \leq M} A_{k,\boldsymbol{m}} \phi_{k,m}(0)$ *exists if and only if the mean value function* $\bar{f}(r)$ *has* $[\frac{(n-3)}{2}]$ *continuous derivatives for* $0 \leq r \leq a$ *which vanish at* $r = a$ ($[\,] =$ *integer part).*

I.5b General case: asymptotic solutions of differential equations. We now analyze the eigenfunction expansions on a general manifold with rotational symmetry. The Riemannian metric is written in geodesic polar coordinates as

$$ds^2 = dr^2 + G(r)^2 d\theta^2$$

where $G(r)$ is a smooth function with $G(r) > 0, \ (r > 0), \ G(0) = 0, \ G'(0) = 1, \ G''(0) = 0$.

The equation satisfied by the radial eigenfunctions is

$$0 = \Delta\phi + \lambda\phi = \phi''(r) + (n-1)\frac{G'}{G}\phi'(r) + \lambda\phi = 0.$$

The asymptotic analysis for $\lambda \to \infty$ has been carried out by Olver [O, Thm. 4.1, p. 444] With some work, these can be expressed asymptotically in terms of Φ_λ, which satisfies the Euclidean equation $\Phi'' + \frac{n-1}{r}\Phi' + \lambda\Phi = 0$ with $\Phi(0) = 1$. With a suitable normalization for ϕ_λ, we have

$$\phi_\lambda(r) = \left(\frac{G}{r}\right)^{\frac{1-n}{2}} \Phi_\lambda(r) + \frac{1}{\sqrt{\lambda^\nu}} \frac{G^{\frac{1-n}{2}}}{\sqrt{r}} \epsilon(r^2, \lambda), \qquad \lambda \to \infty.$$

where $\lim_{\lambda\to\infty} \epsilon = 0$.

In particular $\phi_\lambda(0) = 1 + O(\frac{1}{\sqrt{\lambda}})$, $\lambda \to \infty$. For fixed $r > 0$ the first term is $O(\frac{1}{\sqrt{\lambda}^{(\nu+\frac{1}{2})}})$ while the second term is $O(\frac{1}{\sqrt{\lambda}^{(\nu+1)}})$ when $\lambda \to \infty$. This leads to the following theorem, exactly as for the Fourier–Bessel case.

Theorem 5b. *Let $f(r)$, $0 \le r \le a$, be a piecewise smooth radial function on a geodesic ball of an n-dimensional Riemannian manifold with rotational symmetry. Then the eigenfunction expansion $\sum_k A_k \tilde{\phi}_{\lambda_k}(0)$ converges iff $r \to f(r)$ has $[\frac{n-3}{2}]$ continuous derivatives.*

The complete proofs of the above assertions are contained in [P].

II. Pointwise convergence of Jacobi polynomial expansions

II.1 Statement of results. In the following paragraphs we determine necessary and sufficient conditions for the pointwise convergence of the Jacobi polynomial expansion

$$f(x) \sim \sum_{n\ge 0} c_n P_n^{\alpha,\beta}(x), \qquad -1 \le x \le 1.$$

Here $f(x), -1 \le x \le 1$ is a *piecewise smooth function*; this means that there is a subdivision of points $-1 = a_0 < a_1 < \cdots < a_K = 1$ so that f is C^∞ on each interval $(a_{i-1}, a_i)_{1\le i\le K}$ and at the endpoints there exist the limits $f(a_i \pm 0)$ for f as well as for its higher derivatives. The smoothness index of f is defined by

$$j(f) := \min\{j : f \text{ has } j \text{ continuous derivatives}\}$$

If f is discontinuous, we set $j(f) = -1$. The main result is stated as follows.

Theorem II.1. *At an interior point* $-1 < x < 1$ *the Jacobi polynomial expansion converges without further conditions. At the endpoint* $x = 1$ *the Jacobi polynomial expansion converges if and only if we have the inequality* $j(f) > \alpha - \frac{3}{2}$. *At the endpoint* $x = -1$ *the Jacobi polynomial expansion converges if and only if we have the inequality* $j(f) > \beta - \frac{3}{2}$.

This generalizes the result described in I.4 for the case of ultraspherical polynomial expansions where $\alpha = \beta$ is a half-integer.

II.2 Facts about Jacobi polynomials. We recall the following facts about Jacobi polynomial expansions from [Sz]:

i) The Fourier coefficients are obtained as

$$c_n = \frac{\int_{-1}^{1} f(x) P_n^{\alpha,\beta}(x) w(x)\, dx}{\int_{-1}^{1} P_n^{\alpha,\beta}(x)^2\, w(x)\, dx},$$

where $w(x) = (1-x)^\alpha (1+x)^\beta$.

ii) The normalization coefficients in the denominator are given by [Sz, p. 68]

$$h_n^{\alpha,\beta} = \int_{-1}^{1} P_n^{\alpha,\beta}(x)^2\, w(x)\, dx = \frac{\Gamma(n+\alpha+1)\Gamma(n+\beta+1)}{\Gamma(n+1)\Gamma(n+\alpha+\beta+1)} \frac{2^{\alpha+\beta+1}}{2n+\alpha+\beta+1}.$$

iii) The Jacobi polynomials $y = P_n^{\alpha,\beta}(x)$ satisfy the differential equation [Sz, p. 60]

$$(1-x^2)y'' + (\beta - \alpha - (\alpha+\beta+2)x)y' + n(n+\alpha+\beta+1)y = 0$$

equivalently

$$\frac{d}{dx}[(1-x)^{\alpha+1}(1+x)^{\beta+1}y'] + n(n+\alpha+\beta+1)y = 0.$$

iv) At the endpoints we have [Sz, p. 58]

$$P_n^{\alpha,\beta}(1) = \frac{\Gamma(n+\alpha+1)}{n!\Gamma(\alpha+1)}, \qquad P_n^{\alpha,\beta}(-1) = (-1)^n \frac{\Gamma(n+\beta+1)}{n!\Gamma(\beta+1)}.$$

v) The asymptotic behavior of the Jacobi polynomial at interior points is governed by the uniform asymptotic estimate [Sz, p. 168]

$$P_n^{\alpha,\beta}(\cos\theta) = n^{-\frac{1}{2}} k(\theta)[\cos\big((n+\gamma_1)\theta + \gamma_2\big) + O(n^{-1})], \qquad \epsilon \le \theta \le \pi - \epsilon$$

$$\frac{dP_n^{\alpha,\beta}}{dx}(\cos\theta) = -n^{\frac{1}{2}} k(\theta)[\sin\big((n+\gamma_1)\theta + \gamma_2\big) + O(n^{-1})], \qquad \epsilon \le \theta \le \pi - \epsilon$$

where $k(\theta) = \pi^{-1/2}(\sin\frac{\theta}{2})^{-\alpha-\frac{1}{2}}(\cos\frac{\theta}{2})^{\beta-\frac{1}{2}}$, $\gamma_1 = (\alpha+\beta+1)/2$, $\gamma_2 = -(\alpha+\frac{1}{2})\pi/2$.

II.3 Basic estimates.

Lemma 1. *The normalization coefficients satisfy the asymptotic estimate*

$$h_n^{\alpha,\beta} = \frac{2^{\alpha+\beta}}{n}\left(1 + O\left(\frac{1}{n}\right)\right), \qquad n \to \infty$$

Proof. This is immediate from Stirling's formula applied to the formula for $h_n^{\alpha,\beta}$. □

Lemma 2. *For any piecewise smooth function f,*

$$\int_{-1}^{1} f(x) P_n^{\alpha,\beta}(x) w(x)\, dx = o\left(\frac{1}{\sqrt{n}}\right), \qquad n \to \infty.$$

Proof. The functions $\frac{P_n^{\alpha,\beta}}{\sqrt{h_n}}$ form an orthonormal set, hence the expansion coefficients tend to zero, which translates into the statement that $\int_{-1}^{1} f(x) P_n^{\alpha,\beta}(x)\, w(x)\, dx = o(\sqrt{h_n})$, $n \to \infty$.

Referring to the asymptotic form of h_n from Lemma 1 gives the result.

Lemma 3. *For any piecewise smooth function, we have the identity*

$$n(n+\alpha+\beta+1) \int_{-1}^{1} f(x) P_n^{\alpha,\beta}(x) w(x)\, dx$$

$$= \sum_{1 \le i \le K} (1 - a_i^2) w(a_i) \left[\frac{d P_n^{\alpha,\beta}}{dx}(a_i) \delta f(a_i) - P_n^{\alpha,\beta}(a_i) \delta f'(a_i)\right] - \int_{-1}^{1} P_n^{\alpha,\beta}(x) L F(x) w(x) dx$$

where we have set $\delta f(a) := f(a+0) - f(a-0)$, $Lf(x) := (1-x^2) f'' + [(\beta - \alpha) - x(\alpha+\beta+2)] f'$.

Proof. This results from multiplying the differential equation for $y = P_n^{\alpha,\beta}$ by f, integrating-by-parts twice and using the definition of the δ operator.

II.4 Proof of the Theorem. We first note from Lemma 2 and Lemma 3 the following asymptotic estimate for the numerator of the Fourier coefficient:

$$n(n+\alpha+\beta+1) \int_{-1}^{1} f(x) P_n^{\alpha,\beta}(x) w(x)\, dx$$

$$= \sum_{1 \le i \le K} (1 - a_i^2) w(a_i) \left[\frac{d P_n^{\alpha,\beta}}{dx}(a_i) \delta f(a_i) + O(n^{-\frac{1}{2}})\right]$$

Taking into account the estimate of lemma 1 and the asymptotic behavior of the Jacobi polynomials as expressed by v), we are lead to the following asymptotic evaluation of the Fourier coefficient:

$$c_n = n^{-\frac{1}{2}} \sum_{1 \le i \le K} B_i \sin\big((n + \gamma_1)\theta_i + \gamma_2\big) + O(n^{-1}), \qquad n \to \infty$$

where $\theta_i = \cos^{-1}(a_i) \in (0,\pi)$, $B_i = (1-a_i^2)w(a_i)k(\theta_i)(\delta f)(a_i)\sqrt{2}^{-\alpha-\beta}$.

To prove the convergence at an interior point $-1 < x < 1$ we again use the asymptotic estimate v) and find that the terms of the series behave according to

$$c_n P_n^{\alpha,\beta}(\cos\theta) = n^{-1}k(\theta) \sum_{1\le i\le K} B_i \sin\big((n+\gamma_1)\theta_i + \gamma_2\big)k(\theta_i) \cdot \cos\big((n+\gamma_1)\theta + \gamma_2\big) + O(n^{-2})$$

which is clearly the general term of a convergent numerical series.

In order to prove the convergence at the endpoints, we first consider the case $j(f) = -1$. In this case we have from Lemma 3 and Lemma 2 the following asymptotic estimate for the numerator of the Fourier coefficient:

$$n(n+\alpha+\beta+1)\int_{-1}^{1} f(x)P_n^{\alpha,\beta}(x)w(x)\,dx = \sum_{1\le i\le K}(1-a_i^2)w(a_i)\Big[\frac{dP_n^{\alpha,\beta}}{dx}(a_i)\delta f(a_i) + O(n^{-\frac{1}{2}})\Big]$$

Taking into account the estimate of Lemma 1 and the asymptotic behavior of the Jacobi polynomials as expressed by v), we are lead to the following asymptotic evaluation of the Fourier coefficient:

$$c_n = n^{-\frac{1}{2}} \sum_{1\le i\le K} B_i\big[\sin\big((n+\gamma_1)\theta_i+\gamma_2\big) + O(n^{-1})\big], \qquad n\to\infty.$$

At the endpoint $x = 1$ we have from Stirling's formula $P_n^{\alpha,\beta}(1) = D_\alpha n^\alpha(1+O(1/n))$ where $D_\alpha > 0$. This gives the asymptotic behavior

$$c_n P_n^{\alpha,\beta}(1) = D_\alpha n^{\alpha-\frac{1}{2}} \sum_{1\le i\le K} B_i[\sin\big((n+\gamma_1)\theta_i+\gamma_2\big) + O(n^{-1})].$$

If $\alpha - \frac{1}{2} < 0$ this is the general term of a convergent series, while if $\alpha - \frac{1}{2} \ge 0$ this term doesn't tend to zero, hence a divergent series.

At the endpoint $x = -1$ we have from Stirling's formula $P_n^{\alpha,\beta}(1) = D_\beta(-1)^n n^\beta(1+O(1/n))$ where $D_\beta > 0$. This gives the asymptotic behavior

$$c_n P_n^{\alpha,\beta}(-1) = (-1)^n D_\beta n^{\beta-\frac{1}{2}} \sum_{1\le i\le K} B_i[\sin\big((n+\gamma_1)\theta_i+\gamma_2\big) + O(n^{-1})].$$

If $\beta - \frac{1}{2} < 0$ this is the general term of a convergent series, while if $\beta - \frac{1}{2} \ge 0$ this term doesn't tend to zero, hence a divergent series, which completes the analysis in case $j(f) = -1$.

To prove the theorem in case $j(f) = 0$, we have $\delta f(a_i) = 0$ for all i, $1 \le i \le K$, but $\delta f'(a_i) \ne 0$ for some i, $1 \le i \le K$. The Fourier coefficient is found to be

$$c_n = n^{-\frac{3}{2}} \sum_{1\le i\le K} B_i\big[\cos\big((n+\gamma_1)\theta_i+\gamma_2\big) + O(n^{-1})\big], \qquad n\to\infty$$

where $B_i = (1 - a_i^2)w(a_i)k(\theta_i)(\delta f')(a_i)$. Accordingly the general term of the series is

$$c_n P_n^{\alpha,\beta}(1) = D_\alpha n^{\alpha-\frac{3}{2}} \sum_{1 \le i \le K} B_i[\cos((n+\gamma_1)\theta_i + \gamma_2) + O(n^{-1})].$$

Clearly this is convergent iff $\alpha - \frac{3}{2} < 0$ which was to be proved in this case. The analysis for the endpoint $x = -1$ is entirely similar.

In order to prove the result in case $j(f) \ge 1$ we can appeal to Lemma 3 to reduce this to a previously solved problem. Lemma 3 allows us to express the Fourier coefficient of f in terms of the Fourier coefficient of Lf. whenever the boundary terms are zero, which is precisely the case when $j(f) \ge 1$. We note in passing that, for the series of functions $f,\ f',\ Lf,\ (Lf)',\ \ldots,$ the first time a discontinuity is encountered is exactly the same as for the series of funcction $f,\ f',\ f'',\ \ldots;$ more succinctly $j(f) = j(Lf) + 2$. Combining this with the previous work leads to the following asymptotic expansion of the Fourier coefficient in the general case:

If $j = j(f) = 2k - 1$ is odd, then

$$c_n = n^{-j-\frac{3}{2}} \sum_{1 \le i \le K} B_i\big[\cos((n+\gamma_1)\theta_i + \gamma_2) + O(n^{-1})\big], \qquad n \to \infty$$

where $B_i = (1 - a_i^2)w(a_i)k(\theta_i)\delta(L^k f)(a_i)$.

At the endpoint $x = 1$ the general term of the series is

$$c_n P_n^{\alpha,\beta}(1) = C_\alpha n^{-j+\alpha-\frac{3}{2}} \sum_{1 \le i \le K} B_i\big[\cos((n+\gamma_1)\theta_i + \gamma_2) + O(n^{-1})\big].$$

This is the general term of a convergent series if and only if $-j + \alpha - \frac{3}{2} < 0$, which was to be proved, correspondingly for the endpoint $x = -1$.

If $j = j(f) = 2k$ is even, then

$$c_n = n^{-j-\frac{3}{2}} \sum_{1 \le i \le K} B_i\big[\sin((n+\gamma_1)\theta_i + \gamma_2) + O(n^{-1})\big], \qquad n \to \infty$$

where $B_i = (1 - a_i^2)w(a_i)k(\theta_i)\delta(L^k f)'(a_i)$. This is of the same form as the case of j odd, save for the replacement of cosine by sine, which does not change the proof.

References

[GP1] A. Gray and M. Pinsky, Computer graphics and a new Gibbs phenomenon for Fourier–Bessel Series, Experiment. Math. 1 (1992), 313–316.

[GP2] A. Gray and M. Pinsky, Gibbs' phenomenon for Fourier–Bessel Series, Exposition. Math. 11 (1993), 123–135.

[HL] G.H. Hardy and E. Landau, The lattice points of a circle, Proceedings of the Royal Society, A, 105, (1924), 244–258.

[O] F. W. Olver, Asymptotics and Special Functions, Academic Press, 1974.

[P1] M. Pinsky, Fourier inversion for multidimensional characterstic functions, J. Theoret. Probab. 6 (1993), 187–193.

[P2] M. Pinsky, Pointwise Fourier inversion and related eigenfunction expansions, Comm. Pure Appl. Math. 47 (1994), 653—681.

[PST] M. Pinsky, N. Stanton and P. Trapa, Fourier series of radial functions in several variables, J. Funct. Anal. 116 (1993), 111–132.

[SW] E. M. Stein and G. Weiss, Introduction to Fourier Analysis in Euclidean Spaces, Princeton University Press, 1971.

[Sz] G. Szego, Orthogonal Polynomials, American Mathematical Society, Colloquium Publication vol 23, 1939; Fourth Edition, 1975.

Department of Mathematics
Northwestern University
Evanston, IL 60208, U.S.A.
pinsky@math.nwu.edu

The Distribution of the Particles of a Branching Random Walk

P. Révész

1. Introduction

At time $t = 0$ we have a particle. At time $t = 1$ this particle produces k ($k = 0, 1, 2, \ldots$) offsprings with probability p_k and dies. In case $k = 0$ we say that the process dies out at $t = 1$. In case $k > 0$ each offspring repeats this procedure independently. More formally speaking: let $\{Y(i, j),\ i = 1, 2, \ldots;\ j = 1, 2, \ldots\}$ be an array of i.i.d.r.v.'s with

$$\mathbf{P}\{Y(i, j) = k\} = p_k \qquad (k = 0, 1, 2, \ldots)$$

where

$$p_k \geq 0, \qquad \sum_{k=0}^{\infty} p_k = 1.$$

Define the sequence $\{B_n,\ n = 0, 1, 2, \ldots\}$ as follows:

$$\begin{aligned}
B_0 &= 1,\\
B_1 &= Y(1, 1),\\
B_2 &= Y(2, 1) + Y(2, 2) + \cdots + Y(2, B_1),\\
&\vdots\\
B_n &= Y(n, 1) + Y(n, 2) + \cdots + Y(n, B_{n-1}) \quad (n = 3, 4, \ldots).
\end{aligned}$$

Here $Y(i, j)$ can be considered as the number of offsprings of the j-th particle born at time $t = i$.

From now on we assume that

$$\mathbf{E}Y(i, j) = \sum_{k=0}^{\infty} k p_k = m < \infty$$

and

$$0 < \operatorname{Var} Y(i, j) = \sum_{k=0}^{\infty} (k - m)^2 p_k = \sigma^2 < \infty.$$

The *branching process* $\{B_n\}$ is called subcritical if $m < 1$, critical if $m = 1$, supercritical if $m > 1$. The proof of the next theorem is very simple.

Theorem A.

$$\mathbf{P}\{\lim_{n\to\infty} B_n = 0\} = \begin{cases} 1 & \text{if } m \le 1, \\ q & \text{if } m > 1 \end{cases}$$

where $0 < q < 1$ *and in* $(0,1)$ q *is the unique solution of the equation*

$$g(q) = q, \qquad g(x) = \sum_{k=0}^{\infty} p_k x^k \qquad (0 < x < 1).$$

From now on we assume that

$$m > 1$$

and in this case we cite a more exact result.

Theorem B. (Harris, 1963: Theorems 8.1, 8.3 and (8.4) pp. 13, 16). *There exists a nonnegative r. v.* B *such that*

$$\lim_{n\to\infty} m^{-n} B_n = B \quad a.s., \qquad \mathbf{P}\{B = 0\} = q, \tag{1}$$

$$\mathbf{E}\frac{B_n}{m^n} = \mathbf{E}B = 1, \qquad \operatorname{Var} B = \frac{\sigma^2}{m^2 - m}, \tag{2}$$

$$\mathbf{E}\left|\frac{B_n}{m^n} - B\right| = O(m^{-n/2}), \quad \mathbf{E}\left(\frac{B_n}{m^n} - B\right)^2 = O(m^{-n}) \tag{3}$$

and the distribution of B *is absolutely continuous except for a jump of magnitude* q *at* 0.

We define a *branching random walk* as a branching process on which we superimpose the additional structure of a random walk on $\mathbb{Z}^d$. This model can be described in detail as follows.

At time $t = 0$ a particle located in $0 \in \mathbb{Z}^d$ begins a random walk. It moves at time $t = 1$ with equal probabilities to one of the $2d$ neighbours of 0. Arriving to the new location it produces k offsprings with probability p_k $(k = 0, 1, 2, \dots)$ and dies. Each of the offsprings move independently at time $t = 2$ to one of their neighbours. Arriving to the new locations each of them produce offsprings independently and they die. Repeating this procedure we obtain a branching random walk. Let $\lambda(x, t)$ $(x \in \mathbb{Z}^d,\ t = 0, 1, 2, \dots)$ be the number of particles in x at t. We are interested in the limit properties of $\lambda(x, t)$ as $t \to \infty$.

At first we give a formal definition of $\lambda(x, t)$. Let $e_1, e_2, \dots, e_d$ be the orthogonal unit vectors of $\mathbb{Z}^d$. Let

$$\{X(x, t, \mu), Z(x, t, \mu);\ x \in \mathbb{Z}^d,\ t = 0, 1, 2, \dots,\ \mu = 1, 2, \dots\}$$

be an array of independent r.v.'s with

$$\mathbf{P}\{X(x, t, \mu) = e_i\} = \mathbf{P}\{X(x, t, \mu) = -e_i\} = \frac{1}{2d}$$

$(i = 1, 2, \ldots, d;\ x \in \mathbb{Z}^d,\ t = 0, 1, 2, \ldots,\ \mu = 1, 2, \ldots)$,

$$\mathbf{P}\{Z(x, t, \mu) = k\} = p_k$$

$(k = 0, 1, 2, \ldots,\ x \in \mathbb{Z}^d,\ t = 0, 1, 2, \ldots,\ \mu = 1, 2, \ldots)$, where

$$p_k \geq 0, \qquad \sum_{k=0}^{\infty} p_k = 1, \qquad \sum_{k=0}^{\infty} k p_k = m > 1,$$

$$\sum_{k=0}^{\infty} (k - m)^2 p_k = \sigma^2 < \infty.$$

Further put

$$I_u(v) = \begin{cases} 1 & \text{if } u = v, \\ 0 & \text{if } u \neq v, \end{cases}$$

$$C(x, R) = \{y :\ y \in \mathbb{Z}^d,\ |y - x| \leq R\},$$

$$C(x) = \{y :\ y \in \mathbb{Z}^d,\ |y - x| = 1\}.$$

Then we give the definition of the array

$$\{\lambda(x, t);\ x \in \mathbb{Z}^d,\ t = 0, 1, 2, \ldots\}$$

as follows. Let

$$\lambda(x, 0) = \begin{cases} 1 & \text{if } x = 0, \\ 0 & \text{if } x \neq 0 \end{cases}$$

and

$$\lambda(x, t) = \sum_{y \in C(x)} \sum_{\mu=1}^{\lambda(y, t-1)} I_{x-y}(X(y, t-1, \mu)) Z(y, t-1, \mu)$$

if $x \in \mathbb{Z}^d$ and $t = 1, 2, \ldots$.

The intuitive meaning of the above definition is the following. In order to get the number of the particles located in x at time t, consider the particles located in one of the neighbours of x at time $t - 1$. The μ-th particle located in $y \in C(x)$ at time $t - 1$ is moving to x if $X(y, t-1, \mu) = x - y$ i.e. if $I_{x-y}(X(y, t-1, \mu)) = 1$. If this particle moves to x then it will produce there

$$Z(y, t-1, \mu) = I_{x-y}(X(y, t-1, \mu)) Z(y, t-1, \mu)$$

offsprings.

2. Results

Clearly by (1)

$$\lim_{t\to\infty} m^{-t} \sum_{x\in\mathbb{Z}^d} \lambda(x,t) = \lim_{t\to\infty} m^{-t} B_t = B.$$

It is natural to conjecture that the B_t particles are distributed around the origin according to the normal law i.e. our conjecture is: for any fixed $x \in \mathbb{Z}^d$ as $t \to \infty$

$$m^{-t}\lambda(x,t) \sim \frac{B}{(\pi t)^{1/2}}$$

where $x = (x_1, x_2, \ldots, x_d)$ and $\|x\|^2 = x_1^2 + x_2^2 + \cdots + x_d^2$. Similarly we conjecture that for any fixed $x \in \mathbb{Z}^d$ as $t \to \infty$

$$m^{-t} \sum_{y \le x t^{1/2}} \lambda(y,t) \sim B\Phi(x)$$

where

$$\begin{aligned}\Phi(x) &= \Phi(x_1, x_2, \ldots, x_d)\\ &= \frac{1}{(2\pi)^{d/2}} \int_{-\infty}^{x_1} \cdots \int_{-\infty}^{x_d} \exp\left(-\frac{y_1^2 + y_2^2 + \cdots + y_d^2}{2}\right) dy_1 dy_2 \ldots dy_d.\end{aligned}$$

As we will see later on it is really so. The following result is already known.

$$\lim_{t\to\infty} \mathbf{E}\Big(m^{-t} \sum_{y < x t^{1/2}} \lambda(y,t) - B\Phi(x)\Big)^2 = 0$$

cf. Ney, P. 1991. For further results of these types cf. Asmussen, S. and Kaplan, N. 1976, 1979.

In order to present our results introduce the following notations:

(i) let $\{S_t;\ t = 0, 1, 2, \ldots\}$ be a simple, symmetric, nearest neighbour random walk on $\mathbb{Z}^d$ with $S_0 = 0$,

(ii) $$p(u, v; t) = \mathbf{P}\{S_{s+t} = v \mid S_s = u\},$$

(iii) $$C(0,T) = \{x : x \in \mathbb{Z}^d,\ \|x\| \le T\},$$

(iv) $$D(T) = \{x : x \in C(0,T),\ x \equiv T \pmod 2\}$$

where $x = (x_1, x_2, \ldots, x_d) \equiv T \pmod 2$ means that $x_1 + x_2 + \ldots + x_d \equiv T$ (mod 2).

Then we can formulate our main theorems.

Theorem 1. *For any* $0 < \varepsilon < 1/2$ *there exists a* $C = C(\varepsilon) > 0$ *such that for any* $T = 1, 2, \ldots$ *we have*

$$\mathbf{E} \sum_{x \in \mathbb{Z}^d} \left| \frac{\lambda(x,T)}{m^T} - p(0,x;T)B \right| \le CT^{-(1/2-\varepsilon)}$$

where (cf. Theorem B*)*

$$B = \lim_{T\to\infty} m^{-T} B_T$$

and

$$B_T = \sum_{x\in\mathbb{Z}^d} \lambda(x,T).$$

A strong version of Theorem 1 is:

Theorem 2. *For any* $\varepsilon > 0$

$$\lim_{T\to\infty} T^{1/2-\varepsilon} \sum_{x\in\mathbb{Z}^d} \left| \frac{\lambda(x,T)}{m^T} - p(0,x;T)B \right| = 0 \quad a.s.$$

A simple consequence of Theorem 2 and the simplest form of the central limit theorem is

Theorem 3. *For any* $x \in \mathbb{R}^d$ *and* $\varepsilon > 0$ *we have*

$$\lim_{T\to\infty} T^{1/2-\varepsilon} \left| \sum_{y \le xT^{1/2}} \frac{\lambda(y,T)}{m^T} - B\Phi(x) \right| = 0 \qquad a.s.$$

where

$$\begin{aligned}\Phi(x) &= \Phi(x_1, x_2, \ldots, x_d)\\ &= \frac{1}{(2\pi)^{d/2}} \int_{-\infty}^{x_1} \cdots \int_{-\infty}^{x_d} \exp\left(-\frac{y_1^2 + y_2^2 + \cdots + y_d^2}{2}\right) dy_1 dy_2 \ldots dy_d.\end{aligned}$$

It looks likely that if in Theorems 1, 2 and 3 the sums are taken for a small set then the rate of convergence is better. In fact we prove

Theorem 4. *For any* $0 \le \gamma \le 1/2$, $\varepsilon > 0$ *and* $0 \le a < \infty$ *we have*

$$\mathbf{E}\left(\sum_{x\in D(aT^\gamma)} \left| \frac{\lambda(x,T)}{m^T} - p(0,x;T)B \right| \right) \le CT^{-(d+2-2\gamma(d+1)-\varepsilon)/2}, \tag{4}$$

$$\lim_{T\to\infty} T^{(d+2-2\gamma(d+1)-\varepsilon)/2} \sum_{x\in D(aT^\gamma)} \left| \frac{\lambda(x,T)}{m^T} - p(0,x;T)B \right| = 0 \quad a.s. \tag{5}$$

Theorem 5. *Let $x = x(T) \in D(T^\gamma)$ $(0 \le \gamma \le 1)$ be a sequence of vectors. Then for any $0 < \varepsilon < \gamma$*

$$\mathbf{P}\{T^{(d+2-2\gamma-2\varepsilon)/2}\Big|\frac{\lambda(x,T)}{m^T} - p(0,x;T)B\Big| \ge 1\} \le \exp(-O(T^\varepsilon)) \tag{6}$$

and

$$\lim_{T\to\infty} T^{d+2-2\gamma-\varepsilon}\mathbf{E}\Big(\frac{\lambda(x,T)}{m^T} - p(0,x;T)B\Big)^2 = 0. \tag{7}$$

Consequently for any fixed $x \in \mathbb{Z}^d$ and $0 < \varepsilon < 1$ we have

$$\lim_{T\to\infty} T^{1-\varepsilon}\Big|\frac{\lambda(x,T)}{m^T p(0,x;T)} - B\Big| = \lim_{T\to\infty} T^{1-\varepsilon}\Big|\frac{1}{2}\Big(\frac{2\pi T}{d}\Big)^{d/2}\frac{\lambda(x,T)}{m^T} - B\Big| = 0 \quad a.s. \tag{8}$$

provided that $x \equiv T \pmod 2$.

We also show that (8) gives the strongest possible rate in the following sense

Theorem 6. *Let $d = 1$. Then for any $C > 0$ there exists a $\delta = \delta(C) > 0$ such that*

$$\mathbf{P}\{\Big|(\pi T)^{1/2}\frac{\lambda(0,2T)}{m^{2T}} - B\Big| \ge \frac{C}{T}\} \ge \delta. \tag{9}$$

3. On the moments of $\lambda(x,t)$

Introduce the following notations:

(i)
$$f(x,T,t) = m^{T-t}\sum_{y\in\mathbb{Z}^d}\lambda(y,t)p(y,x;T-t),$$

(ii) let
$$\mathcal{F}(T) = \mathcal{F}\{\lambda(x,t);\ x\in\mathbb{Z}^d,\ t = 0,1,2,\dots,T\}$$
be the smallest σ–algebra with respect to which the array
$$\{\lambda(x,t):\ x\in\mathbb{Z}^d,\ t=0,1,2,\dots,T\}$$
is measurable,

(iii) let
$$\mathcal{G}(T) = \mathcal{G}\{X(y,s,\mu),\ Z(y,s,\mu);\ s\ge T,\ y\in\mathbb{Z}^d,\ \mu\ge 1\}$$
be the smallest σ-algebra with respect to which the r.v.'s in the brackets are measurable,

(iv)
$$\sum_{y\in\mathbb{Z}^d}\lambda(y,t) = B_t.$$

Note that $\mathcal{F}(T)$ and $\mathcal{G}(T)$ are independent σ-algebras. Further by the definition of $\lambda(x,T)$ we have

$$\mathbf{E}\big(\lambda(x,T)\mid \mathcal{F}(T-1)\big)=\frac{m}{2d}\sum_{y\in\mathcal{C}(x)}\lambda(y,T-1).$$

At first we present a number of lemmas.

Lemma 1. (Lawler, G. F. 1991, Theorem 1.2.1) *Assume that*

$$x=(x_1,x_2,\ldots,x_d)\equiv t \pmod 2,$$

i.e.,

$$x_1+x_2+\cdots+x_d\equiv t \pmod 2.$$

Then

$$\Big|p(0,x;t)-2\Big(\frac{d}{2\pi t}\Big)^{d/2}\exp\Big(-\frac{d\|x\|^2}{2t}\Big)\Big|\le O(t^{-(d+2)/2})$$

as $t\to\infty$.

Lemma 2.

$$\sum_{x\in\mathbb{Z}^d}p^2(0,x;t)=p(0,0;2t)\sim 2\Big(\frac{d}{2\pi t}\Big)^{d/2}\qquad (t=1,2,\ldots).$$

Proof.

$$\sum_{x\in\mathbb{Z}^d}p^2(0,x;t)=\sum_{x\in\mathbb{Z}^d}p(0,x;t)p(x,0;t)=p(0,0;2t).$$

The asymptotic relation follows from Lemma 1. □

Lemma 3. *For any* $0\le t\le T$ *and* $x\in\mathbb{Z}^d$ *we have*

$$\mathbf{E}\big(\lambda(x,T)\mid\mathcal{F}(t)\big)=f(x,T,t). \tag{10}$$

Proof. Since

$$\mathbf{E}\big(\lambda(x,t)\mid\mathcal{F}(t)\big)=\lambda(x,t)$$

and

$$f(x,T,T)=\sum_{y\in\mathbb{Z}^d}\lambda(y,T)p(y,x;0)=\lambda(x,T)$$

we have (10) in case $T=t$. It is also easy to see that (10) holds if $T=1$ and $t=0$. Now we give the proof by induction. Assume that (10) holds true for T, for any $0\le t\le T$

and $x \in \mathbb{Z}^d$ then

$$\begin{aligned}
\mathbf{E}\big(\lambda(x, T+1) \mid \mathcal{F}(t)\big) &= \mathbf{E}\Big(\sum_{y \in C(x)} \sum_{\mu=1}^{\lambda(y,T)} I_{x-y}(X(y,T,\mu)) Z(y,T,\mu) \mid \mathcal{F}(t)\Big) \\
&= \frac{m}{2d} \sum_{y \in C(x)} \mathbf{E}\big(\lambda(y,T) \mid \mathcal{F}(t)\big) \\
&= \frac{m}{2d} \sum_{y \in C(x)} f(y,T,t) \\
&= \frac{m}{2d} \sum_{y \in C(x)} m^{T-t} \sum_{z \in \mathbb{Z}^d} \lambda(z,t) p(z,y;T-t) \\
&= m^{T-t+1} \sum_{z \in \mathbb{Z}^d} \lambda(z,t) \frac{1}{2d} \sum_{y \in C(x)} p(z,y;T-t) \\
&= m^{T-t+1} \sum_{z \in \mathbb{Z}^d} \lambda(z,t) p(z,x;T-t+1) = f(x,T+1,t).
\end{aligned}$$

Hence we have (10). □

Note that (10) can be interpreted as follows: $f(x,T,t)$ is an unbiased predictor of $\lambda(x,T)$ observing the process up to t.

Lemma 4.

$$\begin{aligned}
f(x,T,t) = m^{T-t} &\sum_{y \in \mathbb{Z}^d} \sum_{\mu=1}^{\lambda(y,t-1)} Z(y,t-1,\mu) \\
&\times \sum_{u \in C(y)} I_{u-y}(X(y,t-1,\mu)) p(u,x;T-t). \qquad (11)
\end{aligned}$$

Proof. Clearly we have

$$\begin{aligned}
f(x,T,t) &= \frac{m^T}{m^t} \sum_{\substack{y \in \mathbb{Z}^d \\ u \in C(y)}} \sum_{\mu=1}^{\lambda(u,t-1)} I_{y-u}(X(u,t-1,\mu)) Z(u,t-1,\mu) p(y,x;T-t) \\
&= \frac{m^T}{m^t} \sum_{\substack{y \in \mathbb{Z}^d \\ u \in C(y)}} \sum_{\mu=1}^{\lambda(y,t-1)} I_{u-y}(X(y,t-1,\mu)) Z(y,t-1,\mu) p(u,x;T-t) \\
&= \frac{m^T}{m^t} \sum_{y \in \mathbb{Z}^d} \sum_{\mu=1}^{\lambda(y,t-1)} Z(y,t-1,\mu) \sum_{u \in C(y)} I_{u-y}(X(y,t-1,\mu)) p(u,x;T-t).
\end{aligned}$$

Hence we have (11). □

Lemma 5.

$$\mathbf{E}\lambda(x,T) = m^T p(0,x;T), \tag{12}$$

$$\begin{aligned} \mathbf{E}\big(f(x,T,t) \mid \mathfrak{F}(t-1)\big) &= m^{T-t+1} \sum_{y\in\mathbb{Z}^d} \lambda(y,t-1)p(y,x;T-t+1) \\ &= f(x,T,t-1), \end{aligned} \tag{13}$$

$$\mathbf{E}f(x,T,t) = m^T p(0,x;T) \quad (t = 0,1,2,\ldots,T). \tag{14}$$

Proof. (10) with $t = 0$ implies

$$\begin{aligned} \mathbf{E}\lambda(x,T) &= \mathbf{E}\big(\lambda(x,T) \mid \mathfrak{F}(0)\big) = f(x,T,0) \\ &= m^T \sum_{y\in\mathbb{Z}^d} \lambda(y,0)p(y,x;T) = m^T p(0,x;T). \end{aligned}$$

Hence we have (12). By (11) we have

$$\begin{aligned} \mathbf{E}\big(f(x,T,t) \mid \mathfrak{F}(t-1)\big) &= m^{T-t} \sum_{y\in\mathbb{Z}^d} m\lambda(y,t-1)\frac{1}{2d} \sum_{u\in C(y)} p(u,x;T-t) \\ &= m^{T-t+1} \sum_{y\in\mathbb{Z}^d} \lambda(y,t-1)p(y,x;T-t+1) \\ &= f(x,T,t-1). \end{aligned}$$

Hence we have (13).

By (12) and (13)

$$\begin{aligned} \mathbf{E}f(x,T,t) &= \mathbf{E}\big(\mathbf{E}(f(x,T,t) \mid \mathfrak{F}(t-1))\big) \\ &= m^{T-t+1} \sum_{y\in\mathbb{Z}^d} m^{t-1} p(0,y;t-1)p(y,x;T-t+1) \\ &= m^T p(0,x;T) \end{aligned}$$

and we have (14). □

Lemma 6. *Assume that $T - t \to \infty$. Then we have*

$$\begin{aligned} &\sum_{x\in\mathbb{Z}^d} \mathbf{E}\big((f(x,T,t) - f(x,T,t-1))^2 \mid \mathfrak{F}(t-1)\big) \\ &\quad = m^{2T-2t}\big(B_{t-1}((m^2+\sigma^2)p(0,0;2T-2t) - m^2 p(0,0;2T-2t+2))\big) \\ &\quad \sim 2\sigma^2\Big(\frac{d}{4\pi}\Big)^{d/2} m^{2T-2t} B_{t-1} \frac{1}{(T-t)^{d/2}}. \end{aligned}$$

Proof. By (11) we have

$$
\begin{aligned}
f(x,T,t) - f(x,T,t-1) &= m^{T-t} \sum_{y\in\mathbb{Z}^d} \sum_{\mu=1}^{\lambda(y,t-1)} Z(y,t-1,\mu) \\
&\quad \times \sum_{u\in C(y)} I_{u-y}(X(y,t-1,\mu))p(u,x;T-t) \\
&\quad - m^{T-t+1} \sum_{y\in\mathbb{Z}^d} \lambda(y,t-1)p(y,x;T-t+1) \\
&= m^{T-t} \sum_{y\in\mathbb{Z}^d} \sum_{\mu=1}^{\lambda(y,t-1)} Q(y,\mu)
\end{aligned}
$$

where

$$
Q(y,\mu) = Z(y,t-1,\mu) \sum_{u\in C(y)} I_{u-y}(X(y,t-1,\mu))p(u,x;T-t) - mp(y,x;T-t+1).
$$

Observe that

$$
\begin{aligned}
\mathbf{E}\Big(Z(y,t-1,\mu)\sum_{u\in C(y)} I_{u-y}(X(y,t-1,\mu))p(u,x;T-t)\Big) &= \frac{m}{2d} \sum_{u\in C(y)} p(u,x;T-t) \\
&= mp(y,x;T-t+1).
\end{aligned}
$$

Hence

$$
\begin{aligned}
\mathbf{E}\big(Q^2(y,\mu) \mid \mathcal{F}(t-1)\big) &= \mathbf{E}Q^2(y,\mu) \\
&= (m^2+\sigma^2)\frac{1}{2d}\sum_{u\in C(y)} p^2(u,x;T-t) - m^2p^2(y,x;T-t+1)
\end{aligned}
$$

and

$$
\mathbf{E}\big((f(x,T,t) - f(x,T,t-1))^2 \mid \mathcal{F}(t-1)\big) = m^{2T-2t} \sum_{y\in\mathbb{Z}^d} \lambda(y,t-1)\Delta
$$

where

$$
\Delta = \Delta(x,y,t,T) = \frac{m^2+\sigma^2}{2d} \sum_{u\in C(y)} p^2(u,x;T-t) - m^2p^2(y,x;T-t+1).
$$

Consequently by Lemma 2

$$\begin{aligned} m^{2t-2T} & \sum_{x\in\mathbb{Z}^d} \mathbf{E}\big((f(x,T,t)-f(x,T,t-1))^2 \mid \mathcal{F}(t-1)\big) \\ &= \sum_{y\in\mathbb{Z}^d} \lambda(y,t-1) \sum_{x\in\mathbb{Z}^d} \Delta \\ &= \sum_{y\in\mathbb{Z}^d} \lambda(y,t-1)((m^2+\sigma^2)p(0,0;2T-2t) - m^2 p(0,0;2T-2t+2)) \\ &= B_{t-1}\big((m^2+\sigma^2)p(0,0;2T-2t) - m^2 p(0,0;2T-2t+2)\big). \end{aligned}$$

Hence we have Lemma 6 by Lemma 1. □

Lemma 7. *Assume that* $T-t\to\infty$. *Then*

$$\sum_{x\in\mathbb{Z}^d} \mathbf{E}\big((\lambda(x,T)-f(x,T,t))^2 \mid \mathcal{F}(t)\big) \sim 2\sigma^2\Big(\frac{d}{4\pi}\Big)^{d/2} B_t \sum_{i=1}^{T-t} i^{-d/2} m^{T+i-t-2}.$$

Proof. Clearly by (10) we have

$$\begin{aligned} &\mathbf{E}\big((\lambda(x,T)-f(x,T,t))^2 \mid \mathcal{F}(t)\big) \\ &\qquad = \mathbf{E}\big(\mathbf{E}((\lambda(x,T)-f(x,T,t))^2 \mid \mathcal{F}(T-1)) \mid \mathcal{F}(t)\big) \\ &\qquad = \mathbf{E}\big(\mathbf{E}((\lambda(x,T)-f(x,T,T-1))^2 \mid \mathcal{F}(T-1)) \mid \mathcal{F}(t)\big) \\ &\qquad\quad + \mathbf{E}\big((f(x,T,T-1)-f(x,T,t))^2 \mid \mathcal{F}(t)\big). \end{aligned}$$

Since $\lambda(x,T)=f(x,T,T)$ by Lemma 6 with $t=T$ we have

$$\begin{aligned} &\sum_{x\in\mathbb{Z}^d} \mathbf{E}\big((\lambda(x,T)-f(x,T,t))^2 \mid \mathcal{F}(t)\big) \\ &\qquad = \big((m^2+\sigma^2)p(0,0;0) - m^2 p(0,0;2)\big) m^{T-t-1} B_t \\ &\qquad\quad + \sum_{x\in\mathbb{Z}^d} \mathbf{E}\big((f(x,T,T-1)-f(x,T,t))^2 \mid \mathcal{F}(t)\big). \end{aligned}$$

Similarly we get

$$\sum_{x\in\mathbb{Z}^d} \mathbf{E}\big((f(x,T,T-1)-f(x,T,t))^2 \mid \mathcal{F}(t)\big)$$
$$= \sum_{x\in\mathbb{Z}^d} \mathbf{E}\big(\mathbf{E}((f(x,T,T-1)-f(x,T,t))^2 \mid \mathcal{F}(T-2)) \mid \mathcal{F}(t)\big)$$
$$= \sum_{x\in\mathbb{Z}^d} \mathbf{E}\big(\mathbf{E}((f(x,T,T-1)-f(x,T,T-2))^2 \mid \mathcal{F}(T-2) \mid \mathcal{F}(t)\big)$$
$$+ \sum_{x\in\mathbb{Z}^d} \mathbf{E}\big((f(x,T,T-2)-f(x,T,t))^2 \mid \mathcal{F}(t)\big)$$
$$= m^{T-t}B_t((m^2+\sigma^2)p(0,0;2)-m^2p(0,0;4))$$
$$+ \sum_{x\in\mathbb{Z}^d} \mathbf{E}\big((f(x,T,T-2)-f(x,T,t))^2 \mid \mathcal{F}(t)\big).$$

Going step by step by Lemma 1 we obtain

$$\sum_{x\in\mathbb{Z}^d} \mathbf{E}\big((\lambda(x,T)-f(x,T,t))^2 \mid \mathcal{F}(t)\big)$$
$$= m^{T-t-1}\sum_{i=0}^{T-t-1} m^i\big((m^2+\sigma^2)p(0,0;2i)-m^2p(0,0;2i+2)\big)B_t$$
$$\sim m^{T-t-1}B_t\sum_{i=1}^{T-t}\sigma^2 m^{i-1}2\Big(\frac{d}{4\pi i}\Big)^{d/2}$$
$$= m^{T-t-2}\sigma^2 2\Big(\frac{d}{4\pi}\Big)^{d/2}B_t\sum_{i=1}^{T-t}m^i\Big(\frac{1}{i}\Big)^{d/2}$$
$$= 2\sigma^2\Big(\frac{d}{4\pi}\Big)^{d/2}B_t\sum_{i=1}^{T-t}m^{T+i-t-2}\frac{1}{i^{d/2}}.$$

Hence we have Lemma 7.

Lemma 8.

$$\mathbf{E}\Big(\sum_{x\in\mathbb{Z}^d}\Big(\frac{\lambda(x,T)}{m^T}-\frac{f(x,T,t)}{m^T}\Big)^2 \;\Big|\; \mathcal{F}(t)\Big) \le C\frac{1}{m^t(T-t)^{d/2}}\frac{B_t}{m^t}.$$

with an absolute constant $C>0$.

Proof. Lemma 8 is a trivial consequence of Lemma 7. □

Theorem 7. *For any* $0\le\alpha\le 1$ *we have*

$$\mathbf{E}\Big(\sum_{x\in C(0,T^\alpha)}\Big|\frac{\lambda(x,T)}{m^T}-\frac{f(x,T,t)}{m^T}\Big| \;\Big|\; \mathcal{F}(t)\Big) \le C(2T^\alpha)^{d/2}\Big(\frac{1}{m^t(T-t)^{d/2}}\frac{B_t}{m^t}\Big)^{1/2} \quad (15)$$

and

$$\mathbf{E} \sum_{x \in C(0,T^\alpha)} \left| \frac{\lambda(x,T)}{m^T} - \frac{f(x,T,t)}{m^T} \right| \le C(2T^\alpha)^{d/2} (m^t (T-t)^{d/2})^{-1/2} \tag{16}$$

with an absolute constant $C > 0$.

Proof. By Lemma 8 and the Cauchy inequality we have

$$\begin{aligned} &\mathbf{E}\Big(\sum_{x \in C(0,T^\alpha)} \left| \frac{\lambda(x,T)}{m^T} - \frac{f(x,T,t)}{m^T} \right| \;\Big|\; \mathcal{F}(t) \Big) \\ &\qquad \le (2T^\alpha)^{d/2} \Big(\mathbf{E}\Big(\sum_{x \in C(0,T^\alpha)} \Big(\frac{\lambda(x,T)}{m^T} - \frac{f(x,T,t)}{m^T} \Big)^2 \;\Big|\; \mathcal{F}(t) \Big) \Big)^{1/2} \\ &\qquad \le C(2T^\alpha)^{d/2} \Big(\frac{1}{m^t (T-t)^{d/2}} \frac{B_t}{m^t} \Big)^{1/2}. \end{aligned}$$

Hence we have (15). (2) and the Cauchy inequality implies

$$\mathbf{E}(m^{-t} B_t)^{1/2} \le 1$$

which, in turn, by (15) implies (16).

The proofs of Theorems 1 and 2

In order to prove Theorem 1 we prove two lemmas.

Lemma 9. *Let* $1 \le t \le T,\ x \equiv T \pmod 2$ *and*

$$D(T) = \{x : x \in C(0,T),\ x \equiv T \pmod 2\}.$$

Then

$$\begin{aligned} \mathbf{E}\big(\lambda(x,T) \mid \mathcal{F}(t)\big) &\ge \inf_{y \in D(t)} \mathbf{E}\big(\lambda(x,T) \mid \lambda(y,t) = B_t\big) \\ &= m^T \frac{B_t}{m^t} \inf_{y \in D(t)} p(0, x-y; T-t) \end{aligned}$$

and

$$\begin{aligned} \mathbf{E}\big(\lambda(x,T) \mid \mathcal{F}(t)\big) &\le \sup_{y \in D(t)} \mathbf{E}\big(\lambda(x,T) \mid \lambda(y,t) = B_t\big) \\ &= m^T \frac{B_t}{m^t} \sup_{y \in D(t)} p(0, x-y; T-t). \end{aligned}$$

Proof. Observing that $\lambda(y,t) = 0$ if $y \notin D(t)$ we have the inequalities in the above two relations. The equalities in them follow from (11). □

Lemma 10. *Assume the conditions of Lemma* 9 *and let* $xt = o(T)$ *and* $t^2 = o(T)$ *as* $T \to \infty$. *Then we have*

$$\frac{p(0,x;T)}{\inf_{y\in D(t)} p(0,x-y;T-t)} \leq 1 + O\left(\frac{xt+t^2}{T}\right) \tag{17}$$

and

$$\frac{p(0,x;T)}{\sup_{y\in D(t)} p(0,x-y;T-t)} \geq 1 - O\left(\frac{xt+t^2}{T}\right). \tag{18}$$

Further if $1/2 < \alpha < 1$ *then*

$$\sum_{x\in D(T)-D(T^\alpha)} p(0,x;T) = \exp(-O(T^{2\alpha-1})).$$

Proof. At first we consider the case $d = 1$. Then

$$p(0,2x;2T) = \binom{2T}{T-x}\frac{1}{2^{2T}}$$

and by the Stirling formula we get

$$\frac{p(0,2x;2T)}{p(0,2x;2T-2t)} = 1 + O\left(\frac{xt+t^2}{T}\right) \tag{20}$$

provided that

$$xt = o(T) \quad \text{and} \quad t^2 = o(T) \quad \text{as} \quad T \to \infty.$$

Similarly

$$\frac{p(0,2x;2T)}{p(0,2z;2T)} = 1 + O\left(\frac{x^2-z^2}{T}\right) \tag{21}$$

provided that

$$x^2 - z^2 = o(T) \qquad z = o(T) \quad \text{and} \quad x = o(T).$$

Consequently

$$\frac{p(0,2x;2T)}{p(0,2z;2T-2t)} = 1 + O\left(\frac{xt+t^2+x^2-z^2}{T}\right) \tag{22}$$

provided that

$$xt = o(T), \quad z = o(T), \quad x^2 - z^2 = o(T) \text{ and } t^2 = o(T) \text{ as } T \to \infty.$$

Since $|x^2 - (x-y)^2| \leq |x|t + t^2$ if $y \in D(t)$, we have (17) and (18) in case $d = 1$. In case $d \geq 2$ let

$$T_i = \#\{j : 0 \leq j \leq T-1, \ |(e_i, S_{j+1} - S_j)| = 1\}$$

i.e., T_i is the number of those steps in $[0, T]$ when the particle moves in the direction e_i or $-e_i$. Then for any $0 < \varepsilon < 1$ we have

$$\begin{aligned} p(0, x; T) &= \sum{}^{*}\mathbf{P}\{\bigcap_{i=1}^{d}\{S_{t_i}^{(i)} = x_i\} \mid \bigcap_{i=1}^{d}\{T_i = t_i\}\}\mathbf{P}\{\bigcap_{i=1}^{d}\{T_i = t_i\}\} \\ &= \sum{}^{*}\Big(\prod_{i=1}^{d}\mathbf{P}\{S_{t_i}^{(i)} = x_i\}\Big)\mathbf{P}\{\bigcap_{i=1}^{d}\{T_i = t_i\}\} \\ &= \big(1 + o(1)\big)\sum{}_{\varepsilon}\Big(\prod_{i=1}^{d}\mathbf{P}\{S_{t_i}^{(i)} = x_i\}\Big)\mathbf{P}\{\bigcap_{i=1}^{d}\{T_i = t_i\}\} \end{aligned} \tag{23}$$

where

$S_t = (S_t^{(1)}, S_t^{(2)}, \ldots, S_t^{(d)})$,

$x = (x_1, x_2, \ldots, x_d)$,

$\sum^*$ is the summation extended for all d-tuples $(t_1, t_2, \ldots, t_d)$ for which $t_1 + t_2 + \cdots + t_d = T$ and $t_i \equiv x_i \pmod 2$,

$\sum_\varepsilon$ is the summation extended for all d–tuples $(t_1, t_2, \ldots, t_d)$ for which $t_1 + t_2 + \cdots + t_d = T$, $t_i \equiv x_i \pmod 2$ and $|t_i/T - 1/d| \le \varepsilon$ $(i = 1, 2, \ldots, d)$,

(23) and (22) combined imply (17) and (18); (19) follows from Lemma 1.

Now we turn to the proof of Theorem 1.

For any $0 \le t \le T$ and $0 < \alpha < 1$ we have

$$\begin{aligned} \sum_{x \in D(T)} \Big|\frac{\lambda(x, T)}{m^T} - p(0, x; T)B\Big| \le & \sum_{x \in D(T)-D(T^\alpha)} \frac{\lambda(x, T)}{m^T} \\ & + B \sum_{x \in D(T)-D(T^\alpha)} p(0, x; T) \\ & + \sum_{x \in D(T^\alpha)} \Big|\frac{\lambda(x, T)}{m^T} - \frac{\mathbf{E}\big(\lambda(x, T) \mid \mathcal{F}(t)\big)}{m^T}\Big| \\ & + \sum_{x \in D(T^\alpha)} \Big|\frac{\mathbf{E}\big(\lambda(x, T) \mid \mathcal{F}(t)\big)}{m^T} - p(0, x; T)\frac{B_t}{m^t}\Big| \\ & + \sum_{x \in D(T^\alpha)} \Big|p(0, x; T)\frac{B_t}{m^t} - p(0, x; T)B\Big| \end{aligned} \tag{24}$$

Choosing $\alpha > 1/2$ by (19) and (12)

$$\sum_{x \in D(T)-D(T^\alpha)} p(0, x; T) \le \exp(-O(T^{2\alpha-1})), \tag{25}$$

$$\sum_{x \in D(T)-D(T^\alpha)} \mathbf{E}\Big(\frac{\lambda(x, T)}{m^T}\Big) \le \exp(-O(T^{2\alpha-1})). \tag{26}$$

Let $T = [t^\beta]$ with $\beta > 1$ by Theorem 7 and (10) we have

$$\sum_{x \in D(T^\alpha)} \mathbf{E}\Big|\frac{\lambda(x,T)}{m^T} - \frac{\mathbf{E}(\lambda(x,T) \mid \mathcal{F}(t))}{m^T}\Big|$$
$$\leq C(2T^\alpha)^{d/2}(m^t(T-t)^{d/2})^{-1/2} \leq \exp(-O(T^{1/\beta})). \tag{27}$$

Choosing $\alpha > 1/2$ close enough to $1/2$ and β big enough by Lemmas 9 and 10 we have

$$\sum_{x \in D(T^\alpha)} \Big|\frac{\mathbf{E}(\lambda(x,T) \mid \mathcal{F}(t))}{m^T} - p(0,x;T)\frac{B_t}{m^t}\Big| \leq O(T^{1/2-\varepsilon}) \tag{28}$$

for any $\varepsilon > 0$. By (3)

$$\sum_{x \in D(T^\alpha)} \mathbf{E}\Big(\Big|p(0,x;T)\frac{B_t}{m^t} - p(0,x;T)B\Big|\Big) \leq \exp(-O(T^{1/\beta})). \tag{29}$$

By inequalities (24)–(29) we have Theorem 1.

Now, we turn to the proof of Theorem 2. By (26), (27) and (28) respectively for any $K > 0$ we have

$$\lim_{T\to\infty} T^K \sum_{x \in D(T)-D(T^\alpha)} \frac{\lambda(x,T)}{m^T} = 0 \qquad \text{a.s.} \tag{30}$$

resp.

$$\lim_{T\to\infty} T^K \sum_{x \in D(T^\alpha)} \Big|\frac{\lambda(x,T)}{m^T} - \frac{\mathbf{E}(\lambda(x,T) \mid \mathcal{F}(t))}{m^T}\Big| = 0 \qquad \text{a.s.} \tag{31}$$

$$\lim_{T\to\infty} T^K \sum_{x \in D(T^\alpha)} \Big|p(0,x;T)\frac{B_T}{m^T} - p(0,x;T)B\Big| = 0 \qquad \text{a.s.} \tag{32}$$

respectively.

Hence we have Theorem 2 by (24), (25), (30), (31), (28) and (32). □

Remember that Theorem 3 is a simple consequence of Theorem 2.

Proof of Theorem 4. Let

$$T = [t^\beta], \quad \beta = 1/\varepsilon,$$
$$a(x,T) = \frac{\lambda(x,T)}{m^T} - \frac{\mathbf{E}\big(\lambda(x,T) \mid \mathcal{F}(t)\big)}{m^T},$$
$$b(x,T) = \frac{\mathbf{E}\big(\lambda(x,T) \mid \mathcal{F}(t)\big)}{m^T} - p(0,x;T)\frac{B_t}{m^t},$$
$$c(x,T) = p(0,x;T)\Big(\frac{B_t}{m^t} - B\Big).$$

Then by (28) and (29) with $\gamma \le \alpha$ we have

$$\begin{aligned}
\mathbf{E}\Big(\sum_{x \in D(aT^\gamma)} \Big|\frac{\lambda(x,T)}{m^T} - p(0,x;T)B\Big|\Big) &\\
\le \mathbf{E}\Big(\sum_{x \in D(aT^\gamma)} |a(x,T)|\Big) + \mathbf{E}\Big(\sum_{x \in D(aT^\gamma)} |b(x,T)|\Big) + \mathbf{E}\Big(\sum_{x \in D(aT^\gamma)} |c(x,T)|\Big) &\\
\le \exp(-O(T^\varepsilon)) + \mathbf{E}\Big(\sum_{x \in D(aT^\gamma)} |b(x,T)|\Big). &
\end{aligned}$$

By Lemmas 9 and 10 we have

$$\begin{aligned}
\sum_{x \in D(aT^\gamma)} |b(x,T)| &\le \sum_{x \in D(aT^\gamma)} p(0,x;T) O\Big(\frac{xt+t^2}{T}\Big) \\
&\le (aT^\gamma)^d O\Big(\frac{xt+t^2}{T^{1+d/2}}\Big) = O(T^{-(d+2-2\gamma(d+1)-\varepsilon)/2}).
\end{aligned}$$

Hence we have (4) and (5). □

Proof of Theorem 5. Since by Lemmas 9 and 10

$$\begin{aligned}
|b(x,T)| &\le p(0,x;T) O\Big(\frac{xt+t^2}{T}\Big) \\
&\le O\Big(\frac{xt+t^2}{T^{1+d/2}}\Big) \\
&= O(T^{-(d+2-2\gamma-\varepsilon)/2})
\end{aligned}$$

we get (6) and (7) by Markov inequality repeating the proof of Theorem 4.

In order to prove (7) consider

$$\begin{aligned}
\mathbf{E}\Big(\frac{\lambda(x,T)}{m^T} - p(0,x;T)B\Big)^2 &= \mathbf{E}a^2(x,T) + \mathbf{E}b^2(x,T) + \mathbf{E}c^2(x,t) \\
&\quad + 2\mathbf{E}a(x,T)b(x,T) + 2\mathbf{E}a(x,T)c(x,T) \\
&\quad + 2\mathbf{E}b(x,T)c(x,T) \\
&\le \mathbf{E}a^2(x,T) + \mathbf{E}b^2(x,T) \\
&\quad + \mathbf{E}c^2(x,T) + 2(\mathbf{E}a^2(x,T)\mathbf{E}b^2(x,T))^{1/2} \\
&\quad + 2\big(\mathbf{E}a^2(x,T)\mathbf{E}c^2(x,T)\big)^{1/2} \\
&\quad + 2\big(\mathbf{E}b^2(x,T)\mathbf{E}c^2(x,T)\big)^{1/2}.
\end{aligned}$$

Since by Lemma 8

$$\mathbf{E}a^2(x,T) \le C\frac{1}{m^t(T-t)^{d/2}},$$

by Lemmas 9 and 10

$$\mathbf{E}b^2(x,T) \le (p(0,x;T))^2 O\Big(\big(\frac{xt+t^2}{T}\big)^2\Big) \le O\big(\frac{1}{T^d}\cdot\frac{T^{2\gamma+\varepsilon}}{T^2}\big),$$

by (3)

$$\mathbf{E}c^2(x,T) \le O(m^{-t})$$

we have (7). □

Proof of Theorem 6. The proof is based on the following observation. For any fixed t

$$\mathbf{P}\{\lambda(2t,2t) = B_{2t}\} = p(t) > 0.$$

By Lemma 10 we have

$$\begin{aligned}\mathbf{E}\big(\lambda(0,2T) \mid \lambda(2t,2t) = B_{2t}\big) &= m^{2T-2t} p(2t,0;2T-2t)B_{2t}\\ &\sim m^{2T-2t} p(0,0;2T)B_{2t}\big(1+O\big(\frac{t^2}{T}\big)\big). \qquad (33)\end{aligned}$$

Let

$$\begin{aligned}U &= (\pi T)^{1/2}\frac{\lambda(0,2T)}{m^{2T}} - (\pi T)^{1/2}\frac{\mathbf{E}\big(\lambda(0,2T) \mid \lambda(2t,2t) = B_{2t}\big)}{m^{2T}},\\ V &= (\pi T)^{1/2}\frac{\mathbf{E}\big(\lambda(0,2T) \mid \lambda(2t,2t) = B_{2t}\big)}{m^{2T}} - \frac{B_{2t}}{m^{2t}},\\ W &= \frac{B_{2t}}{m^{2t}} - B.\end{aligned}$$

Then

$$(\pi T)^{1/2}\frac{\lambda(0,2T)}{m^{2T}} - B = U+V+W$$

and by (33)

$$\mathbf{P}\big\{V \ge O\big(\frac{t^2}{T}\big) > 0\big\} \ge p(t).$$

It is easy to see that there exist $c_1 > 0,\ c_2 > 0,\ c_3 > 0$ such that for any $0 < t < T < \infty$ we have

$$\begin{aligned}&\mathbf{P}\{U > 0 \mid \lambda(2t,2t) = B_{2t}\} \ge c_1,\\ &\mathbf{P}\{W > 0 \mid \lambda(2t,2t) = B_{2t}\} \ge c_2,\\ &\mathbf{P}\big\{U > 0,\ V \ge O\big(\frac{t^2}{T}\big),\ W > 0 \mid \lambda(2t,2t) = B_{2t}\big\} \ge c_3 p(t)\end{aligned}$$

which easily implies (9).

References

Asmussen, S. and Kaplan, N., Branching random walks I, Stochastic Process. Appl. 4 (1976), 1–13.

Asmussen, S. and Kaplan, N., Branching random walks II, Stochastic Process. Appl. 4 (1979), 15–31.

Harris, T. E., The theory of branching processes, Springer-Verlag, Berlin 1963.

Lawler, G. F., Intersections of random walks, Birkhäuser, Boston 1991.

Ney, P., Branching random walk, Spatial Stochastic Processes (Ed. Alexander, K. S. and Watkins, J. C.), Birkhäuser, Boston 1991.

Technische Universität Wien
Institut für Statistik
Wiedner Hauptstr. 8–10
A-1040 Wien, Austria

On a Random Ergodic Theorem for Bistochastic Operators

Ben-Zion Rubshtein[1]

Abstract. We prove a version of the random ergodic theorem for stationary sequences of random bistochastic operators in L_p, $1 \leq p < \infty$, and find conditions, under which the limit random operator in the theorem does not essentially depend on realizations of the sequence of random operators.

The classical random ergodic theorems for averages of measure preserving transformations were extended on wide classes of random mapping and operators (see **[1]**, ..., **[8]** and the books **[9]** ch. 8 and **[10]** for references).

It has long been noted that the limit random operator of an ergodic stationary sequence of random operators may be essentially depend on the realizations of the random sequence (see for example **[4]**, **[6]**). However, such phenomenon is absent if the random operators are independent, even in the case of general sub-stochastic random operators **[8]**.

We find in this work simple necessary and sufficient conditions, when the limit operator in the random ergodic theorem is not random.

In the other paper **[11]** will be given applications to the convolutions of random measures on compact groups.

1. Stationary sequences of random bistochastic operators (SSRBO)

Let (X, Σ, m) be a measure space with $mX = 1$. A linear operator $T \in L_\infty(m)$ is called bistochastic if it satisfies the following conditions:

(i) $0 \leq f \in L_\infty(m) \Rightarrow 0 \leq Tf$.

(ii) $TI_X = I_X$, (I_A is the indicator of $A \in \Sigma$).

(iii) $\int f dm = \int Tf dm, \ f \in L_\infty(m)$.

Such operator acts also as a contraction in all spaces $L_p(m)$, $1 \leq p \leq \infty$, (**[9]** p. 65) and we denote by M the set of all bistochastic operators on X. The set M is a semigroup with involution, where the conjugate operator T^* of $T \in M$ is defined by $\int f \cdot Tg dm = \int T^* f \cdot g dm$, $T^* \in M$.

The subset of all idempotents of the semigroup M concides with the set $\{E^{\Sigma_1}, \Sigma_1 \subset \Sigma\}$ of the operators of the conditional expectation on σ-sub-algebras Σ_1 of Σ. In particular, $T = T^2 \in M \Rightarrow T = T^*$.

1 Research supported by the Israel Ministries of Science and of Absorption.

We do not distinguish measurable functions or sets from their equivalence classes mod nullsets and assume all σ-algebras are completed with respect to the corresponding measures.

Recall that by the ergodic theorem ([9], ch. 3) for $T \in \boldsymbol{M}$ and $f \in \boldsymbol{L}_p(m)$ there exists the $\lim_{k\to\infty} \frac{1}{k}\sum_{i=0}^{k-1} T^i f = E(T)f$ in the $\boldsymbol{L}_p$-norm, $1 \le p < \infty$, and almost everywhere on X and $E(T)$ is the conditional expectation operator with respect to the σ-algebra $\mathcal{E}(T) = \{A \in \Sigma : TI_A = I_A\}$, $E(T) = E^{\mathcal{E}(T)}$. Moreover $E(T) = E(T^*)$.

Define a random bistochastic operator (RBO) $T = T(\omega)$ as a measurable mapping $\Omega \ni \omega \to T(\omega) \in \boldsymbol{M}$ on a probability space $(\Omega, \mathcal{F}, P)$, where $\boldsymbol{M}$ is alloted by the weak operator topology in $\boldsymbol{L}_p$, $1 \le p < \infty$.

Let $T_0 = T_0(\omega)$ be a RBO. We consider a stationary sequence of RBO as a sequence $T_n = T_n(\omega)$, where $T_n(\omega) = T_0(\theta^n\omega)$, $n \in \mathbb{Z}$ and θ is an invertible measure preserving transformation on $(\Omega, \mathcal{F}, \mathbf{P})$.

The usual way to prove random ergodic theorems is to construct the corresponding skew products on the product space

$$(\overline{\Omega}, \overline{\mathcal{F}}, \bar{\mathbf{P}}) = (\Omega, \mathcal{F}, \mathbf{P}) \times (X, \Sigma, m)$$

as follows. Let $\bar{\boldsymbol{M}}$ be the set of all bistochastic operators on $(\overline{\Omega}, \overline{\mathcal{F}}, \bar{\mathbf{P}})$. A RBO $T = T(\omega)$ defines the operator $\bar{T} \in \bar{\boldsymbol{M}}$ by

$$(\bar{T}\ \bar{f})(\omega, x) = (T(\omega) f_\omega)(x), \quad (\omega, x) \in \overline{\Omega}$$

where $f_\omega(x) = \bar{f}(\omega, x)$ are the sections of the function $\bar{f} \in \boldsymbol{L}_p(\bar{\mathbf{P}})$.

For a given SSRBO $T_n = T_n(\omega)$ we introduce now $\bar{T}_n \in \bar{\boldsymbol{M}}$, $n \in \mathbb{Z}$ and the sequence of the skew products $\hat{T}_n = \bar{T}_n V \in \bar{\boldsymbol{M}}$ with the unitary bistochastic operators $V \in \bar{\boldsymbol{M}}$,

$$(V\bar{f})(\omega, x) = \bar{f}(\theta\omega, x), \quad (\omega, x) \in \overline{\Omega},$$

corresponding to θ on $\overline{\Omega}$.

One has now the equalities

$$V^k \bar{T}_n V^{-k} = \bar{T}_{n+k}, \quad V^k \hat{T}_n V^{-k} = \hat{T}_{n+k}, \quad k, n \in \mathbb{Z}$$

and

$$\hat{T}_n^k = \bar{T}_n \bar{T}_{n+1} \ldots \bar{T}_{n+k-1} V^k = V^k \bar{T}_{n-k} \bar{T}_{n-k+1} \ldots \bar{T}_{n-1}$$

$$\hat{T}_n^{*k} = \bar{T}_{n-1}^* \bar{T}_{n-2}^* \ldots \bar{T}_{n-k}^* V^k = V^k \bar{T}_{n+k-1}^* \ldots \bar{T}_{n+1}^* \bar{T}_n^*$$

or

$$(\hat{T}_n^k \bar{f})(\omega, x) = (T_n(\omega) \ldots T_{n+k-1}(\omega) f_{\theta^k\omega})(x)$$

$$(\hat{T}_n^{*k} \bar{f})(\omega, x) = (T_{n-1}^*(\omega) \ldots T_{n-k}^*(\omega) f_{\theta^{-k}\omega})(x)$$

for $n \in \mathbb{Z}$, $k \in \mathbb{N}$, $(\omega, x) \in \overline{\Omega}$, $\bar{f} \in \boldsymbol{L}_p(\bar{P})$.

In the sequel it will be always assumed that the automorphism θ is ergodic and the σ-algebras $\mathcal{F}$ and Σ are countable generated.

2. Random ergodic theorem

Let $T_n = T_n(\omega)$, $n \in \mathbb{Z}$, $\omega \in \Omega$, be a SSRBO in $\boldsymbol{M}$, and the sequences of the bistochastic operators $\{\bar{T}_n\}$ and $\{\hat{T}_n\}$ are defined as above.

Denote by $S_n = E(\hat{T}_n)$ the fixed point projections of $\hat{T}_n$ and by $G = E^{\mathcal{G}}$ the conditional expectation with respect to the σ-algebra $\mathcal{G}$, where $\mathcal{G}$ is the $\bar{\mathbf{P}}$-completion of the σ-algebra $\{\Omega \times A,\ A \in \Sigma\}$.

Theorem 2.1. *Let* $p \in [1, \infty)$.

a) *There exists a stationary sequence* $E_n = E_n(\omega)$ *of RBO and a subset* $\Omega_1 \subset \Omega$ *with* $\mathbf{P}(\Omega_1) = 1$ *such that* $\overline{E}_n G = S_n G$ *and the following holds:*

$$\lim_{k\to\infty} A^+_{n,k}(\omega) = \lim_{k\to\infty} A^-_{n,k}(\omega) = E_n(\omega), \quad \omega \in \Omega_1$$

where

$$A^+_{n,k}(\omega) = \frac{1}{k}\sum_{i=0}^{k-1} T_n(\omega)\cdot T_{n+1}(\omega)\dots\cdot T_{n+i}(\omega)$$

$$A^-_{n,k}(\omega) = \frac{1}{k}\sum_{i=0}^{k} T_{n-i}(\omega)\cdot T_{n-i+1}(\omega)\dots T_{n-1}(\omega),$$

and the limits exist in the strong operator topology on $\boldsymbol{L}_p(m)$.

b) *There exists* $E \in \boldsymbol{M}$ *such that for* $n \in \mathbb{Z}$ *and a.a.* ω.

$$\lim_{k\to\infty} \frac{1}{k}\sum_{k=0}^{k-1} E_{n+i}(\omega) = \lim_{k\to\infty}\frac{1}{k}\sum_{i=1}^{k} E_{n-i}(\omega)$$

$$= E_n^*(\omega)\cdot E_n(\omega) = E = \int E_n(\omega)\,dP = \int E_n^*(\omega)\,dP,$$

and $\bar{E}G = G\bar{E} = GS_nG$, *where* $\bar{E} = \mathrm{Id}\times E$ *and the limits and the integrals exist in the strong operator topology on* $\boldsymbol{L}_p(m)$.

Proof. The individual ergodic theorem for the bistochastic operator $\hat{T}_n$ implies that for every $\tilde{f} \in \boldsymbol{L}_1(\bar{\mathbf{P}})$ there exists $\Omega_{\tilde{f}} \subset \Omega$, $P(\Omega_{\tilde{f}}) = 1$ such that for $\omega \in \Omega_{\tilde{f}}$:

$$\lim_{k\to\infty} (A_{n,k}(\omega) f_{\theta^k\omega})(x) = (S_n\tilde{f})(\omega, x) \quad \text{a.e. on } X.$$

We choose now a countable subset $H \subset \boldsymbol{L}_\infty(m)$ such that H is a linear space over the rational field Q and H is dense in $\boldsymbol{L}_p(m)$.

For $h \in H$ the equality

$$(E_n(\omega)h)(x) = (S_n\bar{h})(\omega, x), \quad \bar{h}(\omega, x) = h(x),$$

defines for a.a. ω a linear contraction $E_n(\omega)\colon H \to \boldsymbol{L}_\infty(m)$, which is extended on $\boldsymbol{L}_\infty(m)$ as a bistochastic operator, $E_n(\omega) \in \boldsymbol{M}$.

One can find a subset $\Omega_1 \subset \Omega$ with $\mathbf{P}(\Omega_1) = 1$ such that for every $h \in H$ and $\omega \in \Omega_1$ the limits $A_{n,k}(\omega)h \xrightarrow{k\to\infty} E_n(\omega)h$ exist almost everywhere on X and hence in the L_p-norm. As well as the operators $A_{n,k}(\omega)$ and $E_n(\omega)$ are contractions on $L_p(m)$ and H is dense we obtain the convergence $A_{n,k}(\omega) \to E_n(\omega)$, $k \to \infty$, in the strong operator topology on $L_p(m)$ for $\omega \in \Omega_1$.

Using the same arguments to $\hat{T}^*$ and the equality $E(\hat{T}_n^*) = E(\hat{T}_n) = S_n$ we have completed the proof of the part a).

To prove b) we use the vector ergodic theorem (**[9]** §4.2) for the ergodic transformation θ and the vector function $E_n(\omega)f$, $f \in L_p(m)$. Then

$$\lim_{k\to\infty} \frac{1}{k}\sum_{i=0}^{k-1} E_{n+i}(\omega)f = \lim_{k\to\infty} \frac{1}{k}\sum_{i=0}^{k} E_{n-i}(\omega)f = \int E_n(\omega)f\,dP = Ef$$

in the L_p-norm for a.a. ω and the right part of the equality defines $E \in M$.

The equality

$$T_{n-k}(\omega)\ldots T_{n-1}(\omega)E_n(\omega) = E_{n-k}(\omega)$$

implies

$$\lim_{k\to\infty} A^{-}_{n,k}(\omega)E_n(\omega) = E$$

and $E_n^*(\omega)E_n(\omega) = E$ since $A^{-*}_{n,k}(\omega) \to E_n(\omega)$ for a.e. ω.

In particular, $E = E^*$ and thus $\int E_n^*(\omega)\,dP = E$. Taking $\bar{E} = \mathrm{Id} \times E$ one has now $G\bar{E} = \bar{E}G = G\bar{E}G$ and $GS_nG = G\bar{E}_n^*\bar{E}_nG = G\bar{E}_nG$. Thus the proof is completed. □

3. The essential constancy of the limit RBO

We will study conditions when the limit RBO $E_n = E_n(\omega)$ does not essentially depend on $\omega \in \Omega$, i.e., it is non-random.

For SSRBO $T_n = T_n(\omega)$ we define the σ-algebra

$$\Sigma^0 = \{A \in \Sigma : T_n(\omega)I_A = I_A \quad \mathbf{P}\text{-a.e.}\}$$

and the $\mathcal{G}^0$ which is the $\bar{\mathbf{P}}$ completion of the σ-algebra $\{\Omega \times A : A \in \Sigma^0\}$ and let S^0 and G^0 be the corresponding conditional expectation operators. It is easy to see that the σ-algebra $\mathcal{E}(\hat{T}_n) \cap \mathcal{G} = \mathcal{G}^0$ for all n.

Theorem 2.. *The following conditions are equivalent:*

1) *The mapping $\omega \to E_n(\omega)$ is constant a.e. on Ω.*

2) $S_nG = GS_n$.

3) $E_n(\omega) = S^0$ *a.e.*

4) E *is a projection.*

Proof. If the conditions hold for some of $n \in \mathbb{Z}$, they hold also for all n. Thus we can assume $n = 0$.

2) $\Rightarrow$ 1). For $f \in L_1(m)$ denote $\bar{f}(\omega, x) = f(x)$.

Using $\bar{E}_0 G = S_0 G$ and the condition 1) we have

$$G\bar{E}_0 f = G\bar{E}_0 G\bar{f} = GS_0 G\bar{f} = \bar{E}_0 G f = \bar{E}_0 \bar{f}$$

a.e. on $\overline{\Omega}$. Then for every $f \in L_1(m)$ there exists $f^* \in L_1(m)$ such that $\bar{E}_0 \bar{f} = \bar{f}^*$ i.e. $E_0(\omega) f = f^*$ a.e. on Ω.

Taken a countable dense subset $H \subset L_1(m)$ we can find $\Omega_1 \subset \Omega$ with $P(\Omega_1) = 1$ such that $E_0(\omega) h = h^*$ for all $h \in H$, $\omega \in \Omega_1$ and appropriate functions $h^* \in L_1(m)$. This means that all operators $E_0(\omega)$, $\omega \in \Omega_1$, coincide.

4) $\Rightarrow$ 2). By 4) $\bar{E} = \mathrm{Id} \times E$ is a projection and $GS_0G = G\bar{E} = \bar{E}G$ is a projector too. Hence G and S_0 need be commuting since G, S_0 and GS_0G are orthogonal projectors in $L_2(m)$.

1) $\Rightarrow$ 4). If 1) holds, $E_0(\omega) = \int E_0(\omega)\, dP = E$ a.e. on Ω and $E = E_0^*(\omega) E_0(\omega) = E^* E = E^2$ by the part b) of Theorem 1.

2) $\Rightarrow$ 3). The projection G and S_0 are the conditional expectations on the σ-algebras $\mathcal{G}$ and $\mathcal{E}(\hat{T}_0)$, respectively. If they commute their product $S_0 G$ is the conditional expectation on the σ-algebra $\mathcal{G} \cap \mathcal{E}(\hat{T}_0) = \mathcal{G}^0$, i.e. $S_0 G = G^0$.

From the above we have $E_0(\omega) = E$ a.e. on Ω and

$$\begin{aligned}(Ef)(x) &= (\bar{E}G\bar{f})(\omega, x) \\ &= (S_0 G\bar{f})(\omega, x) = (G^0 \bar{f})(\omega, x) = (E^{\mathcal{G}^0} \bar{f})(\omega, x) = (E^{\Sigma^0} f)(x) = (S^0 f)(x)\end{aligned}$$

for every $f \in L_1(m)$ a.e. Thus $E = S^0$.

It is obvious 3) $\Rightarrow$ 1). The theorem is proved. □

Remark. The condition 2) means that the σ-algebras $\mathcal{G}$ and $\mathcal{E}(\hat{T}_n)$ are conditionally independent with respect to their intersection $\mathcal{G}^0$.

The next statements follow immediately from Theorems 1 and 2.

Corollary 1. *The following conditions are equivalent to conditions* 1)–4) *of Theorem* 2.

5) $E_n(\omega) = E$ *a.e.*

6) $E_n(\omega)$ *are projectors a.e.*

7) $E = S^0$.

8) $\bar{E}_n$ *do not depend of* n.

Corollary 2. *The following conditions are equivalent:*

1) $S_n = GS_n$, $(\mathcal{E}(\hat{T}_n) \subset \mathcal{G})$.

2) $S_n = VS_n$.

3) $S_n = \bar{E}_n$.

4) $S_n = \bar{E}$.

5) $S_n = G^0$, $(\mathcal{E}(\hat{T}_0) = \mathcal{G}^0)$.

Remark. One can get the analogous results for

$$B^+_{n,k}(\omega) = \frac{1}{k}\sum_{i=0}^{k-1} T_{n+i}(\omega)\dots T_{n+1}(\omega)\cdot T_n(\omega)$$

$$B^-_{n,k}(\omega) = \frac{1}{k}\sum_{i=0}^{k} T_{n-1}(\omega)T_{n-2}(\omega)\dots T_{n-i}(\omega)$$

taking the skew product $\check{T}_n = \bar{T}^*_n(\omega)V$ instead of $\hat{T}_n$.

Example. Consider the trivial case when $T_n(\omega) = U_n^{-1}(\omega)U_{n+1}(\omega)$, where $U_n = U_n(\omega)$ be a SSRBO with the invertible and hence unitary values $U_n(\omega) \in \boldsymbol{M}$.

Then for a.a. ω

$$A^-_{n,k}(\omega) = U_n^{-1}(\omega)(\frac{1}{k}\sum_{i=1}^{k} U_{n+i}(\omega)) \xrightarrow{k\to\infty} U_n^{-1}(\omega)U$$

in the strong operator topology on $\boldsymbol{L}_p(m)$, $1 \le p < \infty$, where $U = \int U_n(\omega)\,dP$. We have in this situation $E_n(\omega) = U^{-1}(\omega)U$, $E = U^*U$, $S_n = \bar{U}_n^{-1}G\bar{U}_n$, $\mathcal{E}(\hat{T}_n) = \bar{U}_n^{-1}\mathcal{G}$ and $\Sigma^0 = \{A \in \Sigma : U_n(\omega_1)I_A = U_n(\omega_2)I_A \text{ for a.a. } \omega_1, \omega_2\}$.

The conditions of the Theorem 2 hold if $E = U^*U$ is a projection i.e. U is a partial isometry operator in $\boldsymbol{L}_2(m)$.

Choosing the appropriate RBO $U_n = U_n(\omega)$ one can achieve that the σ-algebras $\mathcal{G}$ and $\mathcal{E}(\hat{T}_n) = \bar{U}_n^{-1}\mathcal{G}$ are or are not conditionally independent with respect to their intersection $\mathcal{G}^0$, and hence the conditions of Theorem 2 do or not hold, although $\mathcal{E}(\hat{T}_n) \not\subset \mathcal{G}$. The condition $\mathcal{E}(\bar{T}_n) \subset \mathcal{G}$ may here hold in the only trivial case, when $\mathcal{E}(\bar{T}_n) = \mathcal{G}$, i.e., U_n is not random.

4. More on the condition $\mathcal{E}(\hat{T}_n) \subset \mathcal{G}$

We will give a sufficient condition for a SSRBO $T_n = T_n(\omega)$ in order that the inclusion $\mathcal{E}(\hat{T}_n) \subset \mathcal{G}$ and hence $\mathcal{E}(\hat{T}_n) = \mathcal{G}^0$ hold.

Let $\mathcal{F}_n^-$ and $\mathcal{F}_n^+$ be the σ-algebras of $\mathcal{F}$ generated by the RBO $\{T_k, k \le n\}$ and $\{T_k, k \ge n\}$ respectively.

We will assume in the sequel without loss of generality, that the σ-algebra $\mathcal{F}$ is generated by $\{T_n,\ n \in \mathbb{Z}\}$.

Let also $\bar{\mathcal{F}}_n^\pm$ be the $\bar{P}$-completions of the σ-algebras $\mathcal{F}_n^\pm \otimes \Sigma$ and $\bar{F}_n^\pm$ be conditional expectation operators with respect to $\bar{\mathcal{F}}_n^\pm$.

Theorem 3. a) $\mathcal{E}(\hat{T}_n) \subset \bar{\mathcal{F}}_{n-1}^- \cap \bar{\mathcal{F}}_n^+$, $n \in \mathbb{Z}$.

b) *If the σ-algebra $\mathcal{F}_{n-1}^- \cap \mathcal{F}_n^+$ is P-trivial then $\mathcal{E}(\hat{T}_n) = \mathcal{G}^0 \subset \mathcal{G}$.*

Proof. By definition we have $\bar{T}_{n+k}\bar{F}_n^+ = \bar{F}_n^+\bar{T}_{n+k}$ for $k \geq 0$ and hence

$$\bar{F}_n^+\hat{T}_n^k = \bar{F}_n^+\bar{T}_n \dots \bar{T}_{n+k-1}V^k = \bar{T}_n \dots \bar{T}_{n+k-1}V^k V^{-k}\bar{F}_n^+ V^k = \hat{T}_n^k \bar{F}_{n-k}^+ \text{ for } k \geq 0.$$

If $h \in L_1(\bar{P})$ and $\hat{T}_n h = h$ then

$$\|h - \bar{F}_n^+ h\|_1 = \|\hat{T}_n^k h - \bar{F}_n^+ \hat{T}_n^k h\|_1 = \|\hat{T}_n^k (h - \bar{F}_{n-k}^+ h)\| \leq \|h - \bar{F}_{n-k}^+ h\| \to 0 \text{ as } k \to +\infty,$$

since $\bar{F}_{n-k}^+ = E^{\mathcal{F}_{n-k}^+} \to E^{\mathcal{F}} = \mathrm{Id}$ as $k \to +\infty$ in the strong operator topology on $L_1(\bar{P})$ by the martingale couvergence theorem. Thus $\bar{F}_n^+ h = h$ for every fixed point h of $\hat{T}_n^k$ and hence $\mathcal{E}(\hat{T}_n^k) \subset \overline{\mathcal{F}}_n^+$.

Consider the operator $\hat{T}_n^*$ instead of $\hat{T}_n$ one has in a similar way $\bar{F}_{n-1}\hat{T}_n^{k^*} = \hat{T}_n^{k^*}\bar{F}_{n+k-1}$ for $k \geq 0$ and $\mathcal{E}(\hat{T}_n^*) = \mathcal{E}(\hat{T}_n) \subset \bar{F}_{n-1}$ that completes the proof of the part a).

If the intersection $\mathcal{F}_{n-1}^- \cap \mathcal{F}_n^+$ is P-trivial then $\mathcal{F}_{n-1}^- \cap \mathcal{F}_n^+ = \mathcal{G}$ and $\mathcal{E}(\hat{T}_n) \subset \mathcal{G}$ by the part a). The equality $\mathcal{E}(\hat{T}_n) = \mathcal{G}^0$ follows from the Corollary 2 of Theorem 2.

The theorem is proved. □

If the RBO T_n, $n \in \mathbb{Z}$, are independent the σ-algebras $\mathcal{F}_{n-1}^-$ and $\mathcal{F}_n^+$ are independent too and we obtain the following well known result.

Corollary 1. *For any sequence $\{T_n\}$ of independent identically distributed RBO the equality $\mathcal{E}(\hat{T}_n) = \mathcal{G}^0$ holds.*

Consider now the case when the SSRBO $T_n = T_n(\omega)$ is a Markov one.

Let F_n be the conditional expectation operator on the σ-algebra $\mathcal{F}_n$ generated by $T_n: \Omega \to M$ and $(M, \mathcal{B}, \mu)$ be the state space of the SSRBO $T_n = T_n(\omega)$ with $\mu = P \circ T_n^{-1}$ and $\mathcal{F}_n = T_n^{-1}\mathcal{B}$. One can construct the transition operator $Q: L_p(\mu) \to L_p(\mu)$ of this Markov SSRBO setting $Q = J_{n-1}^{-1} F_{n-1} F_n J_n$, where $J_n: L_p(\mu) \ni f \to f \circ T_n \in L_p(\mu)$ is the coordinated imbedding.

The one-step deterministic σ-algebra $\mathcal{D}(Q)$ is defined as

$$\mathcal{D}(Q) = \mathcal{E}(Q^*Q) = \{A \in \mathcal{B} : QI_A = I_B,\ B \in \mathcal{B}\}.$$

The Markov property of the SSRBO implies $\mathcal{F}_{n-1}^- \cap \mathcal{F}_n^+ = \mathcal{F}_{n-1} \cap \mathcal{F}_n$ and we can deduce from the Theorem 3 the following

Corollary 2. *Let $T_n = T_n(\omega)$ be a Markov SSRBO with the transition operator Q. If the corresponding one-step deterministic σ-algebra $\mathcal{D}(Q)$ is μ-trivial, then $\mathcal{E}(\hat{T}_n) = \mathcal{G}^0 \subset \mathcal{G}$.*

Remark 1. Let $\mathcal{D}_\infty(Q) := \cap_{n=1}^\infty \mathcal{E}(Q^{*n}Q^n)$ and $\mathcal{D}_\infty(Q^*) = \cap_{n=1}^\infty \mathcal{E}(Q^n Q^{*n})$ be the deterministic (tail) σ-algebras of the Markov SSRBO. (**[12]**, **[13]**, **[14]**). It is clear that the triviality of $\mathcal{D}(Q)$ or $\mathcal{D}(Q^*)$ implies the same one for $\mathcal{D}_\infty(Q)$ and $\mathcal{D}_\infty(Q^*)$, but the converse is not true in general even for finite state space Markov chains.

Remark 2. There exist stationary Markov processes with trivial one-step deterministic σ-algebras and non-trivial "past" and "future" asymptotic σ-algebras (**[15]** pp. 113–115). Thus the conditions of the Corollary 2 may be hold even for non-regular Markov SSRBO.

Acknowledgements. I would like to thank Prof. M. Lin for useful discussions.

References

[1] S. M. Ulam, J. v. Neumann. Random ergodic theorems, Bull. Amer. Math. Soc. 51 (1945), 660.

[2] Sh. Kakutani, Random ergodic theorems and Markoff processes with a stable distributions, Proc. Berkley Symp. Math. Stat. Prob. 1950, (1951), 247–261.

[3] C. Ryll-Nardzewski, On the ergodic theorems III - (Random ergodic theorems), Studia Math. 14 (1954), 298–301.

[4] S. Gladysz, Ein Ergodischer Satz. Studia Math. 40 (1956), 148–157.

[5] K. Jacobs, Lecture Notes on Ergodic Theory. (1962-63). Aarhus Univ. Math. Inst.

[6] E. Kin, The general random ergodic theorem. I, II, Z. Wahrscheinlichkeitstheorie verw. Geb. 22 (1972), 120–135, 136–144.

[7] T. Yoshimoto, On the random ergodic theorem. Studia Math. 61 (1977), 231–237.

[8] J. Woś, Random ergodic theorems for sub-Markovian operators. Studia Math. 74 (1982), 191–212.

[9] U. Krengel, Ergodic theorems. De Gruyter Stud. Math. 6, Berlin-New-York 1985.

[10] Y. Kifer, Ergodic theory of random transformations, Birkhäuser. Boston 1986.

[11] B. A. Rubshtein, Convolutions of random measures on compact groups, J. Theoret. Probab. 8 (1995), 523–538.

[12] S. R. Foguel, The ergodic theory of Markov processes. Van Nostrand Reinhold, New York 1969.

[13] V. G. Vinokurov, B. A. Rubshtein, On tail algebras of stationary Markov processes, in: Asymptotic problems for probability distributions, pp. 62–71, FAN, Tashkent 1984 (Russian).

[14] U. Krengel, M. Lin, On the deterministic and asymptotic σ-algebras of a Markov operator. Canad. Math. Bull. 32 (1989), 64–73.

[15] M. Rosenblatt, Markov processes, Structure and asymptotic behaviour, Springer-Verlag, Berlin-Heidelberg 1971.

Dept. of Mathematics and Computer Science
Ben-Gurion University of the Negev
84105 Beer Sheva, Israel

A Central Limit Theorem for Conditional Distributions

Ben-Zion Rubshtein[1]

Abstract. Let $(\xi, \eta) = \{(\xi_n, \eta_n) : n \in \mathbb{Z}\}$, be a two-dimensional ergodic stationary process with $\mathbf{E}\xi_n^2 < \infty$. We find a simple sufficient condition, under which the $\mathbf{P}_y$-conditional distributions of $S_n = \xi_1 + \xi_2 + \cdots + \xi_n$ are asymptotically normal as $n \to \infty$ for almost all realizations $y = \{\eta_n(\omega)\}_{n\in\mathbb{Z}}$ of the process η. Under the same conditions an invariance principle for the corresponding $\mathbf{P}_y$-conditional distributions in $C[0, 1]$ holds too.

1. Introduction

Let $(\xi, \eta) = \{(\xi_n, \eta_n)\}$ be a two-dimensional ergodic stationary process on a probability space $(\Omega, \mathcal{F}, \mathbf{P})$, $\xi_n = \xi_n(\omega)$, $\eta_n = \eta_n(\omega)$, $\omega \in \Omega$, $n \in \mathbb{Z}$, and $\mathbf{E}\xi_n^2 < \infty$.

Consider the conditional distribution functions

$$\mathbf{F}_{n,y}(t) = \mathbf{P}_y\Big\{\frac{S_n - \mathbf{E}_y S_n}{\sqrt{\mathbf{D}_y S_n}} < t\Big\}, \quad t \in \mathbb{R}$$

where $S_n = \xi_1 + \xi_2 + \cdots + \xi_n$ and $\mathbf{P}_y$, $\mathbf{E}_y$, $\mathbf{D}_y$ are the conditional probabilities, expectations and dispersions under a fixed realizations $y = \{\eta_n(\omega)\}$ of the sequence η.

We will find simple sufficient conditions (see (I), (II), (III) below) under which the conditional distributions $\mathbf{F}_{n,y}$ are asymptotically normal for almost all realizations y. The conditions allow the use of [7] (see Theorem 3 below) and the conditionally centered sequences method with bounded L_2-norms [5] and to reduce the problem to non-stationary martingale differences. Then the ergodicity of the process implies the conditions of the martingale limit theorem [3] (see also [4], [8]).

Just the same conditions and arguments entail the invariance principle for conditional distributions. In particular, the result hold in case, when the random variables $\{\xi_n\}$ are conditionally independent with respect to the process $\eta = \{\eta_n\}$.

1 Research supported by the Israel Ministries of Science and of Absorption.

2. Asymptotic normality

One can assume that $\xi_n = \xi_n \circ \Theta^n$ and $\eta_n = \eta_0 \circ \Theta^n$, where Θ is a $\mathbf{P}$-preserving invertible shift transformation of Ω.

Let $\mathcal{B}$ be the σ-algebra generated by the rv's $\{\eta_k, k \in \mathbb{Z}\}$ and $\mathcal{A}_n$ be the σ-algebra generated by $\{\xi_k,\ k \le n\}$ and $\{\eta_k,\ k \in \mathbb{Z}\}$. The conditional expectation with respect to $\mathcal{A}_n$ is denoted by $E_n = \mathbf{E}(\cdot \mid \mathcal{A}_n)$.

Put $S'_n = \xi'_1 + \cdots + \xi'_n = S_n - \mathbf{E}(S_n \mid \mathcal{B})$ where $\xi'_n = \xi_n - \mathbf{E}(\xi_n \mid \mathcal{B})$.

Throughout we assume that the following conditions hold:

(I) The stationary process (ξ, η) is ergodic, i.e., the shift Θ is ergodic.

(II) $\mathbf{E}\xi_n^2 < \infty$.

(III) $\sup_{n\ge 1} \mathbf{E}(E_0 S'_n)^2 < \infty$.

Let $(Y, \mathcal{F}_Y, \mathbf{P}_Y)$ be the realization space of the process $\eta = \{\eta_n\}$ and

$$\eta\colon \Omega \to \eta(\omega) = \{\eta_n(\omega)\} \in Y$$

be the corresponding factor-mapping, $\mathbf{P}_Y = \mathbf{P} \circ \eta^{-1}$.

We can assume without loss of generality, that the probability $\mathbf{P}$ defines the regular conditional probabilities $\{\mathbf{P}_y,\, y \in Y\}$ and conditional expectations $\{\mathbf{E}_y,\, y \in Y\}$ with respect to the mapping $\eta\colon \Omega \to Y$ and the factor-measure $\mathbf{P}_Y$. All conditional probabilities are defined on the σ-algebra $\mathcal{A}$, generated by the rv's $\{\xi_k, \eta_k,\ k \in \mathbb{Z}\}$ and we can consider the process $\{\xi_n\}$ on each of the probability spaces $(\Omega, \mathcal{F}_y, \mathbf{P}_y)$, $y \in Y$, where $\mathcal{F}_y$ is the $\mathbf{P}_y$-completion of $\mathcal{A}$.

Consider the conditional distribution functions

$$\mathbf{F}'_{n,y}(t) = \mathbf{P}_y\Big\{\frac{1}{\sqrt{n}} S'_n < t\Big\}, \quad t \in \mathbb{R},\ y \in Y$$

Theorem 1. *Assume that* (I), (II), *and* (III) *hold. Then for almost all* $y \in Y$*:*

a) $\lim_{n\to\infty} \frac{1}{n}\mathbf{D}_y S_n = \sigma^2$ *a.e., where* $\sigma \ge 0$ *does not depend on* y.

b) $\mathbf{F}'_{n,y} \to \Phi_{0,\sigma}$ *weakly as* $n \to \infty$*, where* $\Phi_{0,\sigma}$ *is the normal law with the mean* 0 *and the variance* σ^2 *and* $\Phi_{0,0}$ *is the degenerate at* 0 *law.*

We define for $n \ge 1$ a random function

$$x_n(t, \omega) = \frac{1}{\sqrt{n}} \left(S'_{[nt]}(\omega) + (nt - [nt])\xi'_{[nt]+1}(\omega)\right), \quad 0 \le t \le 1,\ \omega \in \Omega$$

and let $\mathbf{P}_y^{x_n}$ be the sequence of probability measures on $(C, \mathcal{E})$ determined by the distribution of x_n with respect to the conditional probability $\mathbf{P}_y$, $y \in Y$.

We consider the space $C = C[0, 1]$ with the sup norm topology and $\mathcal{E}$ being the Borel σ-algebra generated by open sets in C (see [2], [10] for the terminology).

We denote by $\mathbf{P}^{\sigma w}$ the probability measure on $(C, \mathcal{E})$, which is the distribution of the random function $\sigma w(t), 0 \leq t \leq 1$, where $\sigma \geq 0$ and $w(t)$ is the standard Wiener process in C (if $\sigma = 0$ the distribution $\mathbf{P}^{\sigma w} = \mathbf{P}^0$ is degenerate at 0).

Theorem 2. *Assume that* (I), (II) *and* (III) *hold and* σ *is defined by Theorem* 1(a). *Then for almost all* $y \in Y$ *the sequence of the conditional distribution* $\mathbf{P}_y^{x_n}$ *converges weakly to* $\mathbf{P}^{\sigma\omega}$ *in* C. *In particular, all finite dimensional distributions of* x_n *with respect to* $\mathbf{P}_y$ *converge weakly to those of the process* $\sigma w(t)$, $0 \leq t \leq 1$.

3. A reduction

The first step of the proof of the above theorems will be a reduction to the martingale case by means of (III) (cf. [5]).

Let $\mathcal{H} = L_2(\Omega, \mathcal{F}, \mathbf{P})$ and for $n \in \mathbb{Z}$

$$\mathcal{H}_n^0 = \{\varphi \in \mathcal{H} : \mathbf{E}(\varphi \mid \mathcal{B}) = 0, \quad E_n\varphi = \varphi\}.$$

We have for all n

$$\mathcal{H}_n^0 \subset \mathcal{H}_{n+1}^0, \quad E_n\mathcal{H}_{n+1}^0 = \mathcal{H}_n^0, \quad V\mathcal{H}_{n+1}^0 = \mathcal{H}_n^0$$

where the unitary operator V in $\mathcal{H}$ is defined by $V\varphi = \varphi \circ \Theta$.

We can now introduce a positive contraction T_0 on $\mathcal{H}_0^0$ as the restriction $T_0 = E_0V|_{\mathcal{H}_0^0}$ of E_0V on $\mathcal{H}_0^0$.

Theorem 3. *For* $\varphi \in \mathcal{H}_0^0$ *the following conditions are equivalent:*

a) *There exists* $\phi \in \mathcal{H}_0^0$ *such that* $\varphi = \phi - T_0\phi$.

b) *There exists* $L_2\text{-}\lim_{n\to\infty} \frac{1}{n}\sum_{k=1}^n \sum_{i=0}^{k-1} T_0^i\varphi = \phi \in \mathcal{H}_0^0$ *and* $\varphi = \phi - T_0\phi$.

c) $\sup_{n\geq 1} \| \sum_{k=0}^{n-1} T_0^k\varphi\|_{L_2} < \infty$.

One can find the proof of the theorem and all corresponding references in [7], where the problem is treated in a more general situation. Applying the theorems for $\varphi = \xi_0' \in \mathcal{H}_0^0$ we get for $n \geq 1$

$$E_0S_n' = E_0(V + \cdots + V^n)\xi_0' = \sum_{k=1}^n T_0^k\xi_0'$$

and hence the condition (III) coincides to (c) in this case.

Under the condition (III) we find

$$\lambda_0 = L_2\text{-}\lim_{n\to\infty} \frac{1}{n} \sum_{k=1}^n \sum_{i=0}^{k-1} T_0^i\xi_0' \in \mathcal{H}_0^0$$

such that $\xi_0' = \lambda_0 - T_0\lambda_0$. Having taken $\lambda_n = \lambda_0 \circ \Theta^n \in \mathcal{H}_n^0$ we get a stationary sequence such that

$$\xi_n' = \lambda_n - E_n\lambda_{n+1} = \lambda_n - \lambda_{n+1} + \zeta_n, \quad n \in \mathbb{Z} \tag{1}$$

where $\zeta_n = \lambda_{n+1} - E_n\lambda_{n+1} \in \mathcal{H}_{n+1}^0$, $\zeta_n = \zeta_0 \circ \Theta^n$ and $E_n\zeta_n = 0$.

This means that ζ_n, $n \in \mathbb{Z}$ are stationary martingale differences with respect to the sequences of the σ-algebras $\{\mathcal{A}_{n+1}, n \in \mathbb{Z}\}$ with $\mathbf{E}(\zeta_n \mid \mathcal{B}) = 0$ and $\mathbf{E}\zeta_n^2 < \infty$. Since $\mathcal{B} \subset \mathcal{A}_n$ for all n we see that $\{\zeta_n\}$ form martingale differences on $(\Omega, \mathcal{F}_y, \mathbf{P}_y)$ for almost all $y \in Y$, but the conditional processes need not be stationary.

4. The martingale differences

We will show now that the martingale differences $\{\zeta_n\}$ satisfy all conditions of the martingale central limit theorem [3] on $(\Omega, \mathcal{F}_y, \mathbf{P}_y)$ for a.a. $y \in Y$.

Denote $\mathbf{E}\zeta_n^2 = b^2$ and $R_n = \zeta_1 + \cdots + \zeta_n$, $R_0 = 0$. Then the individual ergodic theorem and ergodicity of Θ imply that the following limits exist almost everywhere on $(\Omega, \mathcal{F}, \mathbf{P})$:

$$\lim_{n\to\infty} \frac{1}{n}\sum_{k=1}^{n} E_k\zeta_k^2 = \lim_{n\to\infty} \frac{1}{n}\sum_{k=1}^{n} (E_0\zeta_0^2) \circ \Theta^k = E\zeta_0^2 = b^2 \tag{2}$$

$$\lim_{n\to\infty} \frac{1}{n}\mathbf{E}(R_n^2 \mid \mathcal{B}) = \lim_{n\to\infty} \frac{1}{n}\sum_{k=1}^{n} \mathbf{E}(\zeta_k^2 \mid \mathcal{B}) = \lim_{n\to\infty} \frac{1}{n}\sum_{k=1}^{n} (\mathbf{E}(\zeta_k^2 \mid \mathcal{B}))\circ\Theta^k = \mathbf{E}\zeta^2\eta_0^2 = b^2. \tag{3}$$

Let now

$$l_n = \frac{1}{n}\sum_{k=1}^{n} \mathbf{E}\big(\zeta_k^2 I(\zeta_k^2 \geq n) \mid \mathcal{B}\big)$$

where $I(A)$ denotes the indicator function of the set A. Then for any $m \leq n$ one has

$$l_n \leq \frac{1}{n}\sum_{k=1}^{m-1} \mathbf{E}(\zeta_k^2 \mid \mathcal{B}) + \frac{1}{n}\sum_{k=m}^{n} \mathbf{E}\big(\zeta_k^2 I(\zeta_k^2 \geq m) \mid \mathcal{B}\big).$$

By the individual ergodic theorem

$$\frac{1}{n}\sum_{k=1}^{m-1} \mathbf{E}(\zeta_k^2 \mid \mathcal{B}) = \frac{1}{n}\sum_{k=1}^{m-1} \mathbf{E}\big(\zeta^2\eta_0^2 \mid \mathcal{B}\big) \circ \Theta^k \to 0$$

and

$$\frac{1}{n}\sum_{k=m}^{n} \mathbf{E}\big(\zeta_k^2 I(\zeta_k^2 \geq m) \mid \mathcal{B}\big) = \frac{1}{n}\sum_{k=m}^{n} \mathbf{E}\big(\zeta_0^2 I(\zeta_0^2 \geq m) \mid \mathcal{B}\big) \circ \Theta^k \to \mathbf{E}\big(\zeta_o^2 I(\zeta_0^2 \geq m)\big)$$

almost everywhere an $n \to \infty$. Thus for all m

$$\overline{\lim_{n\to\infty}} \, l_n = 0 \text{ a.e.} \tag{4}$$

since $\mathbf{E}\big(\xi_0^2 I(\xi_0^2 \geq m)\big) \to 0$ as $m \to \infty$.

By using (2), (3), (4) we get (under $b \neq 0$) that for a.a. $y \in Y$:

$$\lim_{n\to\infty} \frac{1}{b_n(y)^2} \sum_{k=0}^{n} \mathbf{E}_y(\zeta_k^2 \mid \mathcal{A}_k) = 1 \ \mathbf{P}\text{-a.e.} \tag{5}$$

and for any $\epsilon > 0$

$$\lim_{n\to\infty} \frac{1}{b_n(y)^2} \sum_{k=0}^{n} \mathbf{E}_y\big(\zeta_k^2 I(\zeta_k^2 \geq \epsilon b_n(y)^2)\big) = 0 \tag{6}$$

where

$$b_n(y)^2 = \mathbf{E}_y R_n^2 = \sum_{k=1}^{n} \mathbf{E}_y \zeta_k^2.$$

The conditions (5) and (6) show that the martingale R_n on the probability space $(\Omega, \mathcal{F}_y, \mathbf{P}_y)$ satisfies all conditions of the Brown's limit theorem [3] and we obtain the following results for almost all $y \in Y$ and $b \neq 0$.

A) $\mathbf{P}_y\big(\frac{1}{b_n(y)} R_n < t\big) \to \Phi_{0,1}(t)$, $t \in \mathbb{R}$, as $n \to \infty$.

B) $\mathbf{P}_y^{u_n} \to \mathbf{P}^w$ as $n \to \infty$ weakly in C, where $\mathbf{P}^w$ is the standard Wiener measure and $\mathbf{P}_y^{u_n}$, $n \geq 1$, are the conditional distributions on $(C, \mathcal{E})$ determined by the random function $u_n = u_n(t)$ which is defined as follows:

$$u_n(t) = \frac{1}{b_n(y)}\Big(R_k + \frac{t b_n(y)^2 - b_k(y)^2}{b_{k+1}(y)^2 - b_k(y)^2}\zeta_{k+1}\Big)$$

for $0 \leq t \leq 1$ and $b_k(y)^2 \leq t b_n(y)^2 \leq b_{k+1}(y)^2$, $k = 0, 1, \ldots, n-1$. By (3) we have

$$\lim_{n\to\infty} \frac{1}{n} b_n^2(y) = b^2 = \mathbf{E}\zeta_0^2 \text{ for a.a. } y$$

and hence

C) $\mathbf{P}_y^{v_n} \to \mathbf{P}^{bw}$ as $n \to \infty$ weakly in C where

$$v_n(t) = \frac{1}{\sqrt{n}}\left(R_{[nt]} + (nt - [nt])\zeta_{[nt]+1}\right), \ 0 \leq t \leq 1.$$

5. The error term

We will now complete the proofs of the Theorems 1 and 2 by estimating the difference $S'_n - R_n$.

Let $\mu_n = \lambda_n - \lambda_{n+1}$ and $M_n = S'_n - R_n = \sum_{k=1}^n \mu_k = \lambda_1 - \lambda_{n+1}$, $M_0 = 0$. We have

$$\frac{1}{n}\lambda_1^2 \to 0,\ \frac{1}{n}\lambda_{n+1}^2 \to 0 \ \mathbf{P}\text{-a.e.} \tag{7}$$

by the individual ergodic theorem.

Since

$$\big(\mathbf{E}((S'_n)^2 \mid \mathcal{B})\big)^{\frac{1}{2}} - \big(\mathbf{E}(R_n^2 \mid \mathcal{B})\big)^{\frac{1}{2}} \le \big(\mathbf{E}(\lambda_1^2 \mid \mathcal{B})\big)^{\frac{1}{2}} + \big(\mathbf{E}(\lambda_{n+1}^2 \mid \mathcal{B})\big)^{\frac{1}{2}}$$

there exists the limit

$$\lim_{n\to\infty} \frac{1}{n}\mathbf{E}((S'_n)^2 \mid \mathcal{B}) = b^2 \ \mathbf{P}\text{-a.e.}$$

and hence the limit $\sigma^2 = \lim_{n\to\infty} \frac{1}{n}\mathbf{D}_y S'_n$ in the part (a) of Theorem 1 exists with $\sigma^2 = b^2 = \mathbf{E}\zeta_0^2$.

Since $\frac{1}{\sqrt{n}}(S'_n - R_n) \to 0$ $\mathbf{P}$-a.e. as $n \to \infty$ we have also the $\mathbf{P}_y$-a.e. convergence for a.a. $y \in Y$ and the statement (A) implies the part (b) of Theorem 1.

In order to prove Theorem 2 we will show that conditional distributions $\mathbf{P}_y^{z_n}$ of the random functions

$$z_n(t) = x_n(t) - v_n(t) = \frac{1}{\sqrt{n}}(M_{[nt]} + (nt - [nt])\mu_{[nt]+1})$$

converge weakly to the generate distribution $\mathbf{P}^0$ on $(C, \mathcal{E})$.

Show that for any $\epsilon > 0$ and a.a. y

$$\mathbf{P}_y\Big(\frac{1}{\sqrt{n}} \max_{1\le k\le n} |\lambda_k| > \epsilon\Big) \to 0 \quad \text{as } n \to \infty. \tag{8}$$

We have by the Tchebyshev inequality

$$\mathbf{P}_y\big(\max_{1\le k\le n} \lambda_k^2 > n\epsilon^2\big) \le \sum_{k=1}^n \mathbf{P}_y\{\lambda_k^2 > n\epsilon^2\} \le \frac{1}{\epsilon^2 n}\sum_{k=1}^n \mathbf{E}_y\big(\lambda_k^2 I(\lambda_k^2 > \epsilon^2 n)\big).$$

But the last expression tends to 0 as $n \to \infty$ for a.a. y by the same arguments which we used to prove the above Lindeberg condition (4), (6). Hence

$$\mathbf{P}_y\big(\sup_{0\le t\le 1} |z_n(t)| > \epsilon\big) \le \mathbf{P}_y\big(\max_{1\le k\le n+1} |\lambda_k| > \sqrt{n}\epsilon\big)$$

tends to 0 as $n \to \infty$ for a.a. y. In particular, the $\mathbf{P}_y$-distributions of $z_n(t)$ converge to the generate one $\Phi_{0,0}$ for all $t \in [0, 1]$ and $\mathbf{P}_y\left\{\sup_{0\le s,t\le 1} |z_n(s) - z_n(t)| > \epsilon\right\} \to 0$

as $n \to \infty$ and hence

$$\lim_{h\to 0}\overline{\lim_{n\to\infty}}\, \mathbf{P}_y\Big\{\sup_{0\le s,t\le 1} |z_n(s)-z_n(t)| > \epsilon\Big\} = 0$$

for a.e. $y \in Y$.

Thus the sequence of the conditional distributions $\{\mathbf{P}_y^{z_n}\}$ is tight and $\mathbf{P}_y^{z_n}$ tends weakly to the generate distribution $\mathbf{P}^0$ on $(C, \mathcal{E})$ as $n \to \infty$ for a.e. y.

This completes the proof of Theorem 2.

6. Remarks and consequences

1) The convergence to the degenerate law in Theorems 1, 2 holds iff $\sigma^2 = b^2 = \mathbf{E}\zeta_n^2 = 0$, i.e. when ξ_n' has the form $\xi_n' = \lambda_n - \lambda_{n+1}$ with $\lambda_n \in \mathcal{H}_n^0$.
2) Condition (III) trivially holds if $\mathbf{E}_n\xi_{n+1}' = 0$. In this case S_n', $n \ge 0$, form a martingale. In particular, the above results are fulfilled when the rv's $\{\xi_n\}$ are conditionally independent with respect to the σ-algebra $\mathcal{B}$, generated by η_n, $n \in \mathbb{Z}$.
3) If the spectral radius $\rho = \lim_{n\to\infty} \|T_0^n\|^{\frac{1}{n}}$ of the operator T_0 on $\mathcal{H}_0^0$ is less than 1, the series $\sum_{n=0}^{\infty} T_0^n$ converges in norm, and (III) holds too. One can get generalizations of [9], [11] for conditionally Markov processes in the case under consideration (cf. [5]).
4) We can also get convergence conditions of non-conditional distributions of S_n by (I), (II), (III). Let

$$A_n = \frac{1}{\sqrt{n}}\big(\mathbf{E}(S_n \mid \mathcal{B}) - \mathbf{E}S_n\big), \quad S_n^* = \frac{1}{\sqrt{n}}\big(S_n - \mathbf{E}S_n\big)$$

and $\mathbf{G}_n$, $\mathbf{F}_n^*$ be the distribution functions of A_n and S_n^*, respectively.

Corollary. *Assume that* (I), (II), (III) *hold and* σ *as defined in Theorem* 1(a) *is distinct from* 0. *If* $\mathbf{G}_n$ *converges to a distribution function* $\mathbf{G}$, *then* $\mathbf{F}_n^*$ *converges to* $\Phi_{\mathbf{G},\sigma}$, *where* $\Phi_{\mathbf{G},\sigma} = \Phi_{0,\sigma} * \mathbf{G}$ *is the weighted normal distribution.*

5) If the processes ξ and η are independent of each other we have $\mathbf{P}_y = \mathbf{P}$ for almost all $y \in Y$. In particular we get the following extension of the CLT of Billingsley and Ibragimov ([1], [6]) for stationary ergodic martingales, the same as the Gording–Lifšic theorem for the Markov case [5].

Corollary. *Let* $\xi = \{\xi_n, n \in \mathbb{Z}\}$ *be an ergodic stationary process with* $\mathbf{E}\xi_n^2 < \infty$ *and* $\mathbf{E}\xi_n = 0$.

If

$$\sup_{n\ge 1} \mathbf{E}\big(\mathbf{E}(S_n \mid \xi_k,\ k \le 0)\big)^2 < \infty$$

then $\mathbf{F}_n(t) = \mathbf{P}\{\frac{1}{\sqrt{n}} S_n < t\}$, $t \in \mathbb{R}$, *converges to* $\Phi_{0,\sigma}$ *as* $n \to \infty$, *where* $\sigma = \lim_{n\to\infty} \frac{1}{n}\mathbf{E}S_n^2$.

I wish to thank M. Goldshtein and M. Lin for discussions related to this topic.

References

[1] Billingsley, P., The Lindeberg–Levy theorem for martingales, Proc. Amer. Math. Soc. 12 (1961), 788–792.

[2] Billingsley, P., Convergence of probability measures, Wiley, New York 1968.

[3] Brown, B. M., Martingale central limit theorems, Ann. Math. Stat. 42 (1971), 59–66.

[4] Dvoretsky, A., Central limit theorem for dependent random variables and some applications, Ann. Math. Stat. 40 (1969), 1871.

[5] Gordin, M. I., Lifšic B. A., The central limit theorem for stationary Markov processes, Soviet Math. Dokl. 19 (1978), 392–394.

[6] Ibragimov, I. A., A central limit theorem for a class of dependent random variables, Theory Probab. Appl. 8 (1963), 83–89.

[7] Lin, M., Sine, R., Ergodic theory and the functional equation (I-T) $x = y$, J. Operator Theory 10 (1983), 153–166.

[8] Loynes, R. M., The central limit theorem for backwards martingales, Z. Wahrscheinlichkeitstheorie verw. Geb. 13 (1969), 1–8.

[9] Nagaev, S. V., Some limit theorems for stationary Markov chains, Theory Probab. Appl. 2 (1957), 378-406.

[10] Parthasarathy, K. R., Probability measures on metric spaces, Academic Press, New York 1967.

[11] Rosenblatt, M., Markov processes. Structure and asymptotic behaviour, Springer-Verlag, Berlin-Heidelberg-New York 1971.

Dept. of Mathematics and Computer Science
Ben-Gurion University of the Negev
84105 Beer Sheva, Israel

Complex Methods in the Calculation of Some Distribution Functions

Yoram Sagher and Niandi Xiang

1. Introduction

Given a function $\phi(x)$ defined on $\mathbb{R}^1$, we denote its distribution function by $\phi_*(\alpha)$. Thus,

$$\phi_*(\alpha) = \lambda\{x : |\phi(x)| > \alpha\}, \qquad \forall\alpha > 0$$

where λ denotes the Lebesgue measure.

Distribution functions of some rational functions played an important role in the study of the Hilbert transform. L. Loomis **[1]** computed the distribution function of a certain rational function to prove that the Hilbert transform is of weak type $(1, 1)$. E. Stein and G. Weiss **[2]** used such a calculation to prove that the Hilbert transform is of type (p, p) for $1 < p < \infty$. This calculation also enabled E. Stein and G. Weiss to prove their remarkable result that the distribution function of the Hilbert transform of a characteristic function of a set of finite measure is proportional to the measure of the set. Later A. P. Calderon **[3]** gave a different proof of this property.

In this note we use a simple residue method to give a unified approach to the theorems of L. Loomis and E. Stein – G. Weiss. We also prove a generalization of these results.

It was kindly pointed out to us by the referee that Loomis' result had been anticipated by G. Boole in 1857 **[4]**. G. Boole's analysis of the behavior of a certain rational function is quite similar to Loomis' and Boole's subsequent calculation used an early form of the calculus of residues.

2. The distribution functions of some rational functions

Theorem 1. *Let P_k and Q_n be polynomials of precise degrees k and n, and*

$$\phi(x) = \frac{P_k(x)}{Q_n(x)}.$$

Suppose that $\phi(x)$ has only real poles, these poles are simple, and all residues are positive. For any contour C which encloses all the poles of $\phi(x)$,

(I) *if $k \le n - 1$, then*

$$\phi_*(\alpha) = \frac{2}{\alpha}\frac{1}{2\pi i}\int_C \phi(z)\,dz, \quad \forall\alpha > 0,$$

(II) *if* $k = n$, *and both* $Q_n(x)$ *and* $P_n(x)$ *are monics, then*

$$\phi_*(\alpha) = \begin{cases} \frac{1}{\sinh(\log \alpha)} \frac{1}{2\pi i} \int_C \phi(z)\, dz, & \forall \alpha > 1 \\ \infty & \forall \alpha \le 1 \end{cases} \tag{1}$$

$$\lambda\{x : |\phi(x)| < \alpha\} = \begin{cases} \frac{-1}{\sinh(\log \alpha)} \frac{1}{2\pi i} \int_C \phi(z)\, dz, & \forall \alpha < 1 \\ \infty & \forall \alpha \ge 1. \end{cases} \tag{2}$$

Proof. (I) We may assume that $Q_n(x)$ is monic.

Let $(b_j)_1^n$ be the zeroes of $Q_n(x)$, and

$$\mu_j = \operatorname{Res}(b_j) = \frac{P_k(b_j)}{Q_n'(b)} > 0, \quad 1 \le j \le n,$$

then

$$\phi(x) = \sum_{j=1}^n \frac{\mu_j}{x - b_j}, \quad \phi'(x) = -\sum_{j=1}^n \frac{\mu_j}{(x - b_j)^2},$$

$\phi(x)$ is thus unbounded and strictly decreasing in each intervals (b_j, b_{j+1}) for $0 \le j \le n$, where $b_0 = -\infty$, $b_{n+1} = \infty$. As Loomis observed, this implies

$$I = \lambda\{x : \phi(x) > \alpha\} = \sum \text{zeroes } (\phi - \alpha) - \sum \text{poles } (\phi - \alpha).$$

Clearly,

$$\sum \text{zeroes } (\phi - \alpha) - \sum \text{poles } (\phi - \alpha) = \frac{1}{2\pi i} \int_C z \frac{\phi'(z)}{\phi(z) - \alpha} dz,$$

where C is any contour enclosing all poles and zeroes of $(\phi - \alpha)$.

If ψ is a rational function of order ≤ -2, then

$$\lim_{R \to \infty} \int_{|z|=R} \psi(z)\, dz = 0.$$

Thus if $\{z : |z| < R\}$ contains all poles of ψ,

$$\int_{|z|=R} \psi(z) dz = 0.$$

We apply this to $\psi_k(z) = z\phi'(z)\phi^k(z)$ for $k = 1, 2, \ldots$ in the following calculation.

Let R be large enough so that the contour C is enclosed by the circle $\{z : |z| = R\}$ and $\left|\frac{\phi(z)}{\alpha}\right| < q < 1$ on the circle. Then

$$\begin{aligned} I &= \frac{1}{2\pi i} \int_{|z|=R} z \frac{\phi'(z)}{\phi(z) - \alpha} = -\frac{1}{\alpha} \frac{1}{2\pi i} \int_{|z|=R} z\phi'(z)\big(1 + \frac{\phi}{\alpha} + \frac{\phi^2}{\alpha^2} + \cdots\big)\, dz \\ &= -\frac{1}{\alpha} \frac{1}{2\pi i} \int_{|z|=R} z\phi'\, dz = \frac{1}{2\pi i \alpha} \int_{|z|=R} (\phi - (z\phi)')\, dz \\ &= \frac{1}{2\pi i \alpha} \int_{|z|=R} \phi\, dz = \frac{1}{2\pi i \alpha} \int_C \phi\, dz, \end{aligned}$$

where in the last equality C can be chosen as any contour enclosing all poles of ϕ.

Similarly,

$$\lambda\{x : \phi(x) < -\alpha\} = \frac{1}{\alpha}\frac{1}{2\pi i}\int_C \phi\, dz.$$

(II) Let us prove (1) first.

For $\alpha > 1$, since $(\phi(z) - 1)$ has the same poles and residues as $\phi(z)$, and satisfies the condition in (I), we can apply (I) to $(\phi(z) - 1)$,

$$\begin{aligned}\lambda\{\phi(x) > \alpha\} &= \lambda\{x : \phi(x) - 1 > \alpha - 1\}\\ &= \frac{1}{\alpha - 1}\frac{1}{2\pi i}\int_C (\phi(z) - 1)\, dz\\ &= \frac{1}{\alpha - 1}\frac{1}{2\pi i}\int_C \phi(z)\, dz.\end{aligned}$$

Similarly,

$$\lambda\{x : \phi(x) < -\alpha\} = \lambda\{x : \phi(x) - 1 < -(1 + \alpha)\} = \frac{1}{\alpha + 1}\frac{1}{2\pi i}\int_C \phi(z)\, dz.$$

Finally we have

$$\phi_*(\alpha) = \Big(\frac{1}{e^{\log\alpha} - 1} + \frac{1}{e^{\log\alpha} + 1}\Big)\frac{1}{2\pi i}\int_C \phi(z)\, dz = \frac{1}{\sinh(\log\alpha)}\frac{1}{2\pi i}\int_C \phi(z)\, dz.$$

The proof of $\phi_*(\alpha) = \infty$ for $\alpha \leq 1$ is trivial.

To prove (2), let us observe:

(A) If $\phi(x)$ satisfies the same conditions as before except that all the residues of $\phi(x)$ are negative, then

$$\phi_*(\alpha) = \begin{cases} -\frac{1}{\sinh(\log\alpha)}\frac{1}{2\pi i}\int_C \phi(z)dz, & \forall\alpha > 1\\ \infty, & \forall\alpha \leq 1.\end{cases}$$

(B) If $\psi(z) = \frac{\prod_{j=1}^n (z-a_j)}{\prod_{j=1}^n (z-b_j)}$, where $a_j,\ b_j,\ 1 \leq j \leq n$ are complex numbers, then

$$\frac{1}{2\pi i}\int_C \psi(z)dz = -\frac{1}{2\pi i}\int_C \frac{1}{\psi(z)}\, dz,$$

where C is any contour enclosing all zeroes and poles of ψ. This is so because

$$\frac{1}{2\pi i}\int_C (\psi(z) - 1)\, dz = \sum_{j=1}^n (b_j - a_j) = -\sum_{j=1}^n (a_j - b_j) = -\frac{1}{2\pi i}\int_C \Big(\frac{1}{\psi(z)} - 1\Big)\, dz.$$

By these two observations, notice that all the residues of $\frac{1}{\phi(z)}$ are negative, we have

$$\lambda\{x : |\phi(x)| < \alpha\} = \lambda\Big\{x : |\frac{1}{\psi(x)}| > \alpha^{-1}\Big\} = \begin{cases} \frac{-1}{\sinh(\log\alpha)} \frac{1}{2\pi i} \int_C \phi(z)\, dz, & \forall\alpha < 1 \\ \infty, & \forall\alpha \geq 1. \end{cases}$$

□

Applying the residue theorem to Theorem 1, we immediately get

Corollary 1.1. [1] *If*

$$\phi(x) = \sum_{j=1}^{n} \frac{\mu_j.}{x - b_j}, \quad \mu_j > 0, \quad \forall j,$$

then

$$\phi_*(\alpha) = \frac{2}{\alpha} \sum_{j=1}^{n} \mu_j, \quad \forall\alpha > 0.$$

Corollary 1.2. [2] *If*

$$Q_n(x) = \prod_{i=1}^{n} (x - b_j), \quad P_n(x) = \prod_{i=1}^{n} (x - a_i), \text{ and } \phi(x) = \frac{P_n(x)}{Q_n(x)},$$

where $a_i < b_i < a_{i+1} < b_{i+1}$, $1 \leq i \leq n - 1$, *then*

$$(\log|\phi|)_*(\alpha) = \frac{2}{\sinh(\alpha)} \sum_{i=1}^{n} (b_i - a_i), \quad \forall\alpha > 0. \tag{3}$$

Proof. Let $a > 0$ be given. Since

$$\big(\log|\phi|\big)_*(\alpha) = \lambda\big\{x : |\phi(x)| > e^{\alpha}\big\} + \lambda\big\{x : |\phi(x)| < e^{-\alpha}\big\},$$

and since $\phi(x)$ has only real, simple poles, all residues are positive and both $Q_n(x)$ and $P_n(x)$ monic, part (II) of Theorem 1 holds.

Further, if $r(z) = \frac{P_n}{Q_n} - 1$, then $\frac{1}{z^2} r\left(\frac{1}{z}\right)$ has a simple pole at the origin with residue $\sum_{i=1}^{n} (b_i - a_i)$, and therefore

$$\frac{1}{2\pi i} \int_C \phi(z)\, dz = \frac{1}{2\pi i} \int_{|z|=R} r(z)\, dz = \frac{1}{2\pi i} \int_{|z|=\frac{1}{R}} \frac{1}{z^2} r\left(\frac{1}{z}\right) dz$$

$$= \sum_{i=1}^{n} (b_i - a_i), \qquad \forall\alpha > 1,$$

where C is any contour enclosing all poles of ϕ, and $\{z : |z| = R\}$ is any circle enclosing C, and thus (3) follows. □

3. The distribution functions of limits of some rational functions

Theorem 2. *Let*

$$\phi_n(x) = \frac{P_{k(n)}}{Q_{l(n)}(x)}, \qquad n = 1, 2, \ldots$$

be a sequence of rational functions such that $\phi_n(x)$ has only simple real poles and all its residues are positive. Suppose that there exists some function $\phi(x)$ such that $\phi_n \to \phi$ in measure on $\mathbb{R}^1$, then

(I) *if $k(n) \le \ell(n) - 1$, $\forall n$,*

$$\phi_*(\alpha) = \frac{2}{\alpha} \lim_n \frac{1}{2\pi i} \int_{C_n} \phi_n(z)\, dz, \qquad \forall \alpha > 0,$$

(II) *if $k(n) = l(n)$, $Q_{k(n)}(x)$ and $P_{k(n)}(x)$ are monics, $\forall n$,*

$$\phi_*(\alpha) = \begin{cases} \frac{1}{\sinh(\log \alpha)} \lim_n \frac{1}{2\pi i} \int_{C_n} \phi_n(z)\, dz, & \forall \alpha > 1 \\ \infty, & \forall \alpha \le 1, \end{cases}$$

$$\lambda\{x : |\phi(x)| < \alpha\} = \begin{cases} \frac{-1}{\sinh(\log \alpha)} \lim_n \frac{1}{2\pi i} \int_{C_n} \phi_n(z)\, dz, & \forall \alpha < 1 \\ \infty, & \forall \alpha \ge 1, \end{cases}$$

where C_n are any contours enclosing all the poles of ϕ_n.

Proof. (I) Fix $\alpha > 0$. For all $\delta > 0$, we have

$$\lambda\{x : |\phi| > \alpha + \delta\} \le \liminf_n \lambda\{x : |\phi_n| > \alpha\},$$

$$\limsup_n \lambda\{x : |\phi_n| > \alpha + \delta\} \le \lambda\{x : |\phi| > \alpha\},$$

since $\phi_n \to \phi$ in measure on $\mathbb{R}^1$. Now, let $\delta \to 0$,

$$\lambda\{x : |\phi| > \alpha\} \le \liminf_n \lambda\{x : |\phi_n| > \alpha\}.$$

On the other hand,

$$\begin{aligned} \lambda\{x : |\phi| > \alpha\} &\ge \lim_{\delta \to 0} \limsup_n \lambda\{x : |\phi_n| > \alpha + \delta\} \\ &= \lim_{\delta \to 0} \limsup_n \frac{1}{\alpha + \delta} \frac{1}{2\pi i} \int_{C_n} \phi_n(z)\, dz \\ &= \limsup_n \lambda\{x : |\phi_n| > \alpha\}. \end{aligned}$$

Therefore,

$$\lambda\{x : |\phi| > \alpha\} = \lim_n \lambda\{x : |\phi_n| > \alpha\},$$

and the first claim of Theorem 2 follows.

(II) The second claim of Theorem 2 follows similarly. □

The following corollary is a generalization of Loomis' result.

Corollary 2.1. *Let* $a_k < a_{k+1}$, $\mu_k > 0$, $-\infty < k < \infty$, *and* $\sum_{k=-\infty}^{\infty} \mu_k < \infty$. *Define*

$$\phi(x) = \sum_{k=-\infty}^{\infty} \frac{\mu_k}{x - \alpha_k}$$

then

$$\phi_*(\alpha) = \frac{2}{\alpha} \sum_{k=-\infty}^{\infty} \mu_k, \qquad \forall \alpha > 0.$$

Proof. Let $\phi_{n,m} = \sum_{k=m}^{n} \frac{\mu_k}{x-\alpha_k}$, then $\lim_{\substack{n\to\infty \\ m\to-\infty}} \phi_{n,m}(x) = \phi(x)$ pointwise, and $\phi_{n,m}(x)$ is Cauchy in measure by Corollary 1.1, hence $\phi_{n,m} \to \phi$ in measure. Further, $\phi_{n,m}(x)$ has only real, simple poles and all its residues are positive, thus the result follows from Theorem 2. □

Finally, we derive E. Stein and G. Weiss' result.

Corollary 2.2. *Let* E *be a set of finite measure in the real line. Let* I_E *be the characteristic function of* E *and* HI_E *be its Hilbert transform. Then*

$$(HI_E)_*(\alpha) = \frac{2\lambda(E)}{\sinh(\alpha)}, \qquad \forall \alpha > 0.$$

Proof. Let us choose a sequence of sets $\{E_n\}$ in $\mathbb{R}^1$, where each E_n is the finite union of mutually disjoint intervals,

$$E_n = \bigcup_{j=1}^{l_n} (a_j^{(n)}, b_j^{(n)}),$$

so that $\|I_{E_n} - I_E\|_1 \to 0$.

Since the Hilbert transform is of weak type $(1, 1)$, we also have $HI_{E_n} \to HI_E$ in measure on $\mathbb{R}^1$.

Hence $HI_{E_n} = \log|\phi_n|$ except for finite many points a_j, b_j, $1 \leq j \leq l_n$, where

$$\phi_n(x) = \frac{\prod_{j=1}^{l_n}(x - a_j^{(n)})}{\prod_{j=1}^{l_n}(x - b_j^{(n)})},$$

applying part (II) to Theorem 2 and Corollary 1.2, we immediately have

$$(HI_E)_*(\alpha) = \frac{2}{\sinh(\alpha)} \lim_n \lambda(E_n) = \frac{2\lambda(E)}{\sinh(\alpha)}.$$

□

References

[1] L. Loomis, A note on the Hilbert transform, Bull. Amer. Math. Soc. 52 (1946), 1082–1086.

[2] E. Stein, and G. Weiss, An extension of a theorem of Marcinkiewics and some of its applications, J. Math. Mech. 8, (1959), 263–284.

[3] A. P. Calderón, Singular integrals, Bull. Amer. Math. Soc. 72 (1966), 434.

[4] G. Boole, On the Comparison of Transcendents, with certain applications to the Theory of Definite Integrals, Philos. Trans. Roy. Soc. London 147 (1857), 745–803.

Department of Mathematics, Statistics and Computer Science
University of Illinois at Chicago
Chicago, IL 60607, U.S.A.

Exponential Integrability of Rademacher Series

Yoram Sagher and Kecheng Zhou

Abstract. We prove exponential norm inequalities for Rademacher series.[1]

Rademacher functions are defined as $r_0(t) = 1$ if $0 \le t < \frac{1}{2}$; $r_0(t) = -1$ if $\frac{1}{2} \le t < 1$; $r_0(t+1) = r_0(t)$; and $r_k(t) = r_0(2^k t)$. Throughout the paper, I will stand for a dyadic interval $[(k-1)2^{-n}, k2^{-n}]$, $n = 0, 1, \dots$ and $k = 1, 2, \dots, 2^n$. $f(t) \in \mathrm{BMO}_d$ if and only if $\|f\|_{\mathrm{BMO}_d} = \sup_I \left(\frac{1}{|I|}\int_I |f(t) - f_I|^2\, dt\right)^{\frac{1}{2}} < \infty$, where the supremum is taken over all dyadic intervals and $f_I = \frac{1}{|I|}\int_I f(t)\, dt$. It is well known (**[4]**) that convergent Rademacher series $\sum c_k r_k(t)$ necessarily belong to BMO_d. the John–Nirenberg theorem (**[3]**) as applied to BMO_d states that for some constants b and B, for any dyadic interval $I \subset [0, 1]$, and $\alpha > 0$,

$$|\{t \in I : |f(t) - f_I| > \alpha\}| \le |I| B \cdot \exp\Big\{\frac{-b\alpha}{\|f\|_{\mathrm{BMO}_d}}\Big\}.$$

It follows that if $f \in \mathrm{BMO}_d$ and if $\gamma = \frac{b}{(B+1)\|f\|_{BMO_d}}$, then for all dyadic intervals I,

$$\inf_{a \in \mathbb{R}} \frac{1}{|I|}\int_I e^{\gamma|f(t)-a|}\, dt \le 2. \tag{1}$$

This in turn proves Khinchin's theorem which states that all L^p norms of Rademacher series are equivalent, see (**[4]**).

There are many equivalent norms on BMO_d. An important one is

$$\rho(f) = \inf\Big\{\frac{1}{\lambda} : \sup_I \frac{1}{|I|}\int_I e^{\lambda|f(t)-f_I|}\, dt \le 2\Big\}.$$

This is related to

$$\mu(f) = \inf\Big\{\frac{1}{\lambda} : \sup_I \inf_{a \in \mathbb{R}} \frac{1}{|I|}\int_I e^{\lambda|f(t)-a|}\, dt \le 2\Big\},$$

used by Wolff in his characterization of local BMO (**[1]**, **[2]**, **[6]**). Clearly $\mu(f) \le \rho(f)$. On the other hand, if for some a, $\frac{1}{|I|}\int_I e^{\lambda|f(t)-a|}\, dt \le 2$, then

$$\frac{1}{|I|}\int_I e^{\lambda|f(t)-f_I|}\, dt \le (e^{\lambda|f_I - a|})\frac{1}{|I|}\int_I e^{\lambda|f(t)-a|}\, dt$$

---

1 Mathematics Subject Classification 42A55.

and by Jensen's inequality

$$e^{\lambda|f_I - a|} \leq \frac{1}{|I|} \int_I e^{\lambda|f(t)-a|}\,dt \leq 2,$$

so that $\frac{1}{|I|}\int_I e^{\lambda|f(t)-f_I|}\,dt \leq 4$. Therefore, we have that $\rho(f) \leq 2\mu(f)$.

In this note we show that for a Rademacher series f, $\mu(f) = \rho(f)$. A second related result which we prove here is that given any set E with positive measure, there exists $N = N(E)$ so that $\| \sum_{k=N}^{\infty} c_k r_k(t) \|_{\exp\left(E; \frac{dt}{|E|}\right)} \sim \left(\sum_{k=N}^{\infty} |c_k|^2\right)^{1/2}$ for any c_k's with $\sum_{k=N}^{\infty} |c_k|^2 < \infty$. This is a stronger result than the local version of Khinchin's inequality proved in ([**5**]).

Theorem 1. *Let* $f(t) = \sum_{k=0}^{\infty} c_k r_k(t)$ *with* $\sum_{k=0}^{\infty} |c_k|^2 < \infty$. *Then for any* $\lambda > 0$,

$$\sup_I \inf_{a\in\mathbb{R}} \frac{1}{|I|} \int_I e^{\lambda|f(t)-a|}\,dt = \int_0^1 e^{\lambda|f(t)|}\,dt.$$

Proof. Let us show first that

$$\inf_{a\in\mathbb{R}} \int_0^1 e^{\lambda|f(t)-a|}\,dt = \int_0^1 e^{\lambda|f(t)|}\,dt.$$

Since

$$\int_0^1 e^{\lambda|f(t)-a|}\,dt = \sum_{j=0}^{\infty} \frac{\lambda^j \int_0^1 |f(t)-a|^j\,dt}{j!},$$

it suffices to show that for any $j = 1, 2, \ldots,$

$$\int_0^1 |f(t)-a|^j\,dt \geq \int_0^1 |f(t)|^j\,dt.$$

Let $a > 0$, $q > 1$. Define $\phi_a(a) = \int_0^1 |f(t)-a|^q\,dt$. Because $\left|\sum_{k=0}^{\infty} c_k r_k(t) + a\right|$ and $\left|-\sum_{k=0}^{\infty} c_k r_k(t) + a\right| = \left|\sum_{k=0}^{\infty} c_k r_k(t) - a\right|$ are identically distributed, $\phi_q(a) = \phi_q(-a)$ for all $a \in \mathbb{R}$.

Moreover, since $q > 1$, $\phi_q(a)$ is differentiable at all points and

$$\phi_q'(a) = -q \int_0^1 \left|\sum_{k=0}^{\infty} c_k r_k(t) - a\right|^{q-1} \operatorname{sign}\Big\{\sum_{k=0}^{\infty} c_k r_k(t) - a\Big\}dt$$

$$= -q\left(\int_{\{\sum_{k=0}^{\infty} c_k r_k(t) > a\}} \Big(\sum_{k=0}^{\infty} c_k r_k(t) - a\Big)^{q-1} dt - \int_{\{\sum_{k=0}^{\infty} c_k r_k(t) < a\}} \Big(a - \sum_{k=0}^{\infty} c_k r_k(t)\Big)^{q-1} dt\right).$$

Hence $\phi_q'(a)$ is increasing in $a > 0$. Since $r_k(x) = -r_k(1-x)$, $|\{f > \alpha\}| = |\{f < -\alpha\}|$ for all α. Therefore, we also have that

$$\int_{\{\sum_{k=0}^{\infty} c_k r_k(t) > 0\}} \Big(\sum_{k=0}^{\infty} c_k r_k(t)\Big)^{q-1} dt = \int_{\{\sum_{k=0}^{\infty} c_k r_k(t) < 0\}} \Big(-\sum_{k=0}^{\infty} c_k r_k(t)\Big)^{q-1} dt,$$

that is, $\phi_q'(0) = 0$. Thus $\phi_q'(a) > 0$ for all $a > 0$ and $\phi_q(a)$ is increasing in $a > 0$, and $\phi_q(a) \geq \phi_q(0)$ for all $a \in \mathbb{R}$. Letting $q \to 1$ and applying the dominated convergence theorem, we get $\phi_1(a) \geq \phi_1(0)$. We therefore have

$$\sup_I \inf_{a \in \mathbb{R}} \frac{1}{|I|} \int_I e^{\lambda |f(t) - a|} dt \geq \int_0^1 e^{\lambda |f(t)|} dt.$$

Conversely we show that for any dyadic interval I,

$$\inf_{a \in \mathbb{R}} \frac{1}{|I|} \int_I e^{\lambda |f(t) - a|} dt \leq \int_0^1 e^{\lambda |f(t)|} dt.$$

Let $|I| = 2^{-n}$, then for $k = 0, 1, \ldots, n-1$, $r_k(t)$ are constant on I. Denote these constants by $r_k(I)$, and choose $a = \sum_{k=0}^{n-1} c_k r_k(I)$. Since for $k \geq n$, $r_k(2^{-n}t) = r_{k-n}(t)$, we have

$$\inf_{a \in \mathbb{R}} \frac{1}{|I|} \int_I e^{\lambda |f(t) - a|} dt \leq \int_0^1 e^{\lambda |\sum_{k=n}^{\infty} c_k r_{k-n}(t)|} dt.$$

Noting that $\sum_{k=n}^{\infty} c_k r_{k-n}(t)$ and $\sum_{k=n}^{\infty} c_k r_k(t)$ are identically distributed in $[0, 1)$, we have

$$\int_0^1 e^{\lambda |\sum_{k=n}^{\infty} c_k r_{k-n}(t)|} dt = \int_0^1 e^{\lambda |\sum_{k=n}^{\infty} c_k r_k(t)|} dt.$$

We therefore want to prove that for any $n \geq 0$,

$$\int_0^1 e^{\lambda |\sum_{k=n}^{\infty} c_k r_k(t)|} dt \leq \int_0^1 e^{\lambda |\sum_{k=0}^{\infty} c_k r_k(t)|} dt.$$

Let $r > 1$. Consider functions $f(t)$, defined on $[0, 1]$ such that for some $\lambda > 0$, $\int_0^1 e^{\lambda |f(t)|} dt \leq r$. Then it is clear that the collection of all such functions is a Banach space with norm

$$\|f\|_{\exp(r)} = \inf\Big\{\frac{1}{\lambda} : \int_0^1 e^{\lambda |f(t)|} dt \leq r\Big\}.$$

Let $\varepsilon_k = -1$, if $k < n$; and $\varepsilon_k = 1$, if $k \geq n$. Then, since $\sum_{k=0}^{\infty} c_k r_k(t)$ and $\sum_{k=0}^{\infty} \varepsilon_k c_k r_k(t)$ are identically distributed in $[0, 1]$,

$$\|\sum_{k=n}^{\infty} c_k r_k\|_{\exp(r)} \leq \frac{1}{2}\Big(\|\sum_{k=0}^{\infty} c_k r_k\|_{\exp(r)} + \|\sum_{k=0}^{\infty} \varepsilon_k c_k r_k\|_{\exp(r)}\Big)$$
$$= \|\sum_{k=0}^{\infty} c_k r_k\|_{\exp(r)}.$$

Given $\lambda > 0$, and $f(t) = \sum_{k=0}^{\infty} c_k r_k(t)$. Let $r = \int_0^1 e^{\lambda|f(t)|}\,dt$. Then

$$\frac{1}{\lambda} \geq \|f\|_{\exp(r)} \geq \|\sum_{k=n}^{\infty} c_k r_k\|_{\exp(r)}.$$

It follows that $\int_0^1 e^{\lambda|\sum_{k=n}^{\infty} c_k r_k(t)|}\,dt \leq r$. Therefore,

$$\int_0^1 e^{\lambda|\sum_{k=n}^{\infty} c_k r_k(t)|}\,dt \leq \int_0^1 e^{\lambda|\sum_{k=0}^{\infty} c_k r_k(t)|}\,dt.$$

□

Theorem 2. *If we define*

$$\|f\|_{\exp(E;\frac{dt}{|E|})} = \inf\Big\{\frac{1}{\lambda} : \frac{1}{|E|}\int_E e^{\lambda|f(t)|}\,dt \leq 2\Big\},$$

then, given any $E \subset [0, 1]$, there exists $N = N(E)$ and positive constants α and β so that

$$\alpha\Big(\sum_{k=N}^{\infty} |c_k|^2\Big)^{1/2} \leq \|\sum_{k=N}^{\infty} c_k r_k(t)\|_{\exp(E;\frac{dt}{|E|})} \leq \beta\Big(\sum_{k=N}^{\infty} |c_k|^2\Big)^{1/2}$$

for any c_k's with $\sum_{k=N}^{\infty} |c_k|^2 < \infty$.

Proof. To prove the right-hand side inequality, it suffices to show that given any $E \subset [0, 1]$ with positive measure, there exists $N = N(E) > 0$ so that if $\sum_{k=N}^{\infty} |c_k|^2 < \infty$, and $\frac{1}{\mu} > \frac{6(B+1)}{b}\big(\sum_{k=N}^{\infty} |c_k|^2\big)^{1/2}$, then

$$\frac{1}{|E|}\int_E e^{\mu|\sum_{k=N}^{\infty} c_k r_k(t)|}\,dt \leq 2$$

where b and B are the constants appearing in the John–Nirenberg theorem.

Let $E \subseteq [0,1]$. Using a Calderón–Zygmund decomposition of χ_E we can find dyadic intervals I_j's such that $E \subseteq \cup I_j$ almost everywhere and $|E \cap I_j| > \frac{1}{2}|I_j|$. Let K be such that $|E \cap (\cup_{j>K} I_j)| < |E|^2$. Write $E_1 = E \cap (\cup_{j \le K} I_j)$ and $E_2 = E \cap (\cup_{j>K} I_j)$.

Let $f(t) = \sum_{k=0}^{\infty} c_k r_k(t)$ with $\sum_k |c_k|^2 < \infty$. We have (**[4]**):

$$\|f\|_{\mathrm{BMO}_d} = \Big(\sum_{k=0}^{\infty} |c_k|^2\Big)^{1/2}.$$

For $j \le K$, let $n_j = -\log_2 |I_j|$ and let $N = \max\{n_j : j \le K\}$. By (1) and Theorem 1,

$$\begin{aligned}
2 &\ge \inf_{a \in \mathbb{R}} \frac{1}{|I_j|} \int_{I_j} e^{\gamma |\sum_{k=N}^{\infty} c_k r_k - a|}\, dt \\
&= \inf_{a \in \mathbb{R}} \int_0^1 e^{\gamma |\sum_{k=N}^{\infty} c_k r_{k-n_j} - a|}\, dt \\
&= \int_0^1 e^{\gamma |\sum_{k=N}^{\infty} c_k r_{k-n_j}|}\, dt \\
&= \frac{1}{|I_j|} \int_{I_j} e^{\gamma |\sum_{k=N}^{\infty} c_k r_k|}\, dt.
\end{aligned}$$

For E_1, we have

$$\begin{aligned}
\frac{1}{|E|} \int_{E_1} e^{\gamma |\sum_{k=N}^{\infty} c_k r_k(t)|}\, dt &\le \frac{1}{|E|} \int_{\cup_{j \le K} I_j} e^{\gamma |\sum_{k=N}^{\infty} c_k r_k(t)|}\, dt \\
&= \frac{1}{|E|} \sum_{j \le K} \int_{I_j} e^{\gamma |\sum_{k=N}^{\infty} c_k r_k(t)|}\, dt \\
&\le \frac{\sum_{j \le K} |I_j|}{|E|} \cdot 2 \le 4.
\end{aligned}$$

For E_2, we have, using Hölder's inequality,

$$\frac{1}{|E|} \int_{E_2} e^{\gamma |\sum_{k=N}^{\infty} c_k r_k(t)|/2}\, dt \le \frac{|E_2|^{1/2}}{|E|} \Big(\int_0^1 e^{\gamma |\sum_{k=N}^{\infty} c_k r_k(t)|}\, dt\Big)^{1/2} \le 2,$$

so that

$$\frac{1}{|E|}\int_E e^{\gamma|\sum_{k=N}^{\infty} c_k r_k(t)|/2}\, dt \leq \frac{1}{|E|}\int_{E_1} e^{\gamma|\sum_{k=N}^{\infty} c_k r_k(t)|/2}\, dt + \frac{1}{|E|}\int_{E_2} e^{\gamma|\sum_{k=N}^{\infty} c_k r_k(t)|/2}\, dt \leq 6.$$

From Hölder's inequality we have that, denoting $\mu = \gamma/6$,

$$\frac{1}{|E|}\int_E e^{\mu|\sum_{k=N}^{\infty} c_k r_k(t)|}\, dt \leq 2.$$

The left-hand side inequality follows from the fact that

$$\|\sum_{k=N}^{\infty} c_k r_k(t)\|_{\exp(E;\frac{dt}{|E|})} \geq \frac{1}{2}\|\sum_{k=N}^{\infty} c_k r_k(t)\|_{L^2(E)} \geq \alpha\big(\sum_{k=N}^{\infty} |c_k|^2\big)^{1/2}.$$

□

Theorem 2 clearly implies that for all $0 < p < \infty$, there exists B_p so that for any $E \subset [0, 1]$, $|E| \neq 0$, there exists an $N = N(E)$ so that if $f(t) = \sum_{k=N}^{\infty} c_k r_k(t)$, where $\sum |c_k|^2 < \infty$, then

$$\big(\frac{1}{|E|}\int_E |f(t)|^p\, dt\big)^{1/p} \leq B_p\big(\sum_{k=N}^{\infty} |c_k|^2\big)^{1/2}.$$

Using Hölder's inequality and the equivalence of $\frac{1}{|E|}\int_E |f|^2\, dt$ and $\sum_{k=N}^{\infty} |c_k|^2$, we then have that there exists A_p so that (see **[5]**)

$$A_p\big(\sum_{k=N}^{\infty} |c_k|^2\big)^{1/2} \leq \big(\frac{1}{|E|}\int_E |f(t)|^p\, dt\big)^{1/p} \leq B_p\big(\sum_{k=N}^{\infty} |c_k|^2\big)^{1/2}.$$

References

[1] J. Garcia-Cuerva and J. L. Rubio De Francia, Weighted Norm Inequalities and Related Topics, North-Holland Mathematics Studies, Elsevier Science Publishers B.V., 1985.

[2] Peter J. Holden, Extension Theorems for Functions of Vanishing Mean Oscillation, Pacific J. Math. 142 (1990), 277–295.

[3] F. John and L. Nirenberg, On Functions of Bounded Mean Oscillation, Comm. Pure Appl. Math. 14 (1961), 415–426.

[4] E. Kochneff, Y. Sagher and K. Zhou, BMO Estimates for Lacunary Series, Ark. Mat. 28 (1990), 301–310.

[5] Y. Sagher and K. Zhou, A Local Version of a Theorem of Khinchin, in: Analysis and Partial Differential Equations, Lecture Notes in Pure and Appl. Math., Marcel Dekker, New York 1990, pp. 327–330.

[6] T. Wolff, Restrictions of A_p weights, preprint.

[7] A. Zygmund, Trigonometric Series, Cambridge University Press, 1959.

Department of Mathematics, Statistics and Computer Science
University of Illinois at Chicago
Chicago, IL 60607, U.S.A.

Mathematics & Statistics Dept.
California State University at Sacramento
Sacramento, CA 95819-2694, U.S.A.

Weighted Averages of Contractions Along Subsequences

Dominique Schneider and Michel Weber

I. Introduction — main result

Our purpose is to prove the following theorem:

Theorem 1. *Let $\{p_k,\ k \in \mathbb{N}\}$ be a non-decreasing sequence of positive integers and let $\mathcal{N}$ be a partial index $\mathcal{N} = \{n_k,\ k \in \mathbb{N}\}$. Assume that the following condition is satisfied:*

$$\sum_{N \in \mathcal{N}} \frac{\Phi(N)^2}{N} < \infty, \tag{$\mathcal{H}_1$}$$

where $\Phi(N) = \sqrt{\log[p_N + p + 2]}$, if $N \in I_p$, $p \geq 1$ and $I_p = [2^{c_p}, 2^{c_p+1}[$ denotes the p-th interval of the form $[2^r, 2^{r+1}[$ containing elements of $\mathcal{N}$.

Let further $\{\theta_k,\ k \in \mathbb{N}\}$ be a sequence of independent centered random variables with basic probability space $(\Omega, \mathcal{B}, P)$ and satisfying the following condition

$$\mathbb{E}\Big\{\sup_{N \in \mathcal{N}} \Big(\frac{\sum_{k=1}^N \theta_k^2}{N}\Big)^{1/2}\Big\} < \infty. \tag{$\mathcal{H}_2$}$$

Then, there exists a measurable set Ω^* with $P\{\Omega^*\} = 1$ such that for any $\theta = \{\theta_n,\ n \in \mathbb{N}\} \in \Omega^*$, any other probability space $(X, \mathcal{A}, \mu)$ any contraction T on $L^2(\mu)$, and any element $f \in L^2(\mu)$, the averages

$$A_N^{\theta,T}(F) = \frac{1}{N}\sum_{k=1}^{N} \theta_k f \circ T^{p_k} \tag{1}$$

are converging μ-almost surely to 0 when N tends to infinity along the partial index $\mathcal{N}$.

Remarks. a) A weaker form of the above result is easy to show by using a standard argument when

$$\sup_{k \in \mathbb{N}} \mathbb{E}(\theta_k^2) < \infty.$$

Under this assumption,

$$P \otimes \mu\Big\{\lim_{N\to\infty} A_N^{\theta,T}(f) = 0\Big\} = 1,$$

is always fulfilled whenever $f \in L^2(\mu)$. There are, however, sequences $\{\theta_k,\ k \in \mathbb{N}\}$ and $\mathcal{N}$ satisfying $(\mathcal{H}_2)$ but not the above condition. In that case, there no standard argument known to us to prove that weaker property.

b) If $\mathcal{N} = \{[k^\varepsilon],\ k \geq 1\}$, with $\varepsilon > 1$, then the condition $(\mathcal{H}_1)$ reduces here to

$$\sum_{k\geq 1} \frac{\log(p_{[k^\varepsilon]} + \log k)}{k^\varepsilon} < \infty,$$

which is realized as soon $p_k = O(2^{k^\delta})$ with $\delta > 0$ sufficiently small.

c) If $\mathcal{N} = \{2^k,\ k \geq 1\}$, then the condition $(\mathcal{H}_1)$ reduces to

$$\sum_{k\geq 1} \frac{\log(p_{2^k} + k)}{2} < \infty,$$

which is realized as soon $p_k = O(2^{k^\gamma})$ with $0 \leq \gamma < 1$.

d) The following example illustrates the usefulness of the construction of the function Φ. Let $\varphi, v\colon \mathbb{N} \to \mathbb{N}$ be two increasing maps, and let

$$\mathcal{N} = \Big\{N = 2^{\varphi(k)} + j \ \text{ for some } k \in \mathbb{N} \text{ and } 0 \leq j \leq \frac{2^{\varphi(k)}}{v(k)}\Big\}.$$

Then the condition $(\mathcal{H}_1)$ is realized if and only if

$$\sum_{k\in\mathbb{N}} \frac{\log(p_{2^{\varphi(k)}} + k + 2)}{v(k)} < \infty. \tag{2}$$

We can thus always find a sequence $\{p_k,\ k \in \mathbb{N}\}$ satisfying the above condition and so the conclusion of Theorem 1, no matter how fast is the growth of the function φ.

e) The construction of Φ in Theorem 1 is analogue to the construction of normalizing functions we made in **[We]**.

f) A *speed of convergence* can be exhibited. From assumption $(\mathcal{H}_1)$ there exists $v\colon \mathcal{N} \to \mathbb{N}^*$ decreasing to 0 such that

$$\sum_{n\in\mathcal{N}} \frac{\Phi(N)^2}{v^2(N)N} < \infty.$$

From the very last line of the proof of Theorem 6 in Section II, will follow that

$$\mu\big\{x\colon A_N^{\theta,T}(f) = O(v(N)),\ N \in \mathcal{N}\big\} = 1.$$

II. Proof of the main result

2.1. Preliminaries. The proof relies on the following lemmas.

Lemma 2. ([F3]). *Let $\{G_k,\ k \in \mathbb{N}\}$ be a sequence of Gaussian random vectors with values in a separable Banach space $(B, \|\cdot\|)$. Then,*

$$\mathbb{E}\sup_{k\in\mathbb{N}} \|G_k\| \leq K_1\big\{\sup_{k\in\mathbb{N}} \mathbb{E}\|G_k\| + \mathbb{E}\sup_{k\in\mathbb{N}} |\lambda_k\sigma_k|\big\}, \tag{3}$$

where $\{\lambda_k,\ k \in \mathbb{N}\}$ *is an isonormal sequence,* K_1 *is a universal constant, and for all* $k \in \mathbb{N}$,

$$\sigma_k = \sup_{f \in B',\ \|f\| \le 1} \|\langle G_k, f\rangle\|_2.$$

Lemma 3. ([F1]). *Let* g *be an* $\mathbb{R}$*-valued stationary, separable, continuous in quadratic mean, Gaussian random function defined on* $\mathbb{R}$. *Let us further denote by* m *its associated spectral measure on* $\mathbb{R}^+$ *with*

$$\mathbb{E}\big[|g(s) - g(t)|^2\big] = 2\int \big(1 - \cos 2\pi u(s-t)\big) m\,(du).$$

Then,

$$\mathbb{E} \sup_{\alpha \in [0,1]} g(\alpha) \le K_2 \Big\{ \Big[\int \min(u^2, 1) m\,(du) \Big]^{\frac{1}{2}} + \int \big[m\{]\exp x^2, \infty\}\big]^{\frac{1}{2}}\, dx \Big\}, \tag{4}$$

where K_2 *is a universal constant.*

The proof of the main result will depend very much on the following maximal type inequality.

Theorem 4. *Let* $\{\theta_k,\ k \in \mathbb{N}\}$ *be a sequence of independent random variables. Let further* $\{p_k,\ k \in \mathbb{N}\}$ *be a non-decreasing sequence of positive reals, and finally let* $\mathcal{N}$ *be a partial index* $\mathcal{N} = \{n_k,\ k \in \mathbb{N}\}$. *Assume that condition* $\mathcal{H}_2$ *is satisfied. Then,*

$$\mathbb{E} \sup_{0 \le \alpha \le 1} \sup_{N \in \mathcal{N}} \left| \frac{\sum_{k=1}^{N} \theta_k \exp(2i\pi p_k \alpha)}{\sqrt{N}\Phi(N)} \right| < \infty, \tag{5}$$

where the function Φ *is defined in the Theorem* 1.

Remark. This statement improves an extension due to X. Fernique (**[F2]**) of a work by D. Schneider (**[S]**) based on previous works (**[SZ]**) by Salem and Sygmund.

Proof of Theorem 4. Set for any $p \ge 1$,

$$N_p = \sup\{\mathcal{N} \cap I_p\}, \qquad a_p = \sqrt{N_p \log[pN_p + p + 2]}.$$

It is enough to show

$$\mathbb{E} \sup_{p \ge 1} \sup_{N \in I_p \cap \mathcal{N}} \sup_{0 \le \alpha \le 1} \left| \frac{\sum_{k=1}^{N} \theta_k \exp(2i\pi p_k \alpha)}{a_p} \right| < \infty, \tag{6}$$

and this will follow from

$$\mathbb{E} \sup_{p \ge 1} \sup_{N \in I_p \cap \mathcal{N}} \sup_{0 \le \alpha \le 1} \left| \frac{\sum_{k=1}^{n} \theta_k \cos(2\pi p_k \alpha)}{a_p} \right| < \infty, \tag{6-a}$$

$$\mathbb{E} \sup_{p \ge 1} \sup_{N \in I_p \cap \mathcal{N}} \sup_{0 \le \alpha \le 1} \left| \frac{\sum_{k=1}^{N} \theta_k \sin(2\pi p_k \alpha)}{a_p} \right| < \infty. \tag{6-b}$$

First we show that we can assume in (6-a) and (6-b) that the sequence $\{\theta_k,\ k \in \mathbb{N}\}$ is symmetric.

Let $\{\theta'_k,\ k \in \mathbb{N}\}$ be an independent copy of the sequence $\{\theta_k,\ k \in \mathbb{N}\}$. Then,

$$\begin{aligned}\mathbb{E} \sup_{p\geq 1} \sup_{N\in I_p\cap\mathcal{N}} \sup_{0\leq\alpha\leq 1} \left| \frac{\sum_{k=1}^N \theta_k \cos(2i\pi p_k\alpha)}{a_p} \right| &= \mathbb{E} \sup_{p\geq 1} \sup_{N\in I_p\cap\mathcal{N}} \sup_{0\leq\alpha\leq 1} \left| \frac{\sum_{k=1}^N (\theta_k - \mathbb{E}\theta'_k) \cos(2i\pi p_k\alpha)}{a_p} \right| \\ &\leq \mathbb{E} \sup_{p\geq 1} \sup_{N\in I_p\cap\mathcal{N}} \sup_{0\leq\alpha\leq 1} \left| \frac{\sum_{k=1}^N (\theta_k - \theta'_k) \cos(2i\pi p_k\alpha)}{a_p} \right|.\end{aligned}$$

Since the same argument can be applied to the imaginary part (6-b), we can and do assume that the sequence $\{\theta_k,\ k \in \mathbb{N}\}$ is symmetric. Let $\{\varepsilon_k,\ k \in \mathbb{N}\}$ be a Rademacher sequence independent of the sequence $\{\theta_k,\ k \in \mathbb{N}\}$. By symmetry, it suffices to show

$$\mathbb{E} \sup_{p\geq 1} \sup_{N\in I_p\cap\mathcal{N}} \sup_{0\leq\alpha\leq 1} \left| \frac{\sum_{k=1}^N \theta_k\varepsilon_k \cos(2\pi p_k\alpha)}{a_p} \right| < \infty, \tag{6-a$'$}$$

$$\mathbb{E} \sup_{p\geq 1} \sup_{N\in I_p\cap\mathcal{N}} \sup_{0\leq\alpha\leq 1} \left| \frac{\sum_{k=1}^N \theta_k\varepsilon_k \sin(2\pi p_k\alpha)}{a_p} \right| < \infty. \tag{6-b$'$}$$

Let $\{g_k,\ k \in \mathbb{N}\}$ and $\{g'_k,\ k \in \mathbb{N}\}$ be two independent isonormal sequences, also independent from the sequence $\{\theta_k,\ k \in \mathbb{N}\}$. By virtue of the contraction principle, in order to get (6-a$'$) and (6-b$'$), it is enough to show

$$\mathbb{E} \sup_{p\geq 1} \sup_{N\in I_p\cap\mathcal{N}} \sup_{0\leq\alpha\leq 1} \left| \frac{\sum_{k=1}^N \theta_k g_k \cos(2\pi p_k\alpha)}{a_p} \right| < \infty, \tag{6-a$''$}$$

$$\mathbb{E} \sup_{p\geq 1} \sup_{N\in I_p\cap\mathcal{N}} \sup_{0\leq\alpha\leq 1} \left| \frac{\sum_{k=1}^N \theta_k g'_k \sin(2\pi p_k\alpha)}{a_p} \right| < \infty, \tag{6-b$''$}$$

or equivalently, by Jensen's inequality,

$$\mathbb{E}_\theta\mathbb{E}_g \sup_{0\leq\alpha\leq 1} \sup_{p\geq 1} \sup_{N\in I_p\cap\mathcal{N}} \left| \frac{\sum_{k=1}^N \theta_k \{g_k \cos 2\pi p_k\alpha + g'_k \sin 2\pi p_k\alpha\}}{a_p} \right| < \infty.$$

We set for all $p \geq 1,\ \forall N \in I_p \cap \mathcal{N},\ \forall\alpha \in [0, 1]$,

$$X_p(\alpha, N) = \frac{\sum_{k=1}^N \theta_k \{g_k \cos 2\pi p_k\alpha + g'_k \sin 2\pi p_k\alpha\}}{a_p}.$$

We have, therefore, to consider the following expression

$$\mathbb{E}_g \sup_{p\geq 1} \sup_{\substack{\alpha\in[0,1] \\ N\in I_p\cap\mathcal{N}}} |X_p(\alpha, N)| \tag{7}$$

which can be bounded by virtue of Lemma 2, by

$$\sup_{p\geq 1}\mathbb{E}_g \sup_{\substack{\alpha\in[0,1]\\ N\in I_p\cap\mathcal{N}}} |X_p(\alpha, N)| + \mathbb{E}_\lambda \sup_{p\geq 1} |\lambda_p \sigma_p|,$$

where

$$\forall p \geq 1, \quad \sigma_p = \sup_{\substack{\alpha\in[0,1]\\ N\in I_p\cap\mathcal{N}}} \|X_p(\alpha, N)\|_2.$$

But, by the very construction of the normalizing function Φ,

$$\sigma_p \leq \sup_{p\geq 1}\left[\frac{\sum_{k=1}^{N_p}\theta_k^2}{N_p}\right]^{\frac{1}{2}} \frac{1}{\sqrt{\log p + 2}},$$

so that

$$\mathbb{E}\sup_{p\geq 1}|\lambda_p\sigma_p| < \infty. \tag{8}$$

Examine now the contribution of the second term in (7). By Levy's inequality,

$$\begin{aligned}
&\mathbb{E}_g \sup_{\substack{\alpha\in[0,1]\\ N\in I_p\cap\mathcal{N}}} |X_p(\alpha, N)| \qquad (9)\\
&\leq \frac{1}{a_p}\mathbb{E}_g \sup_{N\leq N_p}\sup_{\alpha\in[0,1]}\left|\sum_{k=1}^{N}\theta_k\{g_k\cos 2\pi p_k\alpha + g_k'\sin 2\pi p_k\alpha\}\right|\\
&\leq \frac{2}{a_p}\mathbb{E}_g \sup_{\alpha\in[0,1]}\left|\sum_{k=1}^{N_p}\theta_k\{g_k\cos 2\pi p_k\alpha + g_k'\sin 2\pi p_k\alpha\}\right|.
\end{aligned}$$

The proof of Theorem 4 is now achieved by applying as in **[F2]**, Lemma 3. It is indeed enough to prove that

$$\mathbb{E}_\theta \sup_{p\geq 1}\sup_{\alpha\in[0,1]}\mathbb{E}_g\frac{1}{a_p}\left|\sum_{k=1}^{N_p}\theta_k\{g_k\cos 2\pi p_k\alpha + g_k'\sin 2\pi p_k\alpha\}\right| < \infty. \tag{10}$$

Conditionally to the sequence $\{\theta_k,\ k\in\mathbb{N}\}$, the spectral measure of the Gaussian random function in (10) is given by $\frac{1}{N_p\Phi(N_p)^2}\sum_{k=1}^{N_p}\theta_k^2\delta_{p_k}$. The property (10) will be then realized as soon as

$$\mathbb{E}\sup_{p\geq 1}\left[\frac{1}{N_p(\log[p_{N_p} + (p+2)]}\right]^{\frac{1}{2}}\left[\sum_{k=1}^{N_p}\theta_k^2[\min(p_k^2, 1)]\right]^{\frac{1}{2}} < \infty, \tag{11}$$

$$\mathbb{E}\sup_{p\geq 1}\left[\frac{1}{N_p(\log[p_{N_p} + (p+2)])}\right]^{\frac{1}{2}}\int_0^\infty\left[\sum_{k=1}^{N_p}\theta_k^2\mathbf{1}_{\{p_k>\exp x^2\}}\right]^{\frac{1}{2}} < \infty. \tag{12}$$

The expression in (11) is easily bounded by

$$\mathbb{E}\sup_{p\geq 1}\Big[\frac{\sum_{k=1}^{N_p}\theta_k^2}{N_p}\Big]^{\frac{1}{2}},$$

which is clearly finite by $(\mathcal{H}_2)$. The second expression in (12) is easy to control. We rewrite this expression as

$$\sup_{p\geq 1}\Big[\frac{1}{N_p(\log p_{N_p}+(p+2))}\Big]^{\frac{1}{2}}\sum_{k=1}^{N_p}\big[(\log^+ p_k)^{\frac{1}{2}}-(\log^+ p_{k-1})^{\frac{1}{2}}\big]\Big[\sum_{j=k}^{N_p}\theta_j^2\Big]^{\frac{1}{2}},$$

where we set $p_{-1}=0$. A straightforward bound is then given by

$$\sup_{p\geq 1}\Big[\frac{\sum_{j=1}^{N_p}\theta_j^2}{N_p}\Big]^{\frac{1}{2}}\Big[\frac{\log^+ p_{N_p}}{\log(p_{N_p}+p+2)}\Big]^{\frac{1}{2}}\leq\Big[\frac{\sum_{j=1}^{N_p}\theta_j^2}{N_p}\Big]^{\frac{1}{2}},$$

which, thanks to assumption $(\mathcal{H}_2)$, again achieves the proof. □

From Theorem 4, we get the following result implying Theorem 1.

Theorem 5. *Let* $\{\theta_k,\ k\in\mathbb{N}\}$, $\{p_k,\ k\in\mathbb{N}\}$, $\mathcal{N}=\{n_k,\ k\in\mathbb{N}\}$ *as in the Theorem* 1, *and assume that conditions* $(\mathcal{H}_1)$ *and* $(\mathcal{H}_2)$ *are verified.*

Then, there exists a positive random variable $C(\theta)$ *such that*

$$\mathbb{E}C(\theta)<\infty, \tag{13}$$

and, such that for any other probability space $(X,\mathcal{A},\mu)$, *any contraction* T *on* $L^2(\mu)$, *and any element* $f\in L^2(\mu)$,

$$\int_X\Big(\sum_{N\in\mathcal{N}}|A_N^{\theta,T}f(x)|^2\Big)d\mu(x)\leq C(\theta)^2\|f\|_{2,\mu}^2. \tag{14}$$

Proof of Theorem 5. Let $f\in L^2(\mu)$ with $\|f\|_{2,\mu}=1$ be fixed and denote by μ_f its spectral measure. By the spectral theorem, we can estimate the left-hand side of (14) as

$$\sum_{N\in\mathcal{N}}\int_X|A_N^{\theta,T}f(x)|^2\,d\mu(x)=\sum_{N\in\mathcal{N}}\int_{[0,1[}\Big|\frac{\sum_{k=1}^N\theta_k\exp(2i\pi p_k\alpha)}{N}\Big|^2d\mu_f(\alpha).$$

By Theorem 4 we have

$$\sup_{0\leq\alpha\leq 1}\Big|\frac{\sum_{k=1}^N\theta_k\exp(2i\pi p_k\alpha)}{N}\Big|\leq B(\theta)\frac{\Phi(N)}{\sqrt{N}},$$

with $\mathbb{E}B(\theta)<\infty$, and we can continue our estimations as

$$\sum_{N\in\mathcal{N}}\int_{[0,1[}\Big|\frac{\sum_{k=1}^N\theta_k\exp(2i\pi p_k\alpha)}{N}\Big|^2d\mu_f(\alpha)\leq\Big(\sum_{N\in\mathcal{N}}B(\theta)^2\frac{\Phi^2(N)}{N}\Big)\cdot\int_{[0,1[}d\mu_f=C^2(\theta),$$

where $\mathbb{E}C(\theta)<\infty$ by assumptions $\mathcal{H}_1$ and $\mathcal{H}_2$. This achieves the proof. □

III. The case $\mathcal{N} = \mathbb{N}$

From our main result and Remark c), the following Theorem can be deduced.

Theorem 6. *Let* $\{p_k,\ k \in \mathbb{N}\}$ *be a non-decreasing sequence of positive integers satisfying*

$$p_k = O(2^{k^\gamma}) \text{ for all } 0 < \gamma < 1. \tag{$\mathcal{H}_3$}$$

Let further $\{Y_k,\ k \in \mathbb{N}\}$ *be a sequence of independent, nonnegative, centered, square integrable, identically distributed random variables with basic probability space* $(\Omega, \mathcal{B}, P)$.

Then there exists a measurable set Ω^* *with* $P\{\Omega^*\} = 1$ *such that for any* $Y = \{Y_n,\ n \in \mathbb{N}\} \in \Omega^*$, *any other probability space* $(X, \mathcal{A}, \mu)$, *any contraction* T *on* $L^2(\mu)$ *and any element* $f \in L^2(\mu)$, *the averages*

$$A_N^{Y,T}(f) = \frac{1}{N}\sum_{k=1}^{N} Y_k f \circ T^{p_k}$$

are converging μ*-almost surely to* 0 *when* N *tends to infinity along* $\mathbb{N}$, *whenever the averages*

$$A_N^{T}(f) = \frac{1}{N}\sum_{k=1}^{N} f \circ T^{p_k}$$

are converging μ*-almost surely when* N *tends to infinity along* $\mathbb{N}$.

The proof will rely on the following Lemmas due to Wierdl (**[Wi]**).

Lemma 7. *(***[Wi]***). Let* $\{f_n,\ n \in \mathbb{N}\}$ *be a sequence of nonnegative numbers. For* $\varrho > 1$ *denote* $I_\varrho = \{t : t = \varrho^k$ *for some positive integer* $k\}$. *Suppose that for each* $\varrho > 1$ *the averages*

$$\frac{1}{t}\sum_{n\le t} f_n$$

converge to some finite limit as t *runs through the sequence* I_ϱ.

Then for each $\varrho > 1$ *this limit is the same, say* L, *and we have*

$$\lim_{t\to\infty} \frac{1}{t}\sum_{n\le t} f_n = L.$$

Note: From the very proof, it is enough to check the convergence for countably many values of ϱ convergent to 1.

Proof of Theorem 6. We apply Theorem 1. According to Remark c) following Theorem 1, assumption $\mathcal{H}_1$ is satisfied for partial index $\mathcal{N} = I_\varrho$ with $\varrho > 1$. Concerning assumption $\mathcal{H}_2$, it is clearly satisfied with the choice $\theta_k = Y_k - \mathbb{E}Y_k$ for all integers k. This follows from the classical maximal inequalities. By Theorem 1, there exists a measurable set Ω^* with $P\{\Omega^*\} = 1$ such that for any $Y = \{Y_n,\ n \in \mathbb{N}\} \in \Omega^*$, any other probability space $(X, \mathcal{A}, \mu)$ any contraction T on $L^2(\mu)$, any element $f \in L^2(\mu)$, and any $\varrho > 1$

$$\mu\{\lim_{I_\varrho \ni N \to \infty} A_N^{\theta,T} = 0\} = 1.$$

Hence

$$\mu\{\lim_{I_\varrho \ni N \to \infty} A_N^{Y,T} \text{ exists}\} = 1.$$

Since the Y_i 's are nonnegative, it suffices to treat the case $\mu\{f \geq 0\} = 1$. But, from Lemma 6 we have

$$\mu\{\lim_{N \to \infty} A_N^{Y,T} \text{ exists}\} = 1.$$

And this achieves the proof. □

Acknowledgment. We are very indebted to Xavier Fernique and Mate Wierdl for several helpful comments.

References

[B] Bourgain, J., Pointwise ergodic theorems for arithmetic sets, with an appendix on return time sequences, jointly with H. Furstenberg, Y. Katznelson, D. Ornstein, Inst. Hautes Études Sci. Publ. Math. 69 (1988).

[F1] Fernique, X., Régularité de fonctions aléatoires gaussiennes stationnaires, Probab. Theory Related Fields 88 (1991), 521–536.

[F2] Fernique, X., Un exemple illustrant l'emploi de méthodes gaussiennes, The Journal of Fourier Analysis and Applications, Kahane special issue (1995), 209–213.

[F3] Fernique, X., C. R. Acad. Sci. Paris 300 (1985), 315–318.

[K] Krengel, U., Ergodic Theorems, de Gruyter Stud. Math. 6, Walter de Gruyter, Berlin 1985.

[S] Schneider, D., Convergence presque sûre de moyennes ergodiques perturbées par des variables aléatoires, Thèse, Publication IRMA 011, Strasbourg 1994.

[SZ] Salem, R., Zygmund A., Some properties of trigonometric series whose terms have random signs, Acta Math. 91 (1954), 245–301.

[We] Weber, M., The law of the iterated logarithm on subsequences - Characterizations, Nagoya Math. J. 118 (1990), 65–97.

[Wi] Wierdl, M., Pointwise ergodic theorem along the primes numbers, Israel J. Math. 64 (1988), 315–336.

Institut de Recherche Mathématique Avancé
Université Louis Pasteur et CNRS
7, rue René Descartes
67084 Strasbourg Cedex
France

On Uniform Distribution of Sequences[1]

Yeneng Sun

Abstract. We consider some general properties of completely uniformly distributed sequences and uniform distribution preserving transformations.

1. Introduction

In this paper we present some properties about completely uniformly distributed sequences and uniform distribution preserving transformations on the interval $[0, 1]$. As applications of the isomorphism theorems of **[14]** (see **[18]** for some further development), we also obtain some results on general spaces by transferring the corresponding results on the interval.

We first collect some immediate consequences of the main theorems of **[14]**. For a Borel measure on a topological space X, the pair (X, μ) is called a topological measure space. A function f from the topological measure space (X, μ) to a topological space Z is said to be μ-continuous if the set of discontinuity points of f is a μ-null set. We shall assume in this paper that any measure on a topological space which we work with is a Borel measure. A bijection F from a topological measure space (X, μ) to another topological measure space (Y, ν) is called an isomorphism if F is Borel measurable, measure preserving, and μ-continuous, and F^{-1} is Borel measurable and ν-continuous. Let μ and ν be probability measures on Polish spaces X and Y respectively. If μ and ν are atomless, then Theorem 1 of **[14]** implies that (X, μ) is isomorphic to (Y, ν). If every atom of μ is a condensation point of X, then the proof of Theorem 2 of **[14]** indicates that there is a probability measure P on the Cantor set C such that (X, μ) is isomorphic to (C, P).

Now recall that a sequence $\omega = \{x_n\}_{n=1}^{\infty}$ in a compact space X is called uniformly distributed with respect to a probability measure μ on X or simply μ-u.d. if

$$\lim_{N\to\infty} \frac{1}{N} \sum_{n=1}^{N} f(x_n) = \int_X f(x)\, d\mu$$

1 Some of the results were reported in the Conference on Convergence in Ergodic Theory and Probability held in Ohio State University in June 1993. The author is very grateful to Professors Joseph Rosenblatt, Dorothy Maharam Stone, Benjamin Weiss, and an anonymous referee for help.

for any continuous function f on X (see [7]). A more special property is complete uniform distribution: let S be the one-sided shift on the product space X^∞ and μ_∞ the product measure on X^∞; the sequence ω is completely uniformly distributed with respect to μ (completely μ-u.d.) in X if the sequence $\{S^n(\omega)\}_{n=1}^\infty$ is μ_∞-u.d. in X^∞. Note that completely u.d. mod 1 sequences are important random number generators (see Chapter 3 of [6]).

A transformation T from X to a compact space Y with a probability measure ν on Y is said to be uniform distribution preserving (u.d.p.) if for every μ-u.d. sequence $\omega = \{x_n\}_{n=1}^\infty$ in X, the sequence $T(\omega) = \{T(x_n)\}_{n=1}^\infty$ is ν-u.d. in Y. For the case that $X = Y = I$, the unit interval $[0, 1]$, and $\mu = \nu = \lambda$, the Lebesgue measure on I, it is shown in [2] and [12] that a transformation T on I is u.d.p. if and only if it is Riemann integrable and measure preserving with respect to λ. Note that Tichy and Winkler have considered in [19] generalizations of the main results of [2] and [12] to a compact metric space X with a probability measure μ on X in Binder's setting which requires μ satisfying that nonempty open sets have positive measures and boundaries of open balls are null sets (see [1]). By using Theorem 1 of [14], we actually generalize in [14] the main results of [1], [2], [3], and [12] to a compact metrizable space with an arbitrary probability measure (see Propositions 23–26 of [14]). We shall present more properties about u.d.p. transformations in this paper.

The paper is organized as follows. Section 2 concerns various properties of complete uniform distribution. In particular, we show that u.d.p. transformations in compact metrizable spaces with arbitrary probability measures can be characterized by complete uniform distribution. A lifting property is also considered for completely u.d. sequences. By some results in ergodic theory, we are able to prove a decomposition property for completely u.d. sequences. Section 3 contains some further results on u.d.p. transformations which include an interpolation property of piecewise linear u.d.p. transformations and some results about piecewise differentiable u.d.p. transformations.

2. Complete uniform distribution of sequences

In this section we present some results involving complete uniform distribution. In the following proposition we obtain a characterization of u.d.p. transformation by complete uniform distribution. It is noted in [19] that the method used there can be used to give a characterization of u.d.p. transformation by complete uniform distribution on a compact metric space X with a probability measure in Binder's setting (see p. 181, [19]). Our method works for an arbitrary probability measure on X.

Proposition 2.1. *Let X and Y be compact metrizable spaces and let μ, ν be probability measures on X and Y respectively. Let T be a mapping from X to Y. Then T maps every completely μ-u.d. sequence in X to a completely ν-u.d. sequence in Y if and only if T is a u.d.p. transformation.*

Proof. Proposition 24 of **[14]** says that T is u.d.p. if and only if T is measure preserving and μ-continuous. Thus if T is u.d.p., then it is easy to see that T preserves complete uniform distribution.

Next we shall show that if T preserves complete uniform distribution, then T is μ-continuous and measure preserving. Note that a real valued function f on X is said to be R-μ-integrable if it is bounded and μ-continuous. Now let f be a real valued function on X which is not R-μ-integrable. By the proof of Proposition 23 of **[14]**, we know that there is a measure preserving and λ-continuous mapping F from the Lebesgue interval (I, λ) to (X, μ) such that $f \circ F$ is not Riemann integrable on I. Similar arguments to **[1]** yield that for a fixed completely u.d. mod 1 sequence $\{a_n\}_{n=1}^{\infty}$, there is a sequence $\{b_n\}_{n=1}^{\infty}$ such that $\lim\limits_{n\to\infty} |b_n - a_n| = 0$ and $\lim\limits_{N\to\infty} \frac{1}{N} \sum_{n=1}^{N} f(F(b_n))$ does not exist. It is easy to see that the sequence $\{b_n\}_{n=1}^{\infty}$ is still completely u.d. mod 1. Now suppose that T is not μ-continuous; then there is an R-ν-integrable function φ on Y such that $\varphi \circ T$ is not R-μ-integrable on X (see the proof of Proposition 24 in **[14]**). Thus the previous argument implies that there is a completely u.d. mod 1 sequence $\{b_n\}_{n=1}^{\infty}$ such that $\lim\limits_{N\to\infty} \frac{1}{N} \sum_{n=1}^{N} \varphi(T(F(b_n)))$ does not exist, which contradicts the assumption that the sequence $\{T(F(b_n))\}_{n=1}^{\infty}$ is completely ν-u.d.. Therefore T is μ-continuous.

To show that T is measure preserving, pick a completely μ-u.d. sequence $\{x_n\}_{n=1}^{\infty}$ in X. Then $\{T(x_n)\}_{n=1}^{\infty}$ is ν-u.d. in Y. For any continuous function f on Y, $f \circ T$ is μ-continuous on X. Hence, we have

$$\int_X f \circ T d\mu = \lim_{N\to\infty} \frac{1}{N} \sum_{n=1}^{N} f \circ T(x_n) = \lim_{N\to\infty} \frac{1}{N} \sum_{n=1}^{N} f(T(x_n)) = \int_Y f d\nu,$$

which implies that T is measure preserving. □

Lemma 1 in **[15]** says that for a given u.d. mod 1 sequence $\{x_n\}_{n=1}^{\infty}$, the sequence $\{(x_n, nx)\}_{n=1}^{\infty}$ is u.d. mod 1 in $\mathbb{R}^2$ for almost all irrational number $x \in [0, 1]$ (see **[4]** for another version of the result). Thus every u.d. mod 1 sequence can be lifted to a u.d. mod 1 sequence in $\mathbb{R}^2$. In the following we consider the lifting property of completely uniformly distributed sequences.

Theorem 2.2. *Let $\omega_0 = \{a_n\}_{n=1}^{\infty}$ be a completely u.d.* mod 1 *sequence. Then for almost all x in* $[0, 1]$, *the sequence $\{(a_n, x \cdot n!)\}_{n=1}^{\infty}$ is completely u.d.* mod 1 *in* $\mathbb{R}^2$.

Proof. Let s be a positive integer. For any given lattice point $(h_1, k_1, \ldots, h_s, k_s) \in \mathbb{Z}^{2s}$ with $(k_1, \ldots, k_s) \neq (0, \ldots, 0)$, define a sequence of functions $\{u_n(x)\}_{n=1}^{\infty}$ such that for every $x \in [0, 1]$,

$$u_n(x) = h_1 a_{n+1} + k_1 x \cdot (n+1)! + \cdots + h_s a_{n+s} + k_s x \cdot (n+s)!.$$

Thus the derivative $u_n'(x) = k_1(n+1)! + \cdots + k_s(n+s)!$. Let $K = \max\{k_i : 1 \leq i \leq s\}$. Without loss of generality, we assume $k_s \neq 0$ and thus $|k_s| \geq 1$. Then, for $n \geq 2 \cdot K \cdot s$, we have

$$|u_n'(x)| \le Ks(n+s)! < (n-ks+s+1+K)(n+s)!$$
$$= (n+s+1)! - K(s-1)(n+s)! \le |u_{n+1}'(x)|$$

Therefore, $|u_n'(x) - u_m'(x)| \ge 1$ for any two different positive integers m and n with $m, n \ge K \cdot s$. Thus the sequence $\{u_n(x)\}_{n=1}^{\infty}$ satisfies the conditions in Koksma's general metric theorem for $[a, b] = [0, 1]$ (see p. 34, **[7]**). Hence, for almost all x in $[0, 1]$, the sequence $\{u_n(x)\}_{n=1}^{\infty}$ is u.d. mod 1. Thus there is a set A in $[0, 1]$ with $\lambda(A) = 1$ such that for every x in A and for every lattice point $(h_1, k_1, \dots, h_s, k_s) \in \mathbb{Z}^{2s}$ with $(k_1, \dots, k_s) \ne (0, \dots, 0)$, the sequence

$$\{h_1 a_{n+1} + k_1 x \cdot (n+1)! + \cdots + h_s a_{n+s} + k_s x \cdot (n+s)!\}_{n=1}^{\infty}$$

is u.d. mod 1. Since ω_0 is completely u.d. mod 1, the sequence

$$\{h_1 a_{n+1} + \cdots + h_s a_{n+s}\}_{n=1}^{\infty}$$

is still u.d. mod 1 for every lattice point $(h_1, \dots, h_s) \ne (0, \dots, 0)$. Hence the sequence

$$\{h_1 a_{n+1} + k_1 x \cdot (n+1)! + \cdots + h_s a_{n+s} + k_s x \cdot (n+s)!\}_{n=1}^{\infty}$$

is u.d. mod 1 for every x in A and for every lattice point

$$(h_1, k_1, \dots, h_s, k_s) \ne (0, 0, \dots, 0, 0).$$

Therefore, Theorem 6.3 (p. 48, **[7]**) yields that

$$\{(a_{n+1}, x \cdot (n+1)!, \dots, a_{n+s}, x \cdot (n+s)!)\}_{n=1}^{\infty}$$

is u.d. mod 1 in $\mathbb{R}^{2s}$ for any x in A. Hence the sequence $\{(a_n, x \cdot n!)\}_{n=1}^{\infty}$ is completely u.d. mod 1 in $\mathbb{R}^2$. □

Remark 2.3. Though a u.d.p. transformation maps u.d. sequences to u.d. sequences, well distributed sequences to well distributed sequences, and completely u.d. sequences to completely u.d. sequences, we shall see that it may also map u.d. sequences which are not well distributed to well distributed sequences and u.d. sequences which are not completely u.d. to completely u.d. sequences. Let f be an isomorphism from (I, λ) to (I^2, λ^2). Let g be the projection from I^2 to I such that $g(x, y) = y$. Let $F = g \circ f$. Then F is a u.d.p. transformation from (I, λ) to itself. Choose two u.d. mod 1 sequences $\omega_1 = \{a_n\}_{n=1}^{\infty}$ and $\omega_2 = \{b_n\}_{n=1}^{\infty}$ such that ω_1 is not well distributed mod 1 and ω_2 is not completely u.d. mod 1. Note that the sequence $\{nx\}_{n=1}^{\infty}$ is well distributed mod 1 for any irrational number x (see **[7]** and also **[17]** for an application of such kind of sequences to the study of compact operators). Then, Lemma 1 of **[15]** yields that there is a well distributed mod 1 sequence $\{a_n'\}_{n=1}^{\infty}$ such that $\omega_1' = \{(a_n, a_n')\}_{n=1}^{\infty}$ is u.d. mod 1 in $\mathbb{R}^2$. By the proof of Lemma 1 of **[15]** and Theorem 2.2 above, we know that $\{(b_n, n!x)\}_{n=1}^{\infty}$ is u.d. mod 1 for almost all $x \in [0, 1]$. It follows from Theorem 2.2 that the sequence $\{n!x\}_{n=1}^{\infty}$ is completely u.d. mod 1 for almost all $x \in [0, 1]$. Hence there is a completely u.d. mod 1 sequence $\{b_n'\}_{n=1}^{\infty}$ such that $\omega_2' = \{(b_n, b_n')\}_{n=1}^{\infty}$ is u.d. mod

1 in $\mathbb{R}^2$. Let $\alpha_1 = f^{-1}(\omega'_1)$ and $\alpha_2 = f^{-1}(\omega'_2)$. Then α_1 is not well distributed mod 1 and α_2 is not completely u.d. mod 1; but $F(\alpha_1)$ is well distributed mod 1 and $F(\alpha_2)$ is completely u.d. mod 1.

By using the isomorphism theorems in **[14]**, we can obtain the following general result.

Proposition 2.4. *Let μ and ν be atomless probability measures on compact metrizable spaces X and Y respectively. For any given countable set $\mathcal{C}$ of completely μ-u.d. sequences in X, we can find a completely ν-u.d. sequence $\omega_0 = \{b_n\}_{n=1}^\infty$ in Y such that for any sequence $\omega = \{a_n\}_{n=1}^\infty$ in $\mathcal{C}$, the sequence $\{(a_n, b_n)\}_{n=1}^\infty$ is completely $\mu \times \nu$-u.d. in $X \times Y$.*

Proof. Let F and G be isomorphisms from (X, μ) and (Y, ν) to the unit Lebesgue interval (I, λ) respectively. Then $F(\mathcal{C}) = \{F(\omega) : \omega \in \mathcal{C}\}$ is a countable set of completely u.d. mod 1 sequences. By Theorem 2.2, we can choose an $x \in [0, 1]$ such that the sequence $\{n!x\}_{n=1}^\infty$ can be paired with any sequence in $F(\mathcal{C})$ to obtain a completely u.d. mod 1 sequence in $\mathbb{R}^2$. For each $n \geq 1$, let $b_n = G^{-1}(\{n!x\})$, where $\{n!x\}$ is the fractional part of $n!x$. Then the sequence $\omega_0 = \{b_n\}_{n=1}^\infty$ has the desired property. □

Remark 2.5. If μ and ν are arbitrary probability measures, a result of Kamae (see **[5]**) yields that any given completely μ-u.d. sequence can be paired with some completely ν-u.d. sequence to form a completely $\mu \times \nu$-u.d. sequence in $X \times Y$. The above proposition tells us that for the atomless case, a completely ν-u.d. sequence can be chosen to pair with countably many given completely μ-u.d. sequences to form completely $\mu \times \nu$-u.d. sequences in the product space $(X \times Y, \mu \times \nu)$ simultaneously.

Let $\omega_0 = \{a_n\}_{n=1}^\infty$ be a completely u.d. mod 1 sequence. Let m be a fixed positive integer. Then, a result of Niven and Zuckerman in **[10]** yields that for any $1 \leq j \leq m$, the sequence $\omega_j = \{a_{mn+j}\}_{n=0}^\infty$ is still completely u.d. mod 1 (see also Section 3.5 of **[6]**). In the following proposition we shall note that the sequences $w_1, \ldots, w_m$ are independent in a certain strong sense. Here we also note that if ω_0 is well distributed mod 1 and thus also u.d. mod 1, then the ω_j's may not be u.d. mod 1. For example, let $a_n = \frac{1}{2}(\{[\frac{n+1}{2}]\sqrt{2}\} + \frac{1-(-1)^n}{2})$, where for a real number x, $\{x\}$ and $[x]$ denote the fractional and integer parts of x respectively. Then the fact that $\{n\sqrt{2}\}_{n=1}^\infty$ is well distributed mod 1 implies that $\omega_0 = \{a_n\}_{n=1}^\infty$ is well distributed mod 1 (see **[16]** for some related results). Take $m = 2$. Then $\omega_2 = \{a_{2n}\}_{n=1}^\infty$ is neither well distributed mod 1 nor u.d. mod 1.

Proposition 2.6. *Let $\omega_0 = \{a_n\}_{n=1}^\infty$ be a sequence in I. Let m be a fixed positive integer. Then, ω_0 is completely u.d.* mod 1 *if and only if the sequence*

$$\omega = \{(a_{mn+1}, a_{mn+2}, \ldots, a_{mn+m})\}_{n=0}^\infty$$

is completely u.d. mod 1 *in $\mathbb{R}^m$.*

Proof. Note that ω is completely u.d. mod 1 in $\mathbb{R}^m$ if and only if for any $s \geq 1$, the following sequence

$$\{(a_{mn+1}, \ldots, a_{mn+m}, a_{m(n+1)+1}, \ldots, a_{m(n+1)+m}, \\ \ldots, a_{m(n+s-1)+1}, \ldots, a_{m(n+s)})\}_{n=0}^{\infty}$$

which we denote as ω_s^m, is u.d. mod 1. Now, if ω_0 is completely u.d. mod 1, then the theorem of Niven and Zuckerman (see **[10]** or **[6]**, p. 134) implies that ω_0 is $(m, m \cdot s)$-distributed (for a definition of $(m, m \cdot s)$-distributed sequences, see **[6]**, p. 134). Thus ω_s^m is u.d. mod 1 and hence ω is completely u.d. mod 1 in $\mathbb{R}^m$. Next, if ω is completely u.d. mod 1 in $\mathbb{R}^m$, then ω_s^m is u.d. mod 1 for every positive integer s. Let k be a positive integer. Then the uniform distribution of ω_k^m yields that

$$\{(a_{mn+j}, a_{mn+j+1}, a_{mn+j+k-1})\}_{n=0}^{\infty}$$

is u.d. mod 1 in $\mathbb{R}^k$ for any $1 \leq j \leq m$. We concatenate those m u.d. mod 1 sequences together to get the sequence $\{(a_{n+1}, a_{n+2}, \ldots, a_{n+k})\}_{n=0}^{\infty}$, which is still u.d. mod 1 in $\mathbb{R}^k$ (see **[7]**, p. 180). Therefore, ω_0 is completely u.d. mod 1. □

As before, we can generalize the above result to a general space.

Proposition 2.7. *Let μ be an atomless probability measure on a compact metrizable space. Let $\omega_0 = \{a_n\}_{n=1}^{\infty}$ be a sequence in X. Let m be a fixed positive integer. Then, ω_0 is completely μ-u.d. if and only if the sequence*

$$\omega = \{(a_{mn+1}, a_{mn+2}, \ldots, a_{mn+m})\}_{n=0}^{\infty}$$

is completely μ^m-u.d. in X^m.

Remark 2.8. Let X and μ be as above and m a fixed positive integer. For each $1 \leq j \leq m$, let $\omega_j = \{a_n^j\}_{n=1}^{\infty}$ be a μ-u.d. sequence in X. Let $\omega = \{a_n\}_{n=1}^{\infty}$ be the sequence such that if $n = m \cdot p + j$ for some integers p, j with $1 \leq j \leq m$, then $a_n = a_{p+1}^j$. Then ω is μ-u.d. in X. But ω is completely μ-u.d. in X if and only if $\{(a_n^1, a_n^2, \ldots, a_n^m)\}_{n=1}^{\infty}$ is completely μ^m-u.d. in X^m.

In Theorem 4 of **[15]**, a decomposition property of u.d. mod 1 sequences is presented. We shall now consider the decomposition property for completely uniformly distributed sequences.

Lemma 2.9. *Let μ be a probability measure on a compact metrizable space X and $\omega_0 = \{a_n\}_{n=1}^{\infty}$ a completely μ-u.d. sequence in X. Let $b \geq 1$. Then the subsequence $\omega = \{a_{[nb]}\}_{n=1}^{\infty}$ is still completely μ-u.d. in X.*

Proof. Define a sequence $\sigma = \{i_n\}_{n=1}^{\infty}$ on $\{0, 1\}$ such that $i_n = 1$ if $n = [mb]$ for some $m \geq 1$, and otherwise $i_n = 0$. Then a result of Weiss **[21]** says that σ is completely deterministic. Since $\lim_{n\to\infty} \frac{[nb]}{n} = b < \infty$, Theorem 4 of **[5]** yields that the subsequence $\omega = \{a_{[nb]}\}_{n=1}^{\infty}$ is completely μ-u.d. in X. □

Proposition 2.10. *Let μ be a probability measure on a compact metrizable space X and $\omega = \{x_n\}_{n=1}^{\infty}$ a completely μ-u.d. sequence in X. Let $\{p_i\}_{i=1}^{s}$ be a finite sequence of positive numbers with $p_1 + p_2 + \cdots + p_s = 1$. Then there is a partition $P_1, P_2, \ldots, P_s$ of the set of positive integers such that*

(1) *the asymptotic density of P_i is p_i for each $1 \leq i \leq s$;*

(2) *if we arrange the elements of P_i as $\{m_{ik}\}_{k=1}^{\infty}$ in natural order and if we let $x_n^i = x_{m_{in}}$ for each $1 \leq i \leq s$ and $n \geq 1$, then each subsequence $\omega_i = \{x_n^i\}_{n=1}^{\infty}$ is completely μ-u.d. in X.*

Proof. We shall use an induction argument on s. When $s = 1$, it is trivial. Now assume that the result holds for $s = k$. To prove the case for $s = k + 1$, let $\{p_i\}_{i=1}^{s+1}$ be a finite sequence of positive numbers with $p_1 + \cdots + p_{s+1} = 1$. Then there is a partition $\overline{P}_1, \overline{P}_2, \ldots, \overline{P}_s$ of the set of positive integers such that

(1) the asymptotic density of $\overline{P}_1$ is $p_1 + p_2$ and that of $\overline{P}_i$ is p_{i+1} for each $2 \leq i \leq s$;

(2) if we arrange the elements of $\overline{P}_i$ as $\{m_{ik}\}_{k=1}^{\infty}$ in natural order and if we let $x_n^i = x_{m_{in}}$ for each $1 \leq i \leq s$ and $n \geq 1$, then each sequence $\overline{w}_i = \{x_n^i\}_{n=1}^{\infty}$ is completely μ-u.d. in X.

Next, let $b = \frac{p_1+p_2}{p_1}$, $P_1 = \{m_{1k} : k = [nb] \text{ for some } n \geq 1\}$ and $P_2 = \overline{P}_1 - P_1$. Then the asymptotic densities of P_1 and P_2 are p_1 and p_2 respectively. The above lemma yields that $\omega_1 = \{x_{[nb]}^1\}_{n=1}^{\infty} = \{x_{m_{1[nb]}}\}_{n=1}^{\infty}$ is completely μ-u.d.. Let $\sigma = \{i_n\}_{n=1}^{\infty}$ be the sequence of $0's$ and $1's$ defined as in the proof of the above lemma. Let $\overline{\sigma} = \{1 - i_n\}_{n=1}^{\infty}$. Since σ is completely deterministic, it is easy to check that $\overline{\sigma}$ is also completely deterministic. Hence if we let $\{h_n\}_{n=1}^{\infty}$ be the list of the elements of $\{k \geq 1 : k \neq [nb] \text{ for any } n \geq 1\}$ in natural order, then by Theorem 4 of **[5]** again, the sequence $\omega_2 = \{x_{h_n}^1\}_{n=1}^{\infty} = \{x_{m_{1h_n}}\}_{n=1}^{\infty}$ is also completely μ-u.d.. Let $P_i = \overline{P}_{i-1}$ for $3 \leq i \leq s + 1$. The rest is clear. □

3. Uniform distribution preserving transformations

In **[12]**, a characterization of simple piecewise linear transformations with a u.d. orbit is given (see **[12]**, p. 468 for the definition of simple piecewise linear transformations). It is noted there that for a simple piecewise linear transformation on I, if one orbit is u.d. mod 1, then almost all orbits are u.d. mod 1. The following well known result indicates that a simple piecewise linear transformation with a u.d. orbit must be ergodic.

Proposition 3.1. *Let T be a measure preserving transformation from a compact metrizable space with respect to a probability measure μ to itself. Then T is ergodic if and only if for almost all $x \in X$, $\{T^n(x)\}_{n=1}^{\infty}$ is a μ-u.d. sequence.*

Proof. See p. 156 of **[20]** for the case that T is continuous. The proof for the general case is the same. □

Now we note that a general u.d.p. transformation with a u.d. orbit is usually not ergodic. To construct a counterexample, let $x_1 = \frac{1}{2}$; choose x_2 from $(0, \frac{1}{2})$ and x_3 from $(\frac{1}{2}, 1)$. Once $\{x_i : 1 \leq i \leq \frac{n(n-1)}{2}\}$ is chosen, we can choose x_i for $\frac{n(n-1)}{2} < i \leq \frac{n(n-1)}{2} + n = \frac{n(n+1)}{2}$ as follows. For $i = \frac{n(n-1)}{2} + j$, pick x_i such that $\frac{(j-1)}{n} < x_i < \frac{j}{n}$ and $x_i \neq x_k$, for any $k < i$. Then it can be checked that $\{x_n\}_{n=1}^{\infty}$ is u.d. mod 1. Define a transformation T on I such that $T(x) = x$ if $x \neq x_n$ for any n and $T(x) = x_{n+1}$ if $x = x_n$ for some n. Then T is a non-ergodic u.d.p. transformation with a u.d. orbit $\{T^n(x_1)\}_{n=1}^{\infty}$. This transformation T is clearly not continuous. Thus if a u.d.p. transformation T has a u.d. orbit and also has certain regularity conditions, for example, continuity, it is natural to ask whether it is ergodic. As shown by the following example, continuity is not enough to guarantee the ergodicity of a u.d.p. transformation with a u.d. orbit.

Example 3.2. Let T be the one sided shift mapping on the product space $X = \{0, 1\}^{\infty}$. Let $\omega_1 = (1, 1, \ldots, 1, \ldots), \omega_2 = (1, 0, \ldots, 1, 0, \ldots)$, and $\omega_3 = T(\omega_2)$. Define a probability measure μ on X such that $\mu(\{\omega_1\}) = \mu(\{\omega_2\}) = \mu(\{\omega_3\}) = \frac{1}{3}$. Then μ is T-invariant. By Kakutani's theorem (see **[11]**), there exists an element ω of X such that the orbit $\{T^n(w)\}_{n=1}^{\infty}$ of ω is μ-uniformly distributed. However, T is not ergodic on (X, μ).

It is shown in **[12]** (see Theorem 1 there) that a map $T: I \longrightarrow I$ is a u.d.p. transformation if and only if for every Riemann integrable function $g: I \longrightarrow \mathbb{R}$, the composition $g \circ T$ is also Riemann integrable and $\int_0^1 g(x)dx = \int_0^1 g(T(x))\,dx$ (see Thereom 2.1 in **[19]** for the generalization to a general space). Then a natural question arises: whether the composition of two almost surely continuous mappings is always almost surely continuous. In Marcus **[9]** it has been shown that every bounded function on $[0, 1]$ is the composition of two Riemann integrable functions, that is, the answer of the question is negative. In **[19]**, Tichy and Winkler extend the result of Marcus to the s-dimensional unit cube and also remark that their construction can be easily generalized to functions on manifolds by using coverings with open sets and coordinate mappings (see p. 180, **[19]**). In the following proposition we shall present a far reaching generalization.

Proposition 3.3. *Let μ and ν be probability measures on separable metric spaces X and Y respectively. Assume that Y is homeomorphic to an uncountable Borel subset of some Polish space. Then there is a μ-continuous mapping F from X into a compact subset of Y such that for any mapping f from X to Y, there is a ν-continuous mapping g from Y to Y with the property $f = g \circ F$.*

Proof. By Lemma 7 of **[14]**, there is a compact subset Z of Y such that $\nu(Z) = 0$ and Z is homeomorphic to the Cantor set C. We embed X in the Hilbert cube I^{∞} and extend μ to I^{∞} in the usual way. Then by Theorem 2 of **[14]**, there is a probability measure P on Z such that (I^{∞}, μ) is isomorphic to (Z, P). Let G be such an isomorphism. Let F be the restriction of G to X. It is clear that F is μ-continuous on X. Now pick a point y_0 in Y. For any mapping f from X into Y, define a mapping g from Y to Y

such that $g(y) = f(F^{-1}(y))$ if $y \in F(X)$ and $g(y) = y_0$ if $y \in Y - F(X)$. It is clear that g is ν-continuous and $f = g \circ F$. □

Remark 3.4. (1) If we let $Y = \mathbb{R}$, then every bounded real valued function on X is the composition of an R-μ-integrable function on X with a Riemann integrable function.

(2) If we assume that X is homeomorphic to an uncountable Borel subset of some Polish space, then by the same method, we can obtain a μ-continuous mapping F from X into a compact subset of X such that for any mapping f from X into Y, there is a μ-continuous mapping g from X to Y with the property $f = g \circ F$.

(3) If both X and Y are countable, the above results may not be true. For example, let $X = Y = \{\frac{1}{n}\}_{n=1}^{\infty} \cup \{0\}$, and let $\mu = \nu$ such that $\mu(\{0\}) = \frac{1}{2}$ and $\mu(\{\frac{1}{n}\}) = \frac{1}{2^{n+1}}$ for $n \geq 1$. Then the mappings which are almost surely continuous on X are always continuous everywhere. Thus, for a mapping f from X to Y which is not continuous at 0, we will not be able to represent it as the composition of two almost surely continuous mappings as above.

Now, let J be an isomorphism from (I^2, λ^2) to (I, λ). We have two functions K and L from I to I such that $J^{-1}(r) = (K(r), L(r))$ for each $r \in I$. Define J_1 to be the identity mapping on I. If J_n has been defined, then J_{n+1} is defined by

$$J_{n+1}(x_1, x_2, \ldots, x_{n+1}) = J(x_1, J_n(x_2, \ldots, x_{n+1})).$$

The inverse of J_n has its components composites of K and L. Then J_n is an isomorphism from (I^n, λ^n) to (I, λ) for each n. Thus for any Riemann integrable function f on I^n, there is a Riemann integrable function h on I such that $f = h \circ J_n$. Also note that for each n, J_n is a superposition of Riemann integrable functions of one and of two variables. Hence we know that every Riemann integrable function of several variables can be expressed as a superposition of Riemann integrable functions of one and of two variables, which could be considered as an analog of Kolmogorov's Superposition Theorem (see **[8]**) on the structures of continuous functions of several variables.

In **[2]** and **[12]**, many conditions are given to guarantee a u.d.p. transformation T on I to be in the simplest form $T(x) = x$ for each $x \in I$ or $T(x) = 1 - x$ for each $x \in I$. In the following, we present a new characterization for these two simple u.d.p. transformations.

Proposition 3.5. *Let T be a continuous u.d.p. transformation from I to I. Assume that for any sequence $\omega = \{x_n\}_{n=1}^{\infty}$, $T(\omega)$ is u.d.* mod 1 *if and only if ω is u.d.* mod 1. *Then either $T(x) = x$ for each $x \in I$ or $T(x) = 1 - x$ for each $x \in I$.*

Proof. We shall show that T is one to one. Suppose not; there are numbers a, b in I with $T(a) = T(b)$ and $a < b$. Let c_1, c_2 be the numbers in $[a, b]$ such that

$$T(c_1) = \inf_{x \in [a,b]} T(x) \quad \text{and} \quad T(c_2) = \sup_{x \in [a,b]} T(x).$$

If $T(c_1) = T(c_2)$, then $T(x) = T(a)$ for every $x \in [a, b]$. Thus, $T^{-1}(\{T(a)\}) \supseteq [a, b]$ and hence $\lambda(T^{-1}(\{T(a)\})) \geq b - a$. On the other hand, since T is measure preserving, we have $\lambda(T^{-1}(\{T(a)\})) = \lambda(\{T(a)\}) = 0$, which is a contradiction. Without loss of generality we assume that $a < c_1 < c_2 \leq b$. Note that $T([a,b]) = T([c_1,c_2]) = [T(c_1), T(c_2)]$ and T is surjective from I to I. Let $\omega' = \{y_n\}_{n=1}^{\infty}$ be a u.d. mod 1 sequence. We shall define a sequence $\omega = \{x_n\}_{n=1}^{\infty}$ as follows. For each $n \geq 1$, if $y_n \in [T(c_1), T(c_2)]$, let x_n be a number in $[c_1, c_2]$ such that $T(x_n) = y_n$; if $y_n \notin [T(c_1), T(c_2)]$, let x_n be a number in $[0,a] \cup (b,1]$ such that $T(x_n) = y_n$. Then $T(\omega) = \omega'$. Since the x_n's never appear in the interval (a, c_1), ω is not a u.d. mod 1 sequence, which contradicts the hypothesis. Therefore, T is one to one. Hence T is monotone. By Proposition 3 of **[12]**, we have $T(x) = x$ or $T(x) = 1 - x$ on I. □

Piecewise linear u.d.p. transformations are extensively studied in **[12]** and used in **[15]** and **[19]** to approximate general u.d.p. transformations. In the following we shall show that piecewise linear continuous u.d.p. transformations have certain interpolation properties, which could be used in the constructions of other piecewise linear u.d.p. transformations.

Lemma 3.6. *For any $b \in I$, there is a piecewise linear continuous u.d.p. transformation T_b such that $T_b(0) = 0$ and $T_b(1) = b$.*

Proof. Define a function T_b from I to I as follows:

$$T_b(x) = \begin{cases} 4x & \text{if } 0 \leq x < \frac{b}{4}, \\ 2x + \frac{b}{2} & \text{if } \frac{b}{4} \leq x < \frac{1}{2} - \frac{b}{4}, \\ -2x + 2 - \frac{b}{2} & \text{if } \frac{1}{2} - \frac{b}{4} \leq x < 1 - \frac{b}{4}, \\ 4x - 4 + b & \text{if } 1 - \frac{b}{4} \leq x \leq 1. \end{cases}$$

It is easy to see that the ordinate and abscissa decompositions of I (see **[12]**, p. 467 for the definitions) with respect to T_b are $\{(0,b), (b,1)\}$ and $\{(0, \frac{b}{4}), (1 - \frac{3b}{4}, 1 - \frac{b}{4}), (1 - \frac{b}{4}, 1), (\frac{b}{4}, \frac{1}{2} - \frac{b}{4}), (\frac{1}{2} - \frac{b}{4}, 1 - \frac{3b}{4})\}$ respectively. Then, by Proposition 7 of **[12]**, T_b is a u.d.p. transformation. The rest is clear. □

Now, let $\{f_i\}_{i=1}^{n}$ be a finite sequence of u.d.p. transformations from I to I. Let $\{a_i\}_{i=0}^{n}$ be a sequence of real numbers such that

$$0 = a_0 < a_1 < \cdots < a_{n-1} < a_n = 1.$$

Define a function T from I to I by letting

$$T(x) = f_i\left(\frac{x - a_{i-1}}{a_i - a_{i-1}}\right) \quad \text{if } x \in [a_{i-1}, a_i) \text{ for some } 1 \leq i \leq n.$$

Then it is easy to see that T is a u.d.p. transformation. □

Lemma 3.7. *For any $b_1, b_2 \in I$ and for any integer $n \geq 2$, there is a piecewise linear continuous u.d.p. transformation $T_{b_1 b_2}^{(n)}$ such that*

(1) $T_{b_1 b_2}^{(n)}(0) = b_1$, $T_{b_1 b_2}^{(n)}(1) = b_2$;

(2) *the absolute value of the slope of each line segment of the graph of* $T_{b_1 b_2}^{(n)}$ *in the unit square* I^2 *is greater than* n.

Proof. Define a function $T_{b_1 b_2}^{(n)}$ from I to I by letting

$$T_{b_1 b_2}^{(n)}(x) = \begin{cases} T_{b_1}(1 - nx) & \text{if } 0 \le x < \frac{1}{n}, \\ T_0(nx - k + 1) & \text{if } \frac{k-1}{n} \le x < \frac{k}{n} \text{ for some } 2 \le k \le n-1, \\ T_{b_2}(nx - n + 1) & \text{if } \frac{n-1}{n} \le x \le 1. \end{cases}$$

Then, $T_{b_1 b_2}^{(n)}$ has the desired properties. □

Proposition 3.8. *Let* $\{a_i\}_{i=0}^n$ *be a finite sequence of real numbers such that*

$$0 = a_0 < a_1 < \cdots < a_{n-1} < a_n = 1.$$

Let $\{b_i\}_{i=0}^n$ *be any finite sequence of real numbers in* I. *Then there exists a piecewise linear continuous u.d.p. transformation* T *from* I *to* I *such that* $T(a_i) = b_i$ *for all* $0 \le i \le n$.

Proof. Define a function T from I to I as follows:

$$T(x) = T_{b_{i-1} b_i}^{(2)}\left(\frac{x - a_{i-1}}{a_i - a_{i-1}}\right) \quad \text{if } x \in [a_{i-1}, a_i) \text{ for some } 1 \le i \le n.$$

It can be checked that T satisfies all the properties. □

The following definitions are due to Bosch (**[2]**).

Definition 3.9. The sequence $(a_i) \subseteq I$, $i = 1, 2, \ldots$, is a partition of I if it satisfies one of the following conditions:

(a) (a_i) is a finite sequence;

(b) (a_i) is an infinite, strictly monotone increasing sequence tending to one;

(c) (a_i) is an infinite, strictly monotone decreasing sequence tending to zero;

(d) (a_i) is a bi-infinite sequence indexed by the set of integers such that $a_i < a_{i+1}$ for all integers i, (a_i) tends to one as i tends to infinity, and (a_i) tends to zero as i tends to negative infinity.

Definition 3.10. A function $f : I \longrightarrow I$ belongs to the class G if and only if there is a partition $B_f = (b_j)$ of I corresponding to f such that

(a) $(c, d) \subseteq (b_j, b_{j+1})$, for any j implies $f^{-1}((c, d)) = \bigcup I_k$, $k = 1, 2, \ldots$, where $\{I_k\}$ is a monotone sequence of disjoint intervals in I (i.e., the left endpoints of the disjoint intervals form a monotone sequence of real numbers); and

(b) for each j, $f^{-1}(b_j)$ is empty or is a partition of I.

Note that Bosch joined the terms zero and one to the partition in Definition 3.9. So we may also assume that $f^{-1}(\{0\})$ and $f^{-1}(\{1\})$ are partition of I. It is shown in **[2]** that a function T in G preserves uniform distribution of sequences if and only if T is measure preserving. By Theorem 4 of **[12]**, such a function must be continuous almost everywhere. We see in the following proposition that every function in the class G is in fact continuous on I except for countably many points. Thus the class G considered in **[2]** is actually quite restricted.

Proposition 3.11. *For any f in G, the discontinuity points of f are at most countably many.*

Proof. Let $f \in G$. Let $D = \bigcup_{n=1}^{\infty}\{x : \omega(x) \geq \frac{1}{n}\}$ be the set of discontinuity points of f, where $\omega(x)$ is the oscillation of f at x. We shall prove that the set $F_n = \{x : \omega(x) \geq \frac{1}{n}\}$ is countable for each $n \geq 1$. Now for any given positive integer n and for any interval (b_j, b_{j+1}) as in the definition of G, there exists finitely many open subintervals $I_1^j, I_2^j, \ldots, I_{n_j}^j$ of (b_j, b_{j+1}) such that the length of each subinterval is less than $\frac{1}{2n}$ and the union of them is (b_j, b_{j+1}). Then for each j and i with $1 \leq i \leq n_j$, $f^{-1}(I_i^j)$ is the union of a sequence of disjoint intervals $\{I_k^{ij}\}$ in I. Let $I_k^{ij} =< a_k^{ij}, b_k^{ij} >$. Since $f(I_k^{ij}) \subseteq I_i^j$, for any $x \in (a_k^{ij}, b_k^{ij})$, $\omega_f(x) \leq \frac{1}{2n}$.

On the other hand, for any $y \in F_n$, if we also assume $f(y)$ is not equal to any of the b_j nor 0, 1, then $f(y) \in (b_j, b_{j+1})$ for some j, i.e. y belongs to some interval $< a_k^{ij}, b_k^{ij} >$. Since $\omega_f(y) \geq \frac{1}{n}$. Then y must be a_k^{ij} or b_k^{ij}. Therefore

$$F_n \subseteq \Big(\bigcup_{i,j,k}\{a_k^{ij}, b_k^{ij}\}\Big)\bigcup\Big(\bigcup_j f^{-1}(\{b_j\})\Big)\bigcup f^{-1}(\{0\})\bigcup f^{-1}(\{1\}).$$

whence F_n is a countable set. □

Note. In the proof of the above proposition, the monotone condition on the sequence $\{I_k\}$ in Definition 3.10 is not used. Thus, if we enlarge the class G by replacing the monotone condition by the condition that the characteristic function of $\bigcup I_k$ be Riemann integrable on [0, 1] as noted by Bosch, any function in G is still continuous on I except countably many points.

In **[2]** Bosch considered functions that are continuously differentiable on I except for a set of points forming a partition of I, where the partition is in the sense of Definition 3.9. In particular, he obtained some necessary or sufficient conditions for functions in this class to be uniform distribution preserving. The proof of Theorems 2.9, 3.1 and 3.4 in **[2]** is complicated. We shall consider a larger class of functions defined in the following way and use relatively higher level analysis to yield stronger results.

Definition 3.12. A function f from I to I is said to be in the class AD if there is an open subset $E = \bigcup_{n=1}^{\infty}(a_n, b_n)$ of I with measure one such that f is continuously differentiable and the derivative $f'(x) \neq 0$ on each (a_n, b_n), where $\{(a_n, b_n)\}_{n=1}^{\infty}$ is a countable collection of disjoint intervals.

Note that if $f \in AD$, then f restricted to (a_n, b_n) is a homeomorphism to some open interval (c_n, d_n). Denote its inverse by h_n. Clearly, h_n is continuously differentiable on (c_n, d_n). Define a function g_n on I for each n as follows:

$$g_n(y) = \begin{cases} |h_n'(y)| & \text{for } c_n < y < d_n, \\ 0 & \text{for } 0 \le y \le c_n \text{ or } d_n \le y \le 1. \end{cases}$$

Now let g_f be the function (depending on f) defined as the infinite sum of all the g_n's, i.e. $g_f(y) = \sum_{n=1}^{\infty} g_n(y)$ for all $y \in I$.

It's clear that each g_n is lower semicontinuous on I. By the results of Problem 48 (see p. 48, **[13]**), g_f is lower semicontinuous. The following theorem characterizes those u.d.p. transformations in the class AD.

Theorem 3.13. *For any given function $f \in AD$, let g_f be the function defined above. Then f is u.d.p. if and only if $g_f(y) = 1$ almost everywhere in I. In this case, $g_f(y) \le 1$ for all $y \in I$.*

Proof. Let $f \in AD$. Let E be the corresponding open set in Definition 3.12. Pick any nonnegative continuous function φ on I. Then

$$\begin{aligned} \int_0^1 \varphi \circ f(x)\,dx &= \int_E \varphi \circ f(x)\,dx + \int_{I-E} \varphi \circ f(x)\,dx \\ &= \sum_{n=1}^{\infty} \int_{(a_n, b_n)} \varphi(f(x))\,dx \\ &= \sum_{n=1}^{\infty} \int_{c_n}^{d_n} \varphi(y)|h_n'(y)|\,dy \\ &= \sum_{n=1}^{\infty} \int_0^1 \varphi(y) g_n(y)\,dy. \end{aligned}$$

Since $\varphi(y)g_n(y) \ge 0$ for all $y \in I$, it follows from the Monotone Convergence Theorem that

$$\sum_{n=1}^{\infty} \int_0^1 \varphi(y) g_n(y)\,dy = \int_0^1 \varphi(y) g_f(y)\,dy.$$

Hence $\int_0^1 \varphi \circ f(x)\,dx = \int_0^1 \varphi(x) g_f(x)\,dx$. Now let $\varphi \equiv 1$; then $1 = \int_0^1 g_f(x)\,dx$. Thus g_f is Lebesgue integrable. Therefore, for any φ in the space $C(I)$ of continuous functions on I, we have $\int_0^1 \varphi \circ f(x)\,dx = \int_0^1 \varphi(x) g_f(x)\,dx$. Since f is Riemann integrable, f is u.d.p. if and only if f is measure preserving, which is equivalent to $\int_0^1 \varphi \circ f(x)\,dx = \int_0^1 \varphi(x)\,dx$ for any $\varphi \in C(I)$. Thus f is u.d.p. if and only if $\int_0^1 \varphi(x)\,dx = \int_0^1 \varphi(x) g_f(x)\,dx$ for any $\varphi \in C(I)$, which is equivalent to $g_f = 1$ almost everywhere by the uniqueness of the representing measure in the Riesz representation theorem. Since $g_f(y)$ is lower semicontinuous, $H = g_f^{-1}((1, \infty))$ is an open set. If $g_f(y) = 1$ a.e., then $\mu(H) = 0$; thus $H = \emptyset$, i.e. $g_f(y) \le 1$ for all y in I. □

Remark 3.14. (1) Theorem 2.9 of **[2]** follows easily from our Theorem 3.13. In that case $E = \bigcup_{n=1}^{m}(a_{n-1}, a_n)$ and $0 = a_0 < a_1 < \cdots < a_m = 1$. Let B be the finite set of all c_n, d_n and $f(a_n)$, where $(c_n, d_n) = f((a_{n-1}, a_n))$. It's clear that g_f is continuous on $I - B$. By Theorem 3.13, we know that $g_f = 1$ almost surely. Hence $g_f(y) = 1$ for every $y \in I - B$. It's easy to check that $g_f(y) = \sum_{f(x)=y} \frac{1}{|f'(x)|}$ for every $y \in I - B$.

(2) When $E = \bigcup_{n=-\infty}^{\infty}(a_{n-1}, a_n)$ and the bi-infinite sequence (a_n) forms a partition in the sense of (d) of Definition 3.9, we define the set B in the same way as above. We still have $g_f(y) = \sum_{f(x)=y} \frac{1}{|f'(x)|}$ for every $y \in I - B$. However, B may not be a partition of I. Under the assumption that B is a partition of I, Bosch obtained his Theorems 3.1 and 3.4. It is easy to see that part (a) of his Theorem 3.1 and his Theorem 3.4 are direct consequences of our Theorem 3.13. For part (b) of Theorem 3.1 in **[2]**, let S be the set of continuity points of g_f in $I - B$. Then, by the result of Problem 31 in (p. 141, **[13]**), $(I - B) - S$ is a set of first category in the topological space $I - B$, thus a set of first category in the usual sense. Since $g_f(y) = 1$ a.e., we must have $g_f(y) = 1$ for any $y \in S$. Clearly, S is a dense subset of I.

References

[1] C. Binder, Über einen Satz von de Bruijn und Post, Sitzungsber. Österr. Akad. Wiss., Math.-Naturw. Kl. Abt. II 179 (1971), 233–251.

[2] W. Bosch, Functions that preserve the uniform distribution of sequences, Trans. Amer. Math. Soc. 307 (1988), 143–152.

[3] N. G. De Bruijn and K. A. Post, A remark on uniformly distributed sequences and Riemann integrability, Indag. Math. 30 (1968), 149–150.

[4] P. Gerl, Relative Gleichverteilung in lokalkompakten Räumen, Math. Z. 121 (1971), 24–50.

[5] T. Kamae, Subsequences of normal sequences, Israel J. Math. 16 (1973), 121–149.

[6] D. E. Knuth, The Art of Computer Programming, Vol. 2: Semi-numerical Algorithms, Addison-Wesley, Reading, Mass., 1969.

[7] L. Kuipers and H. Niederreiter, Uniform Distribution of Sequences, John Wiley & Sons, New York 1974.

[8] G. G. Lorentz, The 13th problem of Hilbert, Proc. Sympos. Pure Math. 28 (1976), 419–430.

[9] S. Marcus, La superposition des fonctions et l' isométrie des certaines classes des fonctions, Bull. Math. Soc. Sci. Math. Phys. Roumanie 1 (1957), 69–76.

[10] I. Niven and H. S. Zuckerman, On the definition of normal numbers, Pacific J. Math. 1 (1951), 103–109.

[11] J. C. Oxtoby, On two theorems of Parthasarathy and Kakutani concerning the shift transformation, in: Ergodic Theory (ed. by F. B. Wright), Academic Press, New York 1962.

[12] S. Porubsky, T. Salát and O. Strauch, Transformations that preserve uniform distribution, Acta Arith. 49 (1988), 459–479.

[13] H. L. Royden, Real Analysis, 2nd ed., Macmillan, New York 1968.

[14] Y. N. Sun, Isomorphisms for convergence structures, Adv. Math. 116 (1995), 322-355.

[15] Y. N. Sun, Some properties of uniformly distributed sequences, J. Number Theory 44 (1993), 273–280.

[16] Y. N. Sun, On metric theorems in the theory of uniform distribution, Compositio Math. 86 (1993), 15–21.

[17] Y. N. Sun, A remark on the trace of some Riesz operators, Arch. Math. 63 (1994), 530–534.

[18] Y. N. Sun, Isomorphisms for convergence structures II: Borel spaces and infinite measures, Adv. Math., to appear.

[19] R. F. Tichy and R. Winkler, Uniform distribution preserving mappings, Acta Arith. 60 (1991), 177–189.

[20] P. Walters, An Introduction to Ergodic Theory, Springer-Verlag, New York 1982.

[21] B. Weiss, Normal numbers as collectives, Proc. of Sympos. on Ergodic Theory, University of Kentucky, 1971.

Department of Mathematics
National University of Singapore
Lower Kent Ridge Road
Singapore 0511, Republic of Singapore

Ergodic Theorems for Exit Laws

Rainer Wittmann[1]

Abstract. Norm and almost everywhere convergence is shown for Cesàro averages of exit laws with respect to linear and nonlinear operators.

1. Introduction

Let T be a mapping on a Banach space X. A sequence $(x_n)_{n\in\mathbb{N}}$ will be called an *exit law* with respect to (X, T), if

$$\sup_{n\in\mathbb{N}} \|x_n\| < \infty, \qquad Tx_{n+1} = x_n \quad (n \in \mathbb{N}).$$

We say that an exit law (x_n) is *trivial*, if $x_n = x_{n+1}$ for any $n \in \mathbb{N}$. Usually X will be an L^p space and T will be a linear positive contraction.

Exit laws appear naturally in the theory of Markov processes (cf. Neveu **[9]**, Proposition V-2-2). In particular, if T is the transition kernel of a Markov process $(\Omega, \mathcal{F}, (P^x)_{x\in E}, X_n)$ and if (f_n) is an exit law with respect to $(T, B(E))$ ($B(E)$ is the Banach space of bounded measurable functions on the state space E), then $f_n \circ X_n$ is a martingale with respect to P^x for any $x \in E$.

Exit laws appear also naturally in the study of the asymptotic behavior of the iterates of linear contractions, as the following theorem of M. Lin **[7]** shows (cf. Derriennic **[2]** for its role in the theory of Markov processes).

Theorem 1.1. *Let T be a linear contraction on a Banach space X and T^* the dual operator on the dual space X^*. Then the following properties are equivalent:*

$$\text{Any exit law with respect to } (X^*, T^*) \text{ is trivial.} \tag{i}$$

$$T^n \text{ is norm convergent} \iff \frac{1}{n}\sum_{i=0}^{n-1} T^i x \text{ is norm convergent.} \tag{ii}$$

$$\{x^* \in X^* : T^*x^* = x^*,\ \|x^*\| \le 1\} = \bigcap_{n=1}^{\infty} T^{*n}\left(\{x^* \in X^* : \|x^*\| \le 1\}\right). \tag{iii}$$

If T is a transition probability then the above three conditions are equivalent to the 0-1 law for the σ-algebra of tail events.

1 Heisenberg Fellow of Deutsche Forschungsgemeinschaft

If an exit law (x_n) is norm convergent and if T^k is equicontinuous, then it is easy to see that (x_n) is already trivial. The situation is different, if we study *averages* of exit laws. This is the subject of this paper. If T is an invertible mapping on X, then $(T^n x)$ is an exit law with respect to (X, T^{-1}) (which also shows that an exit law may converge weakly without being trivial). Thus the study of averages of exit laws with respect to (X, T^{-1}) is equivalent to the study of the averages of $(T^n x)$. In particular, the ergodic theorems below contain the corresponding ergodic theorems for isometries. Sometimes ergodic theorems for exit laws can be reduced to isometries. In other places methods successfull for isometries can be generalized to exit laws. This is especially true for the basic maximal inequality for exit laws in Section 3. The plan of our paper is as follows. Section 2 deals with norm convergence of Cesàro averages of exit laws. In the linear case this problem is very simple. Thus the most interesting results of this section deal with nonlinear operators. Finally in Section 3 we show almost everywhere of Cesàro averages of exit laws with respect to positive linear contractions on L^p $(1 < p < \infty)$.

2. Norm convergence

The proof of the following basic result is obvious.

Theorem 2.1. *Let T be a linear contraction on a Banach space X. Then the space $E_{X,T}$ of exit laws with respect to (X, T) endowed with the pointwise operations and norm $\|(x_n)_{n\in\mathbb{N}}\| = \sup_n \|x_n\| = \lim_n \|x_n\|$ is a Banach space. Moreover*

$$(x_n)_{n\in\mathbb{N}} \longrightarrow (Tx_n)_{n\in\mathbb{N}}$$

defines a linear invertible isometry $S_{X,T}$ on $E_{X,T}$ and

$$(x_n)_{n\in\mathbb{N}} \longrightarrow x_1$$

defines a linear contraction $P_{X,T}$ of $E_{X,T}$ into X.

To obtain mean ergodic theorems, we need information about the Banach space $E_{X,T}$. For this purpose we give the following immediate but useful result.

Proposition 2.2. *Let T be a linear contraction on a uniformly convex Banach space X. Then $E_{X,T}$ is uniformly convex as well. If X is a Hilbert space, then $E_{X,T}$ is a Hilbert space.*

Corollary 2.3. *Let T be a linear contraction on a uniformly convex space X. Then for any $(x_n)_{n\in\mathbb{N}} \in E_{X,T}$ there exists $z \in X$ with $Tz = z$ such that the Cesàro averages $\frac{1}{n}\sum_{i=1}^{n} x_i$ are norm convergent to z.*

Proof. Since $E_{X,T}$ is uniformly convex by Proposition 2.2, by the ordinary mean ergodic theorem, there exists $(z_n)_{n\in\mathbb{N}} \in E_{X,T}$ such that $\frac{1}{n}\sum_{i=0}^{n-1} S^i_{X,T}((x_m)_{m\in\mathbb{N}})$ is norm convergent to (z_n). Since $P_{X,T}$ is a contraction,

$$P_{X,T}\Big(\frac{1}{n}\sum_{i=0}^{n-1} S^i_{X,T}((x_m)_{m\in\mathbb{N}})\Big) = \frac{1}{n}\sum_{i=1}^{n} x_i$$

is norm convergent to $P(z_n) = z_1$. Since

$$\Big\|\frac{1}{n}\sum_{i=1}^{n} x_i - T\Big(\frac{1}{n}\sum_{i=1}^{n} x_i\Big)\Big\| = \frac{1}{n}\|x_n - Tx_1\|,$$

we have $Tz_1 = z_1$. □

Remark 2.4. For the proof of the pointwise ergodic Theorem 3.2 we need slightly more than norm convergence. Namely that

$$\lim_{n\to\infty}\sup_{m\geq 0}\Big\|z - \frac{1}{n}\sum_{i=m+1}^{m+n} x_i\Big\| = 0. \tag{AC}$$

In other words (x_n) is *almost convergent* to z. The concept of almost convergence was introduced by Lorentz **[8]** and, for instance in **[10]**, **[6]**, it was used for nonlinear ergodic theorems. To verify (AC) we introduce the contraction P_n of $E_{X,T}$ into X defined by $P_n((x_m)_{m\in\mathbb{N}}) = x_{n+1}$. Then we have $P_{X,T} = P_0$ and

$$P_m\Big(\frac{1}{n}\sum_{i=0}^{n-1} S^i_{X,T}((x_\ell)_{\ell\in\mathbb{N}})\Big) = \frac{1}{n}\sum_{i=m+1}^{m+n} x_i.$$

Hence

$$\sup_{m\geq 0}\Big\|z - \frac{1}{n}\sum_{i=m+1}^{m+n} x_i\Big\| \leq \Big\|\frac{1}{n}\sum_{i=0}^{n-1} S^i_{X,T}((x_\ell)_{\ell\in\mathbb{N}}) - (z)_{\ell\in\mathbb{N}}\Big\|$$

converges to 0 as n tends to ∞.

We now turn to the nonlinear case. From examples in Krengel, Lin **[4]** we know that one can only expect *weak* convergence of $\frac{1}{n}\sum_{i=1}^{n} T^i x$ if T is a nonexpansive mapping (i.e. $\|Tx - Ty\| \leq \|x-y\|$). Fortunately, for exit laws, one can prove norm convergence. We start in the setting of Hilbert spaces.

Proposition 2.5. *Let* $\varepsilon > 0$ *and* (x_n) *be a sequence in a Hilbert space* H *such that either*

$$\|x_{n+k} + x_{m+k}\|^2 \leq \|x_n + x_m\|^2 + \varepsilon \quad (n,m,k\in\mathbb{N}) \tag{i}$$

$$\|x_n\|^2 \leq \|x_{n+k}\|^2 + \frac{\varepsilon}{2} \quad (n,k\in\mathbb{N}) \tag{ii}$$

or

$$\|x_n - x_m\|^2 \leq \|x_{n+k} - x_{m+k}\|^2 + \varepsilon \quad (n,m,k\in\mathbb{N}) \tag{iii}$$

$$\|x_{n+k}\|^2 \leq \|x_n\|^2 + \frac{\varepsilon}{2} \quad (n,k\in\mathbb{N}). \tag{iv}$$

Then we have

$$\lim_{k\to\infty}\sup_{m,n\geq k}\|A_m - A_n\|^2 \leq 4\varepsilon,$$

where $A_n = \frac{1}{n}\sum_{i=1}^{n} x_i$.

Proof. We use the arguments of the proof of Wittmann **[11]**, Theorem 2.3. Let $\langle\cdot,\cdot\rangle$ be the inner product of the Hilbert space H. If (i) and (ii) hold, then

$$\begin{aligned}2\langle x_{n+k}, x_{m+k}\rangle &= \|x_{n+k}+x_{m+k}\|^2 - \|x_{n+k}\|^2 - \|x_{m+k}\|^2\\ &\leq \|x_n + x_m\|^2 + \varepsilon - \|x_n\|^2 + \frac{\varepsilon}{2} - \|x_m\|^2 + \frac{\varepsilon}{2} = 2\langle x_n, x_m\rangle + 2\varepsilon.\end{aligned}$$

On the other hand, if (iii) and (iv) hold, then

$$\begin{aligned}2\langle x_{n+k}, x_{m+k}\rangle &= \|x_{n+k}\|^2 + \|x_{m+k}\|^2 - \|x_{n+k}-x_{m+k}\|^2\\ &\leq \|x_n\|^2 + \frac{\varepsilon}{2} + \|x_m\|^2 + \frac{\varepsilon}{2} - \|x_n - x_m\|^2 + \varepsilon = 2\langle x_n, x_m\rangle + 2\varepsilon.\end{aligned}$$

Thus under both types of assumptions, we have shown

$$\langle x_{n+k}, x_{m+k}\rangle \leq \langle x_n, x_m\rangle + \varepsilon \quad (n, m, k \in \mathbb{N}). \tag{1}$$

We now define

$$K = \inf\Big\{\Big\|\sum_{n=1}^{p} t_n x_n\Big\|^2 : p \in \mathbb{N},\ 0 \leq t_n \leq 1,\ \sum_{n=1}^{p} t_n = 1\Big\}.$$

We fix $\delta > 0$ for a while and choose $(s_n)_{1\leq n\leq q}$ in $[0, 1]$ such that $\sum_{n=1}^{q} s_n = 1$ and

$$\Big\|\sum_{n=1}^{q} s_n x_n\Big\|^2 \leq K + \delta. \tag{2}$$

Using (1) and (2) we get for any $k \in \mathbb{N}$

$$\begin{aligned}\Big\|\sum_{n=1}^{q} s_n x_{n+k}\Big\|^2 &= \sum_{m,n=1}^{q} s_n s_m \langle x_{n+k}, x_{m+k}\rangle\\ &\leq \sum_{m,n=1}^{q} s_n s_m (\langle x_n, x_m\rangle + \varepsilon)\\ &= \sum_{m,n=1}^{q} s_n s_m \langle x_n, x_m\rangle + \sum_{m,n=1}^{q} s_n s_m \varepsilon\\ &= \Big\|\sum_{n=1}^{q} s_n x_n\Big\|^2 + \varepsilon \leq K + \delta + \varepsilon.\end{aligned}$$

Together with the convexity of $x \to x^2$ this implies

$$\|z_m\|^2 \leq K + \varepsilon + \delta \quad (m \in \mathbb{N}), \tag{3}$$

where

$$z_m = \frac{1}{m}\sum_{k=0}^{m-1}\sum_{n=1}^{q} s_n x_{n+k}.$$

By the definition of K we have $\|\frac{1}{2}(z_m + z_n)\|^2 \geq K$, whence using (3) and the parallelogram identity we get

$$\|z_m - z_n\|^2 = 2\|z_m\|^2 + 2\|z_n\|^2 - \|z_m + z_n\|^2 \leq 4K + 4\varepsilon + 4\delta - 4K = 4\varepsilon + 4\delta. \quad (4)$$

Since

$$z_m = \frac{1}{m}\sum_{n=1}^{q}\sum_{j=n}^{n+m-1} s_n x_j = \frac{1}{m}\sum_{j=1}^{q+m-1}\sum_{n=(j-m+1)\vee 1}^{j\wedge q} s_n x_j$$

and $\sum_{n=1}^{q} s_n = 1$ we have

$$\|z_m - A_m\| \leq \frac{1}{m}\sum_{j=1}^{q-1}\|x_j\|\sum_{n=j+1}^{q} s_n + \frac{1}{m}\sum_{j=m+1}^{m+q-1}\|x_j\|\sum_{n=j-m+1}^{q} s_n \leq 2\frac{q-1}{m}\sup_n \|x_n\|.$$

Together with (4) this implies

$$\lim_{k\to\infty}\sup_{m,n\geq k}\|A_m - A_n\|^2 \leq 4\varepsilon + 4\delta.$$

Since $\delta > 0$ can be made arbitrarily small the assertion follows. □

Corollary 2.6. *Let (x_n) be a sequence in a Hilbert space H such that*

$$\|x_{n+1} - x_{m+1}\| \geq \|x_n - x_m\| \quad (m, n \in \mathbb{N}), \tag{i}$$

$$\|x_{n+1}\| \geq \|x_n\| \quad (n \in \mathbb{N}), \tag{ii}$$

$$\sup_{n\in\mathbb{N}}\|x_n\| < \infty. \tag{iii}$$

Then $\frac{1}{n}\sum_{i=1}^{n} x_i$ is norm convergent.

Proof. Let $\varepsilon > 0$ be given. By (ii) and (iii) there exists $k_\varepsilon \in \mathbb{N}$ such that $\|x_n\| \leq \|x_m\| + \varepsilon/2$ for any $n \geq m \geq k_\varepsilon$. Hence the sequence $(x_{n+k_\varepsilon})_{n\in\mathbb{N}}$ satisfies 2.5(iv). By (i) it also satisfies 2.5(iii) (even with $\varepsilon = 0$). Thus Proposition 2.5 implies

$$\lim_{k\to\infty}\sup_{m,n\geq k}\Big\|\frac{1}{n}\sum_{i=1}^{n} x_i - \frac{1}{m}\sum_{i=1}^{m} x_i\Big\| = \lim_{k\to\infty}\sup_{m,n\geq k}\Big\|\frac{1}{n}\sum_{i=1}^{n} x_{i+k_\varepsilon} - \frac{1}{m}\sum_{i=1}^{m} x_{i+k_\varepsilon}\Big\| \leq 4\varepsilon\,.$$

Thus, since $\varepsilon > 0$ can be made arbitrarily small, the averages $\frac{1}{n}\sum_{i=1}^{n} x_i$ form a Cauchy sequence and the assertion follows. □

Corollary 2.7. *Let T be a nonexpansive mapping on a Hilbert space H with $T0 = 0$. Then $\frac{1}{n}\sum_{i=1}^{n} x_i$ is norm convergent for any exit law (x_n).*

Remark. It can be shown, as in the linear case, that the limit in Corollary 2.7 is a fixed point of T.

From now on let (E, μ) be σ-finite measure space. All L^p spaces are with respect to this measure space.

Theorem 2.8. *Let T be a nonexpansive mapping on L^1, which is order preserving, i.e. $f \leq g$ implies $Tf \leq Tg$, and L^∞ norm decreasing, i.e. $\|Tf\|_\infty \leq \|f\|_\infty$. By Krengel, Lin* **[4]** *we then have $\|Tf\|_p \leq \|f\|_p$ for any $f \in L^1 \cap L^p$ and T can be naturally extended to all of L^p.*

Let $1 < p < \infty$ and (f_n) an exit law with respect to (L^p, T) with $f_n \geq 0$ for any $n \in \mathbb{N}$. Furthermore, we assume that $\lim_{t\to\infty} \sup_n \|(f_n - t)_+\|_p = 0$. Then $\frac{1}{n}\sum_{i=1}^n f_i$ is norm convergent in L^p. Moreover, if the measure μ is finite, then the same holds for $p = 1$.

For averages of iterates *weak* convergence for the same class of operators was shown in Krengel, Lin **[4]** and extended in Krengel, Lin, Wittmann **[6]**. It is important to note that the above mappings T need not be nonexpansive in L^p $(1 < p < \infty)$. In fact there exists a huge class of such mappings, which is explored in Krengel, Lin **[5]** (cf. also Krengel, Lin, Wittmann **[6]**,Proposition 2.9). The pattern of the proof below is similar to the ingenious proof of the above mentioned weak convergence theorem of Krengel, Lin **[4]**. To keep the size of the paper within reasonable limits we omit the proofs of two very important lemmas of **[4]**, which are almost identical in our situation.

Let (f_n) and T as in Theorem 2.8. Then $T(f_{n+1} \wedge t) \leq (Tf_{n+1}) \wedge t = f_n \wedge t$ for any $t \geq 0$, since T is order preserving, whence

$$\begin{aligned}\int (f_n - t)_+ \, d\mu &= \|f_n - (f_n \wedge t)\|_1 = \|Tf_{n+1} - (Tf_{n+1} \wedge t)\|_1 \\ &\leq \|Tf_{n+1} - T(f_{n+1} \wedge t)\|_1 \leq \|f_{n+1} - (f_{n+1} \wedge t)\|_1 \\ &= \int (f_{n+1} - t)_+ \, d\mu.\end{aligned}$$

In particular $\int (f_n - t)_+ \, d\mu$ is convergent for $n \to \infty$. Together with **[4]**, Theorem 3.2 and the proof of **[4]**, Theorem 3.1 we obtain

Proposition 2.9. *Let (f_n) and T as in Theorem 2.8. Let further $D(F)$ be the set of points, where the function $F(t) = \limsup_{n\to\infty} \mu\{f_n \geq t\}$ $(t \in]0, \infty[)$ is discontinuous. Since F is increasing, this set is at most countable. We have*

$$F(t) = \liminf_{n\to\infty} \mu\{f_n > t\} \quad (t \in]0, \infty[\setminus D(F)).$$

Behind Proposition 2.9 is the concept of convergence in distribution, which is somewhat unusual for infinite measures. We refer to Section 3 of **[4]** for more details.

Lemma 2.10. *Let (f_n) and T as in Theorem 2.8 and put $f(n, \alpha, \beta) = (f_n \wedge \beta) - (f_n \wedge \alpha)$ for any $0 < \alpha < \beta < \infty$ and $n \in \mathbb{N}$. Then for any $\eta > 0$ there exists $K \geq 1$ such that*

$i, j \geq K,\ k \geq 0$ imply

$$\|f(i,\alpha,\beta) - f(j,\alpha,\beta)\|_1 \leq \|f(i+k,\alpha,\beta) - f(j+k,\alpha,\beta)\|_1 + \eta \qquad \text{(i)}$$

$$\|f(i+k,\alpha,\beta) \wedge f(j+k,\alpha,\beta)\|_1 \leq \|f(i,\alpha,\beta) \wedge f(j,\alpha,\beta)\|_1 + \eta. \qquad \text{(ii)}$$

The proof is the same as that of **[4]**, Lemma 4.4 except that the "time scale" $\mathbb{N}$ is reversed. More precisely, in **[4]** one has

$$\|f(i+k,\alpha,\beta) - f(j+k,\alpha,\beta)\|_1 \leq \|f(i,\alpha,\beta) - f(j,\alpha,\beta)\|_1 + \eta \qquad \text{(i')}$$

$$\|f(i,\alpha,\beta) \wedge f(j,\alpha,\beta)\|_1 \leq \|f(i+k,\alpha,\beta) \wedge f(j+k,\alpha,\beta)\|_1 + \eta \qquad \text{(ii')}$$

instead of (i) and (ii). Thus the role of $i+k$ and $j+k$ on one side and of i and j on the other side is interchanged. This reversal is necessary, because in **[4]** we have $Tf_n = f_{n+1}$ for $f_n = T^n f$, while for exit laws we have just the opposite, namely $Tf_{n+1} = f_n$. While Lemma 2.10 fits well to Proposition 2.5 yielding norm convergence, (i′) and (ii′) fit well to **[4]**, Lemma 4.3 yielding only weak convergence.

Using Lemma 2.10 instead of **[4]**, Lemma 4.4 the proof of **[4]**, Lemma 4.5 implies

Lemma 2.11. *Let (f_n) be a sequence of nonnegative measurable functions on (E,μ) bounded by $\gamma > 0$. Assume that $\mu\{f_n > 0\}$ tends to a finite limit a, that $\mu\{f_n = \gamma\}$ tends to b and that $\|f_n\|_2$ converges. Moreover, assume that, for any $\eta > 0$, there exists a number $K \in \mathbb{N}$ such that $i, j \geq K,\ k \geq 0$ imply*

$$\|f_{i+k} \wedge f_{j+k}\|_1 \leq \|f_i \wedge f_j\|_1 + \eta.$$

Then, for any $\xi > 0$, there exists a number $K' \in \mathbb{N}$ such that $i, j \geq K',\ k \geq 0$ imply

$$\|f_i - f_j\|_2^2 \leq \|f_{i+k} - f_{j+k}\|_2^2 + \xi + 8(a-b)\gamma^2.$$

We recall from Crandall and Tartar **[1]**:

Proposition 2.12. *Let T be a mapping on L^1_+ such that*

$$\int (Tf - t)_+ \, d\mu \leq \int (f-t)_+ \, d\mu \quad (f \in L^1_+,\ t \in]0,\infty[). \qquad \text{(i)}$$

Then for any increasing convex function $\Phi : \mathbb{R}_+ \to \mathbb{R}_+$ with $\Phi(0) = 0$ we have

$$\int \Phi(Tf)\, d\mu \leq \int \Phi(f)\, d\mu. \qquad \text{(ii)}$$

Lemma 2.13. *Let $1 \leq p < \infty$, (f_n) and T as in Theorem 2.8. Then for any $\varepsilon > 0$ there exists $s > 0$ such that*

$$\sup_n \|f_n \wedge s\|_p^p \leq \varepsilon.$$

Proof. By the argument preceding Proposition 2.9 property (i) of Proposition 2.12 is satisfied. Applying this proposition with $f = f_{n+1}$ and $\Phi(t) = \Phi_{p,s}(t) = (t^p - s^p)_+$

we obtain

$$\int (f_n^p - (f_n \wedge s)^p)\, d\mu = \int \Phi_{p,s}(Tf_{n+1})\, d\mu$$
$$\leq \int \Phi_{p,s}(f_{n+1})\, d\mu = \int (f_{n+1}^p - (f_{n+1} \wedge s)^p)\, d\mu. \quad (1)$$

Since $\|f_n\|_p$ is increasing there exist $K \in \mathbb{N}$ and $s > 0$ such that

$$\sup_n \|f_n\|_p^p \leq \|f_K\|_p^p + \frac{\varepsilon}{2} \quad (2)$$
$$\|f_m \wedge s\|_p^p \leq \frac{\varepsilon}{2} \quad (1 \leq m \leq K). \quad (3)$$

Combining (1), (2) and (3) we obtain for $m \geq K$

$$\|f_m \wedge s\|_p^p = \|f_m\|_p^p - \int (f_m^p - (f_m \wedge s)^p)\, d\mu$$
$$\leq \|f_K\|_p^p + \frac{\varepsilon}{2} - \int (f_K^p - (f_K \wedge s)^p)\, d\mu = \|f_K \wedge s\|_p^p + \frac{\varepsilon}{2} \leq \varepsilon.$$

For $m \leq K$ we use (3) and the proof is complete. □

Proof of Theorem 2.8. This part of the proof is substantially different from the corresponding part of **[4]**, where only the uniqueness of weak limits was to be proved. Let $\varepsilon > 0$ be given. By the assumptions of Theorem 2.8 and by Lemma 2.13 there exists $C > 1$ such that

$$\sup_n \|(f_n - C)_+\|_p \leq \varepsilon, \quad \sup_n \|f_n \wedge C^{-1}\|_p \leq \varepsilon$$

and therefore

$$\sup_n \Big\|\frac{1}{n}\sum_{i=1}^n (f_i - C)_+\Big\|_p \leq \varepsilon, \quad \sup_n \Big\|\frac{1}{n}\sum_{i=1}^n (f_i \wedge C^{-1})\Big\|_p \leq \varepsilon. \quad (1)$$

Let F and $D(F)$ be as in Proposition 2.9. By making C larger, if necessary, we may assume that $C, C^{-1} \in]0,\infty[\setminus D(F)$. We choose $0 = \alpha_0 < C^{-1} = \alpha_1 < \cdots < \alpha_{2M} < \alpha_{2M+1} = C$ in $]0,\infty[\setminus D(F)$ such that

$$\sum_{i=1}^M ((\alpha_{2i+1} - \alpha_{2i})^{p-1}\beta)^{1/p} < \varepsilon \quad (2)$$
$$F(\alpha_{2i-1}) - F(\alpha_{2i}) < \delta^2 \quad (1 \leq i \leq M) \quad (3)$$
$$\alpha_{i+1} - \alpha_i \leq 1 \quad (0 \leq i \leq 2M), \quad (4)$$

where $\beta = \sup_n \|(f_n - C^{-1})_+\|_1 \leq C^{p-1} \sup_n \|f_n\|_p^p$ and $\delta > 0$ is chosen such that $\beta^{2-p}(6\delta C)^{2p-2} = \varepsilon^p$ if $1 < p < 2$, resp. $36\delta^2 C^p = \varepsilon^p$ if $2 \leq p < \infty$, resp. $(6\delta C)\mu(E)^{1/2} = \varepsilon$ if $p = 1$ and $\mu(E) < \infty$. Thus we have placed the big jumps of F

between α_{2i} and α_{2i+1}. Putting $x_{ij} = f(j, \alpha_i, \alpha_{i+1})$ (defined in Lemma 2.10) we have

$$\|x_{2i,j}\|_p^p \leq (\alpha_{2i+1} - \alpha_{2i})^{p-1} \|x_{2i,j}\|_1 \leq (\alpha_{2i+1} - \alpha_{2i})^{p-1} \beta$$

and therefore, by (2),

$$\sup_n \Big\| \frac{1}{n} \sum_{j=1}^{n} \sum_{i=1}^{M} x_{2i,j} \Big\|_p \leq \varepsilon. \tag{5}$$

By Proposition 2.9 $\lim_{j\to\infty} \mu\{x_{ij} > t\}$ exists for any $t \notin \{s - \alpha_i : s \in D(F)\}$ and is finite for any $t \geq 0$ if $i \geq 1$. Hence $\lim_{j\to\infty} \|x_{ij}\|_2$ exists and is finite for any $1 \leq i \leq 2M$. Since $\alpha_i \notin D(F)$ there exists $a_i \in \mathbb{R}_+$ such that

$$\begin{aligned} a_i = \lim_{j\to\infty} \mu\{f_j > \alpha_i\} &= \lim_{j\to\infty} \mu\{x_{ij} > \alpha_i\} \\ &= \lim_{j\to\infty} \mu\{f_j \geq \alpha_i\} = \lim_{j\to\infty} \mu\{x_{i-1,j} = \alpha_i - \alpha_{i-1}\}. \end{aligned}$$

Putting $\gamma = \alpha_{i+1} - \alpha_i$ this can be combined with Lemmas 2.10, 2.11 and we obtain $K \in \mathbb{N}$ such that $m, n \geq K$, $1 \leq i \leq 2M$, $k \geq 0$ implies

$$\|x_{im} - x_{in}\|_2^2 \leq \|x_{i,m+k} - x_{i,n+k}\|_2^2 + \delta^2(\alpha_{i+1} - \alpha_i)^2 + 8(a_i - a_{i+1})(\alpha_{i+1} - \alpha_i)^2.$$

By (3) this implies

$$\|x_{2i-1,m} - x_{2i-1,n}\|_2^2 \leq \|x_{2i-1,m+k} - x_{2i-1,n+k}\|_2^2 + 9\delta^2(\alpha_{2i} - \alpha_{2i-1})^2 \qquad (m, n \geq K,\ 1 \leq i \leq M).$$

Applying now Proposition 2.5 we obtain

$$\lim_{k\to\infty} \sup_{m,n\geq k} \Big\| \frac{1}{m} \sum_{j=1}^{m} x_{2i-1,j} - \frac{1}{n} \sum_{j=1}^{n} x_{2i-1,j} \Big\|_2 \leq 6\delta(\alpha_{2i} - \alpha_{2i-1}) \quad (1 \leq i \leq M)$$

and therefore, putting

$$g_j = \sum_{i=1}^{M} x_{2i-1,j}, \quad h_j = \sum_{i=1}^{M} x_{2i,j}$$

we get

$$\lim_{k\to\infty} \sup_{m,n\geq k} \Big\| \frac{1}{m} \sum_{j=1}^{m} g_j - \frac{1}{n} \sum_{j=1}^{n} g_j \Big\|_2 \leq 6\delta C. \tag{6}$$

Since $\frac{1}{m} \sum_{j=1}^{m} g_j \leq C$ this implies

$$\lim_{k\to\infty} \sup_{m,n\geq k} \Big\| \frac{1}{m} \sum_{j=1}^{m} g_j - \frac{1}{n} \sum_{j=1}^{n} g_j \Big\|_p^p \leq 36\delta^2 C^p = \varepsilon^p,$$

if $p \geq 2$. For $1 < p < 2$ we use that $\sup_n \|g_n\|_1 \leq \beta$, and $\|g\|_p^p \leq \|g\|_2^{2p-2} \|g\|_1^{2-p}$ (a consequence of Hölder's inequality) to obtain

$$\Big\|\frac{1}{m}\sum_{j=1}^{m} g_j\Big\|_p^p \leq \beta^{2-p}\Big\|\frac{1}{m}\sum_{j=1}^{m} g_j\Big\|_2^{2p-2}.$$

Combining this with (6) we get

$$\lim_{k\to\infty}\sup_{m,n\geq k}\Big\|\frac{1}{m}\sum_{j=1}^{m} g_j - \frac{1}{n}\sum_{j=1}^{n} g_j\Big\|_p^p \leq \beta^{2-p}(6\delta C)^{2p-2} = \varepsilon^p.$$

If $p = 1$ and μ finite, then $\|g\|_1 \leq \|g\|_2 \mu(E)^{1/2}$ together with (6) implies

$$\lim_{k\to\infty}\sup_{m,n\geq k}\Big\|\frac{1}{m}\sum_{j=1}^{m} g_j - \frac{1}{n}\sum_{j=1}^{n} g_j\Big\|_1 \leq (6\delta C)\mu(E)^{1/2} = \varepsilon.$$

Thus in all three cases we have shown

$$\lim_{k\to\infty}\sup_{m,n\geq k}\Big\|\frac{1}{m}\sum_{j=1}^{m} g_j - \frac{1}{n}\sum_{j=1}^{n} g_j\Big\|_p \leq \varepsilon. \tag{7}$$

From (5) we get

$$\lim_{k\to\infty}\sup_{m,n\geq k}\Big\|\frac{1}{m}\sum_{j=1}^{m} h_j - \frac{1}{n}\sum_{j=1}^{n} h_j\Big\|_p \leq 2\varepsilon. \tag{8}$$

Since $f_n = g_n + h_n + (f_n \wedge C^{-1}) + (f_n - C)_+$ (7), (8) and (1) we obtain

$$\lim_{k\to\infty}\sup_{m,n\geq k}\Big\|\frac{1}{m}\sum_{j=1}^{m} f_j - \frac{1}{n}\sum_{j=1}^{n} f_j\Big|_p \leq 7\varepsilon.$$

Since $\varepsilon > 0$ can be made arbitrarily small the assertion follows. □

Remarks. It is an open question, whether the assumption

$$\lim_{t\to\infty}\sup_n \|(f_n - t)_+\|_p = 0 \tag{$*$}$$

can be disposed off in the nonlinear case. In the case $p = 1$ and μ finite the answer is negative. We even have a positive linear L^1- L^∞ contraction on a discrete finite measure space and an exit law with divergent Cesàro averages. However, if $1 < p < \infty$, then we have yet no example of an exit law for which $(*)$ fails.

If there exists $p' > p$ such that $\sup_n \|f_n\|_{p'} < \infty$ then $(*)$ is obvious.

If we do not assume $(*)$, then the proof of Theorem 2.8 yields that

$$\lim_{k\to\infty}\sup_{m,n\geq k}\Big\|\frac{1}{m}\sum_{j=1}^{m}(f_j \wedge C) - \frac{1}{n}\sum_{j=1}^{n}(f_j \wedge C)\Big\|_p = 0$$

for any $C > 0$. In particular, we have still convergence in measure, if we do not assume $(*)$.

3. Almost everywhere convergence

In this section (E, μ) will be a σ-finite measure space and all L^p spaces are with respect to this measure space.

Theorem 3.1. *Let* $1 < p < \infty$ *and* T *be a positive linear contraction on* L^p. *For any* $(f_n)_{n\in\mathbb{N}} \in E_{L^p,T}$ *we put*

$$M((f_n)_{n\in\mathbb{N}}) = \sup_n \frac{1}{n}\sum_{i=1}^{n} |f_i|.$$

Then we have

$$\|M((f_n)_{n\in\mathbb{N}})\|_p \leq \frac{p}{p-1} \sup_{n\in\mathbb{N}} \|f_n\|_p.$$

Proof. We fix $N, L \in \mathbb{N}$ with $N < L$ and define $(g_n)_{n\in\mathbb{Z}}$ by

$$g_n = \begin{cases} |f_n| & \text{if } 1 \leq n \leq L \\ 0 & \text{otherwise.} \end{cases}$$

Further we put for any $m \in \mathbb{Z}$

$$M_m = M_m(L, N) = \sup_{1\leq n\leq N} \frac{1}{n}\sum_{i=m+1}^{m+n} g_i.$$

For $1 \leq m \leq L - N$ we have

$$\begin{aligned} M_0 = \sup_{1\leq n\leq N} \frac{1}{n}\sum_{i=1}^{n} |f_i| &\leq \sup_{1\leq n\leq N} \frac{1}{n}\sum_{i=1}^{n} |T^m f_{i+m}| \\ &\leq T^m\Big(\sup_{1\leq n\leq N} \frac{1}{n}\sum_{i=1}^{n} |f_{i+m}|\Big) = T^m(M_m), \end{aligned}$$

because T is positive and linear. Since T is also a contraction this implies

$$\|M_0\|_p \leq \|T^m M_m\|_p \leq \|M_m\|_p \quad (1 \leq m \leq L - N). \tag{1}$$

For each $x \in E$ we apply the Hardy–Littlewood maximal inequality (i.e. the ergodic maximal inequality on Z with the right shift) to the sequence $(g_n(x))_{n\in\mathbb{Z}}$ to obtain

$$\sum_{m=-\infty}^{+\infty} M_m^p(x) \leq \Big(\frac{p}{p-1}\Big)^p \sum_{i=-\infty}^{\infty} |g_i(x)|^p.$$

Integrating over $x \in E$ and using (1) we obtain

$$(L-N+1)\,\|M_0\|_p^p \le \sum_{m=0}^{L-N} \|M_m\|_p^p \le \sum_{m=-\infty}^{\infty} \|M_m\|_p^p$$
$$\le \left(\frac{p}{p-1}\right)^p \sum_{i=-\infty}^{\infty} \|g_i\|_p^p = \left(\frac{p}{p-1}\right)^p \sum_{i=1}^{L} \|f_i\|_p^p$$
$$\le L\left(\frac{p}{p-1}\right)^p \sup_i \|f_i\|_p^p.$$

Thus we have shown

$$\|M_0\|_p \le \left(\frac{L}{L-N+1}\right)^{1/p} \frac{p}{p-1} \sup_i \|f_i\|_p.$$

and therefore

$$\Big\| \sup_{1\le n\le N} \frac{1}{n}\sum_{i=1}^{n} f_i \Big\|_p = \lim_{L\to\infty} \|M_0(L,N)\|_P \le \frac{p}{p-1} \sup_i \|f_i\|_p.$$

As N tends to ∞ the left hand side of this inequality tends to $M((f_n)_{n\in\mathbb{N}})$ and the assertion follows. □

Theorem 3.2. *Let $1 < p < \infty$ and T be a positive linear contraction on L^p. Then for any exit law $(f_n)_{n\in\mathbb{N}}$ with respect to (L^p, T) the Cesàro averages $\frac{1}{n}\sum_{i=1}^{n} f_i$ are almost everywhere convergent.*

Proof. Thanks to the Kakutani–Yosida decomposition (cf. **[3]**, p. 73) for averages of iterates the pointwise convergence theorem is an immediate consequence of the corresponding maximal inequality. Since no such decomposition is available for exit laws, the proof of the pointwise convergence theorem requires more work.

By the improved version of the mean ergodic theorem for exit laws, Remark 2.4, there exists $f \in L^p$ with $Tf = f$ such that

$$\lim_{n\to\infty} \sup_m \|f - A_{mn}\|_p = 0,$$

where $A_{mn} = \frac{1}{n}\sum_{i=m+1}^{m+n} f_i$. We fix $0 < \varepsilon, \delta \le 1$ for a while and choose $n_\varepsilon \in \mathbb{N}$ such that

$$\|f - A_{mn_\varepsilon}\|_p \le \varepsilon \quad (m \ge 0). \tag{1}$$

Then $f'_n = f - A_{n-1,n_\varepsilon}$ is again an exit law and, by (1),

$$\sup_n \|f'_n\| \le \varepsilon. \tag{2}$$

Putting $A'_{mn} = \frac{1}{n}\sum_{i=m+1}^{m+n} f'_i$ we have for $n \ge n_\varepsilon$

$$A'_{mn} = f - \frac{1}{n}\sum_{i=m+1}^{m+n} \frac{1}{n_\varepsilon}\sum_{j=i}^{i+n_\varepsilon-1} f_j = f - \frac{1}{n}\sum_{j=m+1}^{m+n+n_\varepsilon-1} \frac{\min(n_\varepsilon, j-m, m+n+n_\varepsilon-j)}{n_\varepsilon} f_j$$

and therefore

$$\|A'_{mn} - (f - A_{mn})\|_p \le \frac{2n_\varepsilon}{n} \sup_k \|f_k\|_p \quad (n \ge n_\varepsilon). \tag{3}$$

Using Theorem 3.1 we obtain from (2) that

$$\|M\|_p \le \frac{p}{p-1}\varepsilon, \quad M = \sup_n |A'_{0,n}|. \tag{4}$$

Since

$$\Big\{\sup_{n\ge k} |f - A_{0,n}| \ge \delta\Big\} \subset \{M \ge \delta/2\} \cup \bigcup_{n=k}^{\infty} \Big\{|A'_{0n} - (f - A_{0n})| \ge \delta/2\Big\}$$

(3) and (4) imply

$$\begin{aligned}\mu\Big(\Big\{\sup_{n\ge k} |f - A_{0,n}| \ge \delta\Big\}\Big) &\le \mu\left(\{M \ge \delta/2\}\right) \\ &\quad + \sum_{n=k}^{\infty} \mu\left(\{|A'_{0n} - (f - A_{0n})| \ge \delta/2\}\right) \\ &\le (\delta/2)^{-p}\|M\|_p^p + \sum_{n=k}^{\infty} (\delta/2)^{-p}\|A'_{0n} - (f - A_{0n})\|_p^p \\ &\le \Big(\frac{p}{p-1}\frac{2\varepsilon}{\delta}\Big)^p + \sum_{n=k}^{\infty}\Big(\frac{4n_\varepsilon}{\delta n}\Big)^p \sup_m \|f_m\|_p^p.\end{aligned}$$

Letting k tend to ∞ this implies

$$\mu\Big(\Big\{\limsup_{n\to\infty} |f - A_{0,n}| \ge \delta\Big\}\Big) \le \Big(\frac{p}{p-1}\frac{2\varepsilon}{\delta}\Big)^p.$$

Since $\varepsilon > 0$ can be chosen arbitrarily small the assertion follows. □

Remark. We do not know whether almost everywhere convergence of Cesàro averages holds for exit laws in L^1 if T is a positive L^1 - L^∞ contraction. More generally, one can easily formulate an analogue of the Chacon–Ornstein theorem for exit laws. The main problem is that the proof of Hopf's maximal ergodic lemma cannot be used for exit laws.

Acknowledgements. I am indebted to Michael Lin for many valuable comments and corrections of the original manuscript.

References

[1] M. G. Crandall, L. Tartar, Some relations between nonexpansive and order preserving mappings, Proc. Amer. Math. Soc. 78 (1980), 385–390.

[2] Y. Derriennic, Lois ≪zero ou deux≫ pour les processus de Markov. Applications aux marches aléatoires, Ann. Inst. H. Poincaré (B) 12 (1976), 111–129.

[3] U. Krengel, Ergodic Theorems, de Gruyter Stud. Math. 6, de Gruyter, Berlin-New York 1985.

[4] U. Krengel, M. Lin, Order preserving nonexpansive operators in L^1, Israel J. Math. 58 (1987), 170–192.

[5] U. Krengel, M. Lin, An integral representation of disjointly additive order preserving operators in L^1, Stoch. Anal. Appl. 6 (1988), 289–304.

[6] U. Krengel, M. Lin, R. Wittmann, A limit theorem for order preserving nonexpansive operators in L^1, Israel J. Math. 71 (1990), 181–191.

[7] M. Lin, Mixing for Markov operators, Z. Wahrscheinlichkeitstheorie Verw. Geb. 19 (1971), 231–242.

[8] G. G. Lorentz, A contribution to the theory of divergent sequences, Acta Math. 80 (1948), 167–190.

[9] J. Neveu, Bases Mathématiques du calcul de probabilités, Masson, Paris 1964.

[10] S. Reich, Almost convergence and nonlinear ergodic theorems, J. Approx. Th. 24 (1977), 269–272.

[11] R. Wittmann, Mean ergodic theorems for nonlinear operators, Proc. Amer. Math. Soc. 108 (1990), 781–788.

Institut für Mathematische Stochastik der
Universität Göttingen
Lotzestr. 13
D-37083 Göttingen
Germany

On Skew Products of Irrational Rotations with Tori

Qing Zhang

Abstract. Let $f: [0, 1] \to \mathbb{R}^m$ be a piecewise absolutely continuous function. If the vector $\int f'(t)dt$ has rationally independent components, then for any irrational number θ, the map $(t, x) \to (t + \theta, x + f(t))$ (mod 1) is totally ergodic on $\mathbb{T} \times \mathbb{T}^m$.

1. Introduction

Let θ be an irrational number with $0 < \theta < 1$. For any measurable function $f: \mathbb{T} \to \mathbb{R}^m$, define T_f on $\mathbb{T} \times \mathbb{T}^m$ by

$$T_f(t, x) = (t + \theta, x + f(t)) \qquad \text{(mod 1)}$$

for $t \in \mathbb{T}$ and $x \in \mathbb{T}^m$. Then T_f preserves the Lebesgue measure on $x \in \mathbb{T}^m$. (Throughout this note, we will always assume the Lebesgue measure on a torus unless otherwise mentioned.) We shall call T_f a skew product of an irrational rotation with a m-dimensional torus. Ergodicity of T_f is the main question of this note.

Ergodicity of cylinder flows, which are skew products of rational rotations with the line, has been studied by D. Pask **[7]**, I. Oren **[6]**, and P. Hellekalek and G. Larcher **[2]**. Their results directly imply the ergodicity of skew products of irrational rotations with an one-dimensional torus. Recently, A. Iwanik, M. Lemanczyk and D. Rudolph **[5]** proved a result which implies our result in one-dimensional situition. We shall focus on skew products of irrational rotations with multi-dimensional tori and use properties of minimal cocycles to show that if f is piecewise continuous and the vector $\int \frac{d}{dt} f(t)dt$ has rationally independent components, then T_f is totally ergodic. Because of the compactness of a torus in our situation, the common assumption $\int f(t)dt = 0$ for studying cylinder flows is not needed.

In §2, we first state some definitions and properties of cocycles. Then a known theorem relating minimal cocycles to ergodicity will be stated. Several lemmas and propositions will be proved in §3. In §4, we shall prove the main theorem, Theorem 4.6, then several examples will be discussed.

The author would like to thank Professor Tom Ward not only for his valuable suggestions but also for his help in verifying an important fact. Special thanks go to Professor Vitaly Bergelson for his continued help, advice and encouragement.

2. Preliminaries

Let μ be Lebesgue measure on $\mathbb{T} = \mathbb{R}/\mathbb{Z}$ and let $\mathcal{B}_{\mathbb{T}}$ be the σ-algebra generated by all open sets. For any irrational number $\theta \in \mathbb{T}$, define a measure preserving transformation T on $\mathbb{T}$ by $Tt = t + \theta (\text{mod } 1)$. The map T is called an irrational rotation.

Let $\boldsymbol{A}$ be a compact abelian group, $\mu_{\boldsymbol{A}}$ be Haar measure on $\boldsymbol{A}$ and $\mathcal{B}_{\boldsymbol{A}}$ be the σ-algebra generated by all open subsets of $\boldsymbol{A}$. For any measurable function $\alpha: \mathbb{T} \to \boldsymbol{A}$, define a cocycle by

$$\alpha_n(t) = \sum_{k=0}^{n-1} \alpha(T^k t).$$

For brevity, we will sometimes call α a cocycle. For any two cocycles α and β, we say that α is *cohomologous* to β if there exists a measurable function $u: \mathbb{T} \to \boldsymbol{A}$ such that

$$\beta_n(t) = \alpha_n(t) + u(T^n t) - u(t).$$

For any cocycle $\alpha(t)$, let $\boldsymbol{A}_\alpha$ denote the closed subgroup of $\boldsymbol{A}$ generated by $\{\alpha_n(t) : t \in \mathbb{T} \text{ and } n \in \mathbb{Z}\}$. A cocycle $\alpha_n(t)$ is called *minimal* if there is no cocycle $\beta_n(t)$ cohomologous to α such that $\boldsymbol{A}_\beta \subset \boldsymbol{A}_\alpha$ but $\boldsymbol{A}_\beta \neq \boldsymbol{A}_\alpha$.

A cocycle $\alpha_n(t)$ defines a measure preserving transformation T_α on $\mathbb{T} \times \boldsymbol{A}$ by

$$T_\alpha(t, a) = (Tt, \alpha(t) + a).$$

One can show that if $\alpha_n(t)$ is cohomologous to $\beta_n(t)$, then T_α and T_β are isomorphic measure preserving transformations.

A proof of the following theorem can be found in **[8**, page 391**]**.

Theorem 2.1. *With the above definitions,*

any cocycle is cohomologous to a minimal cocycle;

a cocycle $\alpha_n(t)$ is minimal and $\boldsymbol{A}_\alpha = \boldsymbol{A}$ if and only if T_α is ergodic on $\mathbb{T} \times \boldsymbol{A}$;

if $\alpha_n(t)$ and $\beta_n(t)$ are cohomologous, then $\boldsymbol{A}_\alpha = \boldsymbol{A}_\beta$.

For any measurable function $\boldsymbol{f}: \mathbb{T} \to \mathbb{R}^m$, define a $\mathbb{T}^m$-valued cocycle on $\mathbb{T}$ by

$$\alpha_n^f(t) = \boldsymbol{f}_n(t) \ (\text{mod } 1) = \sum_{i=0}^{n-1} \boldsymbol{f}(t + i\theta) \ (\text{mod } 1).$$

Then, for any $t \in \mathbb{T}$ and any $\boldsymbol{x} \in \mathbb{T}^m$

$$T_f(t, \boldsymbol{x}) = (Tt, \boldsymbol{x} + \alpha_1^f(t)) = (t + \theta, \boldsymbol{x} + \boldsymbol{f}(t)) \ (\text{mod } 1)$$

is measure preserving on $\mathbb{T} \times \mathbb{T}^m$. By Theorem 2.1, T_f is ergodic if and only if $\alpha_n^f(t)$ is minimal and $\boldsymbol{A}_{\alpha^f} = \boldsymbol{A}$.

3. Lemmas

In $\mathbb{R}^m$, denote by $|\cdot|$ and $\mathrm{Dist}(\cdot\,,\,\cdot)$ the ordinary Euclidean norm and distance respectively.

Lemma 3.1. *Let $\mathcal{L}$ be an affine subspace in $\mathbb{R}^m$ and $\boldsymbol{k}$ be a vector such that $\boldsymbol{k}+\boldsymbol{v} \notin \mathcal{L}$ for some $\boldsymbol{v} \in \mathcal{L}$. Let $D = \mathrm{Dist}(\boldsymbol{k}+\boldsymbol{v}, \mathcal{L})$ for some $\boldsymbol{v} \in \mathcal{L}$. If $\boldsymbol{f}(t)\colon I \to \mathbb{R}^m$ is absolute continuous such that for some $\lambda > 1$ and $\varepsilon > 0$*

$$\int_I \left|\frac{d\boldsymbol{f}(t)}{dt} - \lambda\boldsymbol{k}\right| dt < \lambda\varepsilon|I|,$$

then

$$\mu\{t \in I : \mathrm{Dist}(\boldsymbol{f}(t), \mathcal{L}) < \varepsilon\} < \frac{(1+\lambda|I|)\varepsilon}{\lambda D}.$$

Proof. It is clear that $D = \mathrm{Dist}(\boldsymbol{k}+\boldsymbol{v}, \mathcal{L})$ does not depend on the choice of $\boldsymbol{v} \in \mathcal{L}$. Without loss of generality, we can assume that $\mathcal{L}$ is a linear subspace. Then $\mathrm{Dist}(\boldsymbol{k}+\boldsymbol{v}, \mathcal{L}) = \mathrm{Dist}(\boldsymbol{k}, \mathcal{L}) = D$. Let $I = [a, b]$. Since

$$\int_I |\frac{d}{dt}\boldsymbol{f}(t) - \lambda\boldsymbol{k}| dt < \lambda\varepsilon|I|,$$

we have that

$$|\boldsymbol{f}(t) - (\lambda\boldsymbol{k}(t-a) + \boldsymbol{f}(a))| < \lambda\varepsilon|I|.$$

Notice that for any $\boldsymbol{v} \in \mathcal{L}$

$$\begin{aligned} |(\lambda\boldsymbol{k}(t-a) + \boldsymbol{f}(a)) - \boldsymbol{v}| &\leq |(\lambda\boldsymbol{k}(t-a) + \boldsymbol{f}(a)) - \boldsymbol{f}(t)| + |\boldsymbol{f}(t) - \boldsymbol{v}| \\ &\leq |\boldsymbol{f}(t) - \boldsymbol{v}| + |I|\lambda\varepsilon. \end{aligned}$$

Therefore

$$\{t \in I : \mathrm{Dist}(\boldsymbol{f}(t), \mathcal{L}) < \varepsilon\} \subset \{t \in \mathbb{R} : \mathrm{Dist}(\lambda\boldsymbol{k}(t-a) + \boldsymbol{f}(a), \mathcal{L}) < (1+|I|\lambda)\varepsilon\}$$

and we only need to show that

$$\mu\{t \in \mathbb{R},\ \mathrm{Dist}(\lambda\boldsymbol{k}(t-a) + \boldsymbol{f}(a), \mathcal{L}) < (1+|I|\lambda)\varepsilon\} < \frac{(1+\lambda|I|)\varepsilon}{\lambda D}.$$

It is clear that

$$\begin{aligned} &\mu\{t \in \mathbb{R},\ \mathrm{Dist}(\lambda\boldsymbol{k}(t-a) + \boldsymbol{f}(a), \mathcal{L}) < (1+|I|\lambda)\varepsilon\} \\ &\quad = \mu\{t \in \mathbb{R},\ \mathrm{Dist}(\lambda\boldsymbol{k}t, \mathcal{L}) < (1+|I|\lambda)\varepsilon\}. \end{aligned}$$

Thus the lemma will follow if

$$\mu\{t \in \mathbb{R},\ \mathrm{Dist}(\lambda\boldsymbol{k}t, \mathcal{L}) < (1+|I|\lambda)\varepsilon\} < \frac{(1+\lambda|I|)\varepsilon}{\lambda D}.$$

and this inequality follows from $\mathrm{Dist}(\lambda\boldsymbol{k}t, \mathcal{L}) = \lambda|t|\,\mathrm{Dist}(\boldsymbol{k}, \mathcal{L})$. $\square$

For any closed subgroup $A \subset \mathbb{R}^m$, the connected component of A containing the origin, A_0, is a linear subspace of $\mathbb{R}^n$. There exists a discrete subgroup Γ such that $\Gamma \times A_0 = A$. It is clear that the distance between two different connected components is always bigger than a fixed positive number δ (which only depends on A). Now we can formulate the following lemma.

Lemma 3.2. *Let $k \notin A_0$ and $D = \mathrm{Dist}(k, A_0)$. If $f(t): I \to \mathbb{R}^m$ is absolute continuous such that, for some $\lambda > 1$ and $\varepsilon > 0$,*

$$\int_I \left|\frac{df(t)}{dt} - \lambda k\right| dt < \lambda\varepsilon|I|,$$

then

$$\mu\{t \in I : \mathrm{Dist}(f(t), A) > \varepsilon\} > |I| - \left(1 + \left[\frac{|f(I)|}{\delta}\right]\right)\frac{1+\lambda|I|)}{\lambda D}\varepsilon.$$

Here δ is the smallest distance between connected components of A and $|f(I)| = \int_I |df(t)/dt|\, dt$.

Proof. Since the distance between any two connected components is δ, it is clear that $f(I)$ intersects at most $1 + \left[\frac{|f(I)|}{\delta}\right]$ many components of A. By Lemma 3.1, we have the lemma immediately. □

A proof and more details of the following form of Kronecker's theorem can be found in **[1**, page 154].

Theorem 3.3 (Kronecker). *Let $\mathbb{T}^m$ be a tourus and let $\theta = (\theta_1, \theta_2, \ldots, \theta_m) \in \mathbb{T}^m$, have the property that $1, \theta_1, \theta_2, \ldots, \theta_m$ are rationally independent. Then, for any $t \in \mathbb{T}^m$, there is a sequence of integers $\{n_k\}$ such that*

$$n_k\theta = (n_k\theta_1, n_k\theta_2, \ldots, n_k\theta_m) \to t.$$

For completeness,, we include a proof of the following known result.

Proposition 3.4. *Let $P: \mathbb{R}^m \to \mathbb{R}^m/\mathbb{Z}^m = \mathbb{T}^m$ be the natural projection and A be a nontrivial closed subgroup of $\mathbb{T}^m$. Let $\tilde{A}_0$ be the connected component of $P^{-1}(A)$ containing the identity 0. Then $\tilde{A}_0$ is a subspace spanned by vectors in $\mathbb{Z}^m$*

Proof. Let $\tilde{A}_0$ be a k-dimensional linear space with $0 < k < m$. There is a $(m-k) \times k$ matrix $A = \left(a_{ij}\right)_{(m-k)\times k}$ such that

$$\tilde{A}_0 = \left\{ (x_1, x_2, \cdots, x_m) \in \mathbb{R}^m : \begin{pmatrix} x_{k+1} \\ \vdots \\ x_m \end{pmatrix} = A \begin{pmatrix} x_1 \\ \vdots \\ x_k \end{pmatrix} \right\}.$$

One can find a matrix $B = \left(b_{ij}\right)_{(m-k)\times(m-k)}$ with rational entries and determinant 1 such that if

$$BA = \begin{pmatrix} c_{11} & \cdots & c_{1k} \\ \vdots & & \vdots \\ c_{i_0 1} & \cdots & c_{i_0 k} \\ \vdots & & \vdots \\ c_{(m-k)1} & \cdots & c_{(m-k)k} \end{pmatrix}$$

then $1, c_{11}, \ldots, c_{i_0 1}$ are rationally independent and $c_{(i_0+1)1}, \ldots, c_{(m-k)1}$ are rational numbers.

We claim that i_0 must be zero. Suppose that this is not the case. Let d be the product of all denominators of fractions $c_{(i_0+1)1}, \ldots, c_{(m-k)1}$ and all denominators of entries in the matrix B^{-1} (note that all entries in B^{-1} are rational). For any $0 \leq \tau < 1$, by Theorem 3.3, there is a sequence $\{t_n\}$ such that, as $t \to \infty$,

$$(dt_n c_{11} - [dt_n c_{11}], dt_n c_{21} - [dt_n c_{21}], \ldots, dt_n c_{i_0 1} - [dt_n c_{i_0 1}]) \to (\tau, 0, \ldots, 0).$$

It is clear that

$$(dt_n, \underbrace{0, \ldots, 0}_{k-1}, dt_n a_{11}, \ldots, dt_n a_{(m-k)1}) \in \tilde{A}_0.$$

Since

$$\begin{aligned}(dt_n a_{11}, \ldots, dt_n a_{(m-k)1}) &= (dt_n c_{11}, \ldots, dt_n c_{(m-k)1})(B^{-1})^t \\ &= (dt_n c_{11}, \ldots, dt_n c_{i_0 1}, 0, \ldots, 0)(B^{-1})^t \\ &\quad + (0, \ldots, 0, dt_n c_{(i_0+1)1}, \ldots, dt_n c_{(m-k)1})(B^{-1})^t\end{aligned}$$

and $(0, \ldots, 0, dt_n c_{(i_0+1)1}, \ldots, dt_n c_{(m-k)1})(B^{-1})^t \in \mathbb{Z}^{m-k}$, where $(\cdot)^t$ means the matrix transpose, we have

$$\begin{aligned}&P\big((dt_n, \underbrace{0, \ldots, 0}_{k-1}, dt_n a_{11}, \ldots, dt_n a_{(m-k)1})\big) \\ &= P\big((\underbrace{0, \ldots, 0}_{k}, (t_n c_{11}, \ldots, t_n c_{i_0 1}, \underbrace{0, \ldots, 0}_{m-k-i})(B^{-1})^t)\big).\end{aligned}$$

This implies

$$\big(\underbrace{0, \ldots, 0}_{k}, (\tau, \underbrace{0, \ldots, 0}_{m-k-1})(B^{-1})^t\big) \in \tilde{A}.$$

Therefore

$$\big(\underbrace{0, \ldots, 0}_{k}, (1, \underbrace{0, \ldots, 0}_{m-k-1})(B^{-1})^t\big) \in \tilde{A}_0,$$

implying that $\tilde{A}_0$ is $k+1$-dimensional. This contradicts the assumption that $\tilde{A}_0$ is k-dimension. So our claim, $i_0 = 0$, is true.

Now we know that all entries in the first column of BA are rational. This implies that all entries in the first column of A are rational. Using the same method, one can show that all entries of A are rational which gives the proposition. □

4. Proof of the main theorem

For an irrational number θ, let $\{p_n/q_n\}$ be the sequence of partial convergents to θ in the continued fraction expansion of θ, so

$$\frac{p_n}{q_n} = a_0 + \cfrac{1}{a_1 + \ddots \cfrac{1}{a_{n-1} + \frac{1}{a_n}}},$$

where

$$\theta = a_0 + \cfrac{1}{a_1 + \cfrac{1}{a_2 + \ddots}}.$$

Basic facts regarding continued fraction will be assumed throughout this paper. Khinchin's book **[4]** provides more than enough for our needs.

For any $x \in \mathbb{T}$, denote by $\|x\|$ the distance from x to the nearest integer. The following lemma is stated in **[7**, page 66**]** without proof. A proof can be found in **[3**, pages 8–9**]**

Proposition 4.1. *Let θ, $0 < \theta < 1/2$, be an irrational number with partial convergents $\{p_n/q_n\}$ and let $\mathcal{P}_n(\theta)$ be the set of right half open partition intervals of $\mathbb{T}$ defined by the points $\{j\theta : j = 0, 1, \dots, q_n - 1\}$. Then each interval of $\mathcal{P}_n(\theta)$ has length $\|q_{n-1}\theta\|$ or $\|q_n\theta\| + \|q_{n-1}\theta\|$.*

A proof of the following lemma can be found in **[7**, pages 67–68**]**.

Lemma 4.2. *Suppose that $g(t)\colon \mathbb{T} \to \mathbb{R}$ has M points of discontinuity. Then for each interval Q of $\mathcal{P}_n(\theta)$, there is a subinterval $J \subset Q$ on which $g_{q_n}(t)$ is continuous, and*

$$\mu(J) \geq \frac{1}{2M-2}\mu(Q).$$

Recall some definitions. For any irrational number θ and any function $\boldsymbol{f}\colon \mathbb{T} \to \mathbb{R}^m$, one can define a cocycle $\boldsymbol{f}_n$ on $\mathbb{R}^m$ by

$$\boldsymbol{f}_n(t) = \sum_{k=0}^{n-1} \boldsymbol{f}(t + k\theta).$$

Then $\alpha_n^{\boldsymbol{f}}(t) = \boldsymbol{f}_n(t) (\mathrm{mod}\ 1)$ defines a cocycle from $\mathbb{T}$ to $\mathbb{R}^m/\mathbb{Z}^m$. For any $t \in \mathbb{T}$ and $\boldsymbol{x} \in \mathbb{T}^m$,

$$T_{\boldsymbol{f}}(t, \boldsymbol{x}) = (t + \theta, \boldsymbol{x} + \alpha_1^{\boldsymbol{f}}(t))$$

is a measure preserving transformation on $\mathbb{T} \times \mathbb{T}^m$.

Throughout this section, we assume that $f: \mathbb{T} \to \mathbb{R}^m$ is piecewise absolutely continuous, that is, there are finitely many points $\omega_i \in \mathbb{T}$ for $i = 1, \ldots, M$ with $0 \leq \omega_1 < \cdots < \omega_M < 1$ such that f is absolutely continuous on each interval $[\omega_i, \omega_{i+1})$ for $i = 1, \ldots, M$, where $[\omega_M, \omega_{M+1})$ should be interpreted as $[\omega_M, \omega_1)$.

Proposition 4.3. *Let $0 < \theta < 1/2$ be an irrational number with partial convergents $\{p_n/q_n\}$ and let $P: \mathbb{R}^m \to \mathbb{R}^m/\mathbb{Z}^m = \mathbb{T}^m$ be the natural projection.Then for any nondiscrete and nontrivial closed subgroup $\boldsymbol{A}$ of $\mathbb{T}^m$, there exists an $\varepsilon > 0$ and a constant $0 < L \leq 1$ such that when n is sufficient large,*

$$\mu\{t \in \mathbb{T} : \mathrm{Dist}(f_{q_n}(t), P^{-1}(\boldsymbol{A})) > \varepsilon\} > L.$$

Proof. Let $\tilde{A} = P^{-1}(\boldsymbol{A})$, $\tilde{\boldsymbol{A}}_0$ be the connected component of $P^{-1}(\boldsymbol{A})$ containing the identity $\mathbf{0}$ and

$$\boldsymbol{k} = \left(\int_0^1 \frac{d}{dt} f_1 dt, \ \ldots, \int_0^1 \frac{d}{dt} f_m dt\right).$$

It is clear that the components of $\boldsymbol{k}$ are rationally independent. We claim that $\boldsymbol{k} \notin \tilde{\boldsymbol{A}}_0$. If this is not the case, by Proposition 3.4, there are vectors $\boldsymbol{v}_1, \boldsymbol{v}_2, \ldots, \boldsymbol{v}_k \in \tilde{\boldsymbol{A}}_0 \cap \mathbb{Q}^m$ such that

$$\boldsymbol{k} = a_1 \boldsymbol{v}_1 + a_2 \boldsymbol{v}_2 + \cdots a_k \boldsymbol{v}_k.$$

So the components of $\boldsymbol{k}$ are rational linear combinations of $a_1, a_2, \ldots, a_k$. Since $m > k$, the components of $\boldsymbol{k}$ must be rationally dependent. This contradiction gives the claim.

Let $\{p_n/q_n\}$ be the sequence of partial convergents for θ and $\mathcal{P}_n(\theta)$ be the set of right half open partition intervals of $\mathbb{T}$ defined by the points of $\{j\theta : j = 0, 1, \ldots, q_n - 1\}$. By Lemma 4.2, we know that there is a constant r, which depends on the number of discontinuous points of f, such that for each interval Q of $\mathcal{P}_n(\theta)$ there is an interval $J \subset Q$ such that f_{q_n} is continuous on J and $|J| > r|Q|$ for some $r > 0$. From results in continued fractions, we know that

$$\left|\theta - \frac{p_{n-1}}{q_{n-1}}\right| = \frac{\gamma_n}{q_{n-1} q_n}$$

with $0 < \gamma_n \leq 1$ and γ_n / q_n decreasing. Actually, we have $1/2 < \gamma_n \leq 1$, since $\|q_{n-1}\theta\| = |q_{n-1}\theta - p_{n-1}|$ and, by Proposition 4.1, $2\|q_{n-1}\theta\| > |Q|$ which implies that $2q_n \|q_{n-1}\theta\| > 1$. Thus we know that

$$r/q_n > |J| > r/2q_n. \tag{1}$$

Let $\delta > 0$ be the smallest distance between any two connected components of $\tilde{A}$ and let $D = \mathrm{Dist}(\boldsymbol{k}, \tilde{\boldsymbol{A}}_0)$. For any $\varepsilon > 0$, let

$$E_n(\varepsilon) = \left\{t : \left|\frac{d}{dt} f_{q_n}(t) - q_n \boldsymbol{k}\right| \geq q_n \varepsilon\right\}.$$

Since

$$\int_0^1 \left|\frac{1}{q_n}\sum_{i=0}^{q_n-1}\frac{d}{dt}f(t+i\theta)-k\right| dt \to 0,$$

there exists $N > 0$ such that when $n > N$,

$$\int_{E_n(\varepsilon)} \left|\frac{d}{dt}f_{q_n}(t) - q_n k\right| dt < q_n\varepsilon. \tag{2}$$

Let

$$\mathcal{P}_n^0(\theta) = \{Q \in \mathcal{P}_n(\theta) : \int_{E_n(\varepsilon)} \left|\frac{d}{dt}f_{q_n}(t) - q_n k\right| dt < 2\varepsilon\}.$$

Then, by (2), $|\mathcal{P}_n^0(\theta)| > q_n/2$. For any $Q \in \mathcal{P}_n^0(\theta)$, there is an interval $J \subset Q$ such that f_{q_n} is continuous on J and $|J| > r|Q|$ for some $r > 0$ (see Lemma 4.2). Noticing (1), we have that

$$\begin{aligned}\int_J \left|\frac{d}{dt}f_{q_n}(t) - q_n k\right| dt &\le \int_{Q\cap E_n(\varepsilon)} \left|\frac{d}{dt}f_{q_n}(t) - q_n k\right| dt - q_n\varepsilon|J| \\ &\le q_n\frac{2\varepsilon}{q_n} + q_n\varepsilon|J| \le q_n\left(\frac{4}{r}+1\right)\varepsilon|J|.\end{aligned} \tag{3}$$

Let $\varepsilon_0 = (1+4/r)\varepsilon$. It follows from Lemma 3.2 that

$$\mu\{t \in J : \mathrm{Dist}(f_{q_n}(t), P^{-1}(A)) > \varepsilon\} > |J| - \left(1 + [\frac{|f_{q_n}(J)|}{\delta}]\right)\frac{1+q_n|J|}{q_n D}\varepsilon_0.$$

Noticing (3) and assuming that $|k| \ge \varepsilon_0$, we have that

$$|f_{q_n}(J)| = \int_J \left|\frac{d}{dt}f_{q_n}(t)\right| dt \le q_n(|k|+\varepsilon_0)|J| \le 2|k|r.$$

Therefore

$$\mu\{t \in J : \mathrm{Dist}(f_{q_n}(t), P^{-1}(A)) > \varepsilon\} > \frac{1}{q_n}\left(\frac{r}{2} - \left(1 + [\frac{2|k|r}{\delta}]\right)\frac{1+r}{D}\varepsilon_0\right).$$

If, at the beginning, we choose $\varepsilon > 0$ such that $|k| \ge (1+4/r)\varepsilon$ and

$$\left(1 + [\frac{2|k|r}{\delta}]\right)\frac{1+r}{D}\left(1+\frac{4}{r}\right)\varepsilon < \frac{r}{4},$$

Then

$$\mu\{t \in J : \mathrm{Dist}(f_{q_n}(t), P^{-1}(A)) > \varepsilon\} > \frac{r}{4q_n}.$$

This implies that

$$\mu\{t \in \mathbb{T} : \mathrm{Dist}(f_{q_n}(t), P^{-1}(A)) > \varepsilon\} > \frac{r}{8}.$$

Corollary 4.4. *For a torus $\mathbb{T}^m$ with $m > 1$, if A is a closed nontriavial discrete subgroup of $\mathbb{T}^m$, then there exists an $\varepsilon > 0$ and a constant $0 < L \leq 1$ such that when n is sufficiently large,*

$$\mu\{t \in \mathbb{T} : \operatorname{Dist}(\boldsymbol{f}_{q_n}(t), P^{-1}(\boldsymbol{A})) > \varepsilon\} > L.$$

Proof. Since A is closed and discrete, then every element in A must have rational components. Otherwise, by Theorem 3.3, A will not be discrete. Taking any $\boldsymbol{a} \in \mathbb{T}^m$, one can let

$$\boldsymbol{A}_1 = \langle \boldsymbol{A}, \{\boldsymbol{a}t : t \in \mathbb{R}\}\rangle.$$

It is clear that $\boldsymbol{A}_1 \supset \boldsymbol{A}$. Then our corollary follows from the Proposition 4.3 immediately. □

Theorem 4.5. *Let $\boldsymbol{f}: \mathbb{T} \to \mathbb{R}^m$ be a piecewise absolutely continuous function. Assume that the vector $\int_0^1 \frac{d}{dt}\boldsymbol{f}(t)dt$ has rationally independent components. Then for any irrational number θ, the measure preserving transformation T_f on $\mathbb{T} \times \mathbb{T}^m$, defined by*

$$T_f(t, \boldsymbol{x}) = (t + \theta, \boldsymbol{f}(t) + \boldsymbol{x}) \pmod 1,$$

is ergodic.

Proof. We first assume $0 < \theta < 1/2$ and $m > 1$. Suppose that T_f is not ergodic. Let $P: \mathbb{R}^m \to \mathbb{T}^m$ be the natural projection. According to Theorem 2.1, there is a nontrivial closed subgroup $\boldsymbol{A}$ of $\mathbb{T}^m$ and a measurable function $\boldsymbol{u}(t): \mathbb{T} \to \mathbb{T}^m$ such that

$$f_n(t) + \boldsymbol{u}(t + n\theta) - \boldsymbol{u}(t) \in P^{-1}(\boldsymbol{A})$$

for almost all $t \in \mathbb{T}$. Let $\{p_n/q_n\}$ be the sequence of partial convergents for θ. Then $q_n\theta \to 0 \pmod 1$ which implies that (here $\boldsymbol{u}$ is bounded)

$$\int_0^1 |\boldsymbol{u}(t + q_n\theta) - \boldsymbol{u}(t)|\, dt \to 0.$$

Hence for any $\tau > 0$, when n is sufficiently large, we have

$$\mu\{t \in \mathbb{T} : |\boldsymbol{u}(t + q_n) - \boldsymbol{u}(t)| > \tau\} < \tau.$$

By Proposition 4.3 and Corollary 4.4, there is an $\varepsilon > 0$ and an L with $0 < L \leq 1$ such that when n is sufficiently large,

$$\mu\{t \in \mathbb{T} : \operatorname{Dist}(\boldsymbol{f}_{q_n}(t), P^{-1}(\boldsymbol{A})) > \varepsilon\} > L.$$

Fix $\tau > 0$ with $\tau < \min\{L/2, \varepsilon/2\}$, and let

$$B = \{t \in \mathbb{T} : \operatorname{Dist}(\boldsymbol{f}_{q_n}(t), P^{-1}(\boldsymbol{A})) > \varepsilon\} \cap \{t \in \mathbb{T} : |\boldsymbol{u}(t + q_n) - \boldsymbol{u}(t)| < \tau\}.$$

Then $\mu(B) > \tau$. For any $t \in B$

$$\operatorname{Dist}(\boldsymbol{f}_{q_n}(t) + \boldsymbol{u}(t + q_n\theta) - \boldsymbol{u}(t), P^{-1}(\boldsymbol{A})) > \operatorname{Dist}(\boldsymbol{f}_{q_n}(t), P^{-1}(\boldsymbol{A})) - \tau > \tau.$$

This contradicts the fact that

$$f_n(t) + \boldsymbol{u}(t+n\theta) - \boldsymbol{u}(t) \in P^{-1}(\boldsymbol{A})$$

for all n and for almost all $t \in \mathbb{T}$.

For a general irrational number θ, there is an integer k such that $k\theta = \theta'$ (mod 1) with $0 < \theta' < 1/2$. Let $S_f = T_f^k$. Then

$$S_f(t, \boldsymbol{x}) = (t + \theta', \boldsymbol{x} + f_k(t))$$

for any $t \in \mathbb{T}$ and $\boldsymbol{x} \in \mathbb{T}^m$. Since

$$\int_0^1 f_k(t)\,dt = \int_0^1 \sum_{i=0}^{k-1} f(t+i\theta)dt = k\int_0^1 f(t)\,dt,$$

$S_f = T_f^k$ must be ergodic, which implies that T_f is ergodic.

Now consider the case for $m = 1$. Suppose that $f: \mathbb{T} \to \mathbb{T}$ is a piecewise absolutely continuous function. Then choose a function (for example, a linear function on $[0, 1]$) $g: \mathbb{T} \to \mathbb{T}$ such that $\int_0^1 g(t)dt$ is rationally independent to $\int_0^1 f(t)dt$. Let $\boldsymbol{f} = (f, g)$. Then $(\mathbb{T} \times \mathbb{T}^2, T_{\boldsymbol{f}})$ is ergodic. Since $(\mathbb{T} \times \mathbb{T}, T_f)$ is a factor of $(\mathbb{T} \times \mathbb{T}^2, T_{\boldsymbol{f}})$, we know that T_f is ergodic. □

Since $\int_0^1 f_k(t)dt = k\int_0^1 \boldsymbol{f}(t)dt$, the following theorem follows immediately from Theorem 4.5.

Theorem 4.6. *Let $\boldsymbol{f}: \mathbb{T} \to \mathbb{R}^m$ be a piecewise absolutely continuous function. Assume that the vector $\int \frac{d}{dt}\boldsymbol{f}(t)dt$ has rationally independent components. Then for any irrational number θ, the measure preserving transformation T_f on $\mathbb{T} \times \mathbb{T}^m$, defined by*

$$T_f(t, \boldsymbol{x}) = (t + \theta, \boldsymbol{f}(t) + \boldsymbol{x}) \pmod 1,$$

is totally ergodic.

At the end of this note, we would like to present several applications. Since $\int t^k dt$ is rational, both known results follow at once from Theorem 4.6

Corollary 4.7. *The measure preserving transformation*

$$(t, x_1, \ldots, x_m) \to (t + \theta, x_1 + \alpha_1 t, \ldots, x_m + \alpha_m t)$$

is totally ergodic on $\mathbb{T}^{m+1}$ if θ is irrational and $\alpha_1, \alpha_2, \ldots, \alpha_m$ are rationally independent.

Corollary 4.8. *Let $p_1(t), \ldots, p_m(t)$ be polynomials, and assume that the coefficients for $p_1(t), \ldots, p_m(t)$ are rationally independent. Then the measure preserving transformation*

$$(t, x_1, \ldots, x_m) \to (t + \theta, x_1 + p_1(t), \ldots, x_m + p_m(t)),$$

is totally ergodic.

References

[1] T. M. Apostol, Modular Functions and Dirichlet Series in Number Theory, Springer-Verlag, New York-Heidelberg-Berlin 1976.

[2] P. Hellekalek and G. Larcher, On ergodicity of a class of skew products, Israel J. Math. 54 (1986), 301–306.

[3] Y. Katznelson, Sigma-finite invariant measures for smooth mappings of the circle, Journal d'Analyse Math. 31 (1977), 1–18.

[4] A. Ya. Khinchin, Continued Fractions, University of Chicago Press, Chicago 1964.

[5] A. Iwanik, M. Lemanczyk and D. Rudolph, Absolutely continuous cocycles over irrational rotations, Israel J. Math. 83 (1993), 73–95.

[6] I. Oren, Ergodicity of cylinder flows arising from irregularities of distribution, Israel J. Math. 44 (1983), 127–138.

[7] D. A. Pask, Skew products over the irrational rotation, Israel J. Math. 69 (1990), 65–74.

[8] R. Zimmer, Extensions of ergodic group actions, Illinois J. Math. 20 (1976), 373–409.

Department of Mathematics
Clark College
1800 McLoughlin Blvd.
Vancouver, WA 98663, U.S.A.